S0-BAN-874

# Medicinal Chemistry

# Medicinal Chemistry
## *A Biochemical Approach*

### SECOND EDITION

## THOMAS NOGRADY
Concordia University, Montreal

New York     Oxford
**OXFORD UNIVERSITY PRESS**
1988

Oxford University Press

Oxford    New York    Toronto
Delhi    Bombay    Calcutta    Madras    Karachi
Petaling Jaya    Singapore    Hong Kong    Tokyo
Nairobi    Dar es Salaam    Cape Town
Melbourne    Auckland

and associated companies in
Berlin    Ibadan

Copyright © 1988 by Oxford University Press, Inc.

Published by Oxford University Press, Inc.,
200 Madison Avenue, New York, New York 10016

Oxford is a registered trademark of Oxford University Press

All rights reserved. No part of this publication may be reproduced,
stored in a retrieval system, or transmitted, in any form or by any means,
electronic, mechanical, photocopying, recording, or otherwise,
without the prior permission of Oxford University Press.

Library of Congress Cataloging-in-Publication Data

Nogrady, Th.
  Medicinal chemistry.

  Bibliography: p.
  Includes index.
  1. Chemistry, Pharmaceutical. I. Title.
  [DNLM: 1. Chemistry, Pharmaceutical. QV 744 N777m]
  RS403.N57   1988      615.7      87-21990
  ISBN 0-19-505368-0
  ISBN 0-19-505369-9 (pbk.)

2 4 6 8 9 7 5 3 1

Printed in the United States of America
on acid-free paper

*To Heather*
*Inspiration, Critic, Sustenance*

# Preface

As a result of the extremely rapid developments in molecular pharmacology and medicinal chemistry, a second edition of this book became necessary sooner than expected. For the same reason, practically all sections have undergone extensive updating. Many topics that were tentative a few years ago have matured and can now be covered in depth. For instance, discussions of the phosphatidylinositol-derived second messengers, tyrosine-phosphorylase, and the role of oncogene products in transmembrane signaling have been added to the description of adenylate cyclase-based mechanisms. Similarly, the section on the rapidly evolving field of co-neurotransmitters and neurohormones has been expanded, and the adenosine receptors and calcium channel blockers have now received more than the previous cursory mention. New topics include atrial natriuretic factors, antiarrhythmic drugs, and the DNA topoisomerase inhibitory mechanism of antitumor and antibacterial agents. To improve the balance of the book, the chapter on drug distribution and metabolism has been expanded, and exciting developments in drug design that utilize numerical techniques and computer graphics have been outlined.

However, as in the first edition, it seemed prudent to treat some of the new ideas with caution to avoid overly speculative concepts that are not quite ready for inclusion in a textbook. The philosophy of the book has remained unchanged: drugs are organized according to their targets. Nevertheless, from a practical point of view, pharmacological classification of drugs will always remain a necessity for the pharmacist, pharmacologist, and physician. The Appendix attending to these needs has therefore been retained and updated.

Many colleagues and reviewers—some of them anonymous—gave generously of their time, offering useful and often extensive suggestions. Professor L. H. M. Jansen (Utrecht) sent detailed remarks on the entire book, and Professor E. J. Ariëns (Nijmegen) helped to update the stereochemical aspects of drug action. To all of them I offer my sincere thanks.

Finally, I am grateful to the Department of Biology, Queen's University, Kingston, Ontario, and its head, Dr. D. T. Dennis, for the generous hospitality extended to me. As always, my wife, Heather, was of great help in graciously enduring the stresses inherent in the review process.

I look forward to a continuing dialogue with readers.

*Kingston, Ontario*                                                                    T. N.
*April 1987*

# Preface to the First Edition

Medicinal chemistry has undergone many changes since ancient herbalist days, but none seems to be as significant and fruitful as the rapid elucidation of the molecular-biochemical mechanisms of drug action achieved in the past twenty five years. Only our full understanding of these mechanisms can lead to a mature, exact, and predictive understanding of molecular pharmacology and rational drug design. This book is an attempt to combine the many possible approaches to the study and teaching of medicinal chemistry into a rational and coherent didactic discipline; it departs from the classical organization and proportions seen elsewhere, in order to promote progressive and creative ways of viewing the chemistry and mode of action of drugs.

The organic chemical approach to drug classification was long ago abandoned as irrational. The widely used pharmacokinetic classification organizes drugs according to their therapeutic action on organs (such as the central nervous system or thyroid gland), pathological syndromes (as with anticonvulsant drugs or antilipidemic agents), or identical therapeutic effects (e.g., local anesthetics or antimalarials). Although this more recent classification is practical for the pharmacist or physician, it is nevertheless as arbitrary as the structural classification. Such an organizational system not only is an obstacle to attempts to correlate the actions of different "classes" of drugs, it also hinders the emergence of much needed leads in rational drug design. The case of the antineoplastic action of many antibiotics comes to mind: while classified differently and used in distinct syndromes, both these and other antineoplastic agents act on DNA.

It is therefore timely that the new generation of medicinal chemists become aware of the fact that a basic understanding of drug action is possible only on a molecular rather than a cellular or organismic level. Using biochemical pharmacology as a framework and unifying structure, the young medicinal chemist should become a truly interdisciplinary practitioner from the start, rather than evolving slowly from a training in organic chemistry or pharmacology, as many of us have done. The molecular approach will nevertheless often overlap with the classical pharmacodynamic organization of drugs. The concept of $H_1$-antihistamines as a group or steroid hormones as a chemically defined family remains valid because their mode of action happens to be cohesive. On the other hand, "tranquilizers" cannot be squeezed into the old categories. They are found in this book among drugs acting on the dopaminergic as well as GABAergic receptors and are also discussed in connection with the biophysical state of neuronal membranes.

To support the main body of a book that discusses drugs on the basis of their molecular targets, considerable space has been devoted to the physicochemical principles of drug action and receptor–effector theories, as well as to an overview of the methods of receptor characterization. The latter is not meant to be a mini-textbook of instrumental techniques; instead it is aimed at giving the student an understanding of the armamentarium of methods that has led to the continuing explosion of molecular insight into drug action. A short chapter on drug distribution and metabolism and another on the principles of drug design underscore the multidisciplinary requirements in the practice of modern medicinal chemistry as a homogeneous discipline. Drug synthesis and immunological aspects of drug action are not covered, and the drugs cited serve as examples rather than as a comprehensive list of relevant pharmaca. Their choice was dictated not by their availability in North America but by their inherent scientific or therapeutic interest. To aid the pharmacy student, the Appendix lists the generic and proprietary names of drugs, their pharmacological actions, and some of their physicochemical characteristics. It also serves as a drug and formula index. Drugs are also grouped by pharmacological activity.

The references at the end of each section are highly selective. Relatively few research papers are listed; instead, the emphasis is on reviews and symposia. To keep in mind parsimony in building library collections, a relatively narrow selection of sources are used. Major standard works are not cited routinely and redundantly but are listed at the end of this preface.

In keeping with the molecular-biochemical bias of this book, a knowledge of basic organic and biochemistry is assumed, but advanced undergraduates as well as graduate students and professionals may find the treatment clear.

Many people have helped in my presumptuous attempt to produce a single-author text. Much of the manuscript was written during a sabbatical stay at the Division of Pharmacology of the University of California at San Diego and during a shorter visit at the Department of Pharmaceutical Chemistry of the University of California at San Francisco. I am most grateful to Palmer Taylor and to Manfred Wolff for their hospitality and help in making available to me the marvelous resources of the University of California.

Pavel Hrdina (University of Ottawa), Leslie Humber (Ayerst Laboratories, Montreal), and Thomas Sandor (Université de Montréal) read individual chapters and supplied valuable information, updating, and corrections. My colleague Mark Doughty read the whole first draft and added his erudition. To all of them, I give my heartfelt thanks. My main editor, grammarian, and critic was, however, my wife, Heather, who immeasurably improved the style and structure of the book as well as the author's disposition after a long day of writing. Doris Tooby in Montreal and Sandra Dutky in La Jolla typed a good part of the initial versions cheerfully and efficiently. The original illustrations are the careful work of Catherine Bata.

I look forward to a dialogue with readers and welcome their additions and criticism.

*Montreal, Quebec*                                                                       T. N.
*September 1983*

# SOME MAJOR WORKS ON MEDICINAL CHEMISTRY

A. Albert (1985). *Selective Toxicity*, 7th ed. Chapman and Hall, London.

E. J. Ariëns (Ed.). *Drug Design*, 10 vols. Academic Press, New York.

J. R. Cooper, F. E. Bloom, and R. H. Roth (1986). *The Biochemical Basis of Neuropharmacology*, 5th ed. Oxford University Press, New York.

R. F. Doerge (Ed.) (1982). *Wilson and Giswold's Textbook of Organic Medicinal and Pharmaceutical Chemistry*, 8th ed. J. B. Lippincott, Philadelphia.

W. O. Foye (Ed.) (1981). *Principles of Medicinal Chemistry*, 2nd ed. Lea and Febiger, Philadelphia.

A. Goodman-Gilman, L. S. Goodman, T. W. Wall, and F. Murad (Eds.) (1985). *Goodman and Gilman's The Pharmacological Basis of Therapeutics*, 7th ed. Macmillan, New York.

N. J. Howe, M. M. Milne, and A. F. Pennel (Eds.) (1978). *Retrieval of Medicinal Chemical Information*, ACS Symposium Series No. 84. American Chemical Society, Washington, DC.

L. L. Iversen, S. D. Iversen, and S. H. Snyder (Eds.) (1975). *Handbook of Psychopharmacology*, 17 vols. Plenum Press, New York.

J. W. Lamble (Ed.) (1981). *Towards Understanding Receptors*. Elsevier Biomedical Press, Amsterdam.

J. W. Lamble (Ed.) (1982). *More About Receptors*. Elsevier Biomedical Press, Amsterdam.

J. W. Lamble and A. C. Abbott (Eds.) (1986). *Receptors, Again*. Elsevier, Amsterdam.

D. Lednitzer and L. A. Mitscher (1976–1984). *Organic Chemistry of Drug Synthesis*, 3 vols. Wiley, New York.

E. E. J. Marler (Ed.) (1985). *Pharmacological and Chemical Synonyms*. 8th ed. Excerpta Medica/Elsevier, Amsterdam.

M. Sittig (1979). *Pharmaceutical Manufacturing Encyclopedia*. Noyes Data, Park Ridge, N. J.

M. Windholz (Ed.) (1984). *The Merck Index*, 10th ed. Merck, Rahway, NJ.

M. E. Wolff (Ed.) (1979–81). *Burger's Medicinal Chemistry*, 4th ed., 3 parts. Wiley-Interscience, New York.

# PERIODICALS OF GENERAL INTEREST

*Advances in Drug Research*. Academic Press, New York.

*Annual Reports of Medicinal Chemistry*. Academic Press, New York.

*Annual Reviews of Biochemistry*. Annual Reviews Inc., Palo Alto, CA.

*Annual Reviews of Neuroscience*. Annual Reviews Inc., Palo Alto, CA.

*Annual Reviews of Pharmacology and Therapeutics*. Annual Reviews Inc., Palo Alto, CA.

*Arzneimittelforschung*. Editio Cantor, Aulendorf, W. Germany.

*Biochemical Pharmacology*. Pergamon Press, New York.

*Drugs of the Future*. Prous S. A., Barcelona.

*European Journal of Medicinal Chemistry*. Elsevier, Amsterdam.

*Journal of Medicinal Chemistry*. American Chemical Society, Washington, DC.

*Medicinal Research Reviews*. Wiley, New York.

*Molecular Pharmacology*. Academic Press, New York.

*Progress in Drug Research*. Birkhäuser, Basel.

*Progress in Medicinal Chemistry*. Elsevier, Amsterdam.

*Trends in Pharmacological Sciences*. Elsevier, Amsterdam.

# Contents

# Medicinal Chemistry

# 1

# Physicochemical Principles of Drug Action

All drug molecules interact with biological structures, such as lipoprotein receptors, enzymes, biomembranes, nucleic acids, or small molecules. This interaction triggers a series of steps that ultimately results in a macroscopic, physiological change which constitutes the drug effect. Only by first unraveling the relatively simple primary interaction between the drug molecule and a macromolecular structure can we understand drug activity at the cellular level. Organs and whole organisms are immensely more complex than individual cells, and therefore require the understanding of many more parameters.

Drug transport from the site of application to the site of action, as well as the drug–stimulus relationship, depends on physicochemical and geometric properties inherent in the structure of the drug molecule. This correlation applies equally to the physicochemical properties of the biological macromolecules with which the drug interacts. However, our knowledge of these macromolecules lags far behind our experience with small compounds. Consequently, in order to achieve rational drug design, the ultimate goal of medicinal chemistry, we have to study the chemical and physical properties of drug molecules and their targets and correlate the sum of these molecular properties with the biological effects of drug–receptor interactions.

This chapter discusses the applications of physicochemical principles to the molecules and modes of action of drugs. Since all biological reactions take place in an aqueous medium or at the interface of water and a lipid, the properties of water and this boundary layer have to be dealt with.

## 1. ROLE AND STRUCTURE OF WATER

Life is based on water, the major constituent of living organisms and their cells. Besides being a universal solvent or dispersing agent, water participates in many reactions, and its role is therefore much more than that of an inert medium: it is a very reactive and unusual chemical compound. Solubility, surface activity, hydrogen bonding, hydrophobic bonding, ionization, and effects on macromolecular conformation (e.g., in drug receptors) all involve water.

## 1.1. The Structure of Bulk Water

Water structure is the consequence of the unique and unusual physical properties of the $H_2O$ molecule. It has a higher melting point, boiling point, and heat of vaporization than such hydrides of related elements as $H_2S$, $H_2Se$, and $H_2Te$, or such isoelectronic compounds as HF, $CH_4$, or $NH_3$. These properties are all a measure of the strong intermolecular forces acting between individual water molecules, which do not let the ice crystal collapse or the molecule leave the surface of the liquid phase easily when heated. These forces result from the high polarity of water caused by the direction of the H—O—H bond angle, which is 104.5°. The

**Fig. 1.1.** Schematic diagram of water structure. (**A**) Tetrahedral hydrogen bonding of a water molecule in ice. (**B**) "Flickering clusters" of liquid water in a two-dimensional diagram.

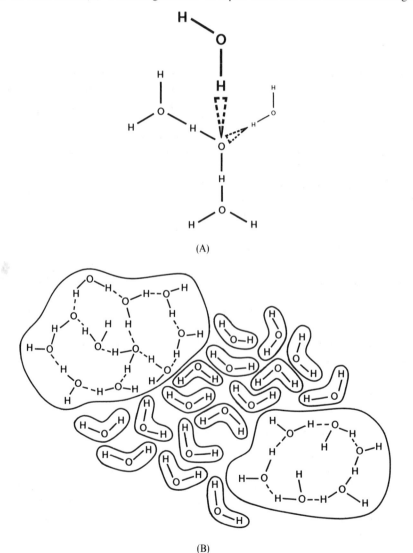

(A)

(B)

more electronegative oxygen attracts the electron of the O—H bond to a considerable extent, leaving the H with a partial positive charge ($\delta^+$), while the O atom acquires a partial negative charge ($\delta^-$). Since the molecule is not linear, $H_2O$ has a *dipole moment*. The partial positive and negative charges of one water molecule will electrostatically attract their opposites in other water molecules, resulting in the formation of *hydrogen bonds* (Fig. 1.1). Such noncovalent bonds can also be formed between water and hydroxyl, carbonyl, or NH groups, as discussed later in Sec. 7.3 of this chapter.

In *ice*, each oxygen atom is bonded to four hydrogen atoms by two covalent and two hydrogen bonds. When ice melts, about 20% of these H bonds are broken, but there is a strong attraction between water molecules even in water vapor. *Liquid water* is therefore highly organized on a localized basis: the hydrogen bonds break and re-form spontaneously, creating and destroying transient structural domains, the so-called flickering clusters. However, because the half-life of any hydrogen bond is only about 0.1 nanosecond ($10^{-10}$ second), the existence of these clusters has statistical validity only; even this has been questioned by some authors who consider water to be a "continuous polymer."

## 1.2. The Solvent Properties of Water

Water will interact with ionic or polar substances and destroy their crystal lattices. Since the resulting hydrated ions are more stable than the crystal lattice, solvation results. Water has a very high dielectric constant (80 Debye units vs. 21 D for acetone), which counteracts the electrostatic attraction of ions, thus favoring further hydration. The dielectric constant of a medium can be defined as a dimensionless ratio of forces: the force acting between two charges in a vacuum and the force between the same two charges in the medium or solvent. According to Coulomb's law,

$$F = \frac{q_1 q_2}{Dr^2}$$

where $F$ is the force, $q_1$ and $q_2$ are the charges, and $r$ is the distance separating them. $D$, the dielectric constant, is a characteristic property of the medium. Since $D$ appears in the denominator, the higher the dielectric constant, the weaker the interaction between the two charges.

Polar functional groups of nonionic organic compounds such as aldehydes, ketones, and amines (possessing free electron pairs) form hydrogen bonds readily with water, and dissolve to a greater or lesser extent, depending on the proportion of polar to apolar moieties in the molecule.

Solutes cause a change in water properties because the hydrate "envelopes" which form around solute ions are more organized and therefore more stable than the flickering clusters of free water. As a result, ions are water-structure *breakers*. The properties of solutions, which depend on solute concentration, are different from those of pure water; the differences can be seen in such phenomena as the freezing-point depression, boiling-point elevation, and increased osmotic pressure of solutions.

Water molecules cannot use all four possible hydrogen bonds when in contact with hydrophobic (literally, "water-hating") molecules. This restriction results in a loss of entropy, a gain in density, and increased organization. So-called icebergs—water domains more stable than the flickering clusters in liquid water—are formed. Such "icebergs" can form around single apolar molecules, producing inclusion compounds called *clathrates*. Apolar molecules are thus water-structure *formers*.

The interaction between a solute and a solid phase—for example, a drug with its lipoprotein receptor—is also influenced by water. Hydrate envelopes or icebergs associated with one or the other phase will be destroyed or created in this interaction and may often contribute to conformational changes in macromolecular drug receptors and, ultimately, to physiological events.

### Selected Readings

F. Franks (Ed.) (1972). *Water. A Comprehensive Treatise*, Vol. 1. *The Physics and Physical Chemistry of Water*. Plenum, New York.

A. L. Lehninger (1982). *Principles of Biochemistry*. Worth, New York, Chap. 4.

L. Packer (Ed.) (1986). *Protons and Water: Structure and Translocation, Methods in Enzymology* (S. P. Colowick and N. O. Kaplan, Eds.), Vol. 127, Part O. Academic Press, New York.

B. Pullman (1978). Aspects of biomolecules in their surroundings: hydration and cation binding. In: *Frontiers in Physicochemical Biology* (B. Pullman, Ed.). Academic Press, New York, pp. 143–163.

M. C. R. Symous (1981). Water structure and reactivity. *Accounts Chem. Res. 14*: 179–187.

## 2. SOLUBILITY

Because a large percentage of all living structures consists of water, all biochemical reactions are based on small molecules dissolved in an aqueous phase (like the cytosol) or on macromolecules dispersed in this phase—usually both. The nonaqueous structures of cells, such as plasma membranes or the membranes of organelles, are of a lipid nature, and can dissolve polar or nonpolar hydrophobic molecules. In either case, a highly significant physical property of all physiologically and pharmacologically important small molecules is their solubility, because only in solution can they interact with the cellular and subcellular structures that carry drug receptors, thus triggering pharmacological reactions. Theoretically, there are no absolutely insoluble compounds; every molecule is soluble in both the aqueous and nonaqueous lipid "compartments" of a cell. The degree of solubility, however, differs in each compartment. The proportion of these concentrations at equilibrium—or the ratio of solubilities—is called the *partition coefficient*, and will be discussed in detail in the next section.

Solubility is a function of many molecular parameters. Ionization, molecular structure and size, stereochemistry, and electronic structure will all influence the basic interaction between a solvent and solute. As we have seen in the previous section, water forms hydrogen bonds with ions or with polar nonionic compounds through —OH, —NH, —SH, and —C=O groups, or with the nonbonding electron pairs of oxygen or nitrogen atoms. The ion or molecule will thus acquire a

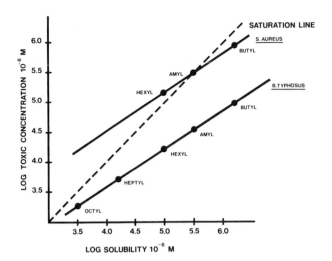

**Fig. 1.2.** Bactericidal concentration of primary aliphatic alcohols versus their water solubility. Since the two axes have the same scale, the broken saturation line has a slope of one. (Data from J. Ferguson (1939), *Proc. Roy. Soc. [B] 127*: 387)

hydrate envelope and separate from the bulk solid; that is, it dissolves. The interaction of nonpolar compounds with lipids is based on a different phenomenon, the hydrophobic interaction (see Sec. 7.2), but the end result is the same: formation of a molecular dispersion of the solute in the solvent. Adrien Albert (1985) discusses these correlations in his excellent book.

There are few examples in which solubility in only one phase correlates with pharmacological activity. One such example is the local anesthetic activity of *p*-aminobenzoic acid esters which is partly proportional to their lipid solubility. Another thoroughly investigated correlation is that between the bactericidal activity of aliphatic alcohols with their solubility (Fig. 1.2).

In the homologous series beginning with *n*-butanol and ending with *n*-octanol, the bactericidal activity increases with increasing molecular weight (i.e., the log toxic concentration decreases) in cultures of the sensitive gram-negative *Salmonella typhi* (formerly called *Bacillus typhosus*). Even the rather water-insoluble octanol is active in a concentration below the saturation point. The "saturation line" in Fig. 1.2 is a diagonal (broken) line with a slope of unity (log solubility vs. log solubility, since the scales of the ordinate and abscissa are identical).

If the same homologous series is tested on cultures of the less sensitive *Staphylococcus aureus* (see Fig. 1.2), the activity line is displaced toward higher concentrations. Whereas *n*-butanol and *n*-pentanol are active, higher members of the series fail to kill the bacteria because the necessary concentration cannot be reached; it is higher than the saturation concentration, thus lying above the saturation line. This interesting interpretation of the "cutoff" point in this homologous series was proposed by J. Ferguson (see Albert, 1985).

Bactericidal aliphatic amines show a cutoff point at dodecylamine, the $C_{12}$ member of the homologous series, even though amines have no solubility problems like the alcohols. However, the next member of the homologous series, the $C_{14}$ amine, is at the critical micelle formation point (see Sec. 4.1), and this compound, or higher members of the series, can contribute fewer and fewer monomeric molecules to the solution. Since monomers are essential to the bactericidal activity, this results in a rapidly diminishing biological effect. Micelle formation can easily be measured by light-scattering or nuclear magnetic resonance relaxation methods.

The effect of solubility on drug action is, however, usually a question of equilibration of the drug between the aqueous phase and the lipid phase of the cell membrane, or even of fat-tissue storage, and leads us to the discussion of partition coefficients.

### 3. PARTITION COEFFICIENTS

The partition coefficient of a drug is defined as the equilibrium constant

$$P = \frac{[\text{drug}]_{\text{lipid}}}{[\text{drug}]_{\text{water}}}$$

of drug concentrations (symbolized by the square brackets) in the two phases. Since partition coefficients are difficult to measure in living systems, they are usually determined *in vitro*, using *n*-octanol as the lipid phase and a phosphate buffer of pH 7.4 as the aqueous phase. This permits standardized measurements of partition coefficients. Because it is a ratio, $P$ is dimensionless. $P$ is also an additive property of a molecule, since each functional group helps determine the polarity and therefore the lipophilic or hydrophilic character of the molecule. These substituent contributions are widely utilized in quantitative structure–activity studies, as discussed later in this chapter.

Partition coefficients thoroughly influence *drug transport characteristics*—the way drugs reach the site of action from the site of application (e.g., injection site, gastrointestinal tract, and so forth). Since drugs are usually distributed by the blood, they must penetrate and traverse many cells to reach the site of action. Hence, the partition coefficient will determine what tissues a given compound can reach. On the one hand, extremely water-soluble drugs may be unable to cross lipid barriers and gain access to organs rich in lipids, such as the brain and other neuronal tissues. However, compounds can cross the "blood–brain barrier" by diffusing from one aqueous phase (blood) to another (cerebrospinal fluid). On the other hand, compounds that are very lipophilic will be trapped in the first "site of loss," like fat tissue, and will be unable to leave this site quickly to reach their target. Naturally, the partition coefficient is only one of several physicochemical parameters influencing drug transport and diffusion, which itself is only one aspect of drug activity. The blood–brain barrier and its role will be discussed in detail in connection with drug transport in Chap. 7, Sec. 1.5.

The partition coefficient and concepts derived from it are particularly important in explaining the mode of action of nonspecific general depressant drugs, such as anesthetics and some barbiturate hypnotics ("sleeping pills"), as well as of structurally nonspecific disinfectants acting on bacterial membranes.

**Table 1.1.** Lipid–water partition coefficients of some depressant compounds

| Substance | Partition coefficient, oleyl alcohol/water | Concentration immobilizing tadpoles, mol/L (water) | Calculated depressant concentration, mol/L (cell lipid) |
|---|---|---|---|
| Ethanol | 0.10 | 0.33 | 0.033 |
| n-Butanol | 0.65 | 0.03 | 0.020 |
| Valeramide | 0.30 | 0.07 | 0.021 |
| Benzamide | 2.50 | 0.013 | 0.033 |
| Salicylamide | 5.90 | 0.0033 | 0.021 |
| o-Nitroaniline | 14.0 | 0.0025 | 0.035 |
| Thymol | 950.0 | $4.7 \times 10^{-5}$ | 0.045 |

Reproduced by permission from Albert (1985), Chapman and Hall, London.

### 3.1. The Overton–Meyer Hypothesis of Anesthetic Activity

Anesthesia refers to the complete lack of somatic sensation. Overton, at the turn of the century, attempted to explain drug-induced anesthesia. He, and later H. H. Meyer, stated that:

1. All neutral lipid-soluble substances have depressant properties on neurons.
2. This activity is most pronounced in lipid-rich cells.
3. The effect increases with increasing partition coefficient, regardless of the structure of the substance.

Although the absolute drug concentration necessary to achieve anesthesia varies greatly, as shown in Table 1.1, the drug concentration in the lipid phase—that is, in the cell membrane—is within one order of magnitude, or 20–50 mM, for all anesthetic agents.

In 1954, Mullins, in a modification to the Overton–Meyer hypothesis, proposed that besides the membrane concentration of the anesthetic, its volume, expressed as its volume fraction (mole fraction × partial molal volume), is important (see Kaufmann, 1977). This reasoning implies that the anesthetic *expands the cell membrane*, and that anesthesia occurs when a critical expansion value is reached, at about 0.3–0.5% of the original volume. The surface area of the membrane will also expand by several percentage points, as measured on red blood cells.

### 3.2. Ferguson's Rule

In 1939, Ferguson extended the usefulness of the Overton–Meyer hypothesis to anesthetics that are administered in the gas phase by inhalation. He discovered that regardless of the nature of the *biophase*—that is, the site of anesthetic action or absolute concentration of the agent in the gas or liquid phase, their effect occurs within a fairly constant range of *thermodynamic activity*. For practical purposes the thermodynamic activity of a substance can be defined as its relative saturation.

For gases,

$$a = \frac{p_t}{p_s}$$

when $p_t$ is the partial vapor pressure of the substance in air, and $p_s$ is the vapor pressure of the substance. For substances dissolved in a liquid, the same type of correlation exists:

$$a = \frac{S_t}{S_0}$$

where $S_t$ is the molar concentration of the dissolved drug necessary for biological activity and $S_0$ is the molar solubility of the drug. The maximum value of the thermodynamic activity is unity, which is the saturation point. For a more rigorous treatment of thermodynamic activity, the reader is referred to Marshall (1978). Table 1.2 gives such a correlation for substances known to be toxic to an agricultural pest. It shows that regardless of their toxic concentrations, which vary by a factor of 4000, all of the substances are toxic at about their half-saturation values in air.

Thermodynamic activity determinations can be useful in distinguishing structurally specific from structurally nonspecific drugs. *Structurally nonspecific drugs* are active at high thermodynamic activities, between 0.01 and 1; that is, they are active only in relatively high doses. Their biological activity is unrelated to their chemical structure, as shown by the ability of dissimilar compounds to show identical biological activities. However, the interaction between a nonspecific drug and its target may be specific insofar as drug binding is concerned.

**Table 1.2.** Agreement of the relative saturation value of various nonspecific toxic substances tested on wireworms

| Substance | Toxic concentration ($\mu$mol/L), lethal in 1000 min at 15°C | $p_s$ (vapor pressure at 15°C, mm) | $p_t/p_s$ (relative saturation of toxic concentration) |
|---|---|---|---|
| Monomethylaniline | 3.7 | 0.22 | 0.3 |
| Dimethylaniline | 6.6 | 0.28 | 0.4 |
| Pyridine | 76 | 10.4 | 0.1 |
| Bromoform | 94 | 3.2 | 0.5 |
| Tetrachloroethane | 141 | 4.2 | 0.6 |
| Chlorobenzene | 200 | 6.8 | 0.5 |
| Toluene | 420 | 17.0 | 0.4 |
| Benzene | 775 | 58 | 0.2 |
| Heptane | 800 | 27 | 0.5 |
| Chloroform | 1040 | 128 | 0.2 |
| Trichloroethylene | 1200 | 52 | 0.4 |
| Carbon tetrachloride | 1600 | 73 | 0.4 |
| Hexane | 3000 | 96 | 0.6 |
| Pentane | 16600 | 320 | 0.9 |

Reproduced by permission from Albert (1985), Chapman and Hall, London.

The majority of compounds used therapeutically and discussed in this book are *structurally specific drugs*. They show pharmacological effects related to their thermodynamic activity and are active at concentrations corresponding to very low thermodynamic activities, normally below 0.001. Groups of drugs that achieve the same effect by identical mechanisms often have similar structures, and variations in their chemical structure (resulting in changes in their physicochemical properties) produce profound alterations in their pharmacological properties. Specific drugs interact with specific and highly selective *drug receptors*, usually macromolecular structures of a lipoprotein or glycoprotein nature. The density of these receptors per unit area of membrane is low, varying from 10 to 1000 receptors per square micrometer for drugs, hormones, and specifically acting neurotransmitters. On the other hand, the average active concentration of 2 mM for nonspecific drugs would assume that millions of molecules of such drugs bind to $\mu m^2$ of membrane, and rather than speak of "receptors" for these agents, we speak simply of their binding sites.

### 3.3. General Anesthetics

The most important group of nonspecific drugs is that of the general anesthetics. These nonspecific compounds, shown in Fig. 1.3, vary greatly in structure, from noble gases such as Ar or Xe to complex steroids.

The pharmacology of anesthesia is complex and extensive; it is concerned with the speed at which narcosis is induced, the rate of recovery, and any possible side effects. The reader is referred to the major pharmacology texts or the treatises of Burger/Wolff (1980) or Doerge (1982) for details (see general bibliography after the Preface to the First Edition).

Among the gaseous anesthetics, **xenon** is active at the low thermodynamic activity of 0.01. Although considered to be a good anesthetic, it is expensive and not readily available. Consequently, it is not used in surgical practice. **Nitrous oxide** (or "laughing gas") is one of the oldest anesthetic agents. By itself it is used only for short operations or in dentistry because it does not produce sufficiently deep anesthesia when administered with the 20% oxygen necessary to maintain normal breathing. The analgesia (loss of pain sensation) produced by nitrous oxide is, however, good even at a 50–60% concentration. **Cyclopropane** is a potent anesthetic agent, but is rarely used because of its highly explosive properties.

The volatile liquids are the most widely used inhalation anesthetics. One of these agents, **diethyl ether**, was known to Paracelsus in the seventeenth century, but its anesthetic activity was discovered only in 1842 by Long and Morton. It causes a slow induction of and slow emergence from narcosis, is irritating, and is also explosive. It gives a deep level of anesthesia, but is no longer used in modern anesthesiology. **Chloroform**, used in 1847 for the first time, is potent and non-explosive, but like most halogenated hydrocarbons it causes liver and kidney damage. This is not a serious side effect of the other three halogenated hydrocarbons shown in Fig. 1.3, which are likewise not explosive. **Halothane** (1-7) is perhaps the most widely used inhalational anesthetic in modern practice because of a combination of favorable properties: rapid induction and emergence, lack of

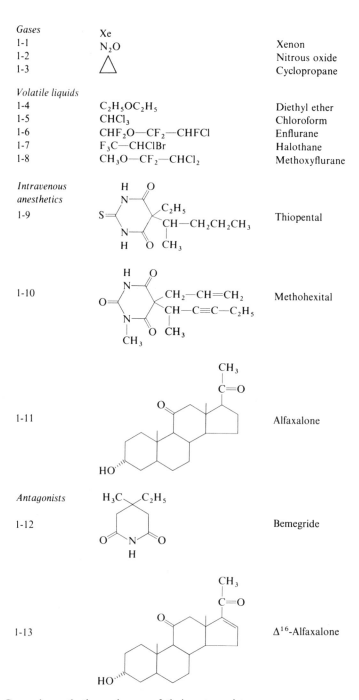

| Gases | | |
|---|---|---|
| 1-1 | Xe | Xenon |
| 1-2 | $N_2O$ | Nitrous oxide |
| 1-3 | △ | Cyclopropane |

| Volatile liquids | | |
|---|---|---|
| 1-4 | $C_2H_5OC_2H_5$ | Diethyl ether |
| 1-5 | $CHCl_3$ | Chloroform |
| 1-6 | $CHF_2O—CF_2—CHFCl$ | Enflurane |
| 1-7 | $F_3C—CHClBr$ | Halothane |
| 1-8 | $CH_3O—CF_2—CHCl_2$ | Methoxyflurane |

| Intravenous anesthetics | | |
|---|---|---|
| 1-9 | | Thiopental |

| 1-10 | | Methohexital |

| 1-11 | | Alfaxalone |

| Antagonists | | |
|---|---|---|
| 1-12 | | Bemegride |

| 1-13 | | $\Delta^{16}$-Alfaxalone |

**Fig. 1.3.** General anesthetics and some of their antagonists.

flammability, and minimal liver damage. **Enflurane** (1-6) shows even less potential liver damage.

Among the intravenous anesthetics, the barbiturates **thiopental** (1-9) and **methohexital** (1-10) are notable. In thiopental, one of the three lactam oxygens of barbituric acid is replaced by sulfur, and the two alkyl side chains impart a lipophilic character to the molecule. Thiopental is known as an "ultrashort"-acting anesthetic because the onset of anesthesia and loss of consciousness occur within seconds of its administration. Barbiturates are therefore extremely useful for short operations, or for the induction of anesthesia before switching to an inhalation anesthetic. The emergence from barbiturate-induced anesthesia is, however, relatively slow, because these compounds must be metabolized from stores in fat tissue in order to be rendered inactive. By contrast, inhalation anesthetics are excreted through the lungs. **Methohexital** (1-10) carries an $N$-methyl group that increases the lipophilic character of the compound. This also prevents its tautomerization to the lactim form of barbituric acid resulting in the formation of an enolate ion. This also increases the lipophilic character because the enolate ion is hydrophilic and thus capable of forming hydrogen bonds with water. The enolate also forms soluble sodium salts. The $N$-methyl group, by maintaining the lipophilic nature of methohexital, therefore prompts the rapid onset of narcosis. Since the unsaturated bonds in the side chains of methohexital increase the rate of drug degradation through biological oxidation, higher doses can be administered safely (see also Chap. 7).

| Lactam | Lactim | Enolate |

Other longer-acting barbiturates, used mainly as sedative-hypnotics ("sleeping pills"), are discussed in detail in Chaps. 4 and 6 because their suspected targets of action are the neurotransmitter $\gamma$-aminobutyric acid, and the neural membranes.

An unusual intravenous anesthetic is **alfaxalone** (1-11), a steroid derivative. Related structurally to the female sex hormone progesterone, alfaxalone, has a 3$\alpha$-OH group that is axial, or perpendicular to the plane of the cyclohexane ring. Interestingly, the 3$\beta$-OH epimer (**betaxolon**), which has an OH that is equatorial, is inactive. If a double bond is introduced between $C_{16}$ and $C_{17}$ ($\Delta^{16}$) in alfaxalone, an antagonist (compound 1-13) is obtained, which is a membrane stabilizer, because its C-17 side chain is equatorial (Chap. 5, Sec. 5.1) and thus easily accommodated between the lipid side chains of the membrane. In **alfaxalone**, this side chain is axial and perturbs the lipid membrane (Fesik and Makriyannis, 1985).

The glutarimide derivative **bemegride** (1-12), which resembles barbiturates in its structure, is a central nervous system (CNS) stimulant (analeptic) and a nonspecific but very useful antagonist of barbiturates, and was used to treat barbiturate poisoning.

The foregoing structure–activity correlations of barbiturates and steroid anesthetics points to an interesting conceptual distinction that must be made. Although the modes of action of barbiturates and steroid anesthetics may indeed be nonspecific, the binding sites of these compounds (presumably in the neuronal membrane) are certainly capable of detecting fine structural differences, such as those between the $\alpha$ or $\beta$ conformation of the 3-OH group of alfaxalone. The other possible explanations for the stereoselectivity of a "nonspecific" drug are stereoselective transport, a different mode of action of intravenous anesthetics, or an assumed, specific "anesthetic receptor." This last possibility will be considered in the next section.

### 3.3.1. Mechanism of Anesthesia

Contemporary hypotheses on the mechanism of anesthesia have been developed steadily since Overton and Meyer first attempted to establish a quantitative relationship between very diverse anesthetics and their narcotic properties. All of those hypotheses that have refined the theory of Overton and Meyer (see Kaufman, 1977) are based on the presumed interaction of anesthetics with the lipids of cell membranes. Ferguson introduced the concept of thermodynamic activity, which permitted dealing with anesthetics in the gas phase, as well as with those in liquid solutes (see above). Mullins pointed out that anesthesia requires a critical volume fraction of the anesthetic in the membrane. He believed that molecules of anesthetic agents are accommodated in the "free volume" space of the membrane. A further extension of this idea was the finding, through biophysical measurement by nuclear magnetic resonance (NMR) and electron paramagnetic resonance (EPR) methods, that anesthetics expand as well as perturb the membrane and result in *fluidization*— a transition of the lipid bilayer of the membrane from a gel to a liquid-crystal form.

Membranes show some selectivity in accepting or rejecting hydrocarbons as anesthetics. Progressing past hexane in the homologous *n*-alkane series, an increasingly large thermodynamic activity is needed for anesthesia until the cutoff point is reached at decane, which is not anesthetic at its saturation concentration. Apparently, such a large molecule does not trigger the necessary structural membrane transformation.

However, the hypotheses relying entirely on membrane fluidization (i.e., "phase transition") are challenged by newer experimental findings. Plotting the potencies of anesthetics against their *n*-octanol/water partition coefficient gives excellent correlation over four orders of magnitude of concentration. Olive oil, and the apolar hexadecane, as organic phase gave much poorer correlations, suggesting somewhat polar anesthetic sites. Moreover, no effects on lipid bilayers were detected by use of x-ray or neutron diffraction methods.

The membrane expansion and critical volume theory is strongly supported by the *pressure reversal of anesthesia*. Anesthetized tadpoles or other aquatic animals, and even mice, will resume normal functioning if subjected to high pressures. However, since high pressure in itself has multiple physiological effects (convulsion, tremors, paralysis), the pressure reversal of anesthesia is probably a complex multineuronal process, and, as discussed below, is not a universal phenomenon.

This discussion ties in well with recent ideas about membrane structure and the fluidization of the lipid bilayer, discussed in detail in Chap. 6. General anesthetics share the effect of membrane fluidization with local anesthetics, alcohols, and barbiturates, although the relationship between membrane fluidity and permeability is clouded. Since it is generally assumed that anesthesia depresses synaptic neuronal transmission, one may draw the implication that it also influences neurotransmitter release through presynaptic membranes. The sensitivity of different membranes to different drugs varies. This accounts for the selective action of anesthetics and results—fortuitiously perhaps—in the selective depression of consciousness with minimal effect on vital functions like respiration and circulation.

Lipids interact with anesthetics by virtue of the hydrophobic (or lipophilic) character. It must not be forgotten that integral membrane proteins also have large hydrophobic areas that interact with the lipid bilayer. Consequently, it is not surprising to find that there are protein–anesthetic interactions involving enzymes, receptors, transport proteins, and such structural proteins as microtubules and microfilaments. Such protein–anesthetic interactions result in conformational changes, or inhibition of the membrane alterations normally necessary to neural function. For instance, anesthetics, by blocking membrane permeability, antagonize the excitation of neurons by acetylcholine (Chap. 4). The interaction of an anesthetic with the ionophore (ion-carrying) part of the acetylcholine receptor, a protein, is therefore an attractive idea, although lacking in experimental evidence. If one adds to this the fact that anesthetic agents also influence $Ca^{2+}$ metabolism in the membrane—which has a profound bearing on neurotransmission—the lack of a coherent and unifying hypothesis of anesthesia is not at all surprising.

Protein involvement in anesthetic binding has been shown by inhibition of firefly luciferase, the enzyme involved in bioluminescence. The enzyme binding site accommodates two molecules of halothane or hexanol, but only one molecule of compounds larger than octanol. The binding of the enzyme substrate, luciferin, was competitively inhibited by most anesthetics at pharmacological concentrations (Franks and Lieb, 1984). These authors speculated that perhaps endogenous ligands (naturally occurring binding substances) are displaced by the anesthetic on central nervous system receptors.

Neurotransmitter involvement in anesthesia has also been suggested on the basis of findings that freshwater crustaceans do not show pressure reversal of anesthesia, although they are just as sensitive to the drug as is any other animal. In other animals, high pressure alone can cause excitation reminiscent of the effect of strychnine, an alkaloid acting through glycine receptors (see Chap. 4, Sec. 8.2). Pressure causes no excitation in these crustaceans, however, and they neither utilize glycine as a neurotransmitter in their CNS nor do they react to strychnine. These observations suggest neurotransmitter involvement in anesthesia based on fairly specific binding; it therefore follows that there must be protein binding sites where anesthetics may enhance glycine binding to its receptors.

Overall, one gains the impression that the lipid phase of the neuronal membrane is the site of most general, nonspecific anesthetic effects, whereas the hydrophobic protein phase of this membrane could be implicated in the specificity and selectivity of some anesthetics. Figure 1.4 summarizes these ideas, which have been put into perspective by the reviews of Kaufman (1977) and Roth and Miller (1986).

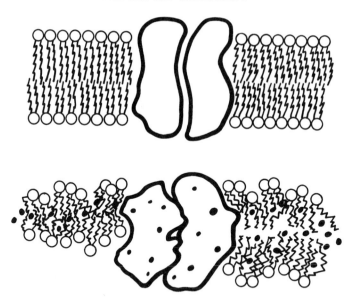

**Fig. 1.4.** Effect of general anesthetics on a lipoprotein membrane. The top shows the normal structure in cross section, with a protein ion channel and the lipid bilayer in the gel state. The bottom schematic shows the lipid bilayer in a disorganized fluid state, and the protein ion channel distorted and inoperative because of protein swelling. The black dots symbolize the anesthetic molecules.

*Hydrophobic fragmental constants* (symbolized by *f* ) were introduced by Nys and Rekker (see Rekker, 1977) to correct for some errors in the determination of partition coefficients, and to simplify the determination of *P* values for small molecules. The fragmental constants are determined statistically by regression analysis; they are additive, and their sum is close to (and usually somewhat higher than) log *P*. Detailed tables of the *f* values for various functional groups have been published in the books of Rekker (1977) and Albert (1985), and are widely used in quantitative structure–activity determinations.

### Selected Readings

A. Albert (1985). *Selective Toxicity*, 7th ed. Chapman and Hall, London.

S. G. Farmer (1985). *Corpora non agunt fixita [sic]. Trends Pharmacol. Sci. 6*: 99–100.

J. Ferguson (1939). *Proc. Roy. Soc. [B] 127*: 384–396.

S. W. Fesik and A. Makriyannis (1985). Geometric requirements for membrane perturbation and anesthetic activity. *Mol. Pharmacol. 27*: 624–629.

B. R. Fink (Ed.) (1980). *Molecular Mechanisms of Anesthesia*. Raven Press, New York.

N. P. Franks and W. R. Lieb (1984). Do general anesthetics act by competitive binding to specific receptors? *Nature 310*: 599–601.

R. D. Kaufman (1977). Biophysical mechanisms of anesthetic action. Historical perspectives and review of current concepts. *Anesthesiology 46*: 49–62.

H. Kubinyi (1979). Lipophilicity and drug activity. In *Progress in Drug Research* (E. Jucker, Ed.), Vol. 23. Birkhäuser, Basel, pp. 97–198.

Y. C. Martin (1978). *Quantitative Drug Design*. Marcel Dekker, New York.

A. G. Marshall (1978). *Biophysical Chemistry*. Wiley, New York.

R. F. Rekker (1977). *The Hydrophobic Fragmental Constant*. Elsevier Science, New York.

S. H. Roth (1979). Physical mechanisms of anesthesia. *Annu. Rev. Pharmacol. Toxicol. 19*: 159–178.

S. H. Roth and K. W. Miller (Eds.) (1986). *Molecular and Cellular Mechanisms of Anesthesia*. Plenum Press, New York.

R. W. Ryall (1979). *Mechanism of Drug Action on the Nervous System*. Cambridge University Press, Cambridge, Chap. 6.

P. Seeman (1975). The action of nervous system drugs on cell membranes. In: *Cell Membranes* (G. Weissman, and R. Claiborne, Eds.). HP Publishing, New York, pp. 239–247.

## 4. SURFACE ACTIVITY AND DRUG EFFECTS

Biological reactions occur in solutions and on the surfaces and interfaces of solids and liquids. The energy situation at a surface differs markedly from that in a solution because special intermolecular forces are at work; therefore, surface reactions require specific consideration. In living organisms, membranes comprise the largest surface, covering all cells (the plasma membrane) as well as many cell organelles (the nucleus, mitochondria, and so forth). Dissolved macromolecules like proteins also account for an enormous surface area (e.g., 1 ml of human blood serum has a protein surface area of $100 \text{ m}^2$). Biological membranes also (1) serve as a scaffold that holds a large variety of enzymes in proper orientation, (2) provide and maintain a sequential order of enzymes that permits great efficiency in multistep reactions, and (3) serve as the boundaries of cells and many tissue compartments. In addition, many drug receptors are membrane bound.

It is therefore easy to see why the physical chemistry of surfaces and the structure and activity of surface-active agents are of interest to the medicinal chemist. Detergents, transport agents for ions, and many disinfectants and antibiotics exert their activity by interacting with biological surfaces.

### 4.1. Surface Interaction and Detergents

All of the molecules in a liquid phase interact with each other and exert a force on neighboring molecules. We have already discussed the hydrogen-bonding interaction of water molecules that creates flickering clusters. The water molecules at a gas–liquid interface, however, are exposed to unequal forces, and are attracted to the bulk water of the liquid phase because no attraction is exerted on them from the direction of the gas phase. This accounts for the surface tension of liquids.

Because the dissolution of a solid is the result of molecular interaction between a solvent and the solid (which, once dissolved, becomes a solute), polar compounds capable of forming hydrogen bonds are water soluble, whereas nonpolar compounds dissolve only in organic solvents as the result of van der Waals and hydrophobic bonds (see Sec. 7). Compounds that are *amphiphilic* (i.e., containing hydrophobic as well as hydrophilic groups) will concentrate at surfaces and thereby influence the surface properties of these interfaces. Only in this way can amphiphilic detergents, through their hydrogen bonding with water and nonpolar interaction

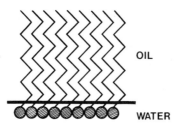

**Fig. 1.5.** Micellar structure of a soap molecule on an oil–water (or air–water) interface. The nonpolar alkyl chains are in the nonpolar phase; the polar carboxylate head groups are in the aqueous phase.

with a nonpolar (organic) phase or with air, maintain an orientation that ensures the lowest potential energy at an interface. A classic example of such behavior is given by soap, a mixture of alkali metal salts of long-chain fatty acids. Figure 1.5 shows the interaction of soap molecules at an oil–water boundary. The circle symbolizes the anionic carboxylate or the polar "head group," and the zig-zag line represents the hydrophobic alkyl chain.

A detergent like soap forms a colloidal solution. At a very low concentration, soap molecules will be dissolved individually. At a higher concentration, the molecules find it more energy-efficient to "remove" their hydrophobic tails from the aqueous phase and let them interact with each other, thus forming a miniature "oil drop" or nonpolar phase, with the polar heads of the soap molecules in the bulk water. At a concentration that is characteristic for a given individual detergent, molecular aggregates, known as *micelles*, are formed. They are often spherical colloidal

**Fig. 1.6.** Different micellar structures: (A) Cross section of a soap micelle in water. (B) Half of a cylindrical soap micelle. (C) Cross section of an "inverted" soap micelle in a nonpolar solvent. (D) Dispersion of a soil particle by soap, making the outside polar and thus water soluble.

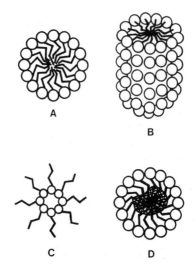

particles, but can also be cylindrical, as shown in Fig. 1.6A,B. The concentration at which such micelles are formed is called the *critical micellar concentration*, and can be determined by measuring the light diffraction of the solution as a function of detergent concentration. The diffraction will show a sudden increase when micelles begin to form.

When soap is dispersed in a nonpolar phase, inverted micelles are formed, in which the nonpolar tails of the soap molecules interact with the bulk solvent while the hydrophilic heads interact with each other, as shown in Fig. 1.6C. This behavior of amphiphilic molecules explains how they can disperse nonpolar particles in water: the hydrocarbon tail of the amphiphile interacts with the particle, such as an oil droplet, dirt, or a lipoprotein membrane fragment, covers the particle, and then presents its hydrophilic head groups to the acqueous phase, as shown in Fig. 1.6D.

### 4.2. Surface-Active Antibacterial Agents

Biological membranes are indispensable to the proper functioning of all cells, including bacteria and fungi. Hence, any agent that disrupts the membrane or otherwise interferes with its integrity or function is a potential threat to the life of the cell.

As discussed in Sec. 2 on solubility, *aliphatic* alcohols are bactericidal (kill bacteria) because they damage the bacterial membrane, resulting in a rapid loss of the cytoplasmic constituents of the bacterium. At high concentrations, such alcohols cause lysis (dissolution) of the bacterial cell.

Because they damage bacterial membranes, **phenol** and **cresol** are also effective disinfectants, and are used in such preparations as Lysol. They not only denature proteins, but also act as a detergents, owing to the polarity of the phenolic hydroxyl group. The activity of phenols can be increased considerably by attaching an alkyl side chain to the benzene ring, as in **n–hexylresorcinol** (1-14), which makes the resulting compound more surface active. **Hexachlorophene** (1-15) and **fentichlor** (1-16) are also very active and are used in disinfectant soaps. Moreover, since fentichlor has a low oral toxicity, it can be used internally for skin infections. Inhibition of the metabolic electron-transport chain and of amino acid uptake also constitutes the bacteriostatic action of fentichlor.

The cationic detergents such as **cetyl-trimethylammonium chloride** (1-17) are more effective than anionic "soaps" such as **sodium dodecylsulfonate** (SDS; 1-18). Nonionic detergents such as **Triton X-100** [octoxynol; (polyethylene glycol)$_{10}$-$p$-isooctylphenyl ether] are very mild, and are used to disperse membranes rather than to kill bacteria. **Chlorhexidine** (1-19), a chlorophenyl-biguanidine derivative, is a very effective compound. It has very low mammalian toxicity and is widely used as a wound and burn antiseptic and surgical disinfectant. Because the imino group of the biguanidine moiety becomes protonated by salt formation, the compound is a cationic detergent. Low concentrations (10–100 $\mu$g/ml) cause a rapid release of cytoplasmic material from the bacterial cell. At concentrations as low as 1 $\mu$g/ml, little leakage is seen, but chlorhexidine is still active because it inhibits the membrane-bound ATPase of bacteria.

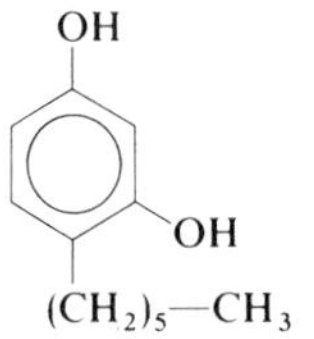

$n$-Hexylresorcinol
1-14

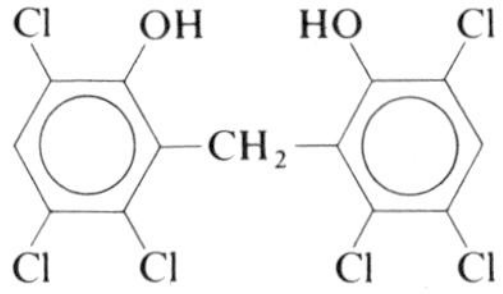

Hexachlorophene
1-15

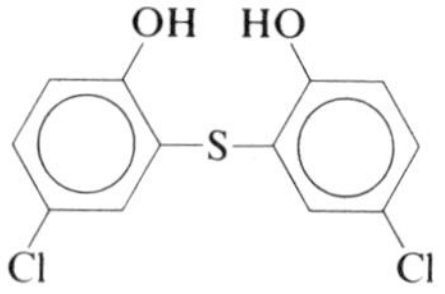

Fentichlor
1-16

$$CH_3-(CH_2)_{15}-\overset{\oplus}{N}(CH_3)_3Cl^{\ominus}$$

Cetyl-trimethyl-ammonium chloride
1-17

$$CH_3-(CH_2)_{11}-SO_3^{\ominus}Na^{\oplus}$$

Sodium dodecylsulfonate
1-18

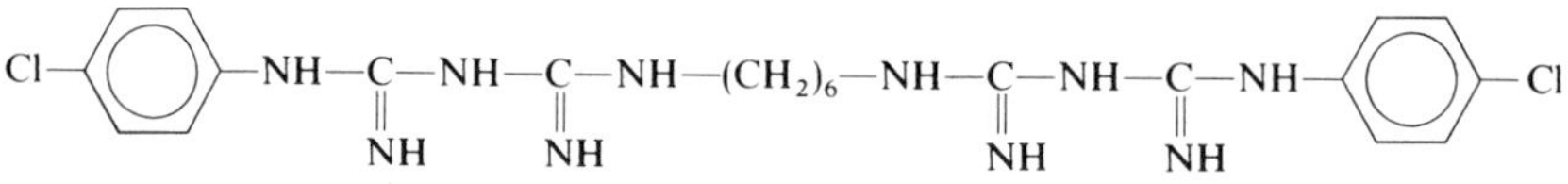

Chlorhexidine
1-19

## 4.3. Membrane-Active Antifungal Agents

Some *Streptomyces* species produce macrocyclic (large-ring) compounds containing numerous (three to seven) double bonds and even more hydroxyl groups, which are usually located on one side of the molecule. Antibiotics of such structure as **amphotericin B** (1-25) and the very similar **nystatin** interact with sterols (see Chap. 5) in the microbial plasma membrane. From five to ten molecules of these antibiotics form a conducting pore or channel through which $K^+$ ions, sugars, and proteins are lost from the microorganism. The structure of such a pore is shown in Fig. 1.7. The inside of the pore is lined with the hydroxyl groups of the antibiotic molecule; the polyene section interacts with the hydrophobic sterol component of the cell membrane. Smaller macrocyclic antibiotics may form aggregates inside the lipoprotein membrane of the cell and cause general membrane disruption. Because these antibiotics interact preferentially with ergosterol, a plant sterol, they have a high affinity for plant membranes, and are therefore fungicidal. With regard to animal membranes, they have a strange selectivity, being lethal to flatworms and snails, not affecting bacteria, and exerting intermediate toxicity in mammals. They are used for such fungal infections as athlete's foot and candidial vaginitis, which are otherwise persistent and difficult to control, and for systemic (internal) fungal infections, which are nearly always fatal if untreated. Thus, the therapeutic value of amphotericin B outweighs the drawback of its toxic side effects.

The recently discovered group of *azole antifungal agents* also act through membrane destabilization by inhibiting ergosterol biosynthesis (see Chap. 5, Sec. 1.1), necessary and specific for fungal membranes. **Ketoconazole** (1-20) can be taken orally, whereas **clotrimazole** (1-21), **miconazole** (1-22), and related compounds are applied topically. The azoles bind very avidly to skin and accumulate in fungi very rapidly. Although contact times may be only 15–20 minutes, the azole remains in the fungus for 120 hours and thus, even in sublethal doses,

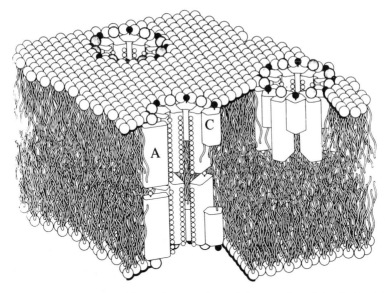

**Fig. 1.7.** Schematic representation of pores formed by amphotericin B (A, 1-25) and cholesterol (C) in a lipid bilayer, showing in cross section the interior of a conducting pore formed from two half pores. (Reproduced by permission from B. de Kruijff and R. A. Demel (1974), *Biochim. Biophys. Acta 339*: 63, Elsevier Biomedical Press, Amsterdam)

leads to a decrease in virulence. Another group of sterol synthesis inhibitors includes the naphthyl-allylamines, represented by **naftifine** (1-23). Morpholines like **tridemorph** (1-24) attack sterol biosynthesis at other points and are valuable agricultural fungicides.

Ketoconazole
1-20

Clotrimazole
1-21

Miconazole
1-22

Naftifine
1-23

Tridemorph
1-24

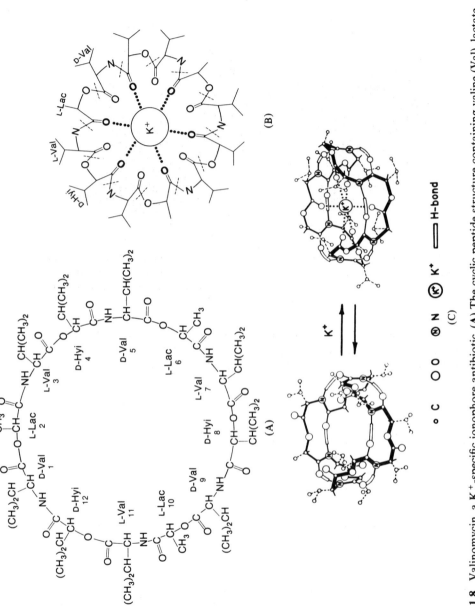

**Fig. 1.8.** Valinomycin, a K$^+$-specific ionophore antibiotic. **(A)** The cyclic peptide structure containing valine (Val), lactate (Lac), and hydroxyisovalerate (Hyi). **(B)** The schematic of the complex: the polar carbonyls hold the K$^+$ ion. The nonpolar side chains are on the outside. **(C)** True conformation of valinomycin in solution and in a complex. (Reproduced by permission from G. Zubay (1983), *Biochemistry*, Addison-Wesley, Reading, MA; and Ovchinnikov (1978), Academic Press, New York)

## 4.4. Ion-Conducting Antibiotics

Some bacteria produce compounds that can become incorporated into lipid membranes and will facilitate the transport of ions, notably $K^+$. Hence, these compounds are called *ionophores*, or ion carriers, in contrast to the polyene antibiotics discussed above, which simply produce leakage through the cell membrane.

Such ionophoric antibiotics can function either as "cage" carriers of an ion or as channel formers. The cage carrier encloses an ion and transports it from one side of the membrane to the other, releasing the ion on the other side. A channel former simply provides a polar tunnel that allows the migration of a polar ion across an otherwise impenetrable lipid layer.

An example of an ionophore that is a cage carrier is **valinomycin** (Fig. 1.8). This cyclic peptide lactone consists of three molecules each of L-valine, D-α-hydroxyisovaleric acid, and L-lactate. The six highly polarized lactone carbonyl oxygens line the inside of the ring, whereas the nonpolar alkyl groups point to the outside of the molecule. Thus, the polar inside can accommodate a nonhydrated potassium ion and surround it with an apolar bracelet. This complex can then be transported through a membrane in an energy-dependent $K^+$–$H^+$ exchange. The selectivity of valinomycin for $K^+$ over $Na^+$ is very high, the ratio being about $10^4$:1. In this way, valinomycin will increase the $K^+$ conductivity of lipid membranes at concentrations as low as $10^{-9}$ M. The high $K^+$ selectivity is due to the relative ease of dehydration of this ion: with its larger diameter, the potassium ion holds hydrate water less firmly than does sodium: consequently, whereas the hydrated sodium ion does not fit the valinomycin "doughnut," the dehydrated $K^+$ does bind easily, with the bonding energy providing a further energy advantage for the selective reaction. The anhydrous and hydrated diameters of $Na^+$ and $K^+$ are, respectively, 0.095 and 0.179 nm for $Na^+$ and 0.122 and 0.133 nm for $K^+$. Clearly, a hydrated sodium ion is larger than a potassium ion with or without a hydrate envelope.

Many antibiotics act similarly to valinomycin, some forming "sandwich" complexes with metal ions. Lambert (1978) and Ovchinnikov (1978) discuss these compounds in detail and give further references.

An example of a channel- or pore-forming antibiotic is **gramicidin A**, a peptide consisting of 15 amino acids (Fig. 1.9). It induces the transmembrane transport of protons, alkali-metal ions, and thallium ions at concentrations as low as $10^{-10}$ M, even though it is unable to complex these ions in solution. Gramicidin also forms several dimers with itself.

There are several hypotheses explaining the channel formation induced by gramicidin A. According to the model shown in Fig. 1.9, two molecules of this compound form a "head-to-head" helix, spanning the total width of the cell membrane. The helix creates a pore lined with hydrophilic groups, permitting ion transport across the otherwise impermeable lipid barrier of the membrane. Such a pore can be as wide as 0.6 nm and can accommodate even large ions as long as they are dehydrated. However, the movement of ions in the pore is not well understood.

**Colicins**, bacteriostatic peptides encoded by bacterial plasmids, have recently been crystallized and investigated by x-ray crystallographic methods. They have a

Val—Gly—Ala—Leu—Ala—Val—Val—Val—Trp—Leu—Trp—Leu—

Trp—Leu—Trp—NH—CH$_2$—CH$_2$—OH

Amino Acid Sequence of Gramicidin A

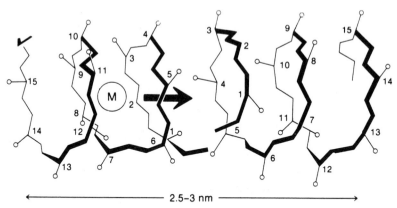

2.5–3 nm

**Fig. 1.9.** Schematic representation of the $\pi^6$-helix of gramicidin A in solution (top) and in the membrane as a pore (bottom). The bold lines are closer to the viewer; arrow represents the direction of ion (M) movement. The side chains are omitted for clarity. (Reproduced by permission from Ovchinnikov (1978), Academic Press, New York)

mass of 79,000 Daltons and an axial ratio of 1:10, giving these peptides a length of about 20 nm. They are capable of forming a transmembrane channel with a diameter so large that glucose molecules can pass through. Of course, this fatally disrupts the membrane potential of the bacterial cell, with consequent bacteriostatic activity.

Ionophoric antibiotics do not distinguish microbial from mammalian membranes, and are therefore therapeutically useless. However, they are excellent tools for studying membrane transport phenomena.

### 4.5. Synthetic Ionophores

The structure of synthetic ionophores is based on a simplified form of "cage" carrier. The synthetic ionophores known as **crown compounds** (e.g., 1-26 and 1-27) are named according to the total number of atoms in the macrocyclic ring and the number of hetero atoms in the ring. The molecules of these compounds are normally flat. The **cryptates** (e.g., 1-28 and 1-29), similar compounds containing nitrogen atoms, are capable of forming three-dimensional cages.

Crown compounds and cryptates can be designed to incorporate ions of any size. These ionophores can even render extremely polar compounds such as KOH or KMnO$_4$ lipophilic and soluble in nonpolar solvents, so that these ions can participate in reactions that would not otherwise take place. Ionophores also have interesting applications in organic synthesis: for example, they can increase the concentration of "naked" anions by trapping the cations normally associated with

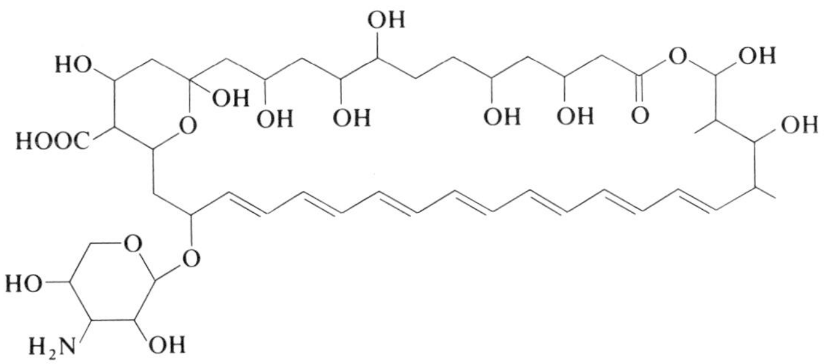

Amphotericin B
1-25

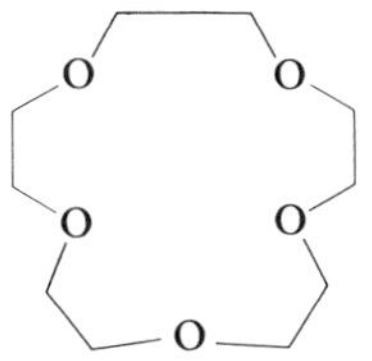

15-Crown-5
1-26

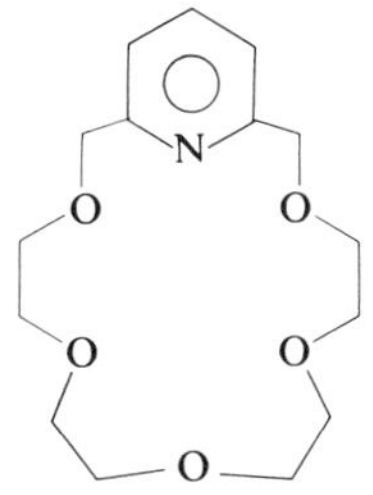

Pyridyl-18-crown-6
1-27

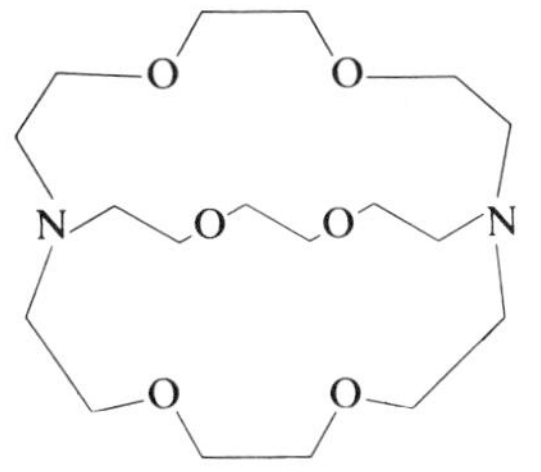

1-28

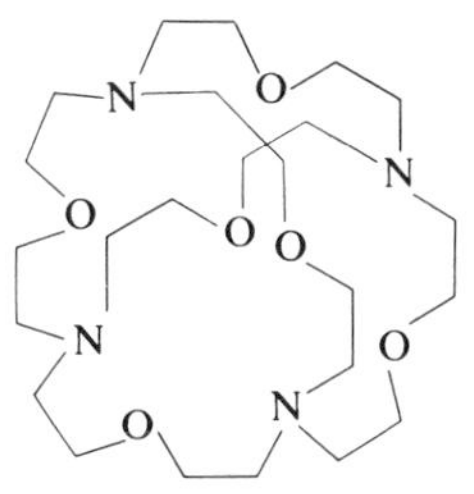

1-29

these anions. Other unusual reactions and their applications are reviewed by Izatt et al. (1979).

Some of the synthetic acidic ionophores have interesting pharmacological properties because they influence $Ca^{2+}$ transport; these compounds cause an increase in cardiac contractility, diuresis, and coronary blood flow, as well as a decrease in peripheral vascular resistance (Pressman, 1976). The acidic ionophores also have a potential application in removing toxic heavy metals or the carcinogenic radioactive $^{90}Sr$—a product of fallout from nuclear explosions which passes, through milk, into the bones of children—from the bones in the presence of large amounts of Ca. This is important because previously it was not possible to remove $Sr^{2+}$ selectively without adversely affecting calcium metabolism and bone structure. However, the considerable toxicity of the available crowns and cryptates restricts their use in humans, and further work seems to be warranted.

### Selected Readings

A. Albert (1985). *Selective Toxicity*, 7th ed. Chapman and Hall, London.

D. Berg, K. H. Büchel, M. Plempel and E. Regel (1986). Antimycotic sterol biosynthesis inhibitors. *Trends Pharmacol. Sci.* 7: 233–238.

M. B. Gravestock and J. F. Ryley (1984). Antifungal chemotherapy. *Annu. Rep. Med. Chem.* 19: 127–136.

R. M. Izatt, J. D. Lamb, D. J. Eatongh, J. J. Christensen, and J. H. Rytting (1979). Design of selective ion binding macrocyclic compounds and their biological application. In: *Drug Design* (E. J. Ariëns, Ed.), Vol. 8. Academic Press, New York, pp. 355–400.

P. A. Lambert (1978). Membrane-active antimicrobial agents. In: *Progress in Medicinal Chemistry* (G. P. Ellis and G. B. West, Eds.), Vol. 15. Elsevier/North Holland, Amsterdam, pp. 87–119.

Y. A. Ovchinnikov (1978). The chemistry of membrane-active peptides and proteins. In: *Frontiers of Physicochemical Biology* (B. Pullman, Ed). Academic Press, New York, pp. 81–113.

B. C. Pressman (1976). Biological application of ionophores. *Annu. Rev. Biochem.* 45: 501–529.

S. H. Yalkowsky, A. A. Sinkula, and S. C. Valvani (Eds.) (1980). *Physical Chemical Properties of Drugs.* Marcel Dekker, New York.

## 5. STEREOCHEMICAL ASPECTS OF DRUG ACTION

Because they interact with optically active, asymmetric biological macromolecules such as proteins, polynucleotides, or glycolipids acting as receptors, it is eminently reasonable that many drugs show stereochemical specificity. This means that there is a difference in action between isomers of the same compound, with one isomer showing pharmacological activity while the other is more or less inactive. Louis Pasteur was the first to show, in 1860, that molds and yeasts can differentiate between (+)- and (–)-tartarates, utilizing only one of the two isomers.

Therefore, complementarity between an asymmetric drug and its asymmetric receptor is often a criterion of drug activity. The effects of highly active or highly specific drugs depend more upon such complementarity than do those of weakly active drugs. Occasionally, the stereoselectivity of a drug is based on a specific and preferential metabolism of one isomer over the other, or on a biotransformation that selectively removes one isomer (see Low and Castagnoli, 1978). Such stereoselective biotransformations may have far-reaching consequences. For instance, microsomal hydroxylation of the tranquilizer **diazepam** (Valium) (1-30) occurs stereoselectively, yielding **(S)-N-methyloxazepam** (1-31). Since this hydroxylated metabolite is pharmacologically active, the stereochemical circumstances of the activation process are crucial, not only for the extent of the activation but also for the rate of elimination of the metabolite.

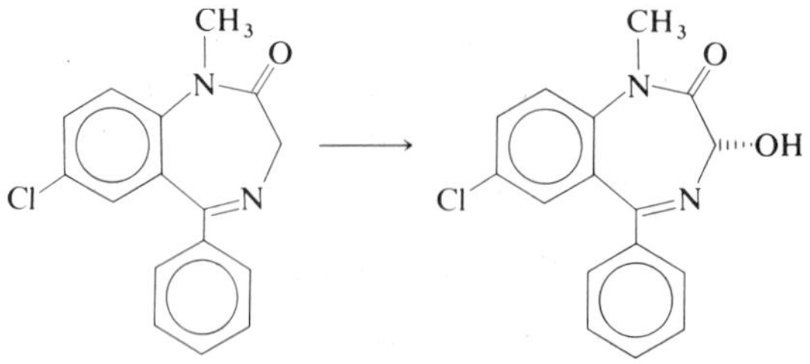

Diazepam
1-30

(S)-N-Methyloxazepam
1-31

The toxicity of environmental carcinogens is also influenced by stereoselective transformations in *vivo* (Fig. 1.10). **Benzo[a]pyrene** (1-32) is transformed to the (–)-*trans*-diol 1-33A, which then undergoes epoxidation to give a 9:1 ratio of

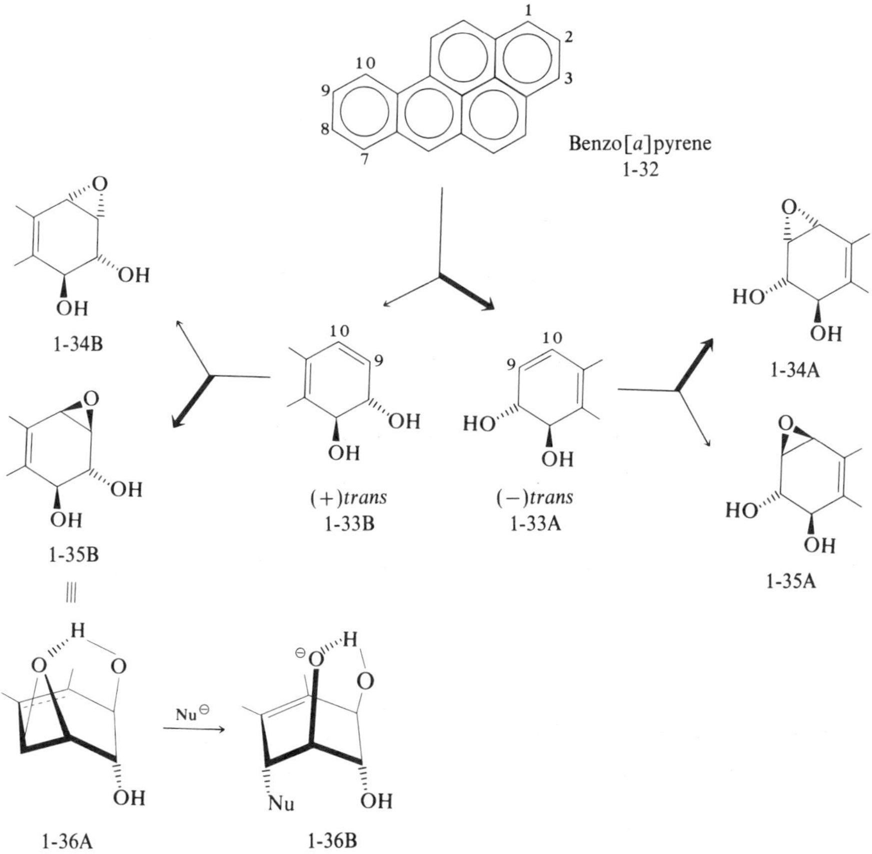

**Fig. 1.10.** Metabolic conversion of benzo [a] pyrene to the carcinogenic diol-epoxides, and their reaction with a nucleophilic biomolecule. (After Low and Castagnoli, 1978)

epoxides 1-34A and 1-35A. The enantiomeric (+)-*trans*-diol 1-33B epoxidizes even more selectively; the ratio of the epoxides is 1:22 for 1-34B and 1-35B. Where the epoxide and 4-OH are both axial (1-35A and 1-35B), anchimeric assistance (as shown in 1-36A and 1-36B) promotes nucleophilic attack on proteins or DNA. However, the (+)-diol-epoxide 1-35B is twice as mutagenic as its (−) epimer.

## 5.1. Optical Isomers

*Optical isomerism* is the result of a dissymmetry in molecular substitution. The basic aspects of optical isomerism are discussed in various textbooks of organic chemistry (see also Tamm, 1982; Rétey and Robinson, 1982), should the reader need to review this area. Optical isomers (enantiomers) *may* have different physiological activities from each other *provided* that their interaction with a receptor or some other effector structure involves the asymmetric carbon atom of the enantiomeric molecule *and* that the three different substituents on this carbon atom interact with the receptor.

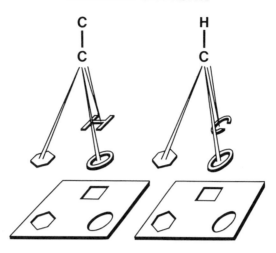

**Fig. 1.11.** Model of the chiral carbon of the two enantiomers of norepinephrine making contact with the receptor. One enantiomer is capable of making a three-point contact, which is necessary for pharmacological activity; the other enantiomer makes a two-point contact only, because the C atom cannot bind to the H bonding site. The figure illustrates the Easson–Stedman hypothesis.

The Easson–Stedman hypothesis assumes that a three-point interaction ensures stereospecificity, since only one of the enantiomers will fit; the other one is capable of a two-point attachment only, as shown in Fig. 1.11 for the reaction with a hypothetical planar receptor. However, it is reasonable to assume that receptor stereospecificity can also undergo a change when the receptor conformation is altered by a receptor–drug interaction.

The difference in pharmacological action between enantiomers can be considerable. (–)-**Levorphanol**, a synthetic analgesic, has a binding equilibrium constant $(K_D)$ of $10^{-9}$ M. ($K_D$ is a dissociation constant, indicating that this drug will occupy half of all accessible morphine receptors at a nanomolar concentration.) (+)-**Dextrorphan**, the optical antipode of (–)-levorphanol, has a $K_D$ of $10^{-2}$ M, reflecting a high and nonphysiological concentration. Qualitatively, dextrorphan is not an analgesic at all, but a very effective antitussive (cough suppressant), an action entirely different from analgesia. (+)-**Muscarine** is about three orders of magnitude more effective as a cholinergic neurotransmitter than (–)-**muscarine**. A very large body of data is available on the selectivity of enantiomeric drugs (cf. Lehman et al., 1976; Stenlake, 1979).

It should be emphasized that the mere sign (+ or –) of the optical rotation produced by an enantiomer is not biochemically decisive to the action of such a molecule. The *absolute* configuration of the compound in question must be considered, and in modern organic chemistry the *Cahn–Ingold–Prelog sequence rules* are followed, which have increasingly replaced the ambiguous and obsolescent D and L designations for absolute configuration. Again, the reader is referred to modern organic chemistry texts for details. The sequence rules relate the absolute configuration of all compounds to (+)-glyceraldehyde, designated as *R* (*rectus* = right-handed) substance.

Even though enantiomeric drug pairs quite often show different potencies, they are seldom antagonists of each other, since the differences in their action are due to differences in their binding properties; antagonists (see Chap. 2, Sec. 3) usually bind more strongly than agonists, and the less active enantiomer of a pair is incapable of displacing the more active one from the receptor. Similarly, such nonspecific drugs as general anesthetics are not stereospecific because they do not act primarily on specific receptors, which are normally dissymmetric macromolecules.

*Diastereomeric drugs*—those having two or more asymmetric centers—are usually active in only one configuration. Unlike enantiomers, which have identical physicochemical properties, the absorption, distribution, receptor binding, metabolism, and every other aspect that influences that influences the pharmacological activity of a drug are different for each diastereomer.

### 5.1.1. Enantiomers and Pharmacological Activity

Lehman et al. (1976) stated the definitions of stereoselectivity in the following manner: the better-fitting enantiomer (the one with higher affinity for the receptor) is called the *eutomer*, whereas the one with the lower affinity is called the *distomer*. The ratio of activity of the eutomer and distomer is called the *eudismic ratio*; the expression of the *eudismic index* is

$$EI = \log \text{affinity}_{Eu} - \log \text{affinity}_{Dist}$$

The correlation of eutomers and distomers with pharmacological activity is illustrated in Fig. 1.12.

**Fig. 1.12.** Plot of affinities of the eutomers and distomers of a series of oxotremorine (4-18) analogues against the affinity of the eutomers. The eutomers (always the *R* isomers in this series), naturally, lie on a line of unit slope, but the distomers form a random pattern. (Reproduced by permission from P. A. Lehman (1983), in *Mechanism of Drug Action* (T. P. Singer et al., Eds.), Academic Press, New York)

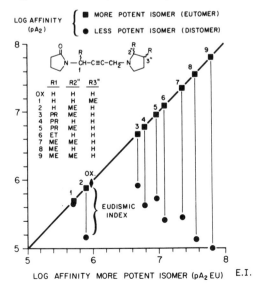

In a series of agonists and antagonists (for definitions, see Chap. 2, Sec. 3), the eudismic affinity quotient can also be defined as a measure of stereoselectivity. Because of widespread misconceptions the distomer of a racemate is often considered "inactive" and of no consequence to pharmacological activity, an idea reinforced by the fact that resolution (i.e., separation) of racemates is economically disadvantageous. Ariëns and his associates published a series of books and papers that show the fallacy of this concept and point out the necessity of using pure enantiomers in therapy and research.

Therefore, the distomer should be viewed as an impurity constituting 50% of the total amount of a drug—an impurity that in the majority of cases is by no means "inert." Soudijn (in Ariëns et al., 1983) lists all the possible unwanted effects of a distomer:

1. It contributes to side effects.
2. It counteracts the pharmacological action of the eutomer.
3. It is metabolized to a compound with unfavorable activity.
4. It is metabolized to a toxic product.

However, there are instances in which the use of a racemate has advantages; sometimes it is more potent than either of the enantiomers used separately (e.g., the antihistamine, isothipendyl), or the distomer is converted into the eutomer *in vivo* (the antiinflammatory drug ibuprofen).

Lately, there has been a tendency to develop drugs with two or more types of action, often with different mechanisms of action. In such a *hybrid drug* (which may be a racemate), the relative proportions of the types of action are predetermined; using two drugs with individual activity instead of a single drug allows finer optimization of therapy, tailored to the needs of the individual patient. The approach can become quite complicated, however, as outlined in the impassioned paper by Ariëns (1984).

Admittedly, the separation of enantiomers is often difficult and expensive. In such instances, we may have no alternative to the use of racemates. However, sometimes an achiral drug can have an effect identical or superior to that of its chiral analogue (e.g., sufentanil vs. morphine; see Chap. 5, Sec. 3.7), in which case its use may be justified on that basis alone.

### Selected Readings

E. J. Ariëns, W. Soudijn, and P. B. M. W. M. Timmermans (Eds.) (1983). *Stereochemistry and Biological Activity of Drugs*. Blackwell, Oxford.

E. J. Ariëns (1984). Stereochemistry, a basis for sophisticated nonsense in pharmacokinetics and clinical pharmacology. *Eur. J. Clin. Pharmacol. 26*: 663–668.

E. J. Ariëns (1986). Chirality in bioactive agents and its pitfalls. *Trends Pharmacol. Sci. 7*: 200–205.

P. A. Lehman, E. J. Ariëns, and J. F. Rodrigues de Miranda (1976). Stereoselectivity and affinity in molecular pharmacology. In: *Drug Research* (E. Jucker, Ed.), Vol. 20. Birkhäuser, Basel, pp. 101–142.

P. A. Lehman (1986) Stereoisomerism and drug action. *Trends Pharmacol. Sci. 7*: 281–285.

L. R. Low and N. Castagnoli (1978). Enantioselectivity in drug metabolism. *Annu. Rep. Med. Chem. 13*: 304–315.

J. Rétey and J. A. Robinson (1982). *Stereospecificity in Organic Chemistry and Enzymology*. Verlag Chemie, Weinheim.

J. B. Stenlake (1979). *Foundations of Molecular Pharmacology*, Vol. 2. The Athlone Press, London, Chap. 3.

C. Tamm (Ed.) (1982). *New Comparative Biochemistry*, Vol. 3: *Stereochemistry*. Elsevier, Amsterdam.

## 5.2. Geometric Isomers

*Cis/trans* isomers are the result of restricted rotation along a chemical bond owing to double bonds or rigid ring systems in the isomeric molecule. *Cis/trans* isomers are not mirror images and have very different physicochemical properties, as reflected in their pharmacological activity. Because the functional groups in these molecules are separated by different distances in the different isomers, they cannot as a rule bind to the same receptor. Therefore, geometric isomerism as such is not of major interest to the medicinal chemist. What does become important as a result of such isomerism is the reactivity and accessibility of substituents on a rigid framework. This aspect will be discussed in conjunction with conformational analysis (see Sec. 5.3).

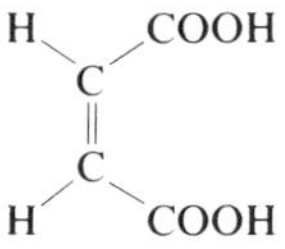

| Maleic acid | Fumaric acid | Decalin | |
| cis | trans | trans | cis |

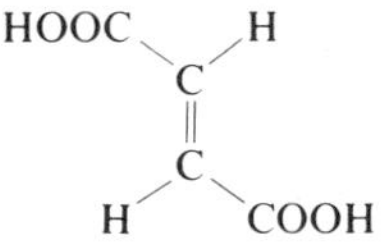

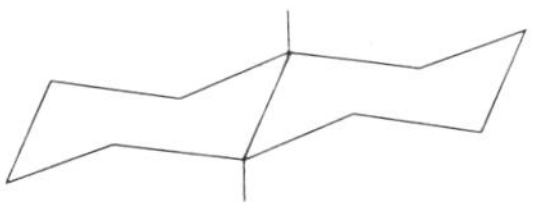

Using the Cahn–Ingold–Prelog sequence rules, Blackwood et al. (1968) designed a system to allow an unambiguous "absolute" assignment of *cis/trans* (or *syn/anti* in the case of C=N bonds) isomerism. For instance, the compound $CHCl=CBrI$ cannot be named unambiguously by the classical rules. However, after the priority of substituents on each carbon atom is determined (using the sequence rules), the configuration in which the two substituents of higher priority lie on the same side is called the Z isomer (for *zusammen*, meaning "together" in German). The configuration in which these substituents lie on opposite sides is designated at the E isomer (for *entgegen*, which means "opposite").

## 5.3. Conformational Isomerism

Isomerism can also be shown by compounds in which the free rotation of atoms around chemical bonds is only slightly hindered. The energy barrier to the transition between different conformations of these isomers is usually very low (on the order of 4–8 kJ/mol), and is easily overcome by thermal motion unless the molecule is made rigid or because nonbonding interactions between functional groups of the molecule favor one conformer over an infinite number of others. The concept and biophysical reality of "preferred" drug conformations and their potential role in receptor binding are currently hotly debated issues among molecular pharmacologists.

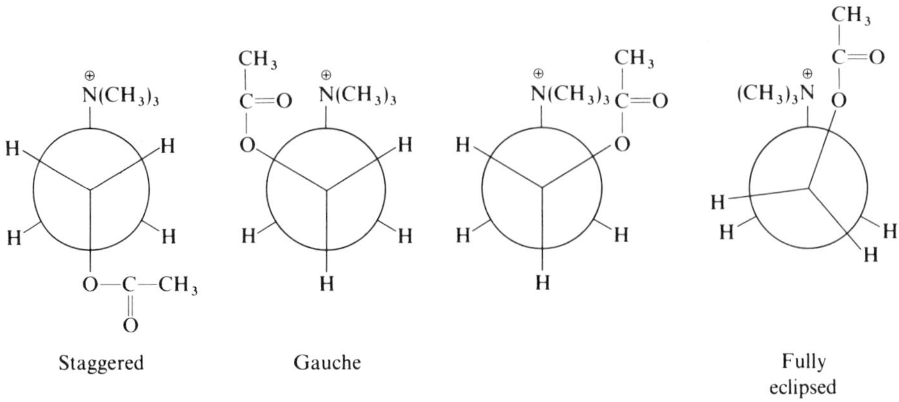

Staggered　　　　　　　　Gauche　　　　　　　　　　　　　　　　Fully
eclipsed

**Fig. 1.13.** Newman projection of acetylcholine conformers.

For aliphatic compounds, the well-known *Newman projection* is used to show the relative position of the substituents on two atoms connected to each other (as in ethane derivatives). For example, Fig. 1.13 shows several possible conformers of acetylcholine. When the trimethylammonium-ion and acetoxy functional groups are as far removed as possible, we speak of a *fully staggered* conformation (erroneously and confusingly also called a *trans* conformation). When the two groups overlap, they are *eclipsed*. Between these two extremes are an infinite number of conformers called *gauche* (or skew) conformers or rotamers (rotational isomers). The potential interaction energy of the trimethylammonium-ion and acetoxy groups is lowest in the staggered conformation and highest when the two groups are eclipsed. The stability of these rotamers is normally opposite. An exception to this exists when two functional groups show a favorable nonbonding interaction (e.g., hydrogen-bond formation).

Because the transition between rotamers occurs very rapidly, the existence of any one conformer can be discussed in statistical terms only. For example, it has been assumed that long hydrocarbon chains exist in the staggered, fully extended, zig-zag conformation. There is, however, a considerable probability of their also existing in skew conformations, effectively reducing the statistical length of the carbon chain. Such considerations become important if one wants to calculate effective intergroup distances in drugs, which play a role in the fit and binding to receptors. For instance, in the anticholinergic drugs **hexamethonium** and **decamethonium** (see Chap. 4, Sec. 2.5), the two quaternary trimethylammonium groups are connected by six and ten $-CH_2-$ groups, respectively. The calculated internitrogen distances based on fully extended six- and ten-carbon chairs in the two drugs, are 0.95 and 1.35 nm, respectively. However, conductimetric experiments show that the effective distances are only 0.63 and 0.95 nm. The latter estimates correlate well with the internitrogen distances seen in rigid neuromuscular blocking like agents **curare** (see Stenlake 1979, p. 132, and Chap. 4, Sec. 2.5).

A low conformational transition barrier can be overcome by the energy of binding to receptors of any kind, including enzymes. This statement is the basis of the *induced fit theory* of Koshland (see Chap. 2). An example of this theory is given by Leger et al. (1980). Using crystallographic data for β-adrenergic agonists, these authors have shown that the distance of the aromatic ring of such an agonist from the charged nitrogen and from the oxygen is critical for pharmacological activity. If necessary, the molecule will fold to preserve this optimum conformation, which is found neither in the solid state nor in a solution of the free drug.

Observations emphasize the need for extreme caution in proposing hypotheses dealing with drug conformations and their correlations with receptor structure or drug design. Since the mid-1970s, many publications have proposed receptor mapping techniques based on the distances between assumed key atoms (usually heteroatoms) or functional groups in drugs, determined by tedious quantum-chemical calculations of "preferred" conformers. Similarly, the design of a number of drugs has been based on shaky assumptions about drug–receptor binding, all founded on conformational analysis. These oversimplifications have been criticized (Martin, 1978, pp. 261–266). Such caveats however, do not detract from the utility of conformational analysis of drugs, of the importance of calculating intergroup distances, or of the potential value of these methods in molecular pharmacology. Another proposed assumption that did not stand the test of time was that the nicotinic and muscarinic cholinergic receptors interacted preferentially with the gauche (folded) and staggered (extended) acetylcholine conformers, respectively (see Chap. 4, Sec. 2). However, the question triggered the development of *rigid analogues* of flexible drugs. The *cis* and *trans* isomers of acetoxycyclopropyl-trimethylammonium iodide are two such analogues of acetylcholine: the additional methylene bridge freezes the staggered and eclipsed acetylcholine conformers in the *trans* and *cis* forms of this cyclopropane derivative. The (+)-*trans* isomer has proven to be almost equipotent with acetylcholine at the muscarinic receptor, but shows little nicotinic activity, and is easily hydrolyzed by acetylcholinesterase, the enzyme that inactivates acetylcholine. The racemic *cis* isomer shows practically no activity at either the nicotinic or muscarinic receptor. This indicates only that acetylcholine probably *assumes* the staggered conformation at the muscarinic receptor.

The conformational analysis of cyclohexane and its derivatives has been very well explored. The cyclohexane ring itself can assume several conformations. The chair conformation is more stable than either the boat or twist form because it permits the maximum number of substituents to exist in a staggered conformation relative to their neighbors. The substituents can assume two conformations relative to the plane of the ring (defined by carbon atoms 2, 3, 5, and 6): *axial* (a), in which they point up or down; and *equatorial* (e), in which they point in the direction of the ring's circumference. As the cyclohexane ring keeps flipping back and forth between many chair forms, the substituents on the ring alternate between axial and equatorial conformations unless stabilized.

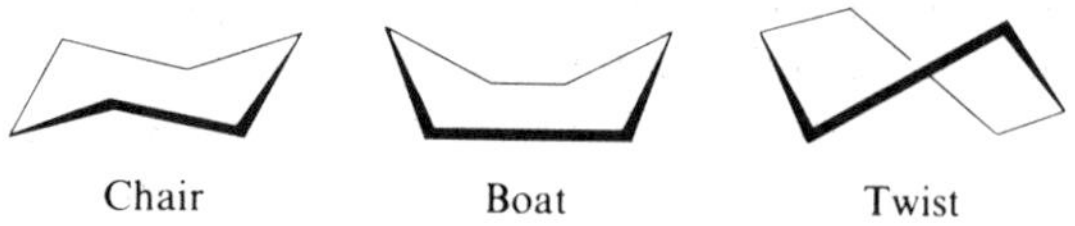

Chair          Boat          Twist

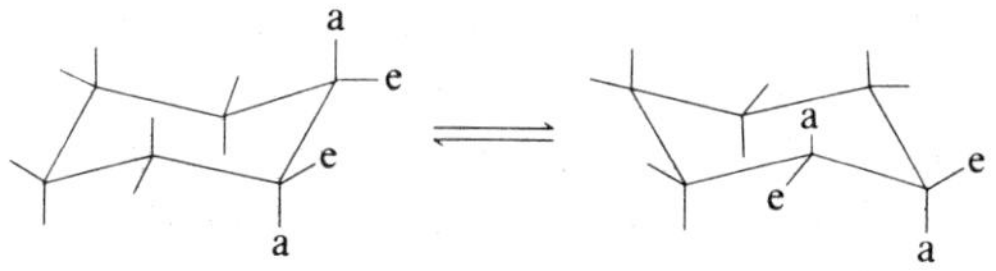

There are several ways to stabilize or "freeze" a given conformation:

1. By electrostatic repulsion of two adjoining substituents (e.g., in 1,2-dichlorocy-clohexane, a diaxial conformation is forced).
2. By steric repulsion.
3. By using a bulky substituent like the *t*-butyl group, which always maintains an equatorial position.

Polycyclic structures like decaline or the steroids are rigid and maintain stable conformations. In such rigid systems, the axial and equatorial substituents can display *cis/trans* isomerism without the presence of a double bond; the restriction on their rotation is ensured by the ring system itself. Diastereomerism can also occur in these molecules. In substituted cyclohexanes or their heterocyclic analogues, 1,2-diaxial or the equivalent diequatorial substituent pairs are considered to be *trans*, while the axial–equatorial pair is regarded as *cis*. 1,3-Diequatorial substituents are, however, *cis*.

The axial or equatorial nature of a substituent has a bearing on its reactivity or ability to interact with its environment. Equatorial substituents are more stable and less reactive than their axial counterparts. For example, equatorial carboxyl groups are stronger acids than axial ones because of the higher stability of the carboxylate ion, whereas equatorial esters are hydrolyzed more slowly than axial ones because they are less accessible to protons or hydroxyl ions during acid- or base-catalyzed hydrolysis.

When contemplating the effect of drug conformation on drug–receptor interactions, one must not forget that the receptor macromolecule also undergoes changes in its molecular geometry, as postulated by the Koshland induced-fit hypothesis. Owing to the enormously more complex nature of macromolecular structure, very little is known about such changes. Figure 1.14 shows the scheme of such a conformational alteration, for a hemoglobin protomer (Albert, 1985). Although the binding of oxygen to hemoglobin cannot be equated to a drug–receptor interaction, the binding of oxygen to the heme moiety happens to be a felicitous example in this case, since oxygen does not undergo any change on binding, just as a drug is not altered by its interaction with a receptor.

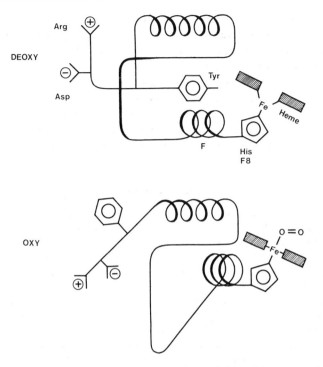

**Fig. 1.14.** Diagram of conformational change in hemoglobin with oxygenation. Oxygenation of the heme ion draws it into the heme plane, which changes the position of helix F, displacing the tyrosine. This, in turn, results in a change in the positions of arginine and aspartate, causing them to break their ionic bonds with neighboring subunits, resulting in changes in the quaternary structure of hemoglobin.

Many examples of conformational changes of enzymes during their reactions with substrates have been well studied and described in the literature, including those of carboxypeptidase, dihydrofolate reductase, and acetylcholinesterase.

### 5.4. Quantitative Assessment of Steric Effects

A knowledge of molecular geometry plays an important role in understanding quantitative structure–activity relationships (see Sec. 9). In this regard, stereochemical factors influence any physicochemical property of a molecule, with the simplest steric parameter being the size of a substituent or other functional group. The first attempt to include steric effects in linear free-energy relationships between the structure and pharmacological activity of a molecule was the *Taft steric parameter* ($E_s$; see Martin, 1978). This parameter was defined as the difference between the logarithm of the relative rate of the acid-catalyzed hydrolysis of a carboxymethyl-substituted compound, and the logarithm of the rate of hydrolysis of methyl acetate as a standard:

$$E_{sx} = \log K_{XCOOCH_3} - \log K_{CH_3COOCH_3}$$

where X is the molecule or molecular fragment in question, to which a carboxy-methyl group has been attached. With some correction suggested by other authors, $E_s$ has proven to be useful in quite a few structure–activity correlations.

Another measure of the molecular geometry of substituents is the *Verloop steric parameter*. This is calculated from bond angles and atomic dimensions — primarily the lengths of substituent groups and several measures of their width. Trivial as this may sound, the consideration of molecular "bulk" is an important and hitherto neglected factor in making multiple quantitative correlations of structure and pharmacological activity. Balaban et al. (1980) devised several related methods.

### Selected Readings

A. Albert (1985). *Selective Toxicity*, 7th ed. Chapman and Hall, London.

A. Balaban, A. Chiriac, J. Motoc, and Z. Simon (1980). *Steric Fit in Quantitative Structure–Activity Relations*. Springer, Berlin.

J. E. Blackwood, C. L. Gladys, K. L. Loening, A. E. Petrarca, and J. E. Rush (1968). Unambiguous specification of stereoisomerism about a double bond. *J. Am. Chem. Soc.* *90*:509–510.

M. Charton and J. Motoc (Eds.) (1983). *Steric Effects in Drug Design*. Springer, Berlin.

J. M. Leger, M. Gadret, and A. Carpy (1980). β-Adrenergic drugs: analysis of crystallographic and theoretical results. *Mol. Pharmacol.* *17*:339–343.

Y. C. Martin (1978). *Quantitative Drug Designs*. Marcel Dekker, New York.

J. B. Stenlake (1979). *Foundations of Molecular Pharmacology*, Vol. 2. The Athlone Press, London, Chap. 3.

## 6. ELECTRONIC STRUCTURE AND ITS EFFECTS ON DRUG ACTIVITY

The chemical structure of a drug and its physicochemical characteristics, chemical reactivity, and ability to interact with receptors ultimately depend on its electronic structure — the arrangement, nature, and interaction of electrons in the molecule. In general, the effect of electron distribution in organic compounds can be direct (short range) or indirect (long range).

### 6.1. Direct Electronic Effects

These effects primarily concern covalent bonding, which involves the overlap of electron orbitals. The "strength" of covalent bonds, the interatomic distances spanned by these bonds, and dissociation constants are all direct consequences of the nature of covalent electrons.

The nonbonding electron pairs of such heteroatoms as O, N, S, and P also play an important role in drug characteristics. They are the basis of such noncovalent interactions as hydrogen bonding (which, as already discussed, has a profound effect on the hydrophilic or lipophilic characteristics of a molecule), charge-transfer complex formation, and ionic bond formation. In all of these phenomena, the nonbonding electron pair participates in a donor–acceptor interaction.

*Indirect electronic effects* occur over a longer range than direct effects, requiring no orbital overlap. Electrostatic ionic interactions fall partly within this category, since the effect of interionic forces decreases by the square of the distance over which they act. Such inductive forces as van der Waals bonds and dipole moments are the results of *polarization* or *polarizability*—the permanent or induced distortion of the electron distribution within a molecule. These forces are especially important in studies of quantitative structure–activity relationships (QSAR) because the electronic effect of a substituent can, by resonance or an inductive or field effect, change the stereoelectronic properties of a molecule and thus influence its biological activity. Examples of different bond types and the details relating to them will be discussed in Sec. 7.

## 6.2. The Hammet Correlations

The Hammet correlations (Hammet, 1970) express quantitatively the relationship between chemical reactivity and the electron-donating or -accepting nature of a substituent. They are perhaps the most widely used electronic indices in QSAR studies of drugs. The Hammet substituent constant ($\sigma$) was originally defined for the purpose of quantifying the effect of a substituent on the dissociation constant of benzoic acid:

$$\log \frac{K_X}{K_H} = \sigma$$

where $K_X$ is the dissociation constant of benzoic acid carrying substituent X; $K_H$ is the dissociation constant of unsubstituted benzoic acid. The effect of the substituent on other reactions (e.g., $E_1$ elimination reactions) has also been investigated in great detail. Electron-attracting substituents (such as $-C\overset{\displaystyle O}{\underset{\displaystyle OH}{\diagup}}$, $-NO_2$, $-\overset{\oplus}{NR_3}$) have a positive $\sigma$ value, while electron-donating substituents ($-OH$, $-OCH_3$, $-NH_2$, $-CH_3$) have a negative $\sigma$. The value of $\sigma$ also varies according to whether the substituent is in the meta or para position. Ortho substituents are subject to too many interferences and are not used in calculating $\sigma$. Detailed tables of $\sigma$ values can be found in the works of Chu (1980), Albert (1985), and Martin (1978).

The Hammet substituent constant includes both inductive and resonance effects (i.e., electronic influences mediated through space and through conjugated bonds). In the case of benzoic acids, direct conjugation is not possible, but in one resonance hybrid, as shown in Fig. 1.15, the electron-withdrawing nitro group puts a positive charge on the C-1 carbon, thus stabilizing the carboxylate ion and decreasing the $pK_a$ of the substituted acid. The electron-donating phenolic hydroxyl group, on the other hand, destabilizes the carboxylate anion by charge repulsion, making the substituted acid weaker. The electronic substituent constants for nonaromatic compounds, introduced by Taft, are related to the acidic and basic hydrolysis rates of substituted acetic acid esters.

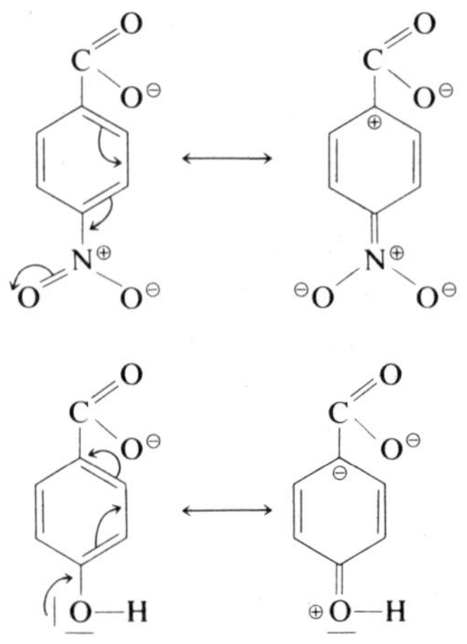

**Fig. 1.15.** Resonance and field effects of the electron-acceptor nitro group and the electron-donor hydroxyl group on the stability (and $pK_a$) of the benzoate ion. Electron acceptors stabilize the anion, whereas electron donors have the opposite effect, increasing the electron density in the vicinity of the carboxylate ion and creating an unfavorable ion–dipole interaction.

### 6.3. Ionization of Drugs

Ionization is another function of the electronic structure of a drug molecule. The $pK_a$ of a drug is important to its pharmacological activity, since it influences both the absorption and passage of the drug through cell membranes. In some cases, only the ionic form of a drug is active under biological conditions.

Drug transport represents a compromise between the increased solubility of the ionized form of a drug and the increased ability of the nonionized form to penetrate the lipid bilayer of cell membranes. A drug must cross many lipid barriers as it travels to the receptor that is its site of action. Yet cell membranes contain many ionic species (phospholipids, proteins) that can repel or bind ionic drugs, and ion channels, usually lined with polar functional groups, can act in an analogous manner. Ionic drugs are also more hydrated, the therefore "bulkier" than nonionic drugs.

As a rule of thumb, drugs pass through membranes in an undissociated form, but act as ions (if ionization is a possibility). A $pK_a$ in the range of 6–8 would therefore seem to be most advantageous, because the nonionized species that passes through lipid membranes has a good probability of becoming ionized and active within this $pK_a$ range. This consideration does not relate to compounds that are actively transported through such membranes.

A high degree of ionization can keep drugs out of cells and decrease their systemic toxicity. This is an advantage in the case of externally applied disinfectants or antibacterial sulfanilamides, which are meant to remain in the intestinal tract to fight infection. Also, some antibacterial aminoacridine derivatives are active only when fully ionized, a phenomenon much studied by A. Albert. These now obsolete bacteriostatic agents intercalate (position or interweave themselves) between the base pairs of DNA. The cations of these drugs, obtained by protonation of the amino groups, then form salts with the DNA phosphate ions, anchoring the drugs firmly in position. Protonation of histamine and its importance in binding is discussed in detail in Chap. 4, Sec. 6.1.

Ionization can also play a role in the electrostatic interaction between ionic drugs and the ionized protein side chains of drug receptors. Therefore, when conducting experiments on drug–receptor binding, it is advisable to regulate protein dissociation by using a buffer. The degree of ionization of any compound can be easily calculated from the Henderson–Hasselbach equation:

$$\% \text{ ionized} = \frac{100}{1 + \text{antilog} \, (\text{pH} - \text{p}K_a)}$$

but the acid or base character of the molecule has to be considered, as is shown in Table 1.3 and is discussed in detail by Albert and Serjeant (1984).

**Table 1.3.** Degree of ionization of acids or bases

|  | % Ionized | |
|---|---|---|
| pH − p$K_a$ | Acid | Base |
| −4.0 | 0.01 | 99.99 |
| −3.0 | 0.10 | 99.90 |
| −2.0 | 0.99 | 99.01 |
| −1.0 | 9.09 | 90.91 |
| −0.5 | 24.03 | 75.97 |
| 0 | 50.00 | 50.00 |
| 0.5 | 75.97 | 24.03 |
| 1.0 | 90.91 | 9.09 |
| 2.0 | 99.01 | 0.99 |
| 3.0 | 99.90 | 0.10 |
| 4.0 | 99.99 | 0.01 |

## Selected Readings

A. Albert (1985). *Selective Toxicity*, 7th ed. Chapman and Hall, London.

A. Albert and E. Serjeant (1984). *The Determination of Ionization Constants*, 3rd ed. Chapman and Hall, London.

K. C. Chu (1980). The quantitative analysis of structure–activity relationships. In: *The Basis of Medicinal Chemistry*, 4th ed. (M. E. Wolff, Ed.), Part 1. Wiley-Interscience, New York, pp. 393–418.

C. R. Ganellin (1977). Chemical constitution and prototropic equilibria in structure–activity analysis. In: *Drug Action at the Molecular Level* (G. C. K. Roberts, Ed.), University Park Press, Baltimore, pp. 1–39.

L. P. Hammett (1970). *Physical Organic Chemistry*, 2nd ed. McGraw-Hill, New York.

Y. C. Martin (1978). *Quantitative Drug Design*. Marcel Dekker, New York.

## 7. CHEMICAL BONDING AND BIOLOGICAL ACTIVITY

In molecular terms, the activity of drugs is initiated by their interaction with some kind of receptor. Since the association of small molecules (e.g., drugs) with macromolecules (e.g., receptors) is promoted and stabilized by bond formation, an understanding of the nature and combination of various chemical bonds is of great interest to the medicinal chemist. As discussed earlier, covalent and noncovalent bonds are both based on electronic interactions but differ greatly in their stability, which is expressed in terms of the bond dissociation energy. Table 1.4 summarizes the various types of bonds and their average bond energies. Although there is no direct correlation between bond energy and drug potency, the energy values give an approximate estimate of the ease of formation and disruption and of the relative strengths of various bond types.

**Table 1.4.** Chemical bonds and average bond energies

| Bond type | Example | Total interaction energy, $-\Delta E$ (kJ/mol) | Electrostatic energy, $-\Delta E_{es}$ (kJ/mol) | Charge-transfer energy, $\Delta E_{ct}$ (kJ/mol) |
|---|---|---|---|---|
| Dispersion (van der Waals) | $Xe \ldots Xe$ | 1.9 | 0 | 0 |
| Hydrophobic | $C_6H_6 \ldots C_6H_6$ | 4.2 | $\neq 0$ | $\neq 0$ |
| Hydrogen | $H_2O \ldots H_2O$ | 37 | 38 | 9 |
| Charge transfer | $\begin{array}{c} NC \quad CN \\ C \\ \Vert \rightarrow H_2O \\ C \\ NC \quad CN \end{array}$ | 17 | 16 | 4 |
| Dipole–dipole | $>C\overset{\frown}{=}O$ $-\overset{..}{N}R_3$ | ~5 | | |
| Ion–dipole | $F^{\ominus} \ldots H_2O$ | 171 | 154 | 75 |
| Ionic | $NH_4^{\oplus}F^{\ominus}$ | 685 | 757 | 149 |
| | $H^{\oplus}Cl^{\ominus}$ | 450 | | |
| Covalent | $>C-C<$ | 346 | | |
| | $-C=C-$ | 614 | | |

Modified from Stenlake (1979) and Kollman (1980).

## 7.1. Dispersion or van der Waals Forces

Van der Waals bonds exist between all atoms, even those of noble gases, and are based on polarizability—the induction of asymmetry in the electron cloud of an atom by a nucleus of a neighboring atom (i.e., a positive charge). This is tantamount to the induced formation of a dipole. However, although the induced dipole–induced dipole interaction sets up a temporary local attraction between the two atoms, this noncovalent interaction decreases very rapidly, in proportion to $1/R^6$, where $R$ is the distance separating the two molecules. Such van der Waals forces operate within an effective distance of about 0.4–0.6 nm and exert an attractive force of less than 2 kJ/mol. Therefore, they are often overshadowed by stronger interactions.

While individual van der Waals bonds make a very low energy contribution to a system, a large number of van der Waals forces can add up to a sizable amount of energy. In the hydrocarbon tails of the lipids in phospholipid membranes, $-CH_2-$ groups attract each other with a force of about 33 kJ/mol, provided they can pack closely. If separation of the phospholipid tails is forced by *cis* double bonds or branched alkyl chains, the attraction drops to 10–12 kJ/mol. Polar lipophilic substituents increase the van der Waals interaction considerably: thus, halogenated hydrocarbons like **halothane** (1-7) or **methoxyflurane** (1-8) (see Fig. 1.3) are more efficient anesthetics than the apolar **xenon** or **cyclopropane** because they bind better to the lipids of neural tissue.

## 7.2. Hydrophobic Interactions

Hydrophobic binding plays an important role in stabilizing the conformation of proteins, in the transport of lipids by plasma proteins, and in the binding of steroids to their receptors, among other examples. The concept of these indirect forces, first introduced by Kauzman in the field of protein chemistry, also explains the low solubility of hydrocarbons in water. Because the nonpolar molecules of a hydrocarbon are not solvated in water owing to their inability to form hydrogen bonds with water molecules, the latter become more ordered around the hydrocarbon molecule, forming an interface, at a molecular level, that is comparable to a gas–liquid boundary. The resulting increase in solvent structure leads to a higher degree of order in the system than exists in bulk water, and therefore a loss of entropy. When the hydrocarbon structures—whether two protein side chains or hexane molecules dispersed in water—come together, they will "squeeze out" the ordered water molecules that lie between them (Fig. 1.16). Since the displaced water is no longer part of a boundary domain, it reverts to a less ordered structure, which results in an entropy gain. This change is sufficient to decrease the free energy of the system by about 3.4 kJ/mol for every methylene group, and is tantamount to a bonding energy because it favors the association of hydrophobic structures. Naturally, once the hydrocarbon chains are in sufficient proximity, van der Waals forces become operative between them. The validity of the concept of hydrophobic binding has been questioned recently (see Albert, 1985, p. 315).

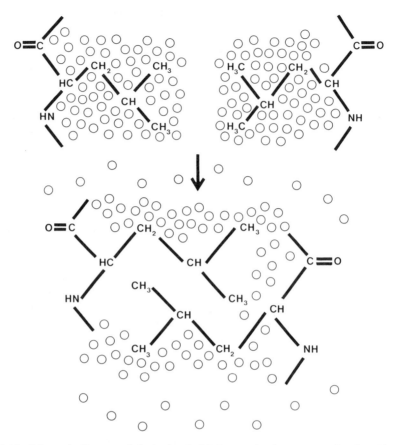

**Fig. 1.16.** Schematic diagram of the hydrophobic interaction between two leucine side chains of a protein. By displacing part of the hydrate envelope, the two alkyl chains occupy the same water "cavity," while many of the water molecules (represented by circles) become randomized. Thus the entropy of the system increases, resulting in a favorable stabilization.

### 7.3. Hydrogen Bonding

Hydrogen bonding has considerable importance in stabilizing structures by *intra*molecular bond formation. Classical examples of such bonding occur in the protein α-helix and the base pairs of DNA. However, hydrogen bonds are probably less important in *inter*molecular bonding between two structures in aqueous solution because the polar groups of such structures form hydrogen bonds with water. Moreover, there is no advantage in exchanging hydrogen bonding with water molecules for hydrogen bonding with another molecule unless additional, stronger bonding brings two molecules into sufficient proximity.

We have seen that hydrogen bonding is based on an electrostatic interaction between the nonbonding electron pair of a heteroatom (N, O, and even S) as the donor, and the electron-deficient hydrogen atom of —OH, —SH, and —NH groups. It is important to remember that —CH hydrogens do not form hydrogen bonds. Hydrogen bonds are strongly directional, and linear hydrogen bonds are energetically preferred to angular bonds. Hydrogen bonds are also weak, having energies ranging from 7 to 40 kJ/mol (see Stenlake, 1979, p. 48).

## 7.4. Charge Transfer

The term "charge transfer" refers to a succession of interactions between two molecules; these can range from very weak donor–acceptor dipolar interactions to interactions that result in the formation of an ion pair, depending on the extent of electron delocalization:

$$D + A \rightleftharpoons DA \rightleftharpoons D^{\delta +}A^{\delta -} \rightleftharpoons D^{\oplus}A^{\ominus} \rightleftharpoons D^{\oplus} + A^{\ominus}$$

Charge transfer (CT) complexes are formed between electron-rich donor molecules and electron-deficient acceptors. Typically, donor molecules are $\pi$-electron-rich heterocycles (furan, pyrrole, thiophene), aromatics with electron-donating substituents, and compounds with free, nonbonding electron pairs. Acceptor molecules are $\pi$-electron-deficient systems such as purines and pyrimidines, aromatics with electron-withdrawing substituents (picric acid), and tetracyanoethylene.

A classic example of CT complex formation occurs in the solution of iodine (an acceptor) in cyclohexene (a donor), with the solution assuming a brown color due to a shift in its absorption spectrum. The brown is not a color in the physical sense, but rather the result of a very broad absorption band encompassing about 200 nm in the visible spectrum and evolving as a result of electronic changes in the CT complex. It will be recalled that, in contrast, a solution of iodine in $CCl_4$—an inert solvent—is purple.

Drug–receptor interactions often involve CT complex formation. Examples include the reactions of antimalarials with their receptors and of some antibiotics that intercalate with DNA (see Chap. 6, Sec. 6); formation of the synaptic storage complexes of such neurotransmitters as norepinephrine and serotonin with ATP; and probably more. The CT energy is proportional to the ionization potential of the donor and the electron affinity of the receptor, but is usually no higher than about 30 kJ/mol.

## 7.5. Dipoles

Molecules in which there is a partial charge separation can interact with each other (a dipole–dipole interaction) or with ions. Dipole moments are bond moments resulting from charge differences and the distance between charges within a molecule; they are vectorial quantities and are expressed in Debye units (about $10^{-20}$ esum, or *e*lectro*s*tatic *u*nits per *m*eter). Linear group moments (as in *p*-dichlorobenzene) can cancel out each other; nonlinear ones (e.g., *m*-dichlorobenzene) are added vectorially. Since so many functional groups have dipole moments, dipole–dipole interactions are frequent. The energy of such interactions can be calculated from the following expression:

$$E = \frac{2\mu_1\mu_2 \cos\theta_1 \cos\theta_2}{Dr^3}$$

where $\mu$ is the dipole moment, $\theta$ is the angle between the two poles of the dipole, $D$ the dielectric constant of the medium, and $r$ the distance between the charges involved in the dipole. Thus, this interaction occurs over a fairly long range, declining only with the third power of the distance between the dipole charges.

Ion–dipole interactions are even more powerful, with energies that can reach 100–150 kJ/mol. The energy of such an interaction can be calculated from

$$E = \frac{e\mu \cos \theta}{D(r^2 - d^2)}$$

where $e$ is the fixed charge and $d$ the length of the dipole. Because the bond energy in this interaction declines only with the square of the distance between the charged entities, it is consequently very important in establishing the initial interaction between two ligands. A classic example of a dipole–ion interaction is that of hydrated ions which, in the process of hydration, become different from the same ions in a crystal lattice.

### 7.6. Ionic Bonds

Ionic bonds are formed between ions of opposite charge. Their electrostatic interaction is very strong:

$$E = \frac{e_1 e_2}{Dr}$$

with a bonding energy $(E)$ that can approach or even exceed the energy of a covalent bond. Ionic bonds are ubiquitous and, since they act across long distances, play an important role in the actions of ionizable drugs.

### 7.7. Covalent Bonds

Although very important generally, covalent bonds are less important in drug–receptor binding than noncovalent interactions. They are seen in the case of antiparasitic agents containing heavy metals (As, Bi, Sb), which inactivate the thiol enzymes of a parasite through bonding of the heavy metal to the sulfur atom of the enzyme thiol groups:

$$\diagdown \!\!\! \diagup \text{Sb—OR} + \text{Enzyme—SH}$$

$$\diagdown \!\!\! \diagup \text{Sb—S—Enzyme} + \text{R—OH}$$

Penicillin (Chap. 6, Sec. 2) acts by acylating a transpeptidase enzyme that is vital to bacterial cell-wall synthesis. Antitumor nitrogen mustards (Chap. 6 , Sec. 5.1) alkylate the amino groups of guanine bases in DNA and cross-link the two strands of the DNA double helix, preventing gene replication and transcription. Most drugs, however, do not attach themselves to their receptors with such permanence.

### Selected Readings

A. Albert (1985). *Selective Toxicity*, 7th ed. Chapman and Hall, London.

P. Andrews (1986). Functional groups, drug–receptor interactions and drug design. *Trends Pharmacol. Sci.* 7:148–151.

A. Ben-Naim (1980). *Hydrophobic Interactions*. Plenum, New York.

P. H. Doukas (1975). The role of charge-transfer processes in the action of bioactive materials. In: *Drug Design* (E. J. Ariëns, Ed.), Vol. 5. Academic Press, New York, pp. 133–167.

P. A. Kollman (1980). The nature of the drug–receptor bond. In: *The Basis of Medicinal Chemistry*, 4th ed. (M. E. Wolff, Ed.), Part 1. Wiley-Interscience, New York, pp. 313–329.

J. B. Stenlake (1979). *Foundations of Molecular Pharmacology*, Vol. 2. The Athlone Press, London, Chap. 2.

## 8. QUANTUM CHEMICAL ASPECTS OF DRUG ACTION

Quantum chemical studies strive toward a Utopian goal in molecular pharmacology: the accurate prediction of the topography of all nuclei and electrons in a drug molecule, as well as the energy of their interaction in a drug–receptor complex. The correlation of these data with the *in vivo* pharmacological activity of a drug would be the ultimate result of a truly quantitative structure–activity correlation analysis. Needless to say, we are very far from attaining these goals, although partial success has been achieved.

The theory and methodology of the calculations is beyond the scope of this book, and the reader is referred to other sources for this information: Marshall (1978; pp. 599–611), Martin (1978), Kier (1971), Weinstein and Green (1981), and Richards (1984).

Quantum chemical calculations, immensely aided by modern computers, can supply two kinds of information:

1. They can describe the electron distribution in a molecule, predict that of unknown compounds, and calculate interatomic distances.
2. Iterative calculations can describe the relative energy contents—and thus the relative stabilities—of all possible conformers of a molecule, as well as indicate the "essential" conformation needed for receptor binding (cf. Richards, 1977).

Two major difficulties frustrate the efforts of quantum pharmacologists. The first is that calculations of molecular orbitals, electron densities, and "preferred" conformations based on solutions to the Schrödinger equation are only approximate even with the most advanced *ab initio* ("from scratch") methods, and even for relatively small drug molecules. The second is that such calculations cannot possibly be done for large macromolecular receptors. This is compounded by the incompleteness of our understanding of nonpolar molecular associations in an aqueous medium—the type of interaction that is all-important in drug–receptor binding. Another, purely practical point is that these extremely complex calculations require substantial computer time and are therefore expensive to perform for any extended series of potential drug molecules.

### 8.1. Electron Distribution in Molecules

The electron distribution in a molecule can be determined experimentally by dipole-moment measurements, NMR methods, and x-ray diffraction. The latter method provides very accurate electron-density maps, but only of molecules in the solid state; it cannot be used to provide maps of the nonequilibrium conformers of a molecule in solution.

Quantum mechanics provides several methods for calculating the orbital energies of atoms, combining the individual atomic orbitals into molecular orbitals and deriving from the latter the probability of finding an electron at any atom in the molecule—which is tantamount to determining the electron density at any atom. There are several methods for doing this, with varying degrees of sophistication, accuracy, and reliability. An example of the histamine monocation is shown in Fig. 1.17. The net positive charge (i.e., the nuclear charge minus the number of electrons in the sphere) is shown for each atom, according to Richards (1977). It is striking that the charge distribution is very uniform and that the protonated side-chain nitrogen, usually written with a positive charge, is not more positive than the other atoms. Richards emphasizes the lack of value of the receptor schematics so often seen in the literature, which show highly localized centers of negative or positive charge that are assumed to attract and bind their electrostatic counterparts. It seems more likely that *overall* electrostatic attractions are the ones that are important to binding, whereas specific short-range binding depends on dispersion forces.

In the same chapter (Richards, 1977), similar calculations for alkylammonium compounds show that current ideas on inductive effects may be too naive. It turns out that the replacement of a hydrogen atom by an alkyl group actually decreases the charge on nitrogen instead of increasing it, as commonly thought. Other authors (Kier, 1971; Kaufman and Koski, 1975) have shown, with histamine and morphine, that charges usually pinpointed on specific atoms should really be represented as

**Fig. 1.17.** Molecular geometry and charges of the histamine monocation. The numbers on the bonds are bond distances in Ångströms; the numbers within the spheres are net positive charges calculated by quantum-chemical methods. (After Richards, 1977)

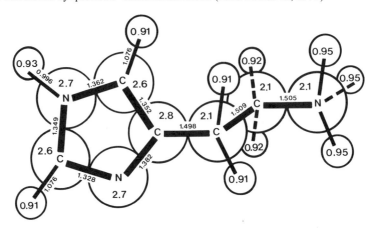

rather wide *areas* around that atom. Advances in quantum theory can be expected to present more such surprises.

Overall $\pi$-electron density of polycyclic hydrocarbons has been assumed to correlate with the carcinogenicity of these compounds (Marshall, 1978, p. 611). According to this hypothesis, defined reactive regions on the molecule undergo metabolism to form reactive intermediates such as epoxides and glycols, which react with cell constituents (see also Chap. 7, Sec. 2). As this model is widely cited in the literature, it is appropriate to warn the reader that however attractive, it is seriously questioned (see Selkirk, 1980, p. 456). However, $\pi$-electron density is very important in the chemical reactivity of aromatic rings, as any modern organic chemistry textbook will affirm.

The *energies of delocalized orbitals* have attracted considerable interest since the early 1960s, when Szent-Györgyi (1960), in his brilliant book on submolecular biology, directed attention to charge-transfer complexes (see Sec. 7.4). The energies of the *h*ighest *o*ccupied *m*olecular *o*rbital (HOMO) and the *l*owest *e*mpty *m*olecular *o*rbital (LEMO) are a measure of electron-donor and -acceptor capacity, respectively, and consequently determine donors and acceptors in charge-transfer reactions. HOMO and LEMO are also reliable estimates of the reducing or oxidizing properties of a molecule. They are expressed in $\beta$ units (a quantum-chemical energy parameter whose value varies from 150 to 300 kJ/mol). The smaller the numerical value of HOMO (a positive number), the better the substance is as an electron donor, since the small number indicates that less energy is required to remove an electron from it. Likewise, the smaller the magnitude of LEMO (a negative number), the more stable the orbital for the incoming electron, which favors electron-acceptor characteristics. Thus, by looking at the numerical values of the HOMO and LEMO of a pair of compounds, one can often decide whether a charge-transfer complex can be formed, and which compound will be the donor and which the acceptor. The most active CT participants have HOMOs or LEMOs of less than plus or minus 0.5.

## 8.2. Conformational Studies by Quantum-Chemical Methods

The bond lengths and bond angles in molecules can be obtained from the crystallographic data generated by x-ray diffraction. These bond lengths and angles, in turn, allow calculation of the conformational potential energies of a large number of conformers. In an iterative (repeating) calculation, individual bond angles, especially those that are seemingly the most crucial to variations in energy, are changed in small (e.g., 5°) steps, and the conformational energy is calculated for each step. This permits the construction of maps of the potential energy of a conformation as a function of two angles (as has been done for the many arylethylamine neurotransmitters—e.g., dopamine, norepinephine, serotonin, histamine, etc.). Conformational probability maps (see Richards, 1977) are even more useful; they circumscribe a space, defined by two bond angles, whose contour defines the region containing a certain percentage of the total number of conformers of a molecule. This approach shows the molecular flexibility and the limits within which pharmacological activity of the molecule can be expected.

The original idea behind these rather tedious calculations was that the "preferred," "lowest-energy," or "essential" conformation of a drug molecule reflected the optimal cooperation of steric, intramolecular-bonding, solvation, and directional forces, and depicted the molecule in its most stable state. Therefore, the reasoning went, this ought to be the most abundant conformer and thus the most likely to bind to the receptor. The interatomic distances between electron-rich groups (e.g., heteroatoms), were considered to be crucial in defining the "pharmacophore" geometry that served as a basis for the design of analogues and derivatives that would bind to the same receptor as a parent drug molecule (cf. Kuntz, 1980, p. 298).

However, although these ideas were widely applied in structure–activity correlations, they seem rather naive in the light of recent concepts of drug binding. As mentioned in Sec. 5.3, a well-reasoned criticism came from Martin (1978, pp. 361–366). Among newer concepts, the "zipper" mechanism of binding proposed by Burgen et al. (1975) assumes that a flexible molecule initiates binding at the molecular subsite that contains the most energetic functional group and one with a long radius of action (e.g., an anion). The molecule then changes its conformation in a stepwise manner, like a zipper, as its other segments bind to the receptor. This concept would also facilitate the conformational modification of large and ponderous receptor macromolecules in their binding to drugs (Fig. 1.18). With this model, there is no *a priori* requirement for a lowest-energy conformation, because

**Fig. 1.18.** Schematic diagram of possible conformational changes during ligand–receptor binding. In the obsolete "lock and key" binding hypothesis (A), only one conformer can bind to a rigid receptor. In the "zipper" model of Burgen et al. (B), any one of the conformers can mold itself to the rigid receptor by a stepwise conformational change that derives the necessary energy from binding itself. In the "induced fit" model (C), both ligand and receptor undergo reversible conformational changes. (Modified from Burgen et al., 1975)

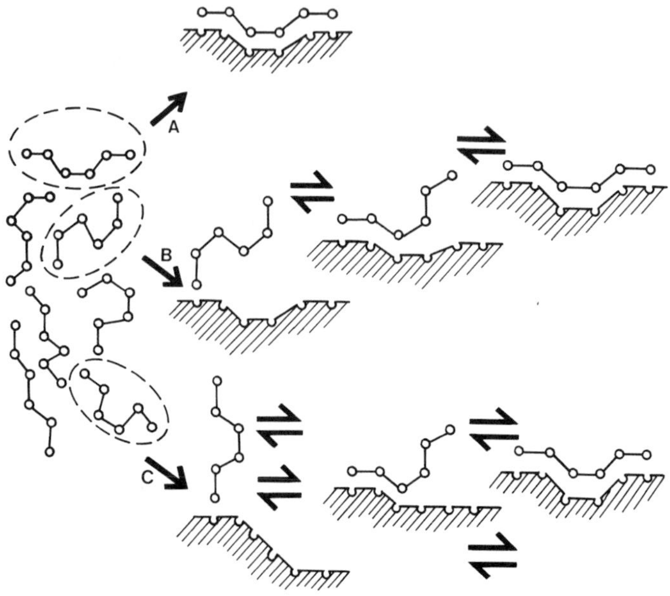

any conformer, within reasonable limits, can be extracted from solution and bound by the receptor. More recently, Marshall and his group (see Weaver et al., 1979) developed computer programs to consider all of the conformers of a drug to permit the generation of receptor site maps. Thus, receptor mapping via small ligand studies may become possible. There is no doubt that conformational studies are essential to the understanding of drug–receptor interactions, but they will require a more flexible approach than that adopted in the past.

### *Selected Readings*

E. Bergman and B. Pullman (Eds.) (1974). *Molecular and Quantum Pharmacology.* Reidel, Dordrecht.

A. S. V. Burgen, J. Feeny, and G. C. K. Roberts (1975). Binding of flexible ligands to macromolecules. *Nature 253*:753–755.

J. J. Kaufman and W. S. Koski (1975). Physicochemical, quantum chemical and other theoretical techniques for the understanding of the mechanism of activation of CNS agents: psychoactive drugs, narcotics and narcotic antagonists and anesthetics. In: *Drug Design* (E. J. Ariëns, Ed.), Vol. 5. Academic Press, New York, pp. 251–340.

L. B. Kier (1971). *Molecular Orbital Theory in Drug Research.* Academic Press, New York.

J. D. Kuntz, Jr. (1980). Drug–receptor geometry. In: *The Basis of Medicinal Chemistry*, 4th ed. (M. E. Wolff, Ed.), Part 1. Wiley-Interscience, New York, pp. 285–312.

A. G. Marshall (1978). *Biophysical Chemistry.* Wiley, New York.

Y. C. Martin (1978). *Quantitative Drug Design.* Marcel Dekker, New York.

W. G. Richards (1977). Calculation of essential drug conformations and electron distributions. In: *Drug Action at the Molecular Level* (G. C. K. Roberts Ed.). University Park Press, Baltimore, pp. 41–54.

W. G. Richards (1984). *Quantum Pharmacology*, 2nd ed. Butterworth, London.

J. K. Selkirk (1980). Chemical carcinogenesis. In: *The Basis of Medicinal Chemistry*, 4th ed. (M. F. Wolff, Ed.), Part 1. Wiley-Interscience, New York, pp. 455–478.

A. Szent-Györgyi (1960). *Introduction to a Submolecular Biology.* Academic Press, New York.

D. C. Weaver, C. D. Barry, M. C. McDaniel, G. R. Marshall, and P. E. Lacy (1979). Molecular requirements for recognition of a glucoreceptor for insulin release. *Mol. Pharmacol. 16*:301–308.

H. Weinstein and J. P. Green (Eds.) (1981). Quantum chemistry in biomedical sciences. *Ann. N.Y. Acad. Sci. 367.*

## 9. QUANTITATIVE STRUCTURE–ACTIVITY RELATIONSHIPS

As we have seen, drug–receptor interactions are determined by the physicochemical parameters of the drug: polarity, ionization, and electron density. All of the electronic properties of a drug are determined by the atomic composition, shape, and size of the drug molecule—in other words, by its chemical structure. Since these physicochemical parameters can be measured and expressed in quantitative terms, the intermolecular binding forces as a function of structure must also have numerical values. Therefore, if the biological actions of the drugs within a series can be measured and the mode of action within the series is identical, it should be possible to calculate a quantitative structure–activity relationship (QSAR). Yet despite great effort and ingenuity, QSARs remain more of a challenge than an achievement. This is understandable if one considers the enormous complexity of physicochemical parameters that relate even to a small drug molecule, let alone a largely unknown macromolecular receptor, or the even more enormous intricacies of an entire living system. The choice of a manageable number of key chemical and biological parameters for QSAR determinations, along with their appropriate statistical juggling, is in itself an intuitive process with rather formidable epistemological aspects. Ganellin (1977) and Barlow (1979) provide thoughtful analyses of the hopes, achievements, difficulties, and pitfalls of QSAR.

Nevertheless, the relationship between chemical structure and biological activity has always been at the center of drug research. In the past, drug structures were modified intuitively and empirically, depending on the imagination and experience of the synthesizing chemist, and based on analogies. Surprisingly, the results were often gratifying, even if obtained only serendipitously, or on the basis of the wrong hypothesis. However, this hit-or-miss approach, practiced even now, is enormously wasteful. Considering that only one of several thousand synthesized compounds will reach the pharmacy shelves, and that the development of a single drug can cost millions of dollars, it is imperative that rational short-cuts to drug design be found. Chapter 8 explores in detail the principles of drug design.

### 9.1. The Hansch Linear Free-Energy Model

This is the most popular mathematical approach to QSAR. Introduced by Corwin Hansch in the early 1960s, and distantly related to the Overton–Meyer concept, it considers (1) the physicochemical aspects of drug transport and distribution from the point of application to the point of effect and (2) the drug–receptor interaction.

In a given group of drugs that have analogous structures and act by the same mechanism, three parameters seem to play a major role:

1. The *substituent hydrophobicity constant*, based on partition coefficients analogous to Hammet constants:

$$\pi_x = \log P_X - \log P_H$$

where $P_X$ is the partition coefficient of the molecule carrying substituent X, and $P_H$ is the partition coefficient of the unsubstituted molecule (i.e., substituted by hydrogen only). More positive $\pi$ values indicate higher lipophilicity of the substituent. Since these values are additive, $P$ values measured on standard molecules permit prediction of hydrophobicity of novel molecules.

2. The *Hammet substituent constant* $\sigma$ (see Sec. 6.2).

3. *Steric effects*, described by the Taft $E_s$ values (Sec. 5.4).

The $\sigma$ and $\pi$ constants of substituents cannot be considered interrelated, but they can be shown in Craig plots, as depicted in Fig. 1.19, in order to avoid choosing a series of substituents where $\pi$ and $\sigma$ are correlated. Extensive tables of these constants for many compounds are given by Martin (1978), who has written the most detailed monograph on the QSAR. They are often useful when correlated to biological activity in the statistical procedure known as multivariate regression analysis. As is well known from pharmacological testing of various drug series, such

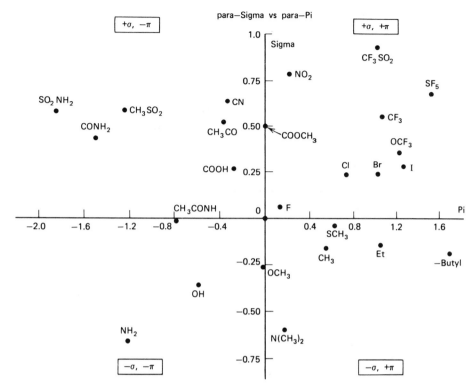

**Fig. 1.19.** Two-dimensional Craig plot of sigma ($\sigma$) substituent constants versus pi ($\pi$) values for aromatic substituents. (Reproduced by permission from Craig (1980), Wiley-Interscience, New York)

correlations can be either linear (Fig. 1.20A) or parabolic (Fig. 1.20B). The linear relationship is described by the equation:

$$\log 1/C = a\pi + bE_s + c\sigma + d$$

where $C$ is the drug concentration for a chosen standard biological effect, and $a, b, c,$ and $d$ are regression coefficients to be determined by iterative curve fitting. The parabolic relationship fits the equation:

$$\log 1/C = -a\pi^2 + b\pi + cE_s + d\sigma + e$$

The coefficients $a, b, c, d,$ and $e$ are fitted to the curve by the least-squares procedure, using regression methods for which computer programs are readily available. The extent of the fit is judged by the correlation coefficient $r$ or the multiple regression coefficient $r^2$, which is proportional to the variance. A perfect fit gives $r^2 = 1.00$. Once the best fit has been achieved and $r$ or $r^2$ has been maximized by using a reasonable number of known compounds (15–20 is an advisable number, depending on the number of variables tested; see Austel, 1984), the curve can be used to predict the biological activity of compounds that have not been tested or, indeed,

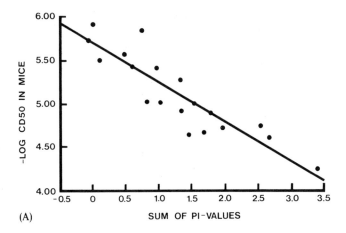

(A)

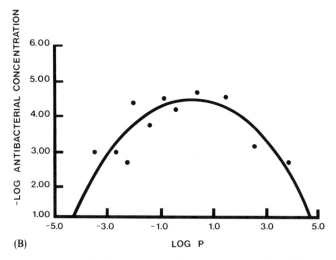

(B)

**Fig. 1.20.** Linear and parabolic Hansch plots. The linear correlation (**A**) shows the dose of substituted penicillins curing 50% of mice infected with *Staphylococcus aureus*, versus the sum of the pi values of the substituents. The parabolic curve (**B**) is the bactericidal concentration of aliphatic fatty acids versus their partition coefficients. (After Martin, 1978)

have not even been synthesized. This requires only the substitution of the optimized regression coefficients constants into the equation, and the use of $\pi$, $\sigma$, and $E_s$ values which are usually available for just about any substituent (see Hansch and Leo, 1979). Naturally, independent variables other than $\pi$ or $\sigma$—including ionization constants, activity coefficients, molar volumes, or molecular orbital parameters—can also be used (see Cramer, 1976).

A regression analysis of the effects of various substituents on a molecule using the Hansch approach is very useful, saving much time and effort in the synthesis and testing of new drugs. Hundreds of examples of such analyses are available in the literature (e.g., Hansch et al., 1977); many show positive predictive values for drug activity, whereas some other drug series cannot be interpreted by this method.

Nevertheless, there are several difficulties and pitfalls in using the Hansch method. First, the inherent disadvantage of regression analysis is that one can obtain good fits ($r^2 > 0.9$) simply by manipulating the constants. Therefore, curve fitting must be done for a relatively large number of compounds to ensure that all predictors are considered. Second, the mode of action may change for drugs within a seemingly continuous series, invalidating the comparison of some compounds in the series with the predictor compounds. The Hansch method cannot anticipate such a change (see Tute, in Yalkowsky et al., 1980).

Other problems with the Hansch method are that biological systems are often too crude as models for its application, or the electronic effects operative in a drug molecule are not sufficiently understood or precise. Finally, the method requires considerable time and expense, even in the hands of an expert. Difficulties notwithstanding, the Hansch approach took both chemists and pharmacologists out of the dark age of pure empiricism and allowed them to consider simultaneously the effects of a large number of variables of drug activity—a feat unattainable with classical methods.

### 9.2. The Free–Wilson Method

This method also assumes that biological activity can be described by the additive properties of the substituents on a basic molecular structure. In the Fujita–Ban modification of this method:

$$\log 1/C = \Sigma a_i X_i + \mu_0$$

where $C$ is the drug concentration for a standardized effect, $a_i$ is the group contribution of the $i$th substituent to the pharmacological activity of the substituted molecule, $X$ is unity if substituent $i$ is present and zero otherwise, and $\mu_0 = 1/C$ for the parent compound. Regression analysis is used to determine $a_i$ and $\mu$ (see Chu, 1980).

In the Fujita–Ban modification of the Free–Wilson method, no assumptions are made about the relevance of the model parameters to the biological activity of the molecule. The effect of each substituent is considered to be independent of any other, and each makes a constant contribution to the overall activity of the molecule. Therefore the method is applicable to compounds with more than one variable group. The result is a table (Table 1.5) showing the contribution of each substituent in each position to the overall biological effect of the molecule. The Free–Wilson equation bears close similarities to the linear Hansch equation, and, as the table shows, the results of the two can be comparable. The Free–Wilson method, however, cannot predict the activities of compounds that have substituents not included in the matrix. Consequently, this method has found only limited application in drug series where many close analogues are already available but physicochemical data are lacking.

**Table 1.5.** An example of a Free–Wilson matrix used for the calculation of group contributions to the adrenergic blocking effect of meta- and para-substituted $N,N$-dimethyl-2-bromophenethylamine

| | Z, meta | | | | | Y, para | | | | | Log $1/C$ Observed | Log $1/C$ Calculated, Fujita–Ban | Log $1/C$ Calculated, Hansch analysis | $ED_{50}$ (nmol/kg body weight, rat) |
|---|---|---|---|---|---|---|---|---|---|---|---|---|---|---|
| | F $a_1$ | Cl $a_2$ | Br $a_3$ | I $a_4$ | CH$_3$ $a_5$ | F $a_6$ | Cl $a_7$ | Br $a_8$ | I $a_9$ | CH$_3$ $a_{10}$ | | | | |
| 1 | | | | | | | | | | | 7.46 | 7.82 | 7.89 | 35.0 |
| 2 | | | | | | 1 | | | | | 8.16 | 8.19 | 7.98 | 7.0 |
| 3 | | | | | | | 1 | | | | 8.68 | 8.59 | 8.38 | 2.1 |
| 4 | | | | | | | | 1 | | | 8.89 | 8.84 | 8.77 | 1.3 |
| 5 | | | | | | | | | 1 | | 9.25 | 9.25 | 8.98 | 0.6 |
| 6 | | | | | | | | | | 1 | 9.30 | 9.08 | 8.79 | 0.5 |
| 7 | 1 | | | | | | | | | | 7.52 | 7.52 | 7.51 | 30.0 |
| 8 | | 1 | | | | | | | | | 8.16 | 8.03 | 8.23 | 7.0 |
| 9 | | | 1 | | | | | | | | 8.30 | 8.25 | 8.42 | 5.0 |
| 10 | | | | 1 | | | | | | | 8.40 | 8.40 | 8.74 | 4.0 |
| 11 | | | | | 1 | | | | | | 8.46 | 8.27 | 8.62 | 3.5 |
| 12 | | 1 | | | | 1 | | | | | 8.19 | 8.37 | 8.32 | 6.4 |
| 13 | | | 1 | | | 1 | | | | | 8.57 | 8.59 | 8.51 | 2.7 |
| 14 | | | | | 1 | 1 | | | | | 8.82 | 8.61 | 8.71 | 1.5 |
| 15 | | 1 | | | | | 1 | | | | 8.89 | 8.80 | 8.72 | 1.3 |
| 16 | | | 1 | | | | 1 | | | | 8.92 | 9.02 | 8.91 | 1.2 |
| 17 | | | | | 1 | | 1 | | | | 8.96 | 9.04 | 9.11 | 1.1 |
| 18 | | 1 | | | | | | 1 | | | 9.00 | 9.05 | 9.11 | 1.0 |
| 19 | | | 1 | | | | | 1 | | | 9.35 | 9.27 | 9.30 | 0.4 |
| 20 | | | | | 1 | | | 1 | | | 9.22 | 9.29 | 9.49 | 0.6 |
| 21 | | | | | 1 | | | | | 1 | 9.30 | 9.53 | 9.53 | 0.5 |
| 22 | | | 1 | | | | | | | 1 | 9.52 | 9.51 | 9.32 | 0.3 |

Reproduced by permission from Chu (1980).

## 9.3. Noncomputer Methods of Analogue Design

These are quicker and easier to use than the Hansch method. The *Topliss scheme* (see Craig, 1980) is an empirical method in which each compound is tested before an analogue is planned, and is compared in terms of its physical properties with analogues already planned. Figure 1.21 shows a Topliss scheme for aromatic ring substitution. If the first derivative, a *p*-chloro analogue, is more active than the parent, unsubstituted compound, then a positive $\pi$ or $\sigma$ correlation with activity is suggested, and the next derivative should be the dichloro compound, etc. A decrease in activity may be due to lipophilic, steric, or electronic effects, and the other branches of the schematic are provided to explore those effects. Schemes for nonaromatic systems are also available (see Martin, 1978, p. 258).

The concept of *molecular connectivity*, introduced by Kier and Hall in 1976, describes compounds in topological terms (see Fig. 1.22). Branching, unsaturation, and molecular shape are all represented in the purely empirical connectivity index $^1\chi$, which correlates surprisingly well with a number of physicochemical properties, including the partition coefficients, molar refractivity, or boiling point. However, higher-order terms are also calculated in this method (cf. Kubinyi, 1979, p. 127). The

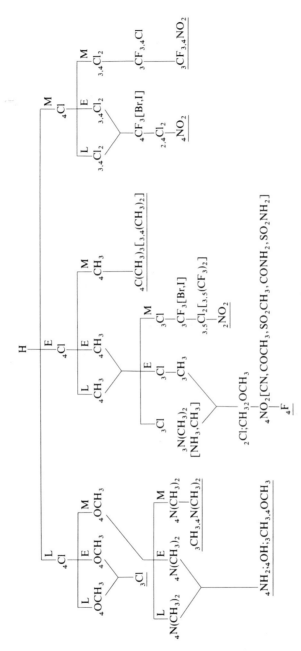

**Fig. 1.21.** A Topliss decision tree for aromatic compounds. The descending lines indicate the sequence of compounds prepared; the brackets indicate alternative compounds. L: less active; E: equiactive; M: more active (Reproduced by permission from Craig (1980), Wiley-Interscience, New York)

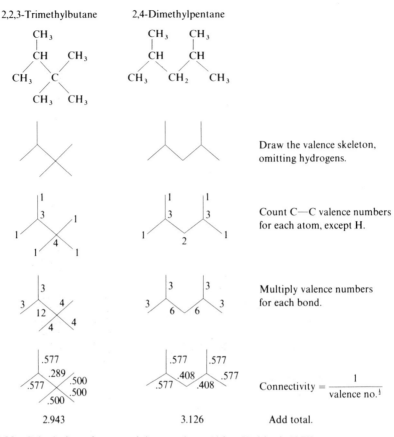

**Fig. 1.22.** Calculation of connectivity numbers. (After Kubinyi, 1979).

use of molecular connectivity data in the Hansch equation gives remarkably high structure–activity correlation coefficients for several series of drugs. However, heteroatoms and steric interactions are not satisfactorily considered, and chance correlations are possible. The method nevertheless offers interesting pragmatic possibilities (see also Henry and Block, 1979; Kier, in Yalkowsky et al., 1980; Kier and Hall, 1986).

Other topological indices based on information theory were developed by Basak, Magnuson, and their co-workers. Many of these are more specific and describe structure–activity relations more precisely than the octanol–water partition co-efficients, but their application is usually beyond the competence of the medicinal chemist. For details, consult Basak et al. (1984) and Dearden (1983).

*Pattern recognition methods,* factor, and cluster analyses are also structure–activity correlative methods, but will not be discussed here. Chu (1980, p. 411) and Martin (1978, p. 261) give some details of these methods. Those interested should become familiar with the very powerful methods of modern statistics. Further aspects of this topic are discussed in Chap. 8, Sec. 3.

## Selected Readings

V. Austel (1984). Design of test series in medicinal chemistry. *Drugs of the Future 9*: 349–365.

R. B. Barlow (1979). Structure–activity relationships. *Trends Pharmacol. Sci. 1*: 109–111.

S. C. Basak, D. K. Harris, and V. R. Magnuson (1984). Comparative study of lipophilicity versus topological molecular descriptors in biological correlations. *J. Pharm. Sci. 73*: 429–437.

K. C. Chu (1980). The quantitative analysis of structure–activity relationships. In: *The Basis of Medicinal Chemistry*. 4th ed. (M. E. Wolff, Ed.), Part 1. Wiley-Interscience, New York, pp. 393–418.

P. N. Craig (1980). Guidelines for drug and analog design. In: *The Basis of Medicinal Chemistry*, 4th ed. (M. E. Wolff, Ed.), Part 1. Wiley-Interscience, New York, pp. 331–348.

R. D. Cramer (1976). Quantitative drug design. *Annu. Rep. Med. Chem. 11*: 301–310.

J. C. Dearden (Ed.) (1983). *Quantitative Approaches to Drug Design*. Elsevier, Amsterdam.

C. R. Ganellin (1977). Chemical constitution and prototropic equilibria in structure–activity analysis. In: *Drug Action at the Molecular Level* (G. C. K. Roberts, Ed.). University Park Press, Baltimore, pp. 1–39.

C. Hansch and A. Leo (1979). *Substituent Constants for Correlation Analysis in Chemistry and Biology*. Wiley-Interscience, New York.

C. Hansch, J. Y. Fukunaga, and P. Y. C. Jow (1977). Quantitative structure–activity relations of antimalarial and dihydrofolate reductase inhibition by quinazolines and 5-substituted benzyl-2, 4-diaminopyridines. *J. Med. Chem. 20*: 96–102.

D. R. Henry and J. H. Block (1979). Classification of drugs by discriminant analysis using fragment molecular connectivity values. *J. Med. Chem. 22*: 465–472.

L. B. Kier and L. H. Hall (1986). *Molecular Connectivity in Structure–Activity Studies*. Res. Studies Press, Lechtworth, England.

H. Kubinyi (1979). Lipophilicity and drug activity. In: *Drug Research*, (E. Jucker, Ed.), Vol. 23. Birkhäuser, Basel, pp. 97–198.

Y. C. Martin (1978). *Quantitative Drug Design*. Marcel Dekker, New York.

A. K. Saxena and S. Ram (1979). Quantitative structure–activity relationships. *Drug Research* (E. Jucker, Ed.), Vol. 23. Birkhäuser, Basel, pp. 199–232.

J. G. Topliss (Ed.) (1983). *Quantitative Structure–Activity Relationships in Drugs*. Academic Press, New York.

J. G. Topliss and J. Y. Fukunaga (1978). QSAR in drug design. *Annu. Rep. Med. Chem. 13*: 292–303.

S. H. Yalkowsky, A. A. Sinkula, and S. C. Valvani (Eds.) (1980). *Physical Chemical Properties of Drugs*. Marcel Dekker, New York.

# 2
# Receptor–Effector Theories

## 1. THE RECEPTOR CONCEPT AND ITS HISTORY

The central theme of molecular pharmacology and medicinal chemistry in the past 25–30 years has been to elucidate the structure and function of drug receptors, an endeavor that goes on unabated. Before attempting to deal with drug receptors in Chaps. 4 and 5, an overview of current concepts and a brief survey of receptor characterization methods will be presented.

The discussion of thermodynamic activities (Chap. 1, Sec. 3.2) led us to the conclusion that specific drugs, those that act at very low concentrations, exert their effects by interacting with a specific macromolecular structure in the living cell. The primary mechanism is the brief formation of a reversible drug–receptor complex. This triggers a secondary mechanism such as the opening of an ion channel, or catalyzes the formation of a "second messenger," often cyclic AMP (cAMP). Other members of this chain reaction, such as kinases, are then activated. This cascade of events finally results in the physiological change attributed to the drug. The same mechanisms also operate with endogenous agents such as hormones and neurotransmitters.

It is generally accepted that endogenous or exogenous agents interact specifically with a *receptor site* on a specialized *receptor molecule*. Drug interaction with this site of binding, which has chemical recognition properties, may or may not trigger the sequence of biochemical events sketched above; therefore, one must distinguish carefully between sites of action (true receptors) and sites of binding ("silent" receptors or, occasionally, separate allosteric antagonist-binding sites).

The receptor concept goes back to 1878. It was formulated by John. N. Langley, a British physiologist who worked on the antagonism of atropine and pilocarpine (see Chap. 4, Secs. 2.4 and 2.5). The term "receptor" was introduced in 1907 by Paul Ehrlich, the famous pioneer of chemotherapy and immunochemistry. His concepts of receptor binding (*Corpora non agunt nisi fixata*—"compounds do not act unless bound"), bioactivation, the therapeutic index, and drug resistance are still valid in principle, though they have undergone considerable expansion and refinement. The early history of the receptor concept is recounted by Parascandola (1980).

## 2. THE NATURE OF RECEPTORS AND THE CRITERIA OF RECEPTOR INDENTITY

In the preceding section, an attempt was made to define the term "receptor." True receptors, those initiating a chain of physicochemical events leading to a pharmacological response, have a diverse chemical nature. Among the receptors known to any extent, several groups can be distinguished.

1. *Lipoproteins* or *glycoproteins* are the most common type of receptor. They are often firmly embedded in the plasma membrane or cell-organelle membrane as intrinsic proteins. This makes their isolation difficult because their structure, and therefore function, is maintained by the surrounding membrane. Isolation of the receptor molecule may cause a deformation or structural collapse, even to the extent that specific binding properties are lost. This was the case during the first attempts to isolate the opiate receptor (Chap. 5, Sec. 3.4). In other cases the situation is much more propitious, as attested for example, by the successful isolation of the nicotinic cholinergic receptor (Chap. 4, Sec. 2.2).

2. *Lipids* in themselves may occasionally be regarded as receptors. The non-specific effect of local anesthetics on the cholinergic ionophore may be attributable to the interaction of these amphiphilic drugs with the lipid "annulus" (ring) of the ionophore protein. Although this lipid layer is only a few molecules thick, it intimately envelops the protein and may have a profound effect on the latter's structure. Recently, a separate local anesthetic binding subunit of the cholinergic receptor complex was proposed.

3. *Pure proteins* are often drug receptors, as in the case of enzymes. As will become evident in Chap. 6, Sec. 3, many drugs exert their effect by specifically affecting enzymes that are vital in biochemical reactions, and thus modify their functions. Receptors transfer the message of the *first messenger*—a neurotransmitter, hormone, or drug—across a cell membrane; the receptor is "coupled" to an *effector system* or *molecule*. As will be shown in Sec. 5.4, three major effector systems are known: adenylate cyclase, the $Ca^{2+}$–phosphatidylinositol system, and tyrosine phosphorylase. These cause a change in concentration of a *second messenger*—cAMP, inositol triphosphate, diacylglycerol, or $Ca^{2+}$ ions—which then activates enzymes or opens ion channels. In general, it is advantageous—given the present stage of our ignorance—to avoid distinguishing too rigorously between drug receptors and enzymes. The similarities of the two are summarized in Table 2.1. The basic difference between drug–receptor and substrate–enzyme binding is that a chemical change "happens" to a substrate after it binds to the enzyme active site, whereas nothing comparable "happens" to a drug, and it will normally dissociate from the receptor unchanged. The cellular repercussions of drug and substrate binding are, of course, very different, and they are considered later in this chapter (see also Macfarlane, 1984).

4. *Nucleic acids* comprise an important category of drug receptors in the widest sense. Chapter 6, Sec. 5, which deals with the details, will show that a number of antibiotics and antitumor agents either interfere directly with DNA replication or transcription, or inhibit translation of the genetic message at the ribosome level. Steroid hormones also have DNA as their acceptor site, and show very high specificity in a

**Table 2.1.** Comparison of enzyme–substrate and drug–receptor relationships

|  | Enzyme–substrate | Drug–receptor |
|---|---|---|
| Symbols | S = substrate<br>E = enzyme<br>ES = enzyme–substrate<br>　　complex<br>P, Q = products<br>$v$ = initial velocity of<br>　　the reaction | D = drug<br>R = receptor<br>RD = drug–receptor complex<br>E = effect of the drug |
| Reaction scheme assumed | $S + E \rightleftharpoons ES \rightarrow P + Q + E$ | $D + R \rightleftharpoons RD \rightarrow E$ |
| Assumptions | 1. Derivation holds only<br>　　for the period for which<br>　　P + Q + E do not form S.<br>2. [ES] ≪ [S].<br>3. $v \propto$ [ES].<br>4. Maximal velocity $V_{max}$ is<br>　　reached when [ES] = $[E_T]$. | 1. One drug molecule<br>　　combines with one receptor.<br><br><br>2. [R] ≪ [D].<br>3. E $\propto$ [RD].<br>4. Maximum effect $E_{max}$ is<br>　　reached when [RD] = $[R_T]$. |
| Measurements | $v$ as a function of [S],<br>　$[E_T]$ constant | E as a function of [D],<br>　$[R_T]$ constant |
| Constants characteristic<br>of the substances<br>studied | $V_{max}$ (maximal velocity)<br>$K_m$ (Michaelis constant, the<br>　concentration of substrate<br>　at half-maximal velocity) | $E_{max}$ (maximum effect)<br>$K_D$ (drug–receptor<br>　dissociation constant, the<br>　dose required to produce<br>　half the maximum effect) |
| Equation | $\dfrac{v}{V_{max}} = \dfrac{[S]}{K_m + [S]}$<br><br>$\dfrac{v}{V_{max}} = \dfrac{1}{2}$ when [S] = $K_m$ | $\dfrac{E}{E_{max}} = \dfrac{[D]}{K_D + [D]}$<br><br>$\dfrac{E}{E_{max}} = \dfrac{1}{2}$ when [D] = $K_D$ |

Reproduced by permission from Y. M. Martin (1978), *Quantitative Drug Design*, Marcel Dekker, New York.

manner that we do not understand at all. Since about 1970, direct experimentation with drug receptors has become a reality, and some of the methods used are discussed in Chap. 3.

Working with a new group of drugs or a new tissue in such experiments is, however, a source of many pitfalls that may lead to hasty or doubtful conclusions. Therefore, a rather extensive set of *criteria* for receptor identification must be followed for *in vivo* as well as *in vitro* investigations (see Burt, 1985; Hollenberg and Cuatrecasas, 1979; Laduron, 1984). These criteria are as follows:

1. The receptor or binding site should be present in the tissue in quantities commensurate with established receptor concentrations; 10–100 pmol/g is the receptor concentration usually found in central or peripheral sites ($B_{max}$).
2. The binding of a drug to the receptor should be saturable, with a binding equilibrium constant ($K_D$) in the nanomolar range. The binding constant is the concentration at which half the receptor sites are occupied; it can be determined by radioisotopic methods, as outlined in Chap. 3, Sec. 3.3. The addition of a nonradioactive ("cold") drug should diminish the degree of saturable binding of

a radiolabeled drug. However, it must be borne in mind that saturability is not identical with specificity.

3. Binding kinetics should be proportional to the rate of the *in vivo* response and should yield an equilibrium constant equal to the dissociation rate constant divided by the association rate constant.

4. Wherever applicable, binding should be stereospecific; but even fulfillment of this criterion is not absolute proof that the site being investigated is a receptor. Opiates, for instance, bind stereospecifically to glass-fiber filters.

5. The receptor should be isolated from an organ or tissue relevant to the activity under investigation. Hallucinogen binding to liver tissue, for example, is unlikely to indicate more than the presence of a metabolizing enzyme.

6. It is desirable that the order of drug binding to a receptor preparation in a related series of drugs be the same as the order of their clinical or at least *in vivo* activity. An example of this can be seen in Fig. 4.31. As a check on methodology, nonspecific drugs should be included in such a series. It is most important to note that failure to meet even one of these criteria jeopardizes the identification of the receptor. Even when all of the criteria are fulfilled, extreme caution in data interpretation is still mandatory.

### 3. DEFINITION OF PHARMACOLOGICAL BINDING TERMS

The findings of classical pharmacology serve as a basis for a discussion of drug—receptor interactions at the molecular level and the biochemical properties of drug receptors. To aid in this discussion, some pharmocological terms are briefly defined.

An *agonist* is a substance that interacts with a specific cellular constituent, the receptor, and elicits an observable response. An agonist may be an endogenous physiological substance like a neurotransmitter or hormone, or it can be a synthetic drug.

There are also *partial agonists* that act on the same receptor as other agonists in a group of ligands (binding molecules) or drugs. However, regardless of their dose they cannot produce the same maximum biological response as a full agonist (Ariëns, 1983).

This behavior necessitates introducing the concept of *intrinsic activity* of an agonist. This is defined as a proportionality constant of the ability of the agonist to activate the receptor as compared to the maximally active compound in the series being studied. The intrinsic activity is a maximum of unity for full agonists and a minimum of zero for antagonists. The intrinsic activity is comparable to the $K_m$ of enzymes.

*Affinity* is the ability of a drug to combine with a receptor; it is proportional to the binding equilibrium constant $K_D$. A ligand of low affinity requires a higher concentration to produce the same effect as a ligand of high affinity. Both agonists and antagonists have affinity for the receptor.

An *antagonist* inhibits the effect of an agonist, but has no biological activity of its own in that particular system. It may compete for the same receptor site that the agonist occupies, or it may act on an allosteric site, which is different from the drug—

receptor site. In *allosteric inhibition*, antagonist binding distorts the receptor, preventing the agonist from binding to it; that is, the antagonist changes the affinity of the receptor for the agonist. In a different system, it may have an independent pharmacological activity.

*Metagonist* is a term introduced by Belleau (DiMaio et al., 1979) to describe a drug that may stabilize an allotropic (variant) form of the receptor that differs from the "active" or "inactive" state. There are few drugs in this class. Some are nonaddictive analgesics that are agonists of morphine but more active as pain relievers, yet also antagonize the narcotic properties of addictive opiates (see Chap. 5, Sec. 3.7).

The *median effective dose* ($ED_{50}$) is the amount of a drug required for half-maximal effect, or producing an effect in 50% of a group of experimental animals. It is usually expressed as mg/kg body weight. The *in vitro* $ED_{50}$ should be expressed as a molar concentration ($EC_{50}$) rather than as an absolute amount. The *median inhibitory concentration* ($IC_{50}$) is the concentration at which an antagonist exerts its half-maximal effect.

The term $pD_2$ refers to the negative logarithm of the molar dose of an agonist necessary for half-maximal effect. It is thus a measure of affinity under ideal conditions (i.e., a linear dose–response relationship).

The $pA_2$ is the negative logarithm of the molar concentration of an antagonist which necessitates the doubling of the agonist dose to counteract the effect of that antagonist and restore the original response.

## 4. CLASSICAL THEORIES OF
## CONCENTRATION–RESPONSE RELATIONSHIPS

The classical theories of drug action were developed by Gaddum and Clark in the 1920s, and were extended to antagonists by Schild. These ideas were expanded by Stephenson (1956) and by Ariëns and his school from 1960 on. Among the numerous reviews in the literature, those of Ariëns et al. (1979), Hollenberg (1985), and Triggle (1978; Triggle and Triggle, 1976) are eminent and cover the details of interest to the pharmacologist. However, it is not possible to appreciate and critically appraise current and rapidly changing ideas on the molecular nature of drug–receptor interactions without reviewing the classical pharmacological theories. Since about 1970, progress in methodology has made direct measurement of drug binding to receptors a routine procedure. The classical theories were of necessity based on measurement of the final effect of drug action—an effect that is many steps removed from the drug–receptor binding process. Therefore, the modern approach more closely follows molecular lines whereas the older pharmacological methodology operates at the cellular and organismic level. Naturally, both avenues have advantages and disadvantages. We shall deal first with the dose–response relationship before reviewing current receptor models.

The classical *occupation theory* of Clark rests on the assumption that drugs interact with identical, independent binding sites and activate them, resulting in a biological response that is proportional to the amount of drug–receptor complex formed. The response ceases when this complex dissociates. Assuming a bimolecular reaction, one can write

$$D + R \rightleftharpoons DR \tag{1}$$

where D = drug and R = receptor.

The dissociation constant at equilibrium is

$$K_D = \frac{[D][R]}{[DR]} \tag{2}$$

The effect ($E$) is directly proportional to the concentration of the drug–receptor complex:

$$E = \alpha[DR] \tag{3}$$

The maximum effect ($E_{max}$) is attained when all of the receptors are occupied:

$$E_{max} = \alpha[R_T] \tag{4}$$

where the total receptor concentration, $[R_T]$, is

$$[R_T] = [R] + [DR] \tag{5}$$

and $\alpha$ is a proportionality factor. Therefore, from (2) and (4):

$$\frac{[DR]}{[R_T]} = \frac{[D]}{K_D + [D]} \tag{6}$$

Dividing (3) by (4):

$$\frac{[DR]}{[R_T]} = \frac{E}{E_{max}} \tag{7}$$

Therefore, from Eqs. (6) and (7):

$$E = \frac{E_{max}[D]}{K_D + [D]} \tag{8}$$

Equation (8) indicates a hyperbolic relationship between the effect and the concentration of free drug. The $ED_{50}$ is therefore equal to $K_D$ (Fig. 2.1). Incidentally, Equation (8) is identical to the Michaelis–Menten relationship in enzyme kinetics, with $E_{max}$ representing $V_{max}$. Dose–response curves (such as those in Fig. 2.2) usually show effect versus the logarithm of the total drug concentration $[D_T]$, assuming that the concentration of bound drug is so small as to be negligible and that $[D_T] \simeq [D]$. However, if the receptor concentration $[R_T]$ becomes large relative to $K_D$, then

$$ED_{50} = K_D + 0.5[R_T] \tag{9}$$

meaning that at a high bound-drug concentration, the total concentration of drug may exceed $K_D$ by an amount equal to one-half the total receptor concentration. It seems that the case $ED_{50} = K_D$ is rather exceptional. If occupation of some of the

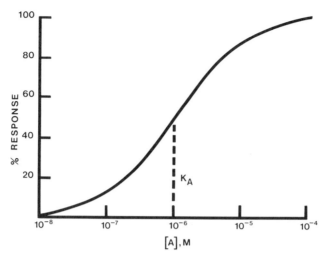

**Fig. 2.1.** A dose–response curve on a semilogarithmic scale, showing the definition of $ED_{50} = K_D$.

receptors is sufficient for a maximal response, as often happens, *spare receptors* will be present and

$$\frac{ED_{50}}{K_D} < 1$$

and the true value of $K_D$ (and thus the affinity of the drug for the receptor) will be underestimated. This case may be an indication that an "induced fit" takes place,

**Fig. 2.2.** Schematic dose–response curves of fractional response versus total drug concentration (log $[D_T]$). Curves *a* and *b* (with different $ED_{50}$ values) show the action of drugs in the same series acting on the same receptor site with different intrinsic activities. Curve *c* represents a partial agonist of the same series. Curve *a'* is the action of *a* in the presence of a competitive antagonist; *d* and *D* are the concentrations of drug necessary for the same response in the absence and presence of competitive antagonist. (After Hollenberg, 1985)

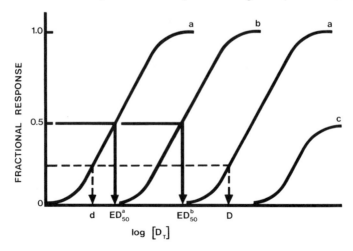

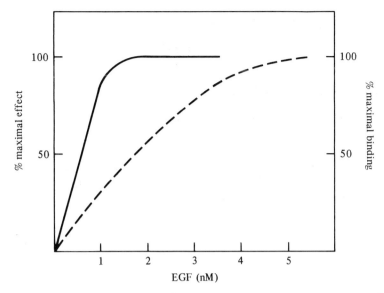

**Fig. 2.3.** Schematic of the biological effect of epidermal growth factor (solid line) and percent maximal binding to (broken line) human fibroblasts. The noncoincidence of the two curves illustrates the spare receptor concept: maximal pharmacological effect (at about 1.8 nM) is attained at less than 100% receptor occupation (at about 5 nM). (Modified from Hollenberg, 1985)

since it seems that a small number of agonist molecules can trigger a conformational change in many receptors, leading to the activation of a larger number of receptors than seems to be warranted (see Colquhoun, 1973, p. 159). The "spare" receptor concept can be tied to the idea of *efficacy* or *intrinsic activity*, meaning that some drugs may have to activate fewer receptors than others to elicit a full pharmacological effect, and are thus said to be more "efficacious."

Such spare receptors are indicated in the dose–response curves for the activity of some peptide hormones, shown in Fig. 2.3. This simplified diagram compares the pharmacological effect of a peptide growth hormone with its binding to cultured human cells on which it acts. One hundred percent effect is reached well before the receptors are saturated. Other systems—for example, insulin in adipocytes (fat storage cells)—also show similar behavior, Such deviation from theoretical predictions demonstrates that receptor concentration can be important in evaluating dose–response data, just as enzyme concentration is important in enzyme kinetics. If a minimum "threshold" occupancy is necessary to observe a biological effect, $ED_{50}$ will overestimate the true $K_D$.

Nevertheless, agonists yielding parallel dose–response curves with the same maximum are assumed to act on the same site but with different affinities (Fig. 2.2). Nonreceptor binding to a "site of loss" (often called a "silent" receptor) can thus be distinguished from relevant binding.

Schild extended these ideas to the description of effects when a *competitive antagonist* (A) is present.

If $Y$ is the proportion of receptors occupied, that is, if

$$Y = \frac{[DR]}{[R_T]} \tag{10}$$

and the AR complex is inactive, then

$$\frac{K_D}{[D]} = \frac{(1 + [A]K_A)Y}{1 - Y} \tag{11}$$

where $K_A$ is the association constant of the antagonist. If the same biological response is achieved at a lower drug concentration [d] in the *absence* of the antagonist, then

$$K_D[d] = \frac{Y}{1 - Y} \tag{12}$$

Dividing (11) by (12):

$$\frac{[D]}{[d]} = 1 + [A]K_A \tag{13}$$

The Schild equation is thus obtained, where $[D]/[d]$ is the "dose ratio," as shown in Fig. 2.2. From Eq. (13), $[A]_2$, the antagonist concentration necessitating a doubling of the agonist concentration to achieve the pure agonist effect is

$$[A]_2 = \frac{1}{K_A}$$

and

$$pA_2 = -\log K_A$$

which provides a convenient experimental method for measuring the "activity" of an antagonist. This is, of course, analogous to the $pD_2 = -\log K_D$ concept (Sec. 3).

The effect of a competitive inhibitor can also be expressed as an *inhibitor affinity constant* $(K_I)$ by plotting the inhibitor concentration versus the reciprocal of the reaction velocity, or versus the reciprocal concentration of the labeled ligand (the isotopically labeled agonist that is displaced by the antagonist). The intersect of the lines so generated is $-K_I$.

$$K_I = \frac{IC_{50}}{1 + \frac{[L^*]}{K^*}}$$

where $IC_{50}$ is the inhibitor concentration that displaces 50% of the labeled ligand, $[L^*]$ is the concentration of the labeled ligand, and $K^*$ is its dissociation constant. This is a method particularly suited for *in vitro* binding experiments; however, it is not suitable for organ preparations or whole-animal studies. Rapidly growing experimental evidence that takes into account the latest *in vitro* binding experiments favors a modified form of the occupation theory of drug activity. There are, however, phenomena that are unexplained by the occupation theory:

1. The inability of partial agonists to elicit a full response while blocking the effect of more active agents.
2. The existence of drugs that first stimulate and then block an effect.
3. *Desensitization* or *tachyphylaxis*—diminution of the effect of an agonist with repeated exposure to or higher concentrations of that antagonist.
4. The concept of spare receptors.

To accommodate some or all of these phenomena, several alternatives to the occupation theory have been proposed. None of them is entirely satisfactory, and some have no physicochemical basis.

The *rate theory* of Paton, as modified by Paton and Rang, rejects the assumption that the response is proportional to the number of occupied receptors, and instead proposes a relationship of response to the *rate* of drug–receptor complex formation. According to this view, the duration of receptor occupation determines whether a molecule is an agonist, partial agonist, or antagonist. Accordingly, the concept of intrinsic activity becomes unnecessary.

The rate theory offers an adequate explanation for the ability of some antagonists to trigger a response before blocking a receptor, and also accounts for desensitization. However, it lacks a plausible physicochemical basis, and conflicts with some experimentally established facts (e.g., the slow dissociation rate of agonists).

The *induced-fit* theory, developed by Koshland primarily for enzymes, states that the morphology of a binding site is not necessarily complementary to the conformation—even the "preferred" conformation— of the ligand. According to this theory, binding produces a mutual plastic molding of both the ligand and receptor as a dynamic process. The conformational change triggered by the mutually induced fit in the receptor macromolecule is then translated into the biological effect. Although this model does not lend itself to the mathematical derivation of binding data, it has altered our ideas on ligand–receptor binding in a revolutionary way, eliminating the rigid and obsolete "lock and key" concept of earlier times.

The *macromolecular perturbation theory* of Belleau is closely related to the induced-fit theory. It proposes that an agonist induces a specific ordering when it binds to a receptor, producing a specific conformational perturbation of the receptor macromolecule. Antagonists, on the other hand, produce a nonspecific disordering effect. Although demonstrated only with alkylammonium cholinergic drugs, Belleau's ideas have a solid physicochemical basis, and his thermodynamic data bear out his hypothesis. The macromolecular perturbation theory also eliminates the need for the concepts of affinity and intrinsic activity, but in its original form is not sufficiently comprehensive.

## 5. MOLECULAR CONCEPTS OF RECEPTOR FUNCTION

The preceding sections have explored classical pharmacological concepts based on the dose–response relationships in tissue or organ preparations. The enormous complexity of living systems and the remoteness of cause from effect (i.e., drug administration and pharmacological action) introduce many complications and artifacts into the study of such relationships.

Molecular pharmacologists and physical scientists have therefore sought to simplify the experimental system as much as possible by stripping it of such unnecessary and irrelevant factors as drug transport and metabolism and putting it on a level accessible to molecular manipulations and precise physicochemical methods. This objective has been increasingly realized as the methodology of quantitative binding experiments on membrane preparations and later on isolated receptors has become more sophisticated, precise, and simple. Isotopically labeled compounds of very high activity have made it possible to work with physiological ligand concentrations down to the picomole level ($10^{-12}$ M). This has allowed direct experimental access to receptor binding sites and has allowed the development of several complementary receptor models. Ariëns, Burgen, Changeux, Colquhoun, Cuatrecasas, Hollenberg, Karlin, Seeman, Snyder, Yamamura, and their many coworkers are among the molecular pharmacologists in the forefront of the spectacular and explosive progress of "receptorology" since the early 1970s. The revolutionary reshaping of our ideas on drug receptors shows no sign of slowing down. The reader is thus warned of the ephemeral nature of some current hypotheses and the inevitable obsolescence of this chapter.

## 5.1. Molecular Properties of Drug Receptors

Early receptor models, based on pharmacological data rather than direct ligand-binding measurements, postulated that agonists and their competitive antagonists became bound to the same receptor site and "completed" for it. This view was partly based on findings in enzymology—where this concept is generally valid for metabolite–antimetabolite competition—as well as on activity studies of vitamins and hormones. The close structural resemblance of agonists and antagonists in these categories constitutes direct proof that they have identical binding sites. The lack of structural correlations between many neurotransmitters and their blocking agents, however, initiated a review of the competitive binding hypothesis.

It is generally accepted that there is a complementarity between a ligand and its receptor site in the sense of the induced-fit concept; this suggests a mutual molding of the drug and macromolecule to take full "advantage" of stereoelectronic interactions. Under optimal conditions, the energies liberated in binding can reach 40–50 kJ/mol, a figure equivalent to binding equilibrium constants of about $10^{-8}$–$10^{-9}$ M, which is considered to represent a high affinity.

Complementarity in the context of induced fit implies a plasticity of the receptor macromolecule, in terms of an ability to undergo conformational changes and associate with ligands. In its activated state (i.e., a different conformation), the receptor can interact with effector molecules, which then transmit a nerve impulse or other signals to other structures. The complementarity also determines the selectivity of the receptor. For stereospecific binding, it is generally assumed that a ligand must have three unequal substituents; this is considered sufficient for great selectivity. Ariëns speculated that if three different binding sites (e.g., a hydrogen bond, an ionic site, and a hydrophobic interaction site) exist, each in five discrete forms in terms of the distance between each pair of bonding sites, the receptor will bind with high affinity only one of a million different compounds. The "discrete forms" of a receptor site are, of course, the result of receptor plasticity.

This capacity of the receptor to assume different molecular geometries without a significant change in function is probably essential to achieving some understanding of the pluralistic nature of many receptors. It is physiologically and structurally unreasonable to assume that a given type of receptor—probably a complex, multisubunit structure that is part of an even more complex membrane framework—is absolutely identical throughout an organism. Mautner pointed out in 1967 (cf. Mautner, 1980), long before the structure of any drug receptor was known in any detail, that the medicinal chemist will have to deal with an "isoreceptor" concept in the same matter-of-fact way that an enzymologist accepts isozymes. Although our present knowledge of receptor structure is still crude, the exact structural and functional identity of, say, an opiate receptor in the central nervous system and in the ileum is rather unlikely. Not only do they have different roles (cf. Chap. 5, Sec 3.4), as participants in neuromodulation and peristaltic regulation, respectively, but they are also probably different in a morphological sense: the neuromodulatory receptors are assumed to be *presynaptic* (see Fig. 4.4 and the review of Starke, 1981), or situated on the presynaptic terminal membrane, *ahead* of the synaptic gap, whereas other receptors are the classical *postsynaptic* receptors, embedded in the postsynaptic membrane of the effector cell or of the next neuron. In the first case, the receptor modulates neurotransmitter release; in the second, it may activate an enzyme such as adenylate cyclase, or trigger an action potential. As we shall later see, almost all neurotransmitters show receptor multiplicity, and medicinal chemists deal with four adrenergic receptor subtypes and three to four different opiate receptors, just to name two examples.

Receptor plasticity could be invoked as the underlying common trait of multiple receptors. For example, although the four adrenergic isoreceptors are similar, they react to the common neurotransmitter, norepinephrine, in a quantitatively different manner. They also show a drug specificity that varies from organ to organ and differs in various species of animals. In subsequent chapters of this book, receptor multiplicity as the rule rather than the exception will become amply evident. It is to be hoped that, in time, the comparison of isoreceptor molecular structures will provide precise criteria for their differentiation. Ariëns and his group (1979, p. 59–64) extend and discuss in detail the ideas expressed above.

The multiplicity of receptor or recognition sites for agonists and antagonists is well documented. One may distinguish (1) agonist binding sites, (2) competitive antagonist binding sites (accessory sites), and (3) noncompetitive antagonist or regulatory binding sites (allosteric sites).

The *agonist binding site* is the subject of continuous discussion throughout this book, ranging from a purely physical approach to the treatment of its biochemical characteristics where these are known. In this discussion, it is implicit that we are dealing with discrete loci on the receptor macromolecule: specific amino acids, lipids, or nucleotides held in just the right geometric configuration by the scaffold of the rest of the molecule, as well as by its supramolecular environment such as a

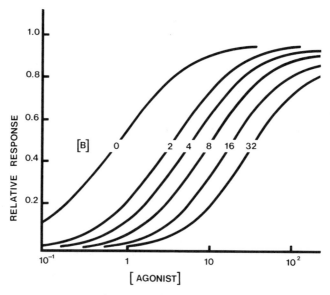

**Fig. 2.4.** Dose–response curves for an agonist in the presence of increasing (i.e., doubled) concentrations of a competitive antagonist, B. (After Triggle, 1978)

membrane. As a matter of fact, most drug receptors are part of the plasma membrane, which is a lipoprotein and thus amphiphilic (see Chap. 6, Sec. 1). There are, of course, notable exceptions to this, such as the steroid hormone receptors, where the steroid binds to a free receptor in the cytosol and the complex undergoes transport.

*Competitive antagonists* were originally assumed to bind to the agonist binding site and in some way displace and exclude the agonist as a result of their very high affinity but lack of intrinsic activity. This would result in displaced but parallel dose–response curves, such as those in Fig. 2.4. Our present views shaped mainly by Ariëns, are at variance with such simplistic older ideas. The mere fact of great chemical dissimilarity between agonists and competitive antagonists in the vast field of neurotransmitters precludes identity of the two receptor sites. Figure 2.5 compares the structures of some selected examples of such substances. It is evident at a glance that although the agonists are structurally similar, a careful analysis is needed to discern correlations between agonist–antagonist pairs or even between antagonists of the same class. As always, there are notable exceptions to this. For example, opiate analgesics and their antagonists are very similar in structure but if one considers the relationship between the endogenous peptide "opiates" known as enkephalins and the opiate antagonists, the conspicuous dissimilarity of the two groups once again is apparent.

The most remarkable property of antagonists is their great receptor affinity, which is often two to four orders of magnitude greater than that of the agonists (Table 2.2). Inspection again reveals that all of the antagonists shown in the table have large nonpolar moieties, usually aromatic rings. Therefore, *accessory binding*

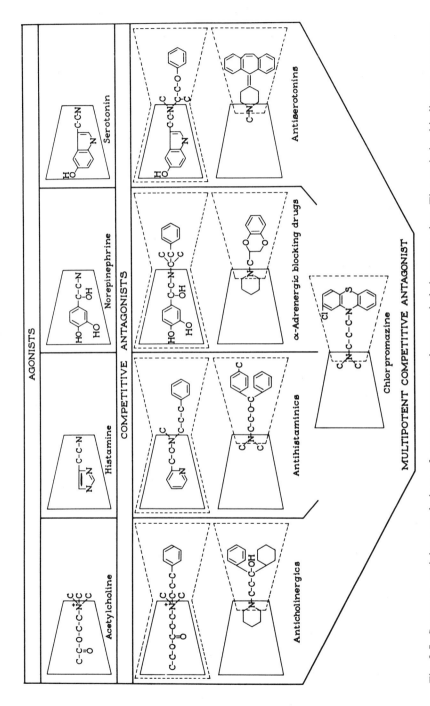

**Fig. 2.5.** Structure–activity correlations of some neurotransmitters and their antagonists. The moieties binding to accessory binding sites are bounded by dotted lines, the agonist portion by solid trapezoids. (Reproduced by permission from Ariëns et al. (1979), Plenum Press, New York)

**Table 2.2.** Comparison of the histamine agonist and antagonist activity of $\beta/\beta$-pyridylethylamine derivatives

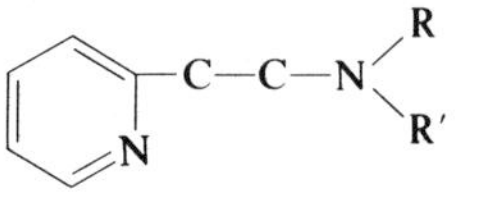

| R | R' | Agonist $pD_2 \pm P_{95}$ | Antagonist $pA_2 \pm P_{95}$ |
|---|---|---|---|
| —H | —H | $5.44 \pm 0.08$ | |
| —C | —H | $5.46 \pm 0.07$ | |
| —C—C | —H | $4.50 \pm 0.10$ | |
| 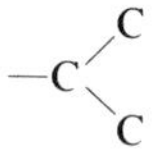 —C(—C)(—C) (isopropyl) | —H | | $4.00 \pm 0.06$ |
| —C—C—C | —H | | $4.19 \pm 0.06$ |
| —C—(C)$_2$—C | —H | | $4.30 \pm 0.14$ |
| —C—(C)$_4$—C | —H | | $4.72 \pm 0.02$ |
| —C—(C)$_6$—C | —H | | $5.90 \pm 0.03$ |
| —C—(C)$_8$—C | —H | | 5.33 |
| 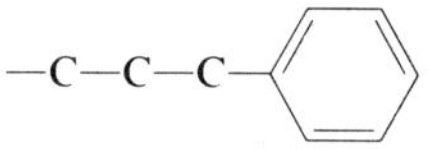 —C—C—C—(phenyl) | —C | | $5.90 \pm 0.13$ |
| 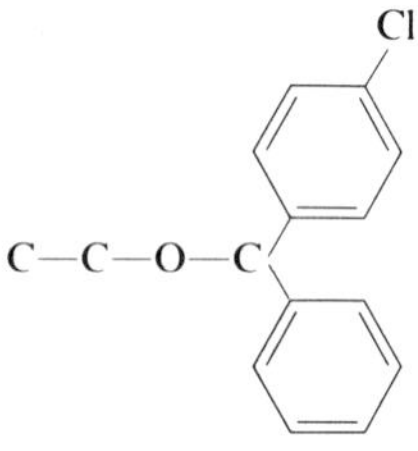 C—C—O—C(—4-Cl-phenyl)(—phenyl) | —C | | $7.80 \pm 0.11$ |
| 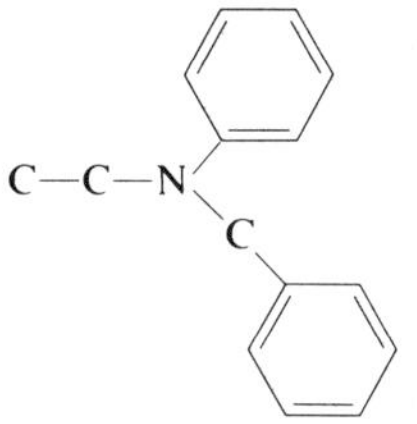 C—C—N(—phenyl)(C—phenyl) | —C | | $7.96 \pm 0.16$ |
| Histamine | | $6.55 \pm 0.13$ | |

*Note*: Derivatives with small N-substituents show agonist activity, whereas derivatives with large N-substituents produce an increasingly antagonist effect.

[a] The $\pm$ figures give the $P_{95}$ for the mean value est./org. = number of estimations per number of organs used. The data were obtained by testing the compounds on the isolated gut of the guinea pig.

Reproduced by permission from Ariëns et al. (1979), Plenum Press, New York.

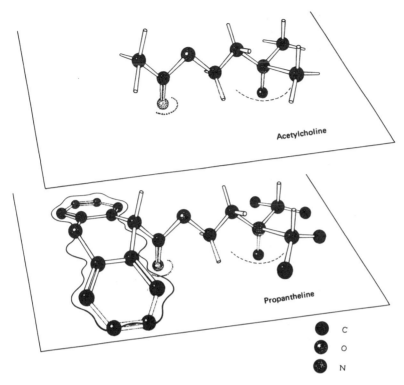

C
O
N

**Fig. 2.6.** Schematic representation of acetylcholine and the anticholinergic drug propantheline binding to the cholinergic receptor. The tricyclic ring system of propantheline binds at
an accessory binding site. (Reproduced by permission from Ariëns et al. (1979), Plenum Press,
New York)

*sites* must exist on the receptor to accommodate these large hydrophobic groups.
These sites are accessory to the agonist binding site, which in most cases is polar;
the situation for acetylcholine and its antagonist propantheline is shown in Fig. 2.6.
The accessory sites are most probably associated with lipid–protein interfaces
*inside* a membrane or with the hydrophobic clefts of a protein. What is even more
remarkable is that there are some compounds that are antagonistic in more than
one system. **Diphenhydramine** (Fig. 4.40), for example, has an antihistaminic as
well as anticholinergic action.

Competitive antagonists can be viewed in two ways. In one of these, the antagonist binding site is considered to be topically close to the agonist site and may
even partially overlap it. The antagonist will therefore interfere with agonist access
to the receptor, even though it need not necessarily occupy both the agonist and
the accessory sites, as shown in Fig. 2.5. On the other hand, the antagonist may
functionally deny agonist accessibility by altering the receptor affinity. This would
be closely analogous to allosteric inhibition, which is discussed below.

The accessory site is therefore the more important binding site for the antagonist,
and the binding sites for the agonist and antagonist are not the same, even though

both are on the same receptor lipoprotein and even though this shows high selectivity for the agonist as well as the antagonist. This fact, combined with different affinities of different forms or states of the receptor, leads to the development of the two-state model of receptors, which is discussed in the next section.

## 5.2. Molecular Models of Receptors

Rodbard and his co-workers (De Lean et al., 1979) have proposed an interesting *"multisubsite" receptor model* which could account for receptor cooperativity, biphasic dose–response curves, competitive and noncompetitive antagonism, and the existence of partial agonists. This model proposes that agonists bind to two subsites and thus trigger the receptor–effector response whereas antagonists bind only to one or the other subsite, including allosteric or accessory sites, and thus do not allow the formation of the R form of the receptor (see below).

*Allosteric sites* are at a distance from the agonist site and may even be on a different receptor protomer in the receptor–effector complex. Their occupation by allosteric inhibitors results in a conformational change that is propagated to the agonist site and changes its affinity. There is thus a mutual exclusion between the agonist and an allosteric antagonist. Moreover, classical pharmacological models cannot distinguish between competitive and allosteric inhibition. Allosteric effectors are not necessarily inhibitors. Just as in enzymology, some may activate whereas others deactivate one or another state of a receptor.

### 5.2.1. The Two-State Receptor Model

The two-state receptor model (also known as the Monod–Wyman–Changeux model) was developed on the basis of the kinetics of competitive and allosteric inhibition as well as through interpretation of the results of direct binding experiments. This model postulates that regardless of the presence or absence of a ligand; the receptor exists in two distinct states the R (relaxed, active, or "on") and T (tense, inactive, or "off") states, which are in equilibrium with each other. An agonist (drug, D) has a high affinity for the R state and will shift the equilibrium to the right; an antagonist (inhibitor, I) will prefer the T state and will stabilize the TI complex:

$$\text{T} \xrightleftharpoons{K_L} \text{R}$$
$$+\text{I} \left\|\, K_{TI} \quad +\text{D} \right\|\, K_{RD}$$
$$\text{TI} \rightleftharpoons \text{RD}$$

Partial agonists have about equal affinity for both forms of the receptor.

Some members of a receptor population are in the R state, even in the absence of any agonist. Thus, the receptor can be thought of having a "tone" like a resting muscle. The ratio of states is defined by the equilibrium constants $K_L$, $K_T$, and $K_R$ (for drug D or inhibitor I), and gives true physicochemical meaning to the concept of

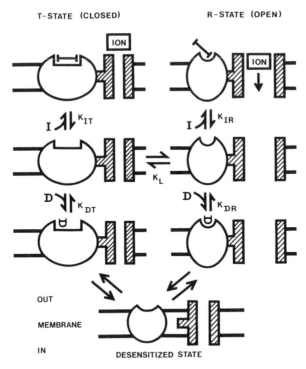

T-STATE (CLOSED)　　　　R-STATE (OPEN)

ION

$K_{IT}$　　　　$K_{IR}$

$K_L$

$K_{DT}$　　　　$K_{DR}$

OUT

MEMBRANE

IN　　　　DESENSITIZED STATE

**Fig. 2.7.** A two-state receptor model connected to an ion channel. The unoccupied receptor (middle row) can exist in a closed T state (left) or an open R state (right). These are in an equilibrium that has an equilibrium constant $K_L$. Both states can bind inhibitor (I) or drug (D), but only proper fit will ensure stabilization of the R or T state. In the desensitized receptor (bottom), the recognition site is detached from the ion channel. (Modified from Hollenberg, 1985)

intrinsic activity. Ariëns et al. (1979) and Triggle (1978) give a detailed derivation of the kinetics of transformation between the two states. Figure 2.7 shows a two-state model of a receptor associated with an ion channel, which regulates membrane conductivity. Besides the R and T states, a desensitized state is also shown. In this latter state, the recognition site is uncoupled from the effector site (in this case, an ionophore), as seen at very high drug concentrations or after repeated stimulation. *Uncoupling* (also called *desensitization, fade,* or, if receptor-specific, *tachyphylaxis*) can also be interpreted in terms of receptor recycling, as discussed in Sec. 5.3.6.

In contrast to the assumption made in the classical occupation theory, the agonist in the two-state model does not "activate" the receptor, but shifts the equilibrium toward the R form. This explains why the number of occupied receptors does not equal the number of activated receptors.

The switch from the R to T form (from agonist to antagonist binding) of a receptor can also be induced by allosteric regulators. The most remarkable example of this occurs with the opiate receptor (see Chap. 5, Sec. 3.4), where $Na^+$ or $Li^+$ (but not $K^+$), acting as such a regulator, shifts the receptor protein to the T (antagonist) form, whereas in a $Na^+$-free buffer the receptor preferentially binds an agonist. Since the effect is measurable at $Na^+$ concentrations as low as 5 mM, which is far below the physiological sodium concentration of 150 mM, the receptor will have a much higher affinity for an antagonist *in vivo* than *in vitro*, and five to ten times more agonist than antagonist will be needed for an $ED_{50}$ because the receptor is in the antagonist conformation.

*Receptor Cooperativity.* Receptor cooperativity, which has largely been studied on hormone receptors and ionophores, is explained by further extension of the two-state model. It is assumed that the cooperation of several receptor protomers is necessary for an effect like the opening of an ion channel, with all of these protomers having to attain an R or a T state to open or close a pore. This means that the binding sites or the receptor protomers on which these sites are situated must interact, and, as they do so, their affinity changes as a function of the proportion of R-state receptors in the assembly. This also means that a drug–receptor complex can trigger the transition of an unoccupied neighboring receptor from the T to the R state. If a ligand facilitates binding or the effect of the receptor, the cooperativity is *positive*; if it hinders these, the cooperativity is *negative* (e.g., in the insulin receptor). Negative cooperativity could also account for the spare receptors (receptor reserve) seen in many systems. As receptors cluster during their own metabolic cycle (see Sec. 5.3.5), low ligand occupancy in such clusters may still lead to a large change in the cluster configuration, resulting in a full effect without a 1:1 ratio of ligand–receptor binding.

The sigmoidal shape of the dose–response curves is the first indication of cooperativity in an oligomeric receptor. Scatchard plots of ligand binding will be concave for positive and convex for negative cooperativity. Hill plots can also indicate the type of cooperativity involved (see Chap. 3, Sec. 4.6).

The cooperative behavior of hemoglobin and some enzymes was treated in the Monod–Wyman–Changeux hypothesis, which assumes that a concerted transition occur simultaneously in all of the subunits of the receptor assembly. In the alternative Koshland–Némethy–Filmer model, only the binding subunit undergoes a conformational change, which alters its interaction with its neighbors and results in a sequential change. The experimental distinction of these models is difficult; as applied to drug receptors, they are discussed in detail by Triggle (1978) and Colquhoun (1973).

As already evident from foregoing discussions, effector or amplifier systems are the parts of the receptor oligomer which convey the fact that a drug has become bound to the receptor (or, to be more precise, that there has been a conformational change to the R or T state), signaling subsequent links in the receptor–effector chain that ultimately trigger the biological effect. Besides initiating the effector step of the drug action, effectors are often amplifiers, magnifying an inconspicuous initial event like the binding of a few thousand ligand molecules at $10^{-9}–10^{-10}$ M concentration. Amplification can take the form of a *cascade*, as in the well-known case of epinephrine or glucagon: these hormones initiate glycogenolysis through a series of enzyme activation steps, as shown in Fig. 2.8, causing the initial effect to be magnified approximately 100 million fold. Indeed, one is struck by the omnipresence

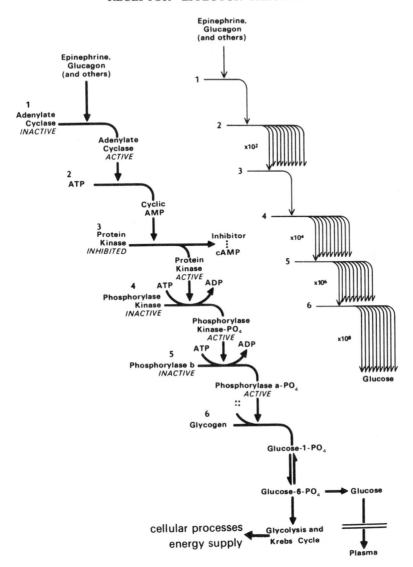

**Fig. 2.8.** The epinephrine-stimulated cascade of glycogenolysis shows the amplification process triggered by minute amounts of this hormone, mediated by cAMP, resulting in a massive mobilization of glucose. (Reproduced by permission from Goldberg, in Weissman and Claiborne (Eds.) (1975), *Cell Membranes: Biochemistry, Cell Biology and Pathology*, H.P. Publishing, New York)

of the enzyme adenylate cyclase (AC), which serves as first amplifier or effector in a large number of drug-initiated cascades as well as in numerous biochemical chains of events in enzymology. The cAMP that is subsequently produced activates kinases, which phosphorylate different proteins acting as final effectors. Since the majority of receptors are localized in cell membranes, this sequence of events constitutes *inter*cellular communication.

Similar cascades are known through the effects of steroid hormones which remove specific repressors from DNA and trigger the transcription of mRNA, resulting in massive protein synthesis. (Chap. 5, Sec. 1).

Another type of effector is the ionophore of an excitable membrane, which in its R (open) conformation allows the passage of about 10,000–20,000 ions in a single impulse, resulting in either membrane depolarization or polarization, and a multitude of possible physiological phenomena (see Chap. 6, Sec. 1 for a discussion of ion channels).

Details on the nature of effector systems are discussed in Sec. 5.3.


### 5.2.2. The Mobile Receptor Model

The mobile receptor model was proposed by Cuatrecasas and by De Haën (cf. Hollenberg, 1985) in an attempt to explain why so many different drugs, hormones, and neurotransmitters can activate adenylate cyclase. According to classical concepts, a recognition site is permanently associated with an effector site, and will regulate its operation on a one-to-one or some other stoichiometric basis. The recognition site is, of course, specific.

If this hypothesis is applied to the case of adenylate cyclase, one of two conditions would have to be assumed, that there are either as many adenylate cyclase isozymes as there are receptors acting through them; or adenylate cyclase would need an enormous variety of specific recognition sites that can answer to many ligands. The latter possibility would imply a lack of selectivity. However, there is no evidence for either assumption.

The mobile receptor concept offers a solution to this problem, in recognizing that the lipid membrane is a two-dimensional liquid, in which the embedded proteins can undergo rapid lateral movement or "translation" at a rate of 5–10 $\mu$m/min, an enormous distance on a molecular scale (Poo, 1985). The recognition protomer of a receptor complex therefore need not be permanently associated with an effector molecule, and thus no stoichiometric relationship is required. Instead, as shown in Fig. 2.10, the recognition protomer can undergo rapid lateral movement, and when activated to the R state can engage in what has been dubbed a "collision coupling." The R state of the receptor has the appropriate conformation to trigger effector activity, which could be the opening of an ionophore or the activation of adenylate cyclase. Therefore, different recognition sites can activate one and the same adenylate cyclase molecule at different times through the same mechanism. By the same token, a single recognition site could activate several adenylate cyclase molecules or other effector systems during its active lifetime. Such multiple collision couplings can be seen as the molecular explanation of positive cooperativity and the concept of receptor reserve. There is no need to invoke multiple recognition sites on the enzyme or a multitude of isoenzymes, only the physical separation of recognition and effector sites and their multipotential association. However, it is probably true that a recognition site that operates an ion gate is more permanently associated with the ionophore than is the drug or hormone receptor that acts through adenylate cyclase or the phosphatidylinositol system.

Alternatives to the collision coupling hypothesis have been proposed by Levitzki (1982, 1986), but kinetic experiments do not support alternative mechanisms.

## Selected Readings

E. J. Ariëns (1983). Intrinsic activity: partial agonists and partial antagonists. *J. Cardiovasc. Pharmacol. 5*: S8–S15.

E. J. Ariëns, A. J. Beld, J. F. Rodrigues de Miranda, and A. M. Simonis (1979). The pharmacon–receptor–effector concept. In: *The Receptors* (R. D. O'Brien, Ed.), Plenum Press, New York, pp. 33–91.

F. J. Barrantes (1979). Endogenous chemical receptors: some physical aspects. *Annu. Rev. Biophys. Bioeng. 8*: 287–321.

J. M. Boeynaems and J. E. Dumont (Eds.) (1980). *Outlines of Receptor Theory*. Elsevier/North Holland, New York.

D. R. Burt (1985). Criteria for receptor identification. In: *Neurotransmitter Receptor Binding*, 2nd ed. (H. T. Yamamura, S. J. Enna, and M. J. Kuhar, Eds.). Raven Press, New York, pp. 41–60.

D. Colquhoun (1973). The relation between classical and cooperative models for drug action. In: *Drug Receptors* (H. P. Rang, Ed.). University Park Press, Baltimore, pp. 149–182.

P. M. Conn (Ed.) (1984–86). *The Receptors*, 3 vols. Academic Press, New York.

A. De Lean and D. Rodbard (1979). Kinetics of cooperative binding. In: *The Receptors* (R. D. O'Brien, Ed.) Plenum Press, New York, pp. 143–192.

A. De Lean, P. J. Munson, and D. Rodbard (1979). Multivalent ligand binding to multisubsite receptors: application to hormone–receptor interactions. *Mol. Pharmacol. 15*: 60–70.

J. DiMaio, F. R. Ahmed, P. Shiller, and B. Belleau (1979). Stereo-electronic control and de-control of the opiate receptor. In: *Recent Advances in Receptor Chemistry* (F. Gualtieri, M. Gianella, and C. Melchiorre, Eds.). Elsevier/North Holland, New York, pp. 221–234.

M. D. Hollenberg (1985). Receptor models and the action of neurotransmitters and hormones. In: *Neurotransmitter Receptor Binding*, 2nd ed. (H. J. Yamamura, S. J. Enna, and M. J. Kuhar, Eds.). Raven Press, New York, pp. 1–39.

M. D. Hollenberg and P. Cuatrecasas (1979). Distinction of receptor from non-receptor interaction in binding studies. In: *The Receptors* (R. D. O'Brien, Ed.). Plenum Press, New York, pp. 193–214.

P. M. Laduron (1984). Criteria for receptor sites in binding studies. *Biochem. Pharmacol. 33*: 833–839.

A. Levitzki (1982). Activation and inhibition of adenylate cyclase by hormones: mechanistic aspects. In: *More About Receptors* (J. W. Lamble Ed.). Elsevier Biomedical Press, Amsterdam.

A. Levitzki (1984). *Receptors: A Quantitative Approach*. Benjamin, Menlo Park.

A. Levitzki (1986). β-Adrenergic receptors and their mode of coupling to adenylate cyclase. *Physiol. Rev. 66*: 819–854.

D. E. Macfarlane (1984). On the enzymatic nature of receptors. *Trends Pharmacol. Sci. 5*: 11–15.

H. G. Mautner (1980). Receptor theories and dose–response relationships. In: *The Basis of Medicinal Chemistry*, 4th ed. (M. E. Wolff, Ed.), Vol. 1. Wiley-Interscience, New York, pp. 271–284.

J. Parascandola (1980). Origins of the receptor theory. *Trends Pharmacol. Sci. 1*: 189–192.

M. Poo (1985). Mobility and localization of proteins in excitable membranes. *Annu. Rev. Neurosci. 8*: 369–406.

D. Rodbard (1980). Agonist versus antagonist. *Trends Pharmacol. Sci. 1*: 222–225.

K. Starke (1981). Presynaptic receptors. *Annu. Rev. Pharmacol. Toxicol. 21*: 7–30.

R. P. Stephenson (1956). A modification of receptor theory. *Br. J. Pharmacol. 11*: 379–393.

D. J. Triggle (1978). Receptor Theory. In: *Receptors in Pharmacology* (J. R. Smithies and R. J. Bradley, Eds.), Marcel Dekker, New York, pp. 1–65.

D. J. Triggle and C. R. Triggle (1976). *Chemical Pharmacology of the Synapse*. Academic Press, New York, Chap. 2.

## 5.3. Receptor Metabolism and Dynamics

Like all proteins, receptors or receptor subunits are coded by appropriate genes, transcribed to mRNA, translated, and further processed in the rough endoplasmic reticulum. The posttranslational processing consists of (1) the removal of peptide leading sequences and (2) N-glycosylation of asparagine. After the receptor protein is packaged in the Golgi apparatus, some carbohydrates are removed and others are added to the branched oligosaccharide "antenna" structures; this process is referred to as "capping" (see Chap. 6, Sec. 1 for a discussion of membrane structure). The oligosaccharides of the receptor glycoprotein seem to serve as recognition units necessary for high-affinity ligand binding, and also as protection from premature proteolytic degradation. The assembled supramolecular receptor is then inserted into the cell membrane as an *intrinsic protein*, that is, one that usually spans the width of the lipid bilayer. It can therefore communicate with the extracellular space as well as with the inside of the cell, thus fulfilling its role as a transmembrane signal transducer.

As a consequence of the mobile receptor concept, recent investigations recognized that membrane-bound receptors undergo dynamic processes that serve as regulatory mechanisms (Hollenberg, 1985a, 1985b; Hanover and Dickson, 1985). This has led to the idea that such receptor regulation is just as important in the overall response of the system as is the response of the target organ (e.g., a muscle cell or secretory cell). There are several categories of regulatory mechanisms, and they differ primarily in the rate of response: some are very fast (milliseconds to seconds), whereas others are much slower and delayed. At this point, we know of the following mechanisms: (1) regulation at the genetic level, (2) regulation by ligand, (3) covalent or noncovalent modification, (4) migration of receptors in the membrane (patch formation), (5) migration of receptors into the cell interior (internalization) and potential recycling to the plasma membrane, and (6) internalization and proteolytic degradation.

### 5.3.1. Regulation at the Genetic Level

Regulation at the genetic level is often observed for hormones that can regulate the rate of synthesis either of their own receptor or of other functionally related receptors (e.g., regulation of oxytocin receptor synthesis in the uterus by estrogens; see Chap. 5, Sec. 2.4).

### 5.3.2. Regulation by Ligand

Ligand regulation of receptors can be either self-regulatory (homospecific) or trans-regulatory (heterospecific). In the first case, ligand binding may initiate internalization of the ligand–receptor complex, thus removing receptors from the cell surface and decreasing the number of available receptors (as in the case of the insulin receptor; see Chap. 5, Sec. 2.7). Heterospecific regulation is shown, for instance by histamine, which at high concentrations can activate acetylcholine receptors, and by benzodiazepine anxiolytics ("tranquilizers"), which regulate GABA receptors (Chap. 4, Sec. 8.1.5).

### 5.3.3. Covalent Modification

Covalent modification of receptors occurs by phosphorylation, sulfhydryl–disulfide redox reactions, and proteolytic cleavage, in the same manner as occurs for many enzymes. Upon ligand binding, the receptor may phosphorylate itself on a tyrosine or serine residue, or the ligand–induced conformational change may make the receptor a substrate for a phosphorylase kinase. The consequences of these reactions are not yet fully understood. Sulfhydryl redox reactions, seen in the nicotinic cholinoceptor (Chap. 4, Sec. 2.2), result in alteration of relative ligand sensitivities: in the insulin receptor, they lead to affinity changes. Thus covalent modifications of receptors, whether homospecific or heterospecific, lend biochemical significance to the pharmacological terms "affinity" and "intrinsic activity."

### 5.3.4. Noncovalent Modification

Noncovalent modifications of receptors can involve interactions with small ligands (ions, nucleotides) or macromolecules (as in the mobile receptor model), leading to allosteric changes. They can also influence the receptor environment, causing a change in membrane potential or receptor distribution (clustering, patching). A notable example is the effect of $Na^+$ ions on the relative affinity of opiate receptors toward agonists and antagonists, or the effect of guanine nucleotides on a number of receptors, as discussed in the next section. The lateral mobility of the ligand–receptor complex, a phenomenon still not well understood, is further regulated by an alteration in membrane fluidity triggered by the ligand–receptor complex itself.

### 5.3.5. Receptor Clustering

Receptor clustering, although a noncovalent interaction, is really an entirely separate regulatory mechanism. Peptide hormone receptors in particular are known to form clusters observable microscopically by use of fluorescent receptor probes. Clustering is a necessary but insufficient prerequisite for the pharmacological effect. Ligand binding to clustered receptors is still necessary for cell activation in such instances as insulin receptor-mediated lipolysis in adipocytes (fat cells). As implied earlier, clustering could explain receptor cooperativity (Sec. 5.2.1) in a positive sense, as well as in a negative sense. Theoretically, dimerization could account for a 30% decrease in binding affinity, and thus influence the response of a whole system.

### 5.3.6. Receptor Internalization

Receptor clustering also accounts for receptor internalization. The basis of this phenomenon is endocytosis via coated pits (Dickson, 1985; Carpentier et al., 1986). These pits are apparent in electron micrographs as membrane invaginations coated on the inner (cytoplasmic) side with a web of the protein clathrine. It has been suggested that certain receptor proteins have structural domains that allow them to react with coated pits (Fig. 2.9). Receptor clusters in coated pits (site 4) are rapidly endocytosed, resulting in the formation of vesicles (endosomes) (site 5) which are then transported to the interior of the cell. As shown in Figure 2.9, internalized receptors undergo one of several fates. Perhaps the most important role of internalization is the removal of receptors from the plasma membrane, the *down-regulation* of a receptor population. As described in Chap. 4, Sec 3.5.4, such down-regulation of adrenoceptors is the mode of action of antidepressant drugs. Some internalized receptors (like the cholesterol-carrying low-density lipoprotein) release their ligand (site 8) and recycle to the cell surface, thus fulfilling the role of a transport protein (site 6). Other receptors undergo dissociation due to the low pH of the endosome, and many will be entirely degraded in lysosomes (site 7), thus

**Fig. 2.9.** Receptor clustering and internalization. Receptors in the plasma membrane ① bind ligand (L) ② and/or are phosphorylated by protein kinase C ③, in preparation for patch formation, which involves clustering in coated pits ④ that are covered on the inside by the protein clathrin (XXX). The coated pits, along with the receptor–ligand complexes, pinch off and form endosomes ⑤, into which the ligand may be released. From the endosome, the receptor can revert to the T form and recycle to the cell surface ⑥, in which case the ligand is released into the cell interior ⑧; alternatively, the receptor can be delivered to the lysosomes ⑦ and degraded, or transported to the nucleus along with the ligand ⑨.

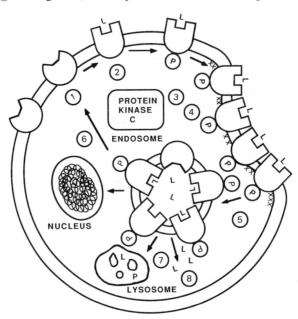

completing the metabolic path of receptors. It has also been hypothesized that endosomes may be the vehicle delivering hormone–receptor complexes to the cell nucleus (site 9), a common process for steroid hormones (Chap. 5, Sec. 1.2), that occurs by an unknown mechanism.

### Selected Readings

J. L. Carpentier, P. Gorden, A. Roberts, and L. Orci (1986). Internalization of polypeptide hormones and receptor cycling. *Experientia 42*: 734–744.

R. B. Dickson (1985). Endocytosis of polypeptides and their receptors. *Trends Pharmacol. Sci. 6*: 164–167.

J. A. Hanover and R. B. Dickson (1985). The possible link between receptor phosphorylation and internalization. *Trends Pharmacol. Sci. 6*: 457–459.

M. D. Hollenberg (1985a). Receptor regulation, Parts 1 and 2. *Trends Pharmacol. Sci. 6*: 242–245; 299–302.

M. D. Hollenberg (1985b). Receptor models and the action of neurotransmitters and hormones: some new perspectives. In: *Neurotransmitter Receptor Binding*, 2nd ed. (H. I. Yamamura, S. J. Enna, and M. J. Kuhar, Eds.). Raven Press, New York.

### 5.4. Transmembrane Signal Transduction

Binding of an agonist or antagonist by a receptor is the first step in a long cascade of events leading to the ultimate, macroscopic physiological effect of the drug or endogenous substance. In the case of receptors that operate on ion channels (e.g., the cholinoceptors and $Na^+$, $K^+$-ATPase), the recognition site and the ion channel are part of the same supramolecular receptor oligomer, and the ion channel will operate in direct response to ligand binding on different parts of the recognition subunits. This arrangement, which is a special case of transmembrane signaling, is discussed in the section on cholinoceptors (Chap. 4, Sec. 2) and ion channels (Chap. 6, Sec. 1.1). A more complex chain of events takes place in the vast majority of receptors—those utilizing chemical signaling. Although the number of steps involved is considerable, we know of only three mechanisms of signal transduction, two of which were discovered since 1982.

The general scheme of signaling begins with the arrival of an extracellular "first messenger"—a neurotransmitter, hormone, or another endogenous substance, or an exogenous ligand such as a drug or bacterial toxin. The receptor–ligand interaction takes place outside the cell, and in most instances the ligand does not enter the cytoplasm. There are, however, some exceptions, as discussed in the previous section on receptor internalization. Generally, the signal delivered by the ligand is conveyed to the cell interior by the receptor–ligand complex, which interacts with a *transducer*. The receptor–ligand–transducer ternary complex then interacts with an *amplifier*, usually an enzyme, which produces a substance that activates an *internal effector* (usually a phosphorylase kinase); the effector kinase then phosphorylates—and thereby activates or deactivates—a site-specific enzyme that regulates the final cellular response. Three systems, using different transducers, are known: (1) the adenylate cyclase system, (2) the guanylate cyclase system, and (3) the inositol triphosphate–diacylglycerol system.

### 5.4.1. The Adenylate Cyclase System

This system has been elucidated by a number of investigators over a relatively long period. Sutherland and Rall discovered cAMP in 1958, Rodbell and co-workers showed the need for GTP in the process in 1971, and the complete sequence of events was mapped more recently by A. G. Gilman and his group. This chain of reactions is shown in Fig. 2.10. There are two kinds of receptors participating in this system: the stimulatory $R_s$ receptors and the inhibitory $R_i$ receptors. Thus membrane receptors that operate through adenylate cyclase can do so either by activating the amplifier (see below) or by inhibiting it. When the receptor is occupied by its ligand, it forms a transient complex with a guanyl nucleotide binding protein, which is occupied by GDP. These protein transducers—likewise either stimulatory $(G_s)$ or inhibitory $(G_i)$—become activated in the binding process. In the ternary ligand–receptor–$G_{GDP}$ complex, the GDP is exchanged for a GTP, which triggers the release of the $\alpha_s$ subunit of the $\alpha\beta\gamma$ trimer $G_s$ protein. The $\beta$ and $\gamma$ subunits are also released. The active $\alpha_s$ subunit then combines with the adenylate cyclase (AC) enzyme (the *amplifier*), which produces cAMP, the second messenger. AC has been purified to homogeneity (Pfeuffer et al., 1985). The active $G_s$ state is terminated by a ligand-activated GTPase which hydrolyzes the bound GTP to GDP. Presumably, the G protein is then reconstituted from the three subunits in the inactive form, ready for the next binding cycle with an occupied receptor. The $G_i$ inhibitory transducer may act directly, in the same way that the $G_s$ protein functions, or perhaps indirectly. In the latter case, the $\beta\gamma$ subunits released from the $G_i$ protein possibly combine with the $\alpha_s$ subunit released from the $G_s$ protein, thus reconstitut-

**Fig. 2.10.** Model of adenylate cyclase activation. ① The receptor, in the tense (off) conformation ($R_T$), binds ligand (L), to form ② the activated ligand–receptor complex ($R_R$–L), which can now undergo collision coupling with the stimulatory guanyl-nucleotide binding protein trimer ($G_s$). ③ The ternary complex (L–$R_R$–$G_s$) is activated by an exchange of the GDP bound on the G protein for a GTP. ④ The ternary complex dissociates into inactive receptor ($R_T$), the ligand (L), the $\beta\gamma$ subunits of the G protein, and ⑤ the activated $\alpha_s$ subunit of the G protein. ⑥ The active $\alpha_s$ subunit binds to adenylate cyclase (AC) and activates it, initiating cAMP synthesis from ATP. ⑦ The $\alpha_s$ subunit is inactivated by hydrolysis of GTP to GDP and inorganic phosphate ($P_i$); the $\alpha_s$ subunit–GDP complex recycles by re-associating with the $\beta\gamma$ subunits.

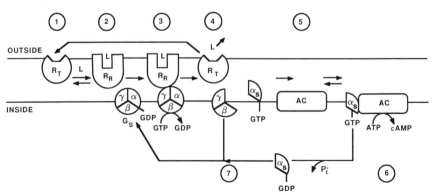

ing the inactive trimer. This idea is supported by the fact that the $G_i$ protein has a higher affinity for GTP than does $G_s$, and therefore the $G_i$ trimer dissociates more readily. It must be kept in mind that the receptors, the G proteins, and the cyclase interact in a mobile system by collision coupling, and thus a large diversity of receptors can activate the same population of G proteins and cyclase.

If the GTP is replaced by a nonhydrolyzable synthetic analogue, guanyl-5'-yl-imidodiphosphate (Gpp(NH)p; the anhydride oxygen is replaced by an NH group), the reaction cannot be terminated, and cAMP will be produced continuously. The GTPase, which normally terminates the active state, can also be inactivated by cholera toxin. The potentially fatal diarrhea and electrolyte loss that occur in cholera reflect the fact that cAMP is an activator of fluid secretion in the intestine. On the other hand, pertussis toxin (causing whooping cough) blocks the activation of $G_i$ by GTP, and probably causes bronchoconstriction.

Another natural compound, the diterpene **forskolin** (2-1), isolated from the roots of *Coleus forskohlii*, activates adenylate cyclase directly (Seamon, 1984). It was used in the folk medicine in India in respiratory disorders, insomnia, and convulsions. It is a potent hypotensive agent (i.e., reduces blood pressure) and a bronchodilator and also reduces intraocular pressure in glaucoma (see Chap. 6, Sec. 3.2).

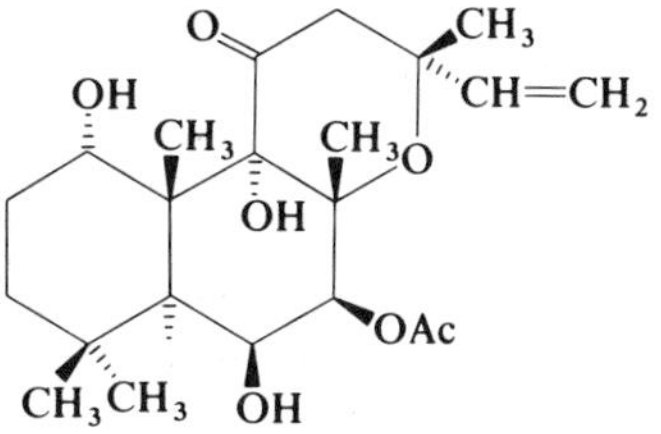

Forskolin
2-1

The most exciting development in signal transduction research was the realization that proteins produced by *oncogenes* (cancer-causing genes) are also GTP-binding proteins. The implications of this finding in carcinogenesis will be discussed in Sec. 5.5.

The final step in signal transduction is the action of cAMP on the regulatory subunit of the enzyme, protein kinase A. This ubiquitous enzyme then phosphorylates and activates enzymes with functions specific to different cells and organs. In fat cells, protein kinase A activates lipase which mobilizes fatty acids; in muscle and liver cells, it regulates glycogenolysis and glycogen synthesis. There are many more examples of activation (Berridge, 1985). On the other hand, if an inhibitory receptor (e.g., the $\alpha_2$-adrenoceptor; see Chap. 4, Sec. 3.3) interacts with a $G_i$ protein, a decrease in lipid degradation results because cAMP production is decreased.

### 5.4.2. The Guanylate Cyclase System

The second type of signal transduction utilizes *cyclic GMP* (cGMP) instead of cAMP as second messenger. It was previously believed that this pathway was not directly connected to a membrane-bound receptor, and could function in the stimulatory mode only, probably by activating protein kinase G. It plays a role in insect behavior, but also in the human retina and in the functioning of atrial natriuretic factor, a newly discovered hormone produced by the heart which regulates blood pressure (Chap. 5, Sec. 5). In the latter case, receptor involvement has to be assumed. It is quite likely that cGMP can also act through $Ca^{2+}$ as a third messenger in activating Ca-dependent protein kinases (Schulman, 1984).

### 5.4.3. The Inositol Triphosphate–Diacylglycerol System

The third and widely utilized signaling pathway is based on *phosphatidylinositol,* a normal constituent of the cell membrane. As shown in Fig. 2.11, the extracellular signal is received by a membrane-bound receptor which interacts with a $G_s$ protein, activating phospholipase C (phosphatidylinositol diphosphate [$PIP_2$] phosphodiesterase), an enzyme that cleaves phosphate diesters. The two products of this cleavage reaction are inositol triphosphate ($IP_3$) and diacylglycerol (DG), both of which act as second messengers, but in different cellular compartments (Berridge, 1985; Hokin, 1985).

**Fig. 2.11.** The phosphatidylinositol signaling system. ① The tense, inactive receptor ($R_T$) binds ligand (L) and changes to the relaxed, active ($R_R$) state. ② It interacts with a G protein and, in a manner probably identical with AC activation, activates phospholipase C (PL-C) or phosphodiesterase. ③. Phosphatidylinositol (PI) is split to yield diacylglycerol (DG), which remains in the membrane and activates protein kinase C (PK-C): ④ PK-C can also be activated directly by phorbol esters. ⑤ Inositol triphosphate ($IP_3$) diffuses into the cytoplasm and liberates $Ca^{2+}$ ions from stores. ⑥ $Ca^{2+}$ then directly, or in combination with calmodulin (CM), activates protein kinases. ⑦ These kinases, as well as PK-C, phosphorylate proteins, initiating physiological processes. The two limbs can operate independently of each other.

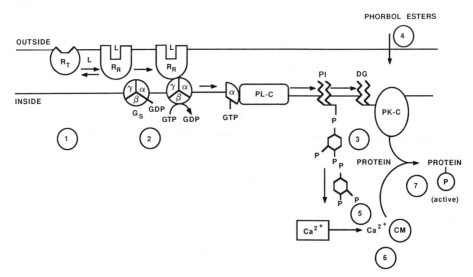

Inositol triphosphate is water soluble and therefore diffuses into the cytoplasm, where it mobilizes calcium from its stores in microsomes or the endoplasmic reticulum. The $Ca^{2+}$ ions then activate Ca-dependent kinases (like troponin C in muscle) directly or bind to the ubiquitous Ca-binding protein calmodulin (Chap. 6, Sec. 1.2), which activates calmodulin-dependent kinases. These kinases, in turn, phosphorylate cell-specific enzymes (Klee et al., 1980; West, 1982; Garrison, 1985).

Diacylglycerol, on the other hand, is lipid soluble and remains in the lipid bilayer of the membrane. There it can activate protein kinase C (PKC), a very important and widely distributed enzyme (Nishizuka, 1986), which serves many systems through phosphorylation: neurotransmitters (acetylcholine, $\alpha_1$- and $\beta$-adrenoceptors, serotonin), peptide hormones (insulin, epidermal growth hormone, somatomedin), and numerous cellular functions (glycogen metabolism, muscle activity, structural proteins, etc.) and also interacts with guanylate cyclase (Garrison, 1985). In addition to diacylglycerol, another normal membrane lipid, phosphatidylserine, is needed for activation of PKC. The $DG-IP_3$ limbs of the pathway usually proceed simultaneously.

Protein kinase C can be activated, independently of the phosphatidylinositol mechanism, by tumor-promoting **phorbol esters** (2-2). These natural compounds and other tumor promoters cause permanent activation of protein kinase C; there is great interest in the mechanism of this activation, as it might shed further light on tumorigenesis: the combined administration of phorbol esters and a calcium ionophore can trigger protein synthesis. Since the PKC activation is permanent in this case, the cellular signaling mechanism may be sufficiently upset to cause uncontrolled growth, that is, cancer.

12-*O*-Tetradecanoyl-phorbol-11-acetate
2-2

The diverse effects of PKC activation can be explained by the fact that this enzyme is produced by several families of genes, presumably as isozymes that have slightly different structures but basically identical mechanisms of action in different cell types (Parker et al., 1986).

The phosphatidylinositol pathway is completed by regeneration of the phospholipid from $IP_3$ and DG, as shown in Fig. 2.12 (see also Majerus et al., 1986). It is remarkable that $IP_3$ is successively dephosphorylated to inositol. The last step of this sequence is inhibited by $Li^+$ ions, which thus block phosphatidylinositol synthesis. Li salts are used to control the symptoms of manic-depressive illness, an affective mental disorder (Chap. 4, Sec. 3.5.4), and it is thus tempting to implicate the last reaction of the PI pathway in the etiology of this disorder.

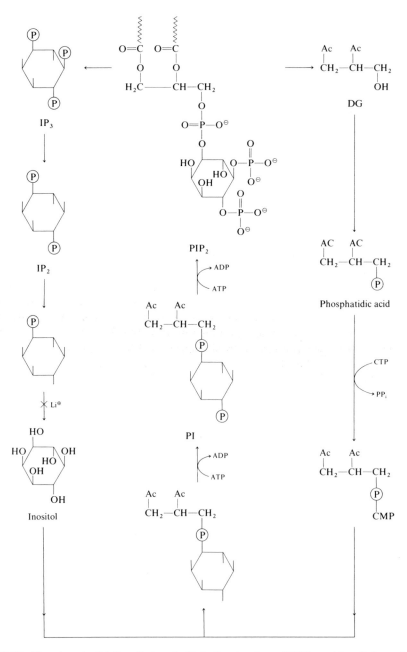

**Fig. 2.12.** The phosphatidylinositol cycle. Only the structure of $PIP_2$ and inositol are shown in detail. Ac, acetate; CTP, cytosine triphosphate; Ⓟ, phosphate group:

$$-O-\overset{\overset{\displaystyle O^{\ominus}}{|}}{\underset{\underset{\displaystyle O}{||}}{P}}-O-$$

## Selected Readings

A. A. Abdel-Latif (1986). Calcium-mobilizing receptors, polyphosphoinositides, and the generation of second messengers. *Pharmacol. Rev. 38*: 227–272.

M. J. Berridge (1985). The molecular basis of communication within the cell. *Sci. Am. 253*(4): 142–152.

J. C. Garrison (1985). Possible roles of protein kinase C in cell function. *Annu. Rep. Med. Chem. 20*: 227–236.

E. J. M. Helmreich and T. Pfeuffer (1985). Regulation of signal transduction by $\beta$-adrenergic hormone receptors. *Trends Pharmacol. Sci. 6*: 438–443.

L. E. Hokin (1985). Receptors and phosphoinositide-generated second messengers. *Annu. Rev. Biochem. 54*: 205–235.

C. B. Klee, T. H. Crouch, and P. G. Richman (1980). Calmodulin. *Annu. Rev. Biochem. 49*: 489–515.

P. W. Majerus, T. M. Conolly, H. Deckmyn, T. S. Ross, T. E. Bross, H. Ishii, V. S. Bansal, and D. B. Wilson (1986). The metabolism of phosphoinositide-derived messenger molecules. *Science 234*: 1519–1526.

S. R. Nahorski, D. A. Kendall, and I. Batty (1986). Receptors and phosphoinositide metabolism in the central nervous system. *Biochem. Pharmacol. 35*: 2447–2453.

Y. Nishizuka (1986). Studies and perspectives of protein kinase C. *Science 233*: 305–312.

P. J. Parker, L. Coussens, N. Totty, L. Rhea, S. Young, E. Chen, S. Stabel, M. D. Waterfield, and A. Ullrich (1986). The complete primary structure of protein kinase C, the major phorbol ester receptor. *Science 233*: 853–859.

E. Pfeuffer, R. M. Dreher, H. Metzger, and T. Pfeuffer (1985). Catalytic unit of adenylate cyclase: purification and identification by affinity cross binding. *Proc. Natl. Acad. Sci. USA 82*: 3086–3090.

H. Schulman (1984). Calcium-dependent protein kinases and neuronal function. *Trends Pharmacol. Sci. 5*: 188–192.

K. B. Seamon (1984). Forskolin and adenylate cyclase: new opportunities in drug design. *Annu. Rep. Med. Chem. 19*: 293–302.

L. W. West (Chairman) (1982). Calmodulin-regulated enzymes: modifications by drugs and disease. *Fed. Proc. 41*: 2251–2299.

### 5.5. Pathophysiological Mechanisms in Signal Transduction

Recent results in the tremendously active field of signal transduction (summarized by Marx, 1987) have begun to reveal some remarkable findings on the role of the PI system in pathophysiology. It has been linked to several *oncogenes*, genes that are normally present, but on activation cause malignant transformation of cells. As mentioned earlier, the protein encoded by the cellular (nonmalignant) *ras* gene may be an integral part of the PI system; it is a G protein (Helmreich and Pfeuffer, 1985) that increases coupling between growth factor receptors and phospholipase C. A point mutation—that is, a substitution of a single amino acid by another—results in oncogenic transformation. If the mutated, malignant *ras* oncogene is now introduced into cells, $IP_3$ is produced despite the absence of a growth factor. This is significant because in the absence of the normally regulated growth factor, the oncogene triggers continuous cell stimulation; such uncontrolled continuous growth is tantamount to malignant transformation of cells. Products of the *fos*

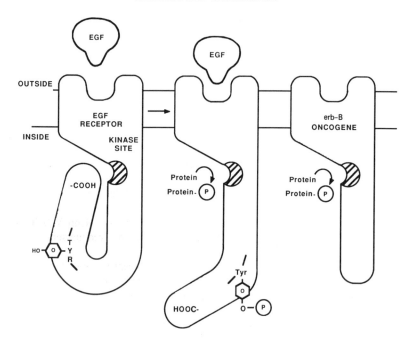

**Fig. 2.13.** Relation between epidermal growth factor receptor and erb-B oncogene structure. The carboxy-terminal end of the EGF receptor blocks the active site of the kinase; the site can function only if the receptor is activated by EGF, undergoes autophosphorylation on a tyrosine, and changes its conformation. The receptor is regulated. The erb-B oncogene is homologous with the EGF receptor, but lacks the C-terminal chain; in the absence of any regulatory substance, the tyrosine kinase remains continually functional. (Modified from R. Schatzman, ACS Medicinal Chemistry Symposium, June 1986, Chapel Hill, NC)

and *myc* oncogenes seem to regulate the expression of other genes and therefore may be capable of initiating permanent alterations in cells that would normally respond to growth factors.

Some other oncogenes linked to the PI system encode *tyrosine kinases*, which activate proteins by phosphorylating their tyrosine hydroxyl groups. Many growth factor receptors (e.g., of epidermal growth factor, EGF; platelet-derived growth factor, PDGF; and insulin-like growth factor) all operate as tyrosine/serine kinases. The interrelation between oncogenes and growth factors has been strengthened by the discovery that oncogenic viruses that produce sarcomas (tumors of connective tissue) code for proteins that are structurally and functionally homologous with EGF and IGF receptors. The difference between the growth-factor receptors and oncogene products might be in the C-terminal portion, as schematically shown in Fig. 2.13. In the EGF receptor molecule, the C-terminus blocks the active site (the phosphorylase) under normal, unactivated conditions. On activation by an EGF molecule, the receptor protein undergoes autophosphorylation and subsequent conformational change, which uncovers its active site and initiates cell proliferation by an unknown mechanism. The product of the *erb-B* oncogene is homologous with the EGF receptor (Antoniades and Pantazis, 1986), except that it lacks the C-terminus of the latter, and thus the active site is continually exposed and functional, and therefore not subject to normal regulation. Thus the uncontrolled tyrosine kinase activity of the viral oncogene distorts normal transmembrane signaling. Other, more complex indirect hormone-related oncogenic changes can also be envisioned (Hollenberg, 1985).

We can now begin to see how several oncogenes (about 25 are known by now) work: some of them (*sis*) code for growth factors (PDGF) that activate other genes (*myc* and *fos*), which specify substances that, through specific functions in the cell nucleus, regulate cell division and growth. When a growth factor is transformed to an oncogene, its regulatory function becomes distorted and uncontrolled, resulting in cancer (Salomon and Perroteau, 1986).

Somatostatin (SS, growth-hormone release inhibitory factor; see Chap. 5, Sec 2.3) seems to inhibit not only the action of growth hormone, but that of EGF-mediated processes (see DeFeudis and Moreau, 1986) at very low (picomolar to nanomolar) concentrations by stopping centrosome separation. It also stimulates phosphoprotein phosphatase (an enzyme that hydrolyzes phosphate esters) activity; that is, it counteracts the effect of tyrosine kinase. Experimental results on various malignancies have been encouraging, but SS has a very short half-life, and therefore more stable analogues must be found.

### Selected Readings

H. N. Antoniades and P. Pantazis (1986). Mitogenic factors as oncogene products. *Annu. Rep. Med. Chem. 21*: 237–245.

M. J. Berridge (1985). The molecular basis of communication within the cell. *Sci. Am. 253*(4): 142–152.

F. V. DeFeudis and J.-P. Moreau (1986). Studies on somatostatin analogues might lead to new therapies for certain types of cancer. *Trends Pharmacol. Sci. 7*: 384–386.

E. J. M. Helmreich and T. Pfeuffer (1985). Regulation of signal transduction by β-adrenergic hormone receptors. *Trends Pharmacol. Sci. 6*: 438–443.

M. D. Hollenberg (1985). Pathophysiological and therapeutic implications of receptor regulation. *Trends Pharmacol. Sci. 6*: 334–337.

J. L. Marx (1987). Polyphosphoinositide research updated. *Science 235*: 974–976.

D. S. Salomon and I. Perroteau (1986). Oncological aspects of growth factors. *Annu. Rep. Med. Chem. 21*: 159–168.

# 3
# Methods of Receptor Characterization

Textbooks present research findings, correlations, hypotheses and their implications in a groomed and necessarily sketchy way. The challenges and frustrations of designing and carrying out experiments, and the skills they require, are rarely revealed.

The reader, especially the student, is often left wondering where all the facts supporting the theory came from. How can one begin to tackle experimentally any of the problems that offer themselves to the critical reader without having to use the voluminous research literature at the outset? How can one begin to collect the armamentarium of methods that change so rapidly but are indispensable to solving increasingly complex and subtle problems?

It assuages the student's insecurity in a new field if the outlines of the methodology are traced even if they can only serve as roadsigns for further study. This chapter is offered with these thoughts in mind.

## 1. METHODOLOGICAL PRINCIPLES

Chemical signals that a drug transmits when it binds to a receptor result in neurophysiological, enzymological, or chromosomal responses. Although these are biochemical effects, drug research differs in two essential points from classical biochemistry. First, the initial event is noncatalytic, and second, the proof of the assumed chain of events requires reconstitution of the original system, disassembled for observation.

The principles underlying the methods currently used in receptor studies can be organized into several categories:

1. Measurements of ligand–macromolecule binding, with a corresponding kinetic analysis
2. The rigorous verification of binding specificity
3. Biochemical isolation of the receptor macromolecule and characterization of its properties
4. Elucidation of the effects secondary to binding

5. Reconstitution of the system by using isolated components and demonstrating its original function

Each of the above tasks is difficult and fraught with pitfalls. Binding experiments are hampered by the extremely low receptor concentration in most tissues, which is normally about 10–100 pmol/g, or less than approximately 1 mg/kg. Only in some specialized tissues such as the electric organs of some fish and eels does the concentration of the nicotinic cholinergic receptor rise to a manageable 10 mg/kg. Low receptor concentrations necessitate the use of low ligand concentrations to ensure that one measures specific binding only. This requires the use of concentrations in the nanomolar ($10^{-9}$ M) range. Since this vanishingly small amount of drug must be detected quantitatively, highly radioactive ligands, of at least 20–30 Ci/mmol activity, are required. Another, equally sensitive method for binding measurements is electron paramagnetic resonance (EPR). Precautions to ensure the specificity of binding are outlined in the next section (see also Kenakin, 1984; Laduron, 1984).

The isolation or solubilization of receptors is even more difficult. Their low concentration requires purification by a factor of many thousands, but the use of extremely mild methods is mandatory to avoid denaturation of the intrinsic membrane-protein receptors. Since the pharmacological properties of the receptor are lost during isolation, binding data are the only criteria for identifying various receptor fractions. Covalently binding "affinity labels" can be useful in isolation attempts, but they furnish an irreversible ligand–receptor complex.

## 2. INDIRECT METHODS OF RECEPTOR CHARACTERIZATION

This chapter is not meant to provide a summary of pharmacological techniques, which, strictly speaking, are the indirect methods of receptor study. However, the molecular pharmacologist is well advised to keep in mind that *in vitro* assays are no substitute for *in vivo* testing to establish a complete pharmacological profile, since both pharmacokinetic phenomena and the bioactivation of drugs precursors are eliminated *in vitro* (cf. Chap. 8, Sec. 4).

Indirect methods employ receptors *in situ*, in tissue or organ preparations. The methods are based on reversible processes and, rather than measuring the properties of the receptor as such, measure only those of its interactions with ligands that are followed by a pharmacological response.

The thermodynamics of binding can be investigated either by equilibrium methods or by studying the kinetic aspects of ligand–receptor interaction when equilibrium is disturbed.

The simplest case of binding deals with a population of single binding sites on every receptor macromolecule when all of these sites are equivalent and do not interact. As already outlined in Chap. 2,

$$D + R \rightleftharpoons DR \rightleftharpoons DR^*$$

Such reactions, leading to receptor "activation" ($R^*$), follow simple Michaelis–

Menten kinetics. The fraction of receptors occupied ($v$) is

$$v = \frac{[DR]}{[DR] + [R]}$$

and the dissociation constant $K_D$ can be determined experimentally.

The principal problem in these determinations is the need to work with very low ligand concentrations (typically lower than micromolar), and with very small amounts of usually impure receptor preparations. Direct measurements of binding are therefore often hopeless, as illustrated by O'Brien (1979), unless one has a reasonably pure receptor and a ligand of high radioactivity.

Specificity of binding, or the lack of it, is the major culprit in kinetic studies. Another is the occurrence of experimental errors inherent in measuring extremely small quantities of ligand and macromolecule. At ligand concentrations in excess of $10^{-6}$ M, and sometimes even $10^{-8}$ M, two kinds of irrelevant binding can be distinguished (see O'Brien, 1979).

The *strictly nonspecific binding* of a ligand refers to absorption that is not in competition with receptor binding and is nonspecific in terms of the properties of the absorbing macromolecule as well as the ligand. For example, the dissolution of a nonpolar drug in the membrane lipid or in fatty tissue would be considered in this category, just as drug degradation products might bind in an unpredictable manner.

The other type of irrelevant binding is *specific nonreceptor binding*, which is sometimes very difficult to distinguish from strictly nonspecific binding. Here the ligand binds specifically; thus stereochemical and structural differences are significant. However, the binding site is located on a macromolecule that is a nonreceptor in the context of the system being investigated. The binding of hallucinogens to liver plasma membranes, the binding of coumarin anticoagulants to site I of human serum albumin, and the affinity of opiates for cerebroside sulfate ($K_D \simeq 10^{-8}$) are some examples of specific nonreceptor binding.

*Specific binding* is therefore defined as the binding that is numerically correlated with the pharmacological action of a drug. As discussed in Chap. 2, Sec. 2, this is also one of the criteria of receptor identity.

Distinguishing between the three kinds of binding function is often very difficult but is essential. One widely used differentiation method is isotopic dilution. The radioligand binding is measured in the absence and then in the presence of a 50- to 100-fold excess of unlabeled ("cold") drug. This should show a difference in radioligand binding proportional to the excess "cold" drug. Usually, strictly nonspecific binding is revealed in such experiments as the fraction of radioligand that cannot be displaced even at high concentrations of the "cold" ligand.

Equilibrium binding methods are the most widely used; these aim to separate excess free ligand from that bound reversibly at the receptor, without disturbing the equilibrium of the ligand–receptor system. Classical *dialysis* is being replaced more and more by faster *filtration* and *ultrafiltration* methods for this separation.

In *equilibrium dialysis*, the binding macromolecule and excess labeled ligand are enclosed in a cellulose acetate bag having an appropriate pore size, and the whole system is immersed in a container of dilute buffer. A large volume of external solution is employed. Ignoring the charge and Donnan effects of its small molecule, the free drug will, in time, equilibrate by diffusion between the inside and outside compartments. Since the pores of the bag do not allow the macromolecule to pass through, the inside of the bag will contain free plus bound ligand whereas the outside compartment will contain only the free ligand. The difference between the concentrations of the two is the amount of bound ligand.

This method is slow and imprecise, because a small difference in ligand concentration is measured against a high background concentration (see O'Brien, 1979). The use of a continuously changing (flowing) outside compartment makes dialysis suitable only for the complete removal of small molecules, as in desalting a protein solution or in the dissociation of tightly bound ligands from receptors.

Ultrafiltration speeds up the passage of free ligand from the inner compartment through application of pressure to the inside solution by means of compressed $N_2$. The two compartments are separated by a porous membrane. Typically, the inside is not filtered to dryness; instead, lost solution is replaced. Therefore, the criteria for equilibrium are not strictly maintained, and the desorption of a bound ligand can occur.

Recently, rapid *filtration* techniques, preferentially using cellulose acetate or glass-fiber filters, have been utilized for receptors on particulate membrane fragments or synaptosomes. The receptor–ligand suspension is first allowed to equilibrate and then filtered very rapidly and washed with cold buffer, to remove all unbound ligand. To avoid losing more than 10% of the bound drug, the method is applied only to systems with a $K_D$ smaller than $10^{-7}$ M. Calculations show very short dissociation half-lives of the receptor–ligand system if the $K_D$ is larger (Table 3.1).

For systems with a $K_D$ higher than $10^{-7}$ M, *centrifugation* in a very rapid minicentrifuge is a good compromise. There is no "separation artifact" in this case, since the pellet stays in contact with the solution, but some free ligand remains trapped. Obviously, washing the pellet is out of the question, since it would result in disequilibration.

**Table 3.1.** Relationship between equilibrium binding constant ($K_D$) and allowable separation time

| $K_D$ (M) | Allowable separation time $(0.15\, t_{1/2})^a$ |
|---|---|
| $10^{-12}$ | 1.2 days |
| $10^{-11}$ | 2.9 hr |
| $10^{-10}$ | 17 min |
| $10^{-9}$ | 1.7 min |
| $10^{-8}$ | 10 sec |
| $10^{-7}$ | 0.10 sec |
| $10^{-6}$ | 0.01 sec |

[a] Calculations of $t_{1/2}$ (half-life for dissociation) assume an association rate constant of $10^6$ $M^{-1} \times sec^{-1}$.

Reproduced by permission from Bennett and Yamamura (1985), Raven Press, New York.

Kinetic binding methods must be used to determine the association rate constants of drug–receptor interactions (i.e., the lifetime of the receptor occupancy). Most experiments based on this concept have been conducted in the area of neurotransmitter–receptor interactions, because the receptor–effector interaction in such cases is direct and measurable by electrical methods. The response is an ion current across the excitable membrane of a neuron or muscle, triggered by the neurotransmitter. The equilibrium or transient change in equilibrium with binding of the ligand is thus expressed as a current (i.e., ion-channel opening per unit of time). Because this current may be very small against a changing background, neuropharmacologists use a "voltage clamp" to overcome this difficulty. In this technique, a microelectrode is used to pass a current through the neuron, which stabilizes the membrane potential at a constant value through feedback regulation. Current flowing in response to the neurotransmitter can then be measured against this steady background, and is proportional to the number of open ion channels. A diagram of the arrangement is shown in Fig. 3.1.

**Fig. 3.1.** Diagrammatic representation of a voltage-clamped cut muscle fiber preparation. The voltage-recording electrode measures the membrane voltage, which, through a feedback circuit, is held constant at a preset level by current passed through the current-passing electrode. The ionic current flow through ion channels opened by acetylcholine (ACh) would tend to lower the membrane voltage (i.e., depolarize the endplate membrane) but this is instantly counteracted by the current-passing electrode. A record of the current passed is therefore exactly equal to the current flow induced by acetylcholine. Such records are the endplate currents (epcs); by convention, they are recorded downward, as illustrated on the right. Endplate currents may be evoked in the two ways illustrated: by stimulating the nerve to the cut fibers or by microiontophoretic application of acetylcholine from the acetylcholine (ACh) pipette to fibers treated with tetrodotoxin. Drug actions that modify both kinds of response are assumed to be exerted postjunctionally, whereas drug effects that are restricted to the epcs evoked by nerve stimulation are assumed to be exerted on the nerve endings. (Reproduced by permission from W. C. Bowman et al. (1984), *Semin. Anesth.* 3: 275–280)

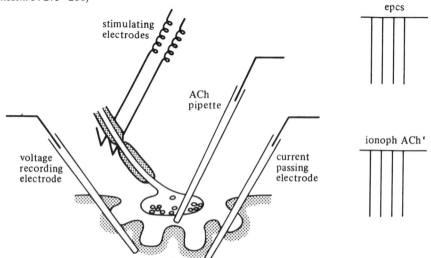

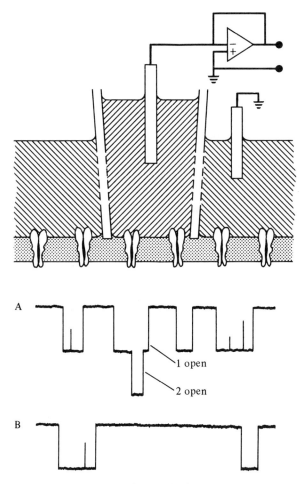

**Fig. 3.2.** Diagrammatic representation of a patch-clamp technique (not to scale). The electrode tip is represented as enclosing two receptor complexes Facsimiles of patch-clamp recordings are shown below. (**A**) With acetylcholine in the pipette. The downward deflections are of uniform amplitude but of different durations. When two receptor complexes are open, the currents summate. (**B**) With acetylcholine and an antagonist (e.g., tubocurarine in the pipette. The shapes of the pulses do not change, but their frequency is diminished. (Reproduced by permission from Bowman (1986), Macmillan, London)

A further refinement of this is the *patch clamp*, consisting of a micropipette-electrode of 0.5- to 6-$\mu$m internal diameter pressed against a cell membrane containing ion channels. The tight closure allows the formation of a high-impedance seal ("*giga*seal," because the impedance is $10^6$ ohm or higher). Under favorable conditions, only a single operating ion channel will be inside the pipette-electrode opening. With the use of the voltage-clamp arrangement, it is possible to measure the interaction of a single ionophore with a ligand (e.g., neurotransmitter). Figure 3.2 shows the principle of a patch-clamp experiment (Bowman, 1986).

A discussion of the powerful and sophisticated techniques of fluctuation analysis, voltage-jump relaxation, and concentration-jump methods is beyond the scope of this book. Colquhoun (1981), Stevens (1980), and McBurney (1985) offer a concise introduction to these kinetic methods.

### Selected Readings

J. P. Bennett, Jr., and H. I. Yamamura (1985). Neurotransmitter, hormone, or drug receptor binding methods. In: *Neurotransmitter Receptor Binding*, 2nd ed. (H. I. Yamamura, S. J. Enna, and M. J. Kuhar, Eds.). Raven Press, New York.

W. C. Bowman (1986). Mechanism of action of neuromuscular blocking drugs. In: *Mechanism of Drug Action*, Vol. 1 (G. N. Woodruff, Ed.). Macmillan, London.

D. Colquhoun (1979). The link between drug binding and response: theories and observations. In: *The Receptors* (R. D. O'Brien, Ed.), Vol. 1. Plenum Press, New York, pp. 93–142.

D. Colquhoun (1981). How fast do drugs work? *Trends Pharmacol. Sci. 2*: 212–217.

J. R. Cooper, F. E. Bloom, and R. H. Roth (1986). *The Biochemical Basis of Neuropharmacology*, 5th ed. Oxford University Press, New York.

T. Heidmann and J. P. Changeux (1978). Structural and functional properties of the acetylcholine receptor protein in its purified and membrane-bound states. *Annu. Rev. Biochem. 47*: 317–357.

M. D. Hollenberg and P. Cuatrecasas (1975). Biochemical identification of membrane receptors: principles and techniques. In: *Handbook of Psychopharmacology* (L. L. Iversen, S. D. Iversen, and S. H. Snyder, Eds.), Vol. 2. Academic Press, New York, pp. 129–177.

M. D. Hollenberg and P. Cuatrecasas (1979). Distinction of receptor from non-receptor interactions in binding studies. In: *The Receptors* (R. D. O'Brien, Ed.), Vol. 1. Plenum Press, New York, pp. 193–214.

T. P. Kenakin (1984). The classification of drugs and drug receptors in isolated tissues. *Pharmacol. Rev. 36*: 165–222.

P. M. Laduron (1984). Criteria for receptor sites in binding studies. *Biochem. Pharmacol. 33*: 833–839.

R. N. McBurney (1985). New approaches to the study of rapid events underlying neurotransmitter action. In: *Neurotransmitters in Action* (D. Bousfield, Ed.). Elsevier, Amsterdam, pp. 47–54.

R. D. O'Brien (1979). Problems and approaches in noncatalytic biochemistry. In: *The Receptors* (R. D. O'Brien, Ed.), Vol. 1. Plenum Press, New York, pp. 311–335.

C. F. Stevens (1980). Biophysical analysis of the function of receptors. *Annu. Rev. Physiol, 42*: 643–652.

J. C. Venter and L. C. Harrison (Eds.) (1984). *Receptor Biochemistry and Methodology*, 5 vols. Vol. 1: *Membranes, Detergents and Receptor Solubilization*; Vol. 2: *Receptor Purification Procedures*; Vol. 3: *Molecular and Chemical Characterization of Membrane Receptors*. A. R. Liss, New York.

### Periodicals of General Interest

*Journal of Pharmacological Methods*. Elsevier/North Holland, New York, 1978–.
*Journal of Biochemical and Biophysical Methods*. Elsevier/North Holland, New York. 1979–.

## 3. DIRECT METHODS OF RECEPTOR CHARACTERIZATION

The previous chapter reviewed methods that can give indirect but valuable evidence on the characteristics of receptors. Many of these techniques have been borrowed from biochemistry and physiology. However, a true understanding of the molecular characteristics of drug receptors will come only from the isolation, purification, and full characterization of these complex proteolipids, including the mapping of their membrane environment. There is, unfortunately, an inherent pitfall in this effort: since most drug receptors are membrane-bound, their isolation from the membrane scaffold alters their properties and conformation. Even if isolation does succeed, the isolated "receptor" is not the same as the native functional entity in its membrane constraint because the function of the amplification site (ion channel or coupler to adenylate cyclase) is necessarily lost.

Nevertheless, many receptors can withstand the rigors of biochemical manipulation. For instance, the nicotinic acetylcholine receptor (Chap. 4, Sec. 2.2), some of the cytoplasmic steroid receptors (Chap. 5, Sec. 1), the insulin receptor (Chap. 5, Sec. 2.7), and some receptors of an enzymatic nature can be studied in fairly pure state.

As in enzyme chemistry, the isolation, characterization, and protein sequencing of all receptors seems unlikely in the near future. Nevertheless, many receptors can be purified or enriched on membrane fragments and studied directly by biophysical methods. Such studies utilize purification techniques, isotopic methods, spectroscopic tools (e.g., fluorescence, magnetic resonance), and receptor reconstitution experiments.

### 3.1. Solubilization of Membrane-Bound Receptors

This is the first step required in receptor purification. Since most receptors are tightly bound intrinsic proteins, deeply embedded in the membrane (see Fig. 6.1), the membrane lipid organization, in which the protein "icebergs" float, must be destroyed. This can be achieved by detergents which attach themselves, with their lipophilic ends, to the lipophilic portion of the receptor protein. When the protein-bound detergent presents its polar side to the aqueous medium, the entire protein–detergent assembly becomes hydrophilic and turns into a colloidal dispersion. The detergents most widely used for this purpose are **Na-dodecyl sulfonate** (**SDS**; 1-18, an anionic detergent), Na-deoxycholate (a natural anionic bile acid derivative), a large number of very gentle nonionic detergents (Tween, Lubrol, Triton, Brig, etc.), and some specific glycoprotein-solubilizers like Li-diiodosalicylate. All of these compounds are widely used in membrane biochemistry (Lindstrom, 1985).

For all practical purposes, all of the protein characteristics are masked when the "solubilized" receptor is surrounded by detergent molecules. The only type of purification profitable at this stage is gel chromatography, which separates compounds according to their molecular size. To retain the adsorbed detergent on the protein, detergent is often added to the gel. Electrophoresis can sometimes be used, although electric charges on the protein may also be masked.

The next step in the purification process is removal of the detergent, usually by gel filtration on Biobeads SM2 or dialysis. The last traces of detergent are all but impossible to remove; this can be a blessing in disguise, since the intrinsic membrane proteins are by definition water insoluble and often aggregate and precipitate at this step. If all of the detergent is removed, this makes further work difficult.

At this point, the purification of the receptor is still only partial, because the general methods described above can reveal only the size or charge of a macromolecule, and not its function. Specific separation techniques utilize the only receptor property still available at this stage: the affinity of the receptor to a specific ligand—an agonist or antagonist.

## 3.2. Affinity Methods

### 3.2.1. Affinity Chromatography

Affinity chromatography is a highly efficient separation method based on the *function* of the substance to be separated, whether an enzyme, antibody, receptor protein, or even a membrane fragment or cell. Since this technique is independent of the size, polarity, or charge of the molecules to be separated, it is distinct from all other methods of chromatography. A specific ligand (substrate, drug, antagonist, antigen) is attached to an insoluble, porous, inert support (agarose, polyacrylamide, dextran, porous glass beads) by a long side chain. In this way, a small molecule acting as "bait" is attached by a "fishing line" (side chain) to a "fishing rod" (the support). Since the bait is highly specific, only a complementary macromolecule will bind to it efficiently when the usual column chromatography system is set up.

The arrangement of such an affinity column is shown in Fig. 3.3. A long amide side chain is attached to an insoluble polysaccharide and carries a quaternary anilinium compound, a specific ligand for the acetylcholine receptor. Only a long chain allows satisfactory ligand-receptor binding, since a short one may not reach binding sites buried in large membrane fragments or proteins. After the specific binding is established, all of the unbound protein contaminants are washed off the column. The bound macromolecule or membrane fragment can then be eluted in a pure state by washing the column with an agent that binds very strongly to the receptor (e.g., an enzyme substrate or a drug antagonist), or by otherwise breaking the ligand–receptor bond, such as through ionic interactions with a salt or pH gradient. Examples of affinity purifications, shown in Table 3.2, attest to the high efficiency of this method, which can purify a protein by a factor of many thousands in a single passage.

### 3.2.2. Affinity Labeling

Affinity labeling is a method for marking specifically and permanently the binding sites of enzymes, drug receptors, neurotransmitter receptors, ribosomes, antibodies, or hormone receptors. The method was introduced by Cuatrecasas and Anfinsen in 1968, and is simple in principle. A site-specific ligand is modified by the incorporation of a functional group that is capable of *covalently* binding to a protein —OH, —NH, or —CH group. Binding specificity is provided by the ligand

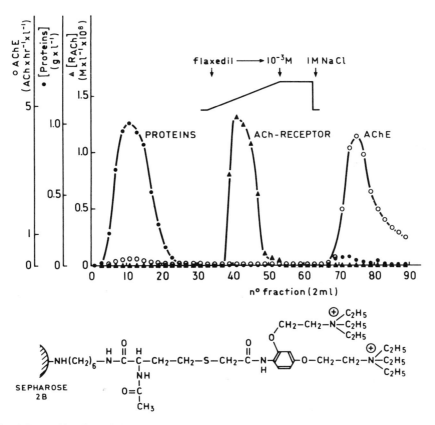

**Fig. 3.3.** Purification of the nicotinic cholinergic receptor by affinity chromatography. The diagram at the bottom shows the affinity "bait" side chain; the top the elution pattern of the chromatogram. The receptor fraction was eluted by a gradient of flaxedil, a quaternary neuromuscular blocking agent that dissociates the receptor from the column. (Reproduced by permission from Olsen et al. (1972), *FEBS Lett 28*: 96)

**Table 3.2.** Some examples of protein purification by affinity chromatography

| Protein | Source | $K_D$ | Recovery (%) | Purification (fold) |
|---|---|---|---|---|
| Dihydrofolate reductase | Phage T$_4$ | $10^{-7}$ | 80 | 6,000[a] |
| | Mammalian skin | $10^{-8}$ | 33 | 3,700[a] |
| Avidin | Egg white | $10^{-15}$ | 90 | 4,000[a] |
| Insulin receptor | Liver membrane | $10^{-11}$ | 70 | 250,000[a] |
| Acetylcholine receptor | *Torpedo* electroplax | $10^{-9}$ | 50 | 6,000[b] |

[a] Lowe and Dean (1974), *Affinity Chromatography*, Wiley, New York.

[b] Eldefrawi et al. (1972), *Proc. Natl. Acad. Sci. USA 69*: 1776.

(substrate, drug, etc.) itself, directing it to the receptor recognition site. There, after noncovalent binding of the ligand molecule, the anchoring group will react with any conveniently located protein side group and, by forming a covalent bond, will permanently anchor the ligand to its receptor site. This powerful method is variously used in the elucidation of receptor composition or topology, in the masking or isolation of receptors, and in the design of long-acting drugs or antagonists. It can also be used for the introduction and anchoring of other molecular probes on the receptor; spin labels, fluorescent markers, and isotopically marked ligands can all be attached by making them part of an affinity label.

However, there are some drawbacks to this method. Normally, the attached label cannot be split off easily, and therefore only covalent receptor–label complexes can be isolated and studied. Furthermore, the specificity of many alkylating or acylating affinity labels is often poor, or the binding inefficient, resulting in only partial labeling. Nevertheless, the method is enjoying wide popularity (see Jacoby and Wilchek, 1977).

For efficient use, the following prerequisites for affinity labeling have to be observed:

1. The reagent must first form a reversible, specific complex with the binding site.
2. The reaction should show saturation kinetics.
3. Competing ligands that cannot covalently bind to the receptor should protect the active site from labeling and reduce the *rate* of labeling; however, the *degree* of labeling remains the same.
4. The active site or receptor should show inactivation proportional to the extent of labeling.

The nature of labeling can be followed kinetically. In specific affinity labeling, the reaction is first order:

$$R + L \rightleftharpoons R \ldots L \rightarrow R\!-\!L$$

The receptor (R) and the ligand (L) are in a reversible equilibrium; covalent binding takes place *within* the complex and is therefore first order with regard to the ligand. Unfortunately, nonspecific labeling can also proceed in the same fashion. The other type of nonspecific binding is second order, without an equilibrium:

$$R + L \rightarrow R\!-\!L$$

and can be detected by kinetic measurements.

The "classical" affinity labels depend on alkylation or acylation. Some of the widely used labels are shown in Table 3.3. These classical agents are not very specific, since they alkylate or acylate any appropriate nucleophile on their way to the target. Better and more selective agents have been found in the *photoaffinity labels*. Upon ultraviolet irradiation, these compounds produce a highly reactive carbene, a nitrene, or an excited $-\!C\!\!=\!\!O$ or $-\!C\!\!=\!\!S$ triplet which binds to any electrophilic site and is even capable of insertion into aromatic rings.

The strategy applied in photoaffinity labeling rests on the principle that the label will equilibrate (in the dark) with both specific and nonspecific binding sites. However, the binding to nonspecific sites is by definition much weaker. If the system

**Table 3.3.** Some widely used affinity labels and their reactions

---

"Classical" labels

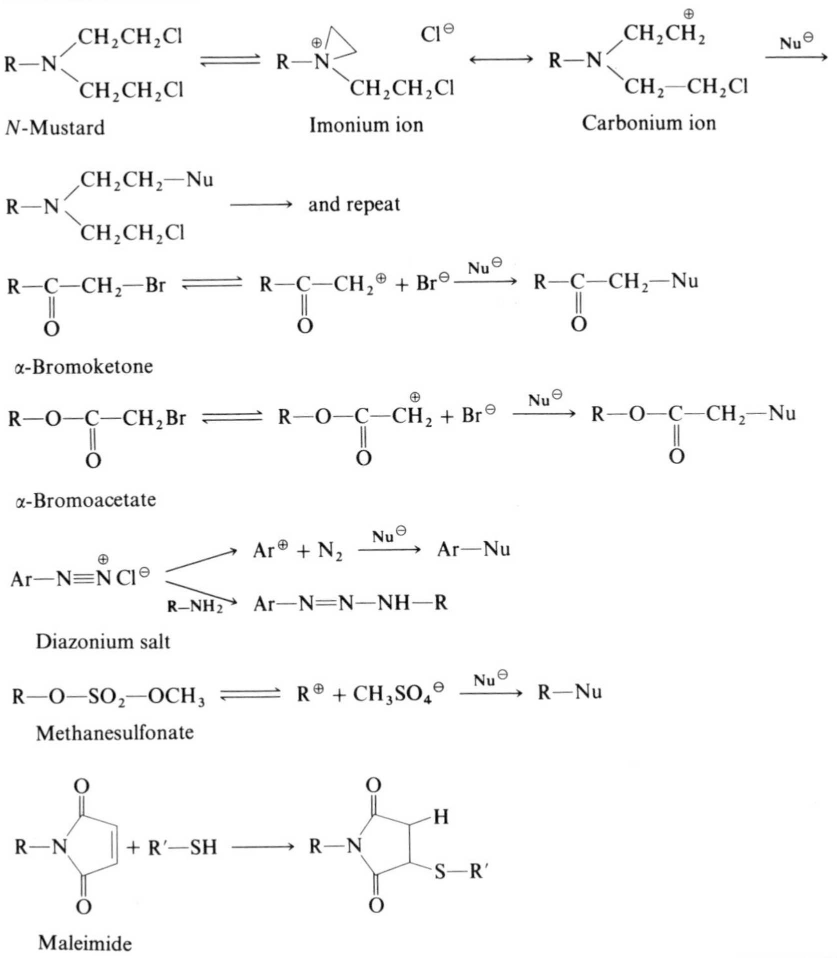

---

$$R-N\begin{smallmatrix}CH_2CH_2Cl\\CH_2CH_2Cl\end{smallmatrix} \rightleftharpoons R-\overset{\oplus}{N}\underset{CH_2CH_2Cl}{\diagup\!\!\!\triangle}\ Cl^{\ominus} \longleftrightarrow R-N\begin{smallmatrix}\overset{\oplus}{CH_2CH_2}\\CH_2-CH_2Cl\end{smallmatrix}\ \overset{Nu^{\ominus}}{\longrightarrow}$$

N-Mustard          Imonium ion          Carbonium ion

$$R-N\begin{smallmatrix}CH_2CH_2-Nu\\CH_2CH_2Cl\end{smallmatrix} \longrightarrow \text{and repeat}$$

$$\underset{O}{R-\overset{\|}{C}-CH_2-Br} \rightleftharpoons \underset{O}{R-\overset{\|}{C}-CH_2^{\oplus}} + Br^{\ominus} \overset{Nu^{\ominus}}{\longrightarrow} \underset{O}{R-\overset{\|}{C}-CH_2-Nu}$$

α-Bromoketone

$$\underset{O}{R-O-\overset{\|}{C}-CH_2Br} \rightleftharpoons \underset{O}{R-O-\overset{\|}{C}-CH_2^{\oplus}} + Br^{\ominus} \overset{Nu^{\ominus}}{\longrightarrow} \underset{O}{R-O-\overset{\|}{C}-CH_2-Nu}$$

α-Bromoacetate

$$Ar-\overset{\oplus}{N}\equiv N\ Cl^{\ominus} \begin{cases} \longrightarrow Ar^{\oplus} + N_2 \overset{Nu^{\ominus}}{\longrightarrow} Ar-Nu \\ \overset{R-NH_2}{\longrightarrow} Ar-N=N-NH-R \end{cases}$$

Diazonium salt

$$R-O-SO_2-OCH_3 \rightleftharpoons R^{\oplus} + CH_3SO_4^{\ominus} \overset{Nu^{\ominus}}{\longrightarrow} R-Nu$$

Methanesulfonate

$$R-N\begin{smallmatrix}O\\\text{(maleimide ring)}\\O\end{smallmatrix} + R'-SH \longrightarrow R-N\begin{smallmatrix}O\\\text{(succinimide ring)}\,H\\S-R'\\O\end{smallmatrix}$$

Maleimide

---

Nu = nucleophile.

is disequilibrated by removing excess free label by adsorption or ultrafiltration, the label will dissociate preferentially from the low-affinity, nonspecific sites. At this point, the system is irradiated by ultraviolet light at a wavelength absorbed by the label, photoactivation takes place, and the carbene or nitrene will bind very rapidly to just about any functional group in its immediate vicinity. The addition of "scavenger" compounds, which remove free carbenes or nitrenes that could label at random, further increases the selectivity of binding.

The literature describes numerous examples of affinity labels based on steroid hormones (Katzenellenbogen, 1977), protein hormones, enzyme substrates, nucleotides (specifically cAMP), and neurotransmitters (Cavalla and Neff, 1985).

**Table 3.3.** (*continued*)

Photoaffinity labels

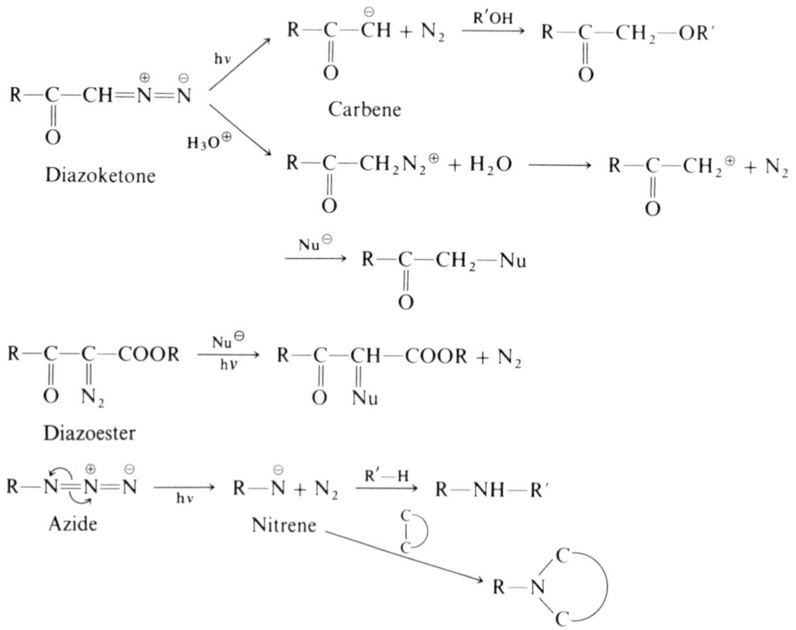

$Nu$ = nucleophile.

Two novel methods of affinity labeling have been developed recently (Simons and Thompson, 1982) (Fig. 3.4). *Chemoaffinity labeling* utilizes a cross-linker like phthalaldehyde, to link an amine label with an —SH group on the receptor protein, forming an isoindole tautomer that has the advantage of being fluorescent. In the other method, known as *electrophilic transfer affinity labeling*, a xanthate label is attached to an affinity carrier (in this example, a steroid). A strategically placed nucleophile in the appropriate position can cleave the steroid xanthate, which now labels the receptor. The steroid-thiol dissociates, leaving behind a functional, unoccupied binding site, capable of normal activity but carrying a radiolabel (tritium or another marker) close to the active site.

### 3.3. Radioisotopic and Fluorescence Methods

These methods are so widely used in modern biochemistry, and there are so many excellent introductory texts on their use (e.g., Cooper, 1977), that the principles and methods of radiotracer use will be assumed to be familiar to the reader, and only the principles of its application to drug–receptor studies will be dealt with. The first and foremost radiotracer technique is the isotopic labeling of ligands, as used in biochemistry and enzymology. In this application, a relatively stable isotopic element ($^3H$, $^{14}C$, $^{32}P$, $^{35}S$, or the $\gamma$-emitting $^{125}I$ or $^{131}I$) is attached covalently by

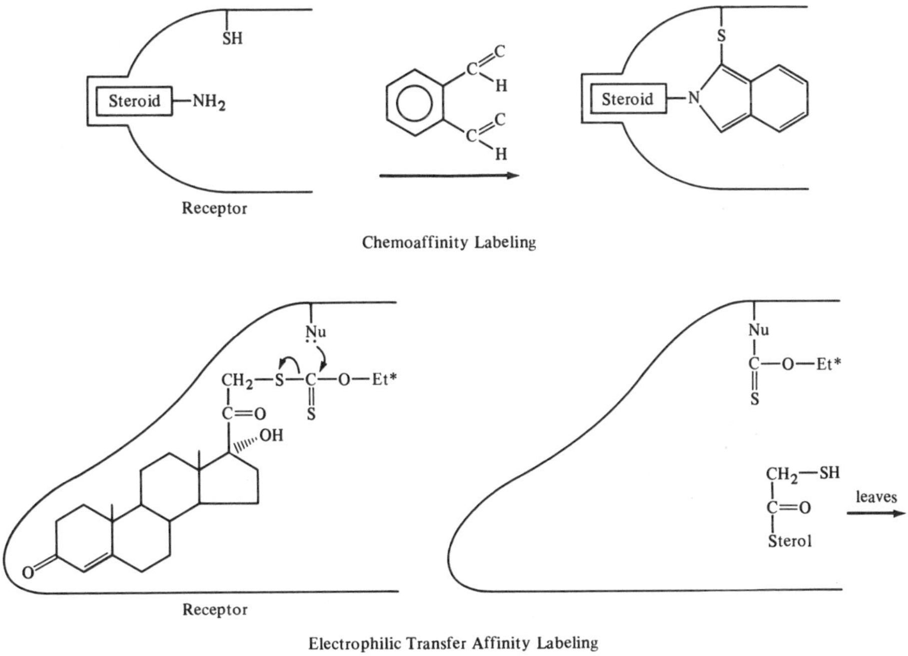

Chemoaffinity Labeling

Electrophilic Transfer Affinity Labeling

**Fig. 3.4.** Chemoaffinity and transfer affinity labeling. In chemoaffinity labeling, the phthal-aldehyde cross-linker connects the bound amino-steroid and a thiol group on the receptor protein. The resulting isoindole tautomer is fluorescent. Electrophilic transfer affinity labeling utilizes a xanthate derivative of a corticosteroid bound to its receptor. A strategically situated nucleophile (Nu; e.g., —OH, —NH$_2$) on the receptor molecule cleaves the xanthate derivative, and the sterol-thioketone leaves the receptor site. What remains is a vacant, functional corticosteroid receptor labeled in the immediate vicinity by a radioactively labeled (*) thiol ester. (Modified from Simons and Thompson, 1982)

synthetic means to a small molecule like a drug or to a peptide in a way that leaves the pharmacological activity of the molecule intact. Replacing an H, C, or P atom by an isotope usually fulfills this criterion; the iodination of proteins is sometimes a more difficult matter and may impair their activity. When very high levels of radioactivity are required, such as in the study of insulin receptors, which are present only in the vanishingly small amount of 10,000 receptors per fat cell, only the use of $^{125}$I will yield the necessary specific activity of 2000–3000 Ci/mmol for obtaining significant counts. Tritiation can usually achieve activities of only 20–30 Ci/mmol; however, in many drug binding studies, that level may be sufficient.

After the binding equilibrium of isotopically labeled ligands is established, the detection of bound species (normally on membrane-bound or macromolecular receptors) is achieved by isolation of the receptor–radioligand complex by centrifugation or rapid filtration through cellulose or glass-fiber filters. As in most binding experiments in which a noncovalent equilibrium is measured, artifacts (like filter binding of free radioligand) must be prevented or carefully corrected, and the criteria of receptor–ligand interaction which were discussed in previous sections still apply. The counting of radiolabeled ligands is usually done in liquid scintillation counters (a $\gamma$-counter is used for $^{125}$I), in which dually labeled preparations containing $^3$H and $^{14}$C can also be handled simultaneously. As always, the statistical treatment of data is all-important in obtaining reliable results suitable for evaluation of the transport, storage, binding specificity, or affinity of drugs and biomolecules. Elucidation of the number of binding sites, allosteric interactions, displacement phenomena, or stereoelectronic effects is among the instances in which the application of radioligands is extremely useful and universally employed.

For the study of the anatomical or histological distribution of receptors or the accumulation of drugs, the method of *autoradiography* is often used. In this technique, high-affinity ligands with a high level of radioactivity are administered to an intact animal, organ, or cell. After allowing a suitable time for distribution and specific binding equilibration, the organ or tissue is fixed by standard histological methods (preferably flash-frozen) and sectioned for optical or electron microscopy. The sections are then covered with a photographic emulsion (in the dark) and allowed to self-expose for 2–6 weeks. Such a long period is necessary for the formation of detectably dense silver grains in the emulsion. The emulsion is then developed with the finest grain possible, the tissue section is stained, and the section–photographic emulsion sandwich is examined microscopically or electron-microscopically. Where the ligand has accumulated, silver grains will be superimposed on histological features. By counting the silver grains, receptor quantification is also possible, provided the ligand was able to reach all of the receptor sites and had a sufficiently high affinity to resist dissociation and diffusion during the long exposure period. Figure 3.5 shows an electron micrograph of a neuromuscular endplate labeled with $^{125}$I-bungarotoxin (see Chap. 4, Sec. 2.2); Barnard (1979) has published a long list of receptors that have been localized in this way, and Kuhar et al. (1986) have published a review on autoradiography and immunohistochemistry.

Perhaps the most sensitive isotopic method is *radioimmunoassay*, which is applicable to almost any molecule that can act as a hapten (i.e., part of a synthetic antigen). The drug or ligand is first coupled to a protein (e.g., serum albumin), forming an antigen. Antibody to this antigen is then raised in an appropriate animal, and purified to the extent possible. A solution of this antibody is then equilibrated with an isotopically labeled ("hot") analogue of the compound under investigation, after which the free radioactive ligand is removed. Upon the addition of a "cold" ligand, a reequilibration occurs and some of the bound, radiolabeled ligand is displaced and becomes free. After the removal of both "hot" and "cold" antigen–antibody complexes, the increase in free radioactive ligand is measured. This gives an extremely sensitive measure of the "cold" ligand, detectable in amounts that would be impossible to quantify otherwise. An analogous immunoassay based on the electron spin resonance (ESR) spectra of spin-labeled ligands is described in Sec. 3.4.

*Fluorescence probes* can have multiple uses in medicinal chemistry and receptor characterization, just as in other areas of biochemistry. For example, many biomolecules as well as drugs show *native fluorescence*, having fluorescent groups. In proteins this property is usually due to aromatic moieties, such as in tryptophan,

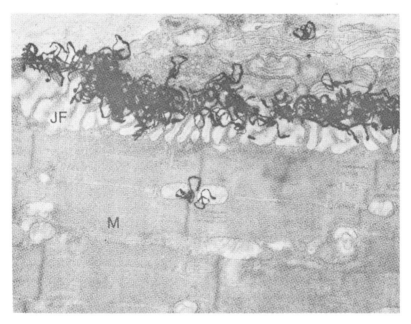

**Fig. 3.5.** Neuromuscular endplate labeled with [$^{125}$I]bungarotoxin. The autoradiogram shows the accumulation of cholinoceptors in the synapse. (Reproduced by permission from Tertuk and Salpeter (1974), *PNAS 71*: 1376)

phenylalanine, or tyrosine. A very large number of small molecules show native fluorescence, and good tables of these can be found in the monograph of Udenfriend (1972). In addition, fluorescent analogues of many small biomolecules can be prepared. Such compounds, as well as other entirely synthetic fluorescent molecules, can be used as *fluorescence labels*, which act as markers that report on their environment through changes in fluorescence behavior. The two most widely used labels are shown on page 108.

Dansyl chloride, which has been used for a long time, reacts in many nucleophilic substitution reactions; however, fluorescamine is much more selective and sensitive. Other fluorescent labels can be selectively attached to macromolecules through affinity labeling techniques (see previous section).

In fluorescence, the absorbed radiation at the excitation frequency will increase the energy of an electron in a compound, resulting in an excited singlet or triplet state. This energy can be lost as radiation at a longer wavelength through fluorescence, or can be transmitted to another molecule or another part of the same molecule in a radiationless energy transfer called *quenching*. This will, of course, decrease the *quantum yield* of fluorescence—the proportion of light quanta participating in excitation and the resulting light emission. This proportion has a maximum of unity (Lakowicz, 1983). The fluorescence of a molecule may also depend on its enviroment, and many compounds, such as fluorescamine, will fluoresce in nonpolar media only (e.g., when bound to the hydrophobic part of a protein or dissolved in a biological membrane). If a conformational change or other

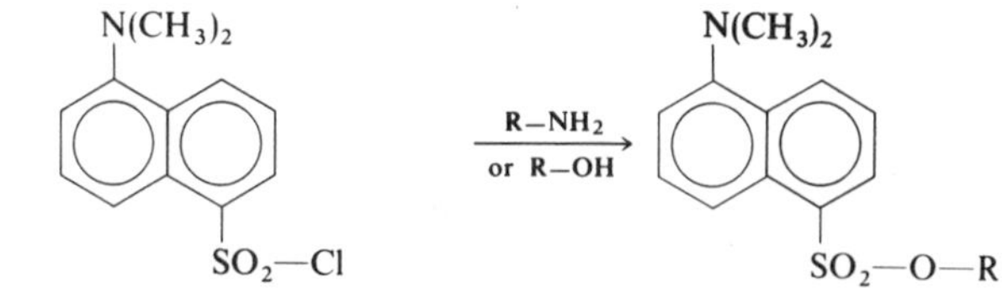

5-Dimethylamino-1-naphthalene-
sulfonyl chloride
(Dansyl chloride)

Dansyl label

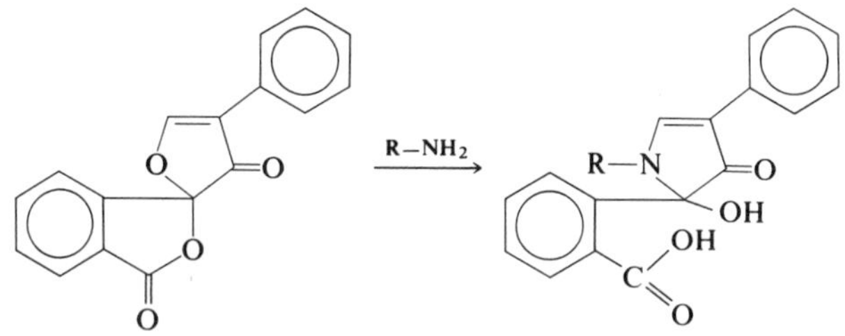

Fluorescamine
(nonfluorescent)

Fluorescent derivative

interaction exposes the fluorescent label to a more hydrophilic environment, the quantum yield of fluorescence will change, serving as a measure of the environmental alteration. Fluorescence labels bear a considerable similarity to spin labels in this respect.

Sometimes, both the energy donor (fluorescing singlet) and acceptor (an excitable group) in a fluorescent process can be part of a single molecule or a tight complex, and the efficiency of energy transfer can be used to calculate the distance between the two groups. The galactose receptor protein of bacteria, an important factor in bacterial chemotaxis, was mapped in this way (Zukin et al. 1977).

In dilute solution, *fluorescence polarization* is a measure of rotational diffusion during the lifetime of an excited singlet. With advanced techniques, including laser and photon counting fluorometers, the lifetime and degree of polarization of the phenol fluorescence of opiates can be measured, and the binding of morphine to receptors studied, among other applications of such techniques.

### 3.4. Magnetic Resonance Methods

Magnetic resonance—the absorption of electromagnetic radiation by an atom or electron spinning in an applied magnetic field—is a technique widely used in chemistry and physics. It has also found increasing application in the biological sciences since it can furnish information on the structural features of complex molecules, as well as on their dynamic interactions and environment. Because of the high selectivity of this method, individual atoms or functional groups can often be followed during reactions, binding equilibria, or phase transitions. Each of these phenomena is of great interest to the molecular pharmacologist.

### 3.4.1. Nuclear Magnetic Resonance

Nuclear magnetic resonance (NMR) is based on the fact that a number of important *nuclei* (e.g., $^1H$, $^2H$, $^{13}C$, $^{19}F$, $^{23}Na$, $^{31}P$, $^{35}Cl$) show the atomic property called *magnetic momentum*; their nuclear spin quantum number, $I$, is larger than zero. (For $^1H$, $^{13}C$, $^{19}F$, and $^{31}P$, $I = 1/2$.) When such a nucleus (or an unpaired electron) is put into a strong magnetic field, the axis of the rotating atom will describe a precessional movement, like that of a spinning top. The precessional frequency, $\omega_0$, is proportional to the applied magnetic field $H_0$: $\omega_0 = \gamma H_0$, where $\gamma$ is the magnetogyric ratio, which is different for each nucleus or isotope. Since the spin quantum number of the nucleus can be either $+1/2$ or $-1/2$, there are two populations of nuclei in any given sample, one with a higher energy than the other. These populations are not equal: the lower-energy population is slightly more abundant. The sample is then irradiated with the appropriate radiofrequency. At a certain frequency, the atom population with the lower energy will absorb the energy of the radiofrequency and be promoted to the higher energy level, and will be *in resonance* with the irradiating frequency. The energy absorption can be measured with a radio receiver (just as in the case of any other electromagnetic radiation such as ultraviolet or infrared, using the appropriate detectors) and can be displayed in the form of a spectrum of absorption versus the irradiating frequency. The great information content of this spectrum derives from the fact that each nucleus of a molecule (e.g., each proton) will have a slightly different resonance frequency, depending on its "environment" (the atoms and electrons that surround it). In other words, its magnetic momentum will be "shielded" differently in different functional groups. This makes it easy to distinguish, for example, the protons on a $C—CH_3$ group from an $O—CH_3$ group or an $N—CH_3$ group; aliphatic or aromatic protons; carboxylic acid or aldehyde protons; and so on, because they absorb at different frequencies. In the same fashion, every carbon atom in a molecule can be distinguished by $^{13}C$ magnetic resonance spectroscopy. Figure 3.6 shows a comparison (and assignment) of a proton with a carbon magnetic resonance spectrum of cholesterol, demonstrating the capability of these methods for structure elucidation.

The only drawback to NMR is its low sensitivity. Concentrations in the millimolar range are normally required, although with computer enhancement techniques (like Fourier transform), signals at $10^{-6}$–$10^{-5}$ M concentrations can be detected. This is especially important for nuclei that have a low natural abundance, such as $^{13}C$ (1.1%) or deuterium, $^2H$ (0.015%).

Fourier-transform (pulsed) proton NMR techniques allow an even more sophisticated assignment of resonances to specific protons. If the single high-frequency pulse is replaced by *two* pulses of variable pulse separation, the introduction of a second time parameter yields a *two-dimensional* NMR spectrum, with two frequency axes. Resonances on the diagonal are the normal, one-dimensional spectrum, but off-diagonal resonances show the mutual interaction of protons through several bonds. This allows the assignment of all protons even in very large molecules; recently, the three-dimensional spectrum of a small protein has been deduced by use of a three-pulse method. Kabsch and Rösch (1986) summarize this technique succinctly . *In vivo* NMR (Mildvan, 1984; MacKenzie, 1985), by which various

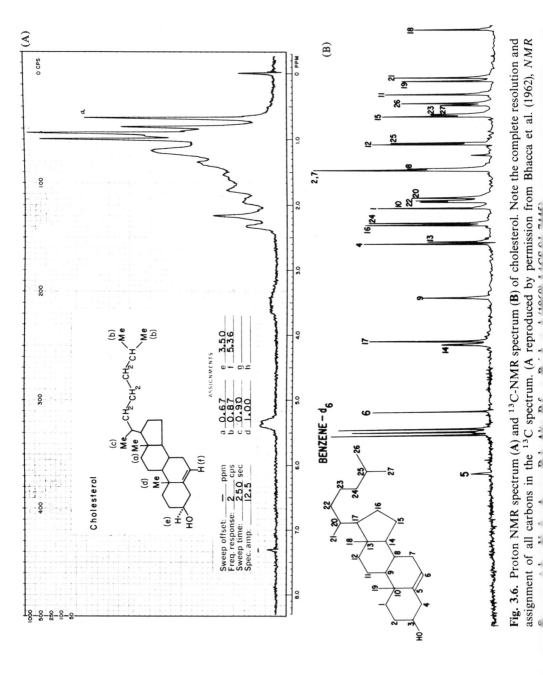

**Fig. 3.6.** Proton NMR spectrum (**A**) and $^{13}$C-NMR spectrum (**B**) of cholesterol. Note the complete resolution and assignment of all carbons in the $^{13}$C spectrum. (A reproduced by permission from Bhacca et al. (1962), *NMR*

nuclei can be followed in intact organs to pinpoint metabolic processes, is also undergoing rapid development.

*Applications.* Nuclear magnetic resonance permits *counting* of the protons in a molecule. The area under each NMR resonance peak is proportional to the protons contained in that functional group. One of the easily identifiable groups in the spectrum is used as a relative standard; electronic integration of the peak areas will give the number of protons in each group of signals, clarifying the assignment of resonances to specific structural features.

The detection of *relaxation rates* is a further application of NMR spectroscopy. When a particular nucleus, like a methyl proton, is irradiated by a strong radio-frequency and absorbs it, the populations of protons in the high- and low-spin states are equalized, and the signal disappears after a while. It will be recalled that the NMR signal is based on energy absorption; if all of the nuclei of a given type are in the high-spin state, absorption is not possible, and "saturation" occurs. Upon removal of the strong irradiating frequency, the high- and low-spin populations will once again become unequal by transferring energy either to the solvent (spin–lattice relaxation, $T_1$) or to another spin in the molecule (spin–spin relaxation, $T_2$), and the appropriate spectrum line will assume its original amplitude. The time necessary for this recovery is called the *relaxation time*, whereas its reciprocal is the *relaxation rate*. We shall see in some later examples how relaxation rates can be used in elucidating molecular interactions.

Another tool in NMR spectral analysis is the observation of slight *shifts* of the various peaks. Hydrogen bonding and charge-transfer complex formation will shift resonances downfield (to lower frequencies) and upfield, respectively. On the other hand, the *coupling constant*, or separation distance between the sublines of doublets or triplets, is a result of line splitting by neighboring protons. Thus, line multiplicity (in addition to line position) is used in determining the nature of a proton and its neighbor. An ethyl group, for instance, gives a triplet (—$CH_3$ split by the adjacent —$CH_2$— group into three peaks) and a quartet (—$CH_2$— split into four peaks by the —$CH_3$). The magnitude of the coupling constant for two protons is also influenced by the dihedral angle of the X—Y bond in an H—X—Y—H structure, and can be used in conformational analysis.

In peptides, the coupling constants of the —CH— and —NH— protons show a correlation with the dihedral angle (Fig. 3.7). This, however, can be ambiguous, since some coupling constants (e.g., at 4 ppm in Fig. 3.7) can be assigned to four different dihedral angles. Additional structural information can be obtained from the coupling constant of the H—$^{13}C$—N—H structure or H—C—C—$^{15}N$ arrangement, giving correlations that do not overlap with the H—X—Y—H curve.

Nuclear magnetic resonance spectroscopy is also an attractive tool for studying the *interaction of small molecules with one another.* In one of the author's studies, formation of the serotonin neurotransmitter complex (Chap. 4, Sec. 5) with ATP was investigated in this way. The neurotransmitter is stored in this complexed form in nerve endings and blood platelets. In the serotonin (5-HT) NMR spectrum in Fig. 3.8, the side chain —$CH_2$— groups are seen as two slightly asymmet-

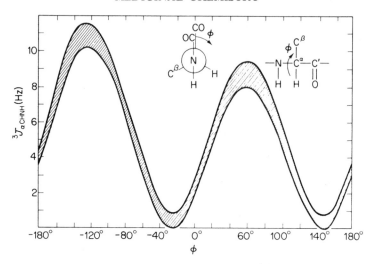

**Fig. 3.7.** Dependence of the CH—NH coupling constant on the dihedral $\phi_L$ of the two protons. (Reproduced by permission from V. F. Bystrov (1976), *Prog. NMR Spectrosc.* *10*: 41)

rical triplets. When serotonin interacts with ATP (Fig. 3.9), the increase in ATP concentration increases the line widths of these triplets. This indicates an increase in the relaxation rate $(T_2)$, due to immobilization of the side chain through the formation of a salt of the side chain —$NH_3^+$ with the phosphate group of the ATP. Observation of the aromatic proton resonance lines of ATP (Fig. 3.10) shows that the H-2 line is shifted upfield as the relative 5-HT concentration increases, indicating a charge-transfer interaction between the indole ring of serotonin and the pyrimidine part of the ATP purine ring. Additionally, the rather broad H-2

**Fig. 3.8.** Serotonin NMR spectrum at 220 MHz.

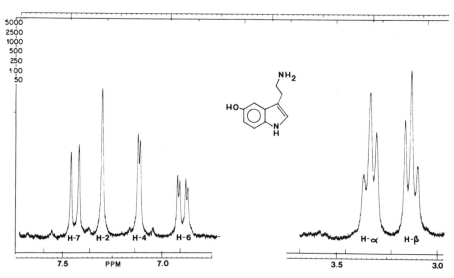

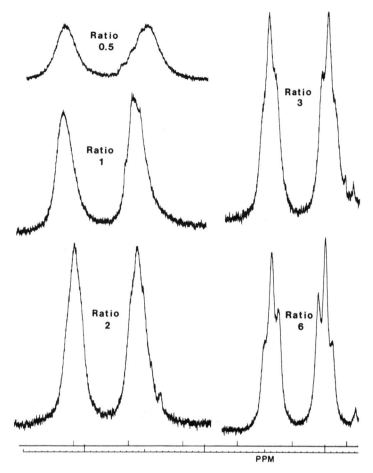

**Fig. 3.9.** Serotonin side-chain resonances at different serotonin: ATP ratios at 220 MHz. Note the line broadening at high ATP concentrations. (From Nogrady et al. (1972), *Mol. Pharmacol. 8*: 565)

resonance becomes sharper. This effect was interpreted as representing the disruption of ATP "stacks" by complex formation; it is known that ATP forms micellar stacks (consisting of molecules piled on top of each other), which leads to line widening.

Much NMR work has been done on the *interaction of small molecules* with *macromolecules*, obviously of great interest in drug–receptor binding studies as well as in enzymology. In principle, the small-molecule resonances are easy to follow, provided they are not overlapped by the very complex and broad spectra of the macromolecules in the same solution. This technique was used to gain information on drug binding to serum albumin, and in some cases the binding moieties of the small molecule could be recognized by increased relaxation rates of some of the protons. It is much more difficult to obtain data on the dynamics of the binding of a macromolecule, such as an enzyme. (For the details of this procedure, see Roberts, 1977.)

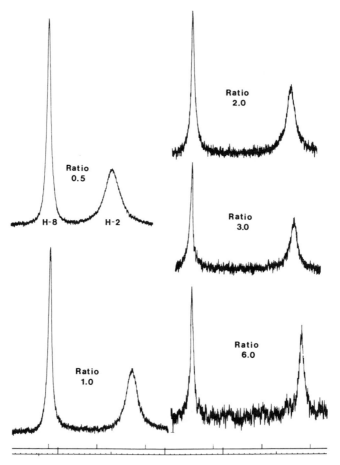

**Fig. 3.10.** Adenine proton resonances in ATP in the presence of increasing amounts of serotonin. The H-2 resonance is shifted upfield, indicating a charge-transfer interaction between the two ring systems. (From Nogrady et al. (1972), *Mol. Pharmacol. 8*: 566)

### 3.4.2. Electron Paramagnetic Resonance

Electron paramagnetic resonance (EPR) or electron spin resonance (ESR) spectroscopy is based on the same principles as NMR, since the electron also has a spin. However, the irradiating frequency is higher ($\sim 10^{10}$ Hz, in the microwave region) than in NMR ($10^7$–$10^8$ Hz, which, as already noted, is in the radiofrequency region). Only unpaired electrons can be detected in this way, but since the magnetic moment of the electron is several orders of magnitude larger than that of a proton, the method is much more sensitive. With EPR, electron concentrations as low as $10^{-9}$ M can be easily detected. The mathematical theory of EPR is, however, distinct from that of NMR.

The information contained in an EPR spectrum can be derived from the line intensity, line width (a measure of interaction between the electron and its environment), and the "hyperfine splitting" of lines, which results in multiplets. This multiplicity stems from the interaction of the electron with nuclei having a spin $I > 0$ (i.e., those that give NMR spectra).

There can be more than one absorption peak for a single unpaired electron as the result of its interaction with a nucleus. In nitroxyls, the line is a triplet, because the electron interacts with nitrogen. For instrumental reasons, ESR spectra are recorded as dispersion lines (i.e., the first derivative of an absorption line of Lorentzian line shape). Further splitting can be seen as a result of the orientation of electrons in the magnetic field. Indeed, in the spectrum of a free radical frozen into a solid matrix, three different spectra with different line widths can be recorded. They are in the $X$, $Y$, and $Z$ axial directions of the magnetic field. In solution, an average of these is seen, since the molecule tumbles rapidly. By slowing this tumbling (averaging) motion, one can retard the electron reorientation and cause the line width to broaden. This can be achieved, for example, by recording the spectra in viscous solutions, such as glycerol, and can be enhanced by cooling these solutions (Fig. 3.11). Such line broadening is also seen if a free radical is forced into a decreased rate of reorientation; it is therefore a useful tool for measuring molecular motion and the restraints of the environment on such motion, like that caused by a conformational change, immobilizing the spin label.

*Applications.* In the majority of biochemical applications, *stable free radicals* like nitroxyls (or oxyls) of the structure $>N \overset{\cdot}{\rightarrow} O$ are used. The dot on the coordinate valence line between the nitrogen and oxygen indicates a resonance hybridization of the electron, which "belongs" equally to N and O.

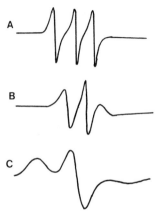

**Fig. 3.11.** Line broadening of a spin label on increasing immobilization: **(A)** in methanol solution, tumbling freely; **(B)** in glycerol at 25°C; **(C)** in glycerol at −150°C.

Since free radicals are notoriously unstable, they must be stabilized by structural constraints such as steric hindrance, as in the piperidine, pyrrolidine, or oxazine derivatives. These groups can be used either by themselves or attached to an appropriate biomolecule. Such compounds, called *spin labels*, are used in a number of applications for measuring molecular mobility, conformational change, and the binding of drugs, as well as serving as molecular "yardsticks." They are also useful in immunoassays.

Molecular motion can be measured in two ways. As an example of one of these, spin-labeled stearic acid analogues were prepared and dissolved in lipid membranes. A comparison of the spectra of these compounds shows (Fig. 3.12) that when the label is close to the polar head of the fatty acid, it is almost completely immobilized (strongly broadened spectrum), whereas when it is near the end of the alkyl chain (i.e., in the middle portion of the lipid bilayer), it is very mobile, indicating membrane fluidity. Such an experimental arrangement was also used to show that general anesthetics increase cell membrane fluidity, whereas cholesterol acts as a membrane stabilizing agent.

As noted above, EPR labels can also be used as *molecular "yardsticks."* Sulfanilamide spin labels such as that shown here, have been used to measure the depth of

$$O \overset{\ominus}{\leftarrow} N \text{—(CH}_2)_n\text{—}\bigcirc\text{—SO}_2\text{NH}_2$$

(with $CH_3CH_3$ groups above and $CH_3CH_3$ groups below the ring nitrogen)

the active-site "pocket" of the enzyme carbonic anhydrase (Chap. 6, Sec. 3.4). The sulfanilamide binds directly to the Zn atom at the bottom of the active-site pocket. The nitroxyl label, on the other end of the molecule, is restricted in its mobility by the sides of this pocket, and shows a broadened spectrum. If one gradually increases the side-chain length, the nitroxyl will eventually be able to protrude from the enzyme pocket, regain its mobility, and hence revert to its normal spectrum. By this method, the active-site crevice was found to be 1.4 nm deep.

**Fig. 3.12.** Electron spin resonance spectra of spin-labeled stearic acid derivatives in a biological membrane. The derivative labeled near the end of the alkyl chain shows high mobility (dashed line), whereas the one labeled near the carboxyl group is highly immobilized (solid line). (Reproduced by permission from Marshall (1978), Wiley, New York)

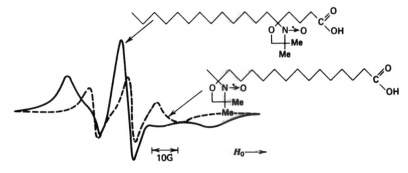

Intramolecular distances can be measured quite accurately by spin labeling specific sites in a molecule with a paramagnetic probe, like a nitroxyl. The unpaired electron of the spin label perturbs the relaxation of protons in other nuclei in the vicinity. By measuring the relaxation rates of protons as far removed as 2 nm from the label, the method can be used to calculate the distances of these protons from the reference point of the free radical, and thus from each other. Several enzyme active sites have been mapped in this way, by use of combined EPR and NMR techniques.

Finally, EPR has been ingeniously used in developing *spin immunoassays* for detecting minute amounts of morphine or other narcotics(Chap. 5, Sec. 3.6) in the urine of addicts and in forensic material (Leute et al., 1972). Morphine is first coupled covalently to bovine serum albumin, and the product is used to prepare a morphine antibody in rabbits. When spin-labeled morphine binds to this anti-morphine antibody, it binds specifically and strongly, and its EPR spectrum becomes considerably broadened. If unlabeled morphine is now added to this system (as in a 1-ml urine specimen from a heroin addict), the spin-labeled morphine is displaced from the antibody and shows the sharp, symmetrical spectrum of the free spin label. The method is semiquantitative and quite specific; other drugs often abused by drug addicts (methadone, barbiturates, and amphetamines) do not react. It is also cheaper than the well-established radioimmunoassay for morphine, in which isotopically labeled ligands are used in a similar experimental setup.

*Receptor function reconstitution* is the culmination of successful receptor isolation and purification, as well as proof of an understanding of the molecular architecture of the receptor environment. Several molecular transport systems have been successfully reconstituted (see Miller and Racker, 1979); the methodology for such reconstitution is available and has been employed successfully in other systems as well (Lindstrom, 1985).

### Selected Readings

E. A. Barnard (1979). Visualization and counting of receptors. In: *The Receptors* (R. D. O'Brien, Ed.), Vol. 1. Plenum Press, New York, pp. 247–310.

J. P. Berliner (1976). *Spin Labeling*. Academic Press, New York.

L. J. Berliner and J. Reuben (Eds.) (1978–) *Biological Magnetic Resonance*, 7 vols. Plenum Press, New York.

R. Blumenthal and A. E. Shamoo (1979). Incorporation of transport molecules into black lipid membranes. In: *The Receptors* (R. D. O'Brien, Ed.), Vol. 1. Plenum Press, New York, pp. 215–245.

D. Cavalla, and N. H. Neff (1985). Chemical mechanisms for photoaffinity labeling of receptors. *Biochem. Pharmacol. 16*: 2821–2826.

V. Chowdhry and F. H. Westheimer (1979). Photoaffinity labeling of biological systems. *Annu. Rev. Biochem. 48*: 293–325.

T. G. Cooper (1977). *The Tools of Biochemistry*. Wiley-Interscience, New York.

W. B. Jacoby and M. Wilchek (Eds.) (1974). Affinity techniques. In: *Methods in Enzymology* (S. P. Colowick and N. D. Kaplan, Eds.), Vol. 34B. Academic Press, New York.

W. B. Jacoby and M. Wilchek (Eds.) (1977). Affinity labeling. In: *Methods in Enzymology* (S. P. Colowick and N. D. Kaplan, Eds.), Vol. 36. Academic Press, New York.

W. Kabsch and P. Rösch (1986). Protein structure determination. *Nature 321*: 469–470.

J. A. Katzenellenbogen (1977). Affinity labeling as a technique in determining hormone mechanisms. In: *Biochemical Action of Hormones* (G. Litwack, Ed.), Vol. 4. Academic Press, New York, pp. 1–84.

M. J. Kuhar (1985). Receptor localization with the microscope. In: *Neurotransmitter Receptor Binding*, 2nd ed. (H. I. Yamamura, S. J. Enna, and M. J. Kuhar, Eds.). Raven Press, New York.

M. J. Kuhar, E. B. de Souza, and J. R. Unerstall (1986). Neurotransmitter receptor mapping by autoradiography and other methods. *Annu. Rev. Neurosci. 9*: 27–59.

J. R. Lakowicz (1983). *Principles of Fluorescence Spectroscopy.* Plenum Press, New York.

A. G. Lee (1978). Fluorescence and NMR studies on membranes. In: *Receptors and Recognition* (P. Cuatrecasas and F. M. Greaves, Eds.), Ser. A, Vol. 5. Chapman and Hall, London, pp. 81–131.

R. K. Leute, E. F. Ullman, A. Goldstein, and L. A. Herzberg (1972). Spin immunoassay technique for determination of morphine. *Nature New Biol. 236*: 93–94.

J. M. Lindstrom (1985). Techniques for studying the biochemistry and cell biology of receptors. In: *Neurotransmitter Receptor Binding*, 2nd ed. (H. I. Yamamura, S. J. Enna, and M. J. Kuhar, Eds.). Raven Press, New York.

C. R. Lowe and P. D. G. Dean (1974). *Affinity Chromatography.* Wiley-Interscience, New York.

N. E. Mackenzie (1985). NMR spectroscopy in biological systems. *Annu. Rep. Med. Chem. 20*: 267–276.

A. G. Marshall (1978). *Biophysical Chemistry.* Wiley-Interscience, New York, pp. 419–434.

A. S. Mildvan (Chairman) (1984). NMR approaches to biochemical problems. *Fed. Proc. 43*: 2633–2670.

C. Miller and E. Racker (1979). Reconstitution of transport functions. In: *The Receptors* (R. D. O'Brien, Ed.), Vol. 1. Academic Press, New York, pp. 1–31.

A. J. Moss, G. V. Dalrymple, and C. M. Boyd (1976). *Practical Radio-immunoassay.* C. V. Mosby, St. Louis.

G. C. K. Roberts (1977). Substrate and inhibitor binding to dihydrofolate reductase. In: *Drug Action on the Molecular Level* (G. C. K. Roberts, Ed.). University Park Press, Baltimore, pp. 127–150.

P. J. Sadler (1975). NMR spectroscopy biological sciences. In: *Progress in Medicinal Chemistry* (P. G. Ellis and G. B. West, Eds.), Vol. 12. Elsevier/North Holland, Amsterdam, pp. 159–190.

R. R. Sharp (1976). Magnetic resonance probes in drug binding. *Annu. Rep. Med. Chem. 11*: 311–320.

S. S. Simons, Jr., and E. B. Thompson (1982). Affinity labeling of glucocorticoid receptors: new methods in affinity labeling. In: *Biochemical Action of Hormones* (G. Litwack, Ed.), Vol. 9. Academic Press, New York, pp. 221–254.

C. F. Stevens (1980). Biophysical analysis of the function of receptors. *Annu. Rev. Physiol. 42*: 643–652.

L. Stryer (1978). Fluorescence energy transfer as a spectroscopic ruler. *Annu. Rev. Biochem. 47*: 819–846.

S. Udenfriend (1972). *Fluorescence Assay in Biology and Medicine*, 2 vols. Academic Press, New York.

D. L. William-Smith and S. J. Wyward (1975). Electron paramagnetic resonance in medicinal chemistry. In: *Progress in Medicinal Chemistry* (G. P. Ellis and G. B. West, Eds.), Vol. 12. Elsevier/North Holland, Amsterdam, pp. 191–246.

M. Williams and D. C. U'Pritchard (1984). Drug discovery at the molecular level: a decade of radioligand binding in retrospect. *Annu. Rep. Med. Chem. 19*: 283–292.

R. S. Yalow (1980). Radioimmunoassay. *Annu. Rev. Biophys. Bioeng. 9*: 327–375.

R. S. Zukin, P. R. Hartig, and D. E. Koshland, Jr. (1977). Use of a distant reporter group as evidence for a conformational change in a sensory receptor. *Proc. Natl. Acad. Sci. USA 74*: 1932–1936.

## 4. DATA TREATMENT

*Binding constants* are derived from *in vivo* pharmacological experiments or from the *in vitro* use of labeled ligands.

As shown in Chap. 2, Sec. 4, in the reaction of a drug with a single population of noninteracting sites,

$$D + R \rightleftharpoons DR \tag{1}$$

where

$$K_D = \frac{[D][R]}{[DR]} \tag{2}$$

the fraction of occupied sites $v$ is

$$v = \frac{[DR]}{[DR] + [R]} \tag{3}$$

### 4.1. Direct Plot

In the direct plot, Eq. (2) is solved for $[DR]$, and its value substituted into Eq. (3):

$$v = \frac{[D]}{K_D + [D]} \tag{4}$$

$K_D$ can then be obtained from a plot of $v$ against $[D]$ (Fig. 3.13A), if the receptor concentration is constant. This is, of course, the same as the direct plot of enzyme activity shown in every biochemistry textbook. As with all hyperbolic relationships, there are several drawbacks to this technique: many data points are needed at the beginning of the curve, at low $[D]$ values, where accuracy is limited. Also, determination of the maximum effect is almost impossible, since we are dealing with an asymptotic curve.

### 4.2. Titration Plot

In the titration plot, Eq. (4) is solved for $K_D$:

$$K_D = [D] \frac{1 - v}{v} \tag{5}$$

and, obtaining $\log_{10}$ of both sides,

$$\log K_D = \log[D] + \log \frac{1 - v}{v} \tag{6}$$

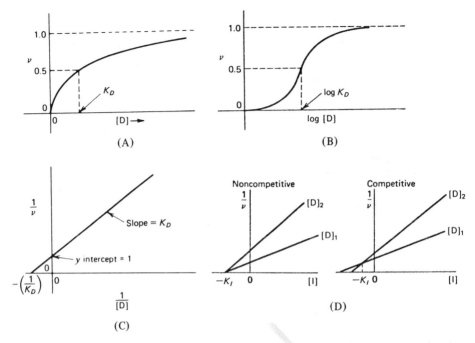

**Fig. 3.13.** Different graphical displays of binding constants: **(A)** Direct plot. **(B)** Titration plot. **(C)** Double reciprocal plot. **(D)** Dixon plot. (Modified by permission from Marshall (1978), Wiley, New York)

where one can use the negative logarithms

$$-\log[D] = pD$$

and

$$-\log K_D = pK_D$$

and arrive at

$$pD = pK_D + \log\left(\frac{1-v}{v}\right) \tag{7}$$

By plotting $v$ against $\log[D]$, the curve in Fig. 3.13B is obtained. In acid–base titrations, pD is pH and $pK_D$ is $pK_a$, because $[D] = [H^+]$. This is the reason for the name of the curve, well known from analytical chemistry. The drawback of this plot is that many points are needed in the vicinity of the inflection point.

### 4.3. Double Reciprocal Plot

The hyperbolic direct plot can easily be straightened out, as analogies from classical enzymology teach us. The most popular data treatment for yielding straight lines is the double reciprocal plot, also known as the Lineweaver–Burke or Benesi–

Hildebrand plot. Here, we take the reciprocal of Eq. (4):

$$\frac{1}{v} = 1 + \frac{K_D}{[D]} \tag{8}$$

If one plots $1/v$ against $1/[D]$, as in Fig. 3.13C, $K_D$ and $v$ can be obtained directly with good precision.

## 4.4. Dixon Plot

Another useful method of data reduction is the Dixon plot, where $1/v$ is plotted against $[I]$ (Fig. 3.13D), the inhibitor concentration, at a fixed $[D]$. This allows for the determination of $K_I$ without the need to determine the absolute concentration of $[D]$—a great advantage in cases in which the substrate is a polynucleotide or a protein, as is often the case in chemotherapy.

## 4.5. Scatchard Plot

Perhaps the most widely used method for extracting binding data is the Scatchard plot. This is obtained from

$$\frac{v}{[D]} = \frac{1}{K_D} - \frac{v}{K_D} \tag{9}$$

Thus, plotting $v/[D]$ against $v$ or, alternatively, $[D]_{bound}/[D]_{free}$ against $[D]_{bound}$ gives a straight line. The slope is $1/K_D$ (the *binding* constant); the abscissa intercept is the number of binding sites if $v$ is shown as mol $v$/mol R. An example of the actual treatment of experimental results of serotonin binding to brain receptors is shown in Table 3.4 and Fig. 3.14, starting with crude data (cpm of [$^3$H]serotonin) and resulting in a plot.

The great advantage of the Scatchard plot is its linearity (i.e., all data are weighted equally). Errors in measurement register on both axes, and are therefore eliminated.

The Scatchard plot is most useful when there are *multiple binding sites*; in this case, however, the plot is not linear. This most commonly occurs in the case involving a small population of receptors with a high affinity, accompanied by a large population with a low affinity. Figure 3.15 shows such a case in which there are two kinds of binding sites: one with a $K_D = 10^{-3}$ M and a number $(n_1) = 10$; the other with a $K_D = 4 \times 10^{-2}$ and a number $(n_2) = 30$ per receptor macromolecule. The number of low-affinity sites is sometimes difficult to determine because of the asymptotic nature of the curve, and numerical solutions are advisable. Nevertheless, it is easy to see that the Scatchard plot is an excellent way of extracting binding-site constants and numbers from experimental findings. However, it must be remembered that these properties of the Scatchard plot are valid only if the binding sites are independent, that is, if there is no cooperative interaction between them. Indeed, Cuatrecasas and Hollenberg (1976) have warned that downward curvature of the Scatchard plot may result from the underestimation of nonspecific binding, simultaneous binding on two sites, and some other factors, besides negative cooperativity and differences in affinity.

**Table 3.4.** Serotonin binding to rat brain membranes[a]

| [³H]5-HT added | | Scatchard plot data [³H]5-HT bound (cpm) | | | [³H]5-HT bound/free | |
|---|---|---|---|---|---|---|
| cpm/2 ml × 10⁻⁴ | nM[b] | Total | Blank | Specific | pmol/g tissue[c] | pmol/g-nM × 10 |
| 4.06 | 1.21 | 1,358 | 321 | 1,037 | 3.24 | 26.8 |
| 8.12 | 2.11 | 2,208 | 541 | 1,667 | 5.21 | 21.6 |
| 12.2 | 3.62 | 3,097 | 719 | 2,378 | 7.43 | 20.5 |
| 16.2 | 4.83 | 3,910 | 1,041 | 2,869 | 8.96 | 18.6 |
| 24.4 | 7.24 | 5,294 | 1,524 | 3,770 | 11.8 | 16.3 |
| 61.0 | 18.1 | 9,665 | 4,358 | 5,306 | 16.6 | 9.17 |
| 122 | 36.2 | 15,843 | 9,148 | 6,695 | 20.9 | 5.77 |

| [³H]5-HT (nM) | log | Hill plot data $B$ (pmol/g) | $\dfrac{B}{B_{max} - B}$ | $\log \dfrac{B}{(B_{max} - B)}$ |
|---|---|---|---|---|
| 1.21 | 0.083 | 3.24 | 0.146[d] | −0.836[d] |
| 2.41 | 0.382 | 5.21 | 0.258 | −0.588 |
| 3.62 | 0.559 | 3.43 | 0.413 | −0.384 |
| 4.83 | 0.684 | 8.96 | 0.546 | −0.263 |
| 7.24 | 0.860 | 11.8 | 0.868 | −0.0616 |
| 18.1 | 1.258 | 16.6 | 1.89 | 0.276 |
| 3.62 | 0.559 | 20.9 | 4.64 | 0.667 |

[a] [³H]5-HT (16.2 Ci/mmol) was incubated with 2-ml aliquots of rat hippocampal membranes containing 20 mg wet weight of tissue per sample. Blank samples represent binding in the presence of $10^{-5}$ M unlabeled 5-HT. Particulate-bound [³H]5-HT trapped on glass filters was counted at 44% efficiency.

[b] Because less than 3% of added [³H]5-HT was bound, total added [³H]5-HT is set equal to "free" [³H]5-HT in these calculations.

[c] $\dfrac{\text{cpm specific [³H]5-HT}}{44\% \text{ efficiency}} \times \dfrac{1 \text{ nCi}}{2.24 \times 10^3 \text{ dpm}} \times \dfrac{1 \text{ pmol[³H]5-HT}}{16.2 \text{ nCi}} \times \dfrac{1}{0.02 \text{ g tissue}}$.

[d] These data are calculated using $B_{max}$ of 25.4 pmol/g, determined from the Scatchard plot (Fig. 3.14).

Reproduced by permission from Bennett (1978), Raven Press, New York.

The interpretation of biphasic Scatchard plots seems to be fraught with pitfalls. Laduron (1982) has warned that the low-affinity limb of such a plot may indicate a binding site that does not modulate a physiological response and is therefore not a receptor in the strict sense. This would indicate that the burgeoning number of multiple receptor sites should be reviewed by measuring nonspecific (blank) binding, using a displacing agent that has the same pharmacological properties as the drug but does not belong to the same chemical class. Light (1984) discusses the analysis of nonlinear Scatchard plots.

## 4.6. Hill Plot

The *cooperativity of receptor sites* can be recognized from binding data. Cooperativity means that binding to one receptor site facilitates binding to subsequent receptor sites in the same population. The classical example is oxygen binding by

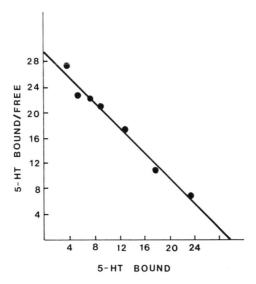

**Fig. 3.14.** Scatchard plot of serotonin binding to rat brain membranes. Data for construction of the plot are shown in Table 3.4. (Reproduced by permission from Bennett (1978), Raven Press, New York)

**Fig. 3.15.** Biphasic Scatchard plot showing two populations of receptors: one with a $K_D = 10^{-3}$ and $n = 10$ binding sites, and the other a low-affinity population with a $K_D = 10^{-2}$ and $n = 30$.

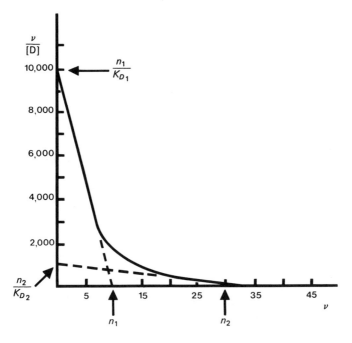

123

hemoglobin, treated in every biochemistry text. In a direct plot, the curve of cooperative binding is sigmoidal instead of hyperbolic.

If Eq. (4) is modified to incorporate $n$ theoretical sites on the receptor, then

$$v = \frac{n[D]^n}{K_D + [D]^n} \tag{10}$$

from which

$$\frac{v}{n - v} = \frac{[D]^n}{K_D} \tag{11}$$

If $F$ is the fraction of occupied active sites, then the number of occupied sites becomes

$$v = nF \tag{12}$$

and therefore

$$\frac{v}{n - v} = \frac{F}{1 - F} = \frac{[D]^n}{K_D} \tag{13}$$

In logarithmic form, this becomes:

$$\log \frac{F}{1 - F} = n \log [D] - \log K_D$$

**Fig. 3.16.** Cooperative binding of oxygen to myoglobin and hemoglobin shown in a "direct" plot and a Hill plot. Binding is not equally cooperative over the whole range, as shown in the deviation from linearity in the Hill plot for hemoglobin at high and low $pO_2$ values. The slope indicates the number of binding sites. (Reproduced by permission from Marshall (1978), Wiley, New York)

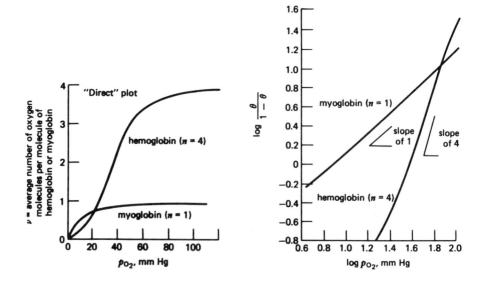

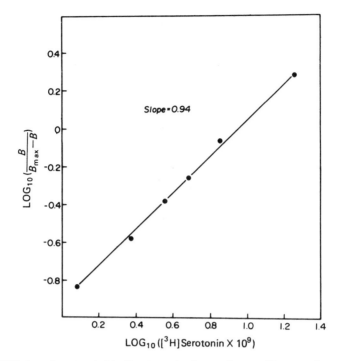

**Fig. 3.17.** Hill plot of serotonin binding to rat brain membranes. Data are the same as those used in constructing Fig. 3.14, and are shown in Table 3.4. (Reproduced by permission from Bennett (1978), Raven Press, New York)

and we get a straight line, the *Hill plot*, if $\log F/1 - F$ is plotted against $\log [D]$. The slope is $n$. Thus, the Hill plot gives, approximately, the number of interacting sites. One example is shown in Fig. 3.16, using the myoglobin and hemoglobin examples. Another example (for noncooperative bindings) is shown in Fig. 3.17; this figure uses the data in Table 3.4 that were employed for the Scatchard plot in Fig. 3.14, which shows [$^3$H] serotonin binding to brain membranes.

If the slope of the Hill plot is less than unity, negative cooperativity is suspected (i.e., the binding of the first ligand inhibits subsequent binding). The insulin receptor (Chap. 5, Sec. 2.7) shows such behavior. Positive or negative cooperativity would indicate a conformational change that increases or decreases the affinity of the receptor site for the drug.

Positive cooperativity can sometimes be noticed on Scatchard plots that become convex at the beginning instead of concave, as shown in Fig. 3.15. De Lean and Rodbard (1979) discuss the theoretical implications of such cases.

### Selected Readings

J. P. Bennett, Jr. (1978). Methods in binding studies. In: *Neurotransmitter Receptor Binding* (H. J. Yamamura, S. J. Enna, and M. J. Kuhar, Eds.). Raven Press, New York, pp. 57–90.

P. Cuatrecasas and M. D. Hollenberg (1976). Membrane receptors and hormone action. *Adv. Protein Chem. 30*: 251–451.

A. De Lean and D. Rodbard (1979). Kinetics of cooperative binding. In: *The Receptors* (R. D. O'Brien, Ed.). Plenum Press, New York, pp. 140–192.

L. Iversen, S. D. Iversen, and S. H. Snyder (1982). *Handbook of Psychopharmacology*, Vol. 15: *New Techniques in Psychopharmacology*. Plenum Press, New York.

J. M. Klotz and D. L. Hunston (1971). Properties of graphical representation of multiple classes of binding sites. *Biochemistry 10*: 3065–3069.

P. M. Laduron (1982). Towards a unitary concept of opiate receptor. *Trends Pharmacol. Sci. 3*: 351–352.

K. E. Light (1984). Analysing nonlinear Scatchard plots, *Science 223*: 76–77.

A. G. Marshall (1978). *Biophysical Chemistry*. Wiley, New York.

# 4

# Drugs Acting on Neurotransmitters
# and Their Receptors

The preceding chapters, dealing with the physicochemical principles of drug action and the theory and practice of receptor studies, have served to lay the groundwork for the rest of the book. We are now equipped to deal with drugs and their modes of action in a manner as rational as present-day concepts permit. In view of the contemporary quest for a basic understanding of drug action at the molecular rather than the cellular or organismic level, our organizational framework rests on bio-chemical molecular mechanisms, as outlined in the Preface to the First Edition.

## 1. OUTLINE OF NEUROANATOMY AND NEUROPHYSIOLOGY

This chapter deals with endogenous messengers, their targets, and the drugs that affect them. Since most of these messengers act on nerve cells, it is appropriate to review the anatomy, histology, and physiology of neurons, and to discuss briefly those neuronal networks that can be manipulated therapeutically.

### 1.1. The Neuron

The nerve cells are highly specialized cells that conduct and trigger bioelectric impulses, communicate with each other in intricate networks, and regulate all tissues and organs. The membrane of the nerve cell is "excitable" because it can undergo changes in its permeability, triggered by small, endogenous neurotrans-mitter molecules or by drugs.

Figure 4.1 shows the organization of a nerve cell (a motoneuron), with all of its components. The cell body carries short, branching *dendrites* which receive and transfer incoming signals to the cell; these signals are then transmitted to the next neuron (or to a tissue) by the long *axon*. The axon of a motoneuron, shown in the figure, is insulated by the lipid *myelin sheath*, interrupted by the *nodes of Ranvier*. These gaps allow the exchange of ions between the axon and its surroundings. The axon terminates in a nerve ending, in this case a *neuromuscular endplate* which communicates with the membranes of muscle cells. In other neurons, the nerve

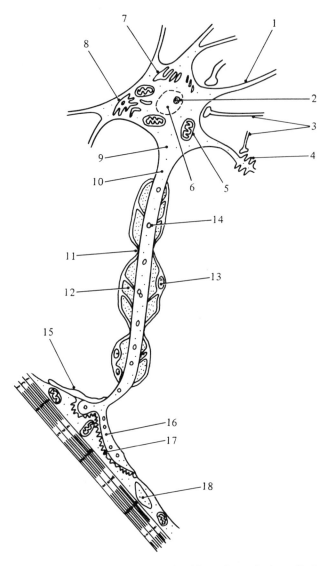

**Fig. 4.1.** Structure of a motor neuron: (1) dendrite; (2) nucleolus; (3) incoming nerve endings; (4) dendritic spines; (5) mitochondrion; (6) nucleus; (7) endoplasmic reticulum; (8) Golgi apparatus; (9) axon hillock; (10) axon; (11) node of Ranvier; (12) myelin sheath; (13) Schwann cell nucleus; (14) neurotransmitter vesicle; (15) dendroglia cell; (16) neuromuscular endplate; (17) endplate junctional folds; (18) muscle cell nucleus. (Modified from Schadé and Ford, 1973)

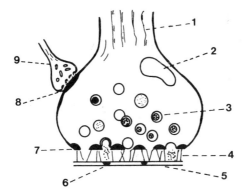

**Fig. 4.2.** Synaptic button: (1) axon; (2) mitochondrion; (3) synaptic vesicles; (4) synaptic gap; (5) postsynaptic membrane; (6) postsynaptic receptor; (7) synaptopore grid; (8) presynaptic receptor; (9) incoming neuron showing axoaxonal communication.

ending can be a knoblike *synaptic button* (Fig. 4.2) in contact with the dendrites, axon, or cell body itself of another nerve cell, with chemical signals rather than electric impulses being used for transmission. The synapse contains mitochondria and one or more types of *synaptic vesicles*—spheres of 0.3–0.9 μm in diameter, surrounded by a membrane and filled with a neurotransmitter often complexed with protein and ATP. The presynaptic membrane seems to have an inner grid composed of *synaptopores*, which are assumed to direct the synaptic vesicles to the membrane when they are about to discharge the neurotransmitter. However, there are other mechanisms of neurotransmitter release. The *synaptic gap* separates two interconnected neurons which only rarely communicate electrically with one another. Normally, the neurotransmitter, released into the synaptic gap, is guided by filaments to the *postsynaptic membrane* and its receptors, which are really parts of the next neuron.

### 1.2. Nerve Conduction

All cells show transmembrane electric potential. A microelectrode placed into a cell will indicate a potential that is 50–80 mV more negative than the potential recorded by an electrode outside the cell. This is a result of ion imbalance. Inside the cell there is a high $K^+$ ion concentration (about 120 mM) and low $Na^+$ concentration (about 20 mM); the reverse is true outside the cell. Since the hydrated potassium ion has a smaller diameter than the hydrated sodium ion, it can diffuse out of the cell along the intra- to extracellular concentration gradient, which sodium ions are incapable of doing. This leaves a negative charge inside the cell because the protein anions of the cytosol are not counterbalanced. The buildup of this negative charge eventually prevents the loss of more $K^+$ ions, and an equilibrium is reached; the cell becomes polarized and the transmembrane potential (*resting potential*) stabilizes.

The difference between an ordinary cell and an excitable cell becomes evident when a depolarizing current is applied. In an ordinary cell, such as an erythrocyte,

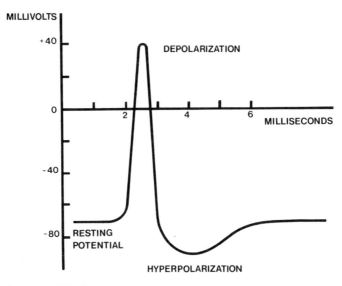

**Fig. 4.3.** Action potential of a neuron.

the membrane potential merely equalizes to zero; in a neuron, however, an explosive, self-limiting process allows the potential to overshoot zero and become about 30 mV more positive within the cell than outside it. This depolarization is called an *action potential*, and is carried first by sodium ions and then by potassium ions. It lasts only about a millisecond, during which time sodium rushes in and potassium rushes out through ion channels opened by conformational changes in the membrane. The original ionic disequilibrium is then reestablished through the rapid elimination of $Na^+$ ions. Figure 4.3 shows the sequence. In myelinated nerves, such ion exchange can occur only at the nodes of Ranvier, and the action potential jumps very rapidly from node to node without a loss of potential. This wave of depolarization passes along the axon to the nerve ending and can be repeated several hundred times per second.

### 1.3. Synaptic Transmission

Synaptic transmission is not electrical but chemical, and is triggered by the arrival of the action potential at the nerve ending. This causes a $Ca^{2+}$ ion influx across the membrane and results in the release of the neurotransmitter characteristic of that particular nerve. There seem to be several different release mechanisms, although none is well understood. When released, the neurotransmitter crosses the synaptic gap and binds momentarily to the receptor on the postsynaptic membrane. This receptor occupation initiates the electrical axonal wave of depolarization of the next (postsynaptic) neuron; alternatively, it can trigger the activation of an enzyme like adenylate cyclase and the formation of cAMP as a second messenger. The released neurotransmitter is then either destroyed enzymatically or taken back into the synapse and recycled. *Inhibitory neurotransmitters*, on the other hand, activate $Cl^-$

ion uptake through the postsynaptic neuronal membrane. This effect makes the intracellular potential more negative than the original resting potential and thus *hyperpolarizes* the neuronal membrane. Naturally, a greater than normal impulse will be necessary to fire such a hyperpolarized neuron, since the threshold value of the action potential remains the same. Both excitatory and inhibitory impulses summate and trigger an all-or-none response of a particular neuron, on which hundreds of other neurons may synapse.

Recently, both polarizing and depolarizing postsynaptic potentials have been discovered that persist for several seconds instead of the millisecond duration of classical potentials; this finding indicates that some other forms of neuronal communication exist. In the case of the release of many newly discovered neurohormones (see Sec. 1.4 and also Chap. 5, Sec. 2), no depolarizing current or other change in membrane properties can be observed. These neurohormones, besides being capable of triggering classical responses in postsynaptic cells, also seem to affect their target cells by modifying their response to "classical" neurotransmitters. This effect, which is not always easily observable, has been termed an "enabling" (or "disenabling") activity (for details, see Cooper, Bloom, and Roth, 1986).

Besides binding to postsynaptic receptors, a released neurotransmitter also diffuses to *presynaptic receptors* or autoreceptors, fulfilling an important feedback regulatory function by facilitating or inhibiting transmitter release (Langer, 1981). The concept of presynaptic autoreceptors has been questioned, however, especially in case of the adrenergic neuronal system (Laduron, 1985; Göthert, 1985). It has been suggested that these presynaptic receptors are *heteroreceptors*—that is, they respond to cotransmitters produced by the same neuron. For instance, it is known that neurotensin regulates the release of norepinephrine, its cotransmitter.) As we realize the fallacy of the "one neuron–one transmitter" dogma, (see next section), it makes sense to assume that cotransmitters must regulate each other's release and metabolism, and that there may be considerable overlap among presynaptic auto- and heteroreceptor functions. However, the autoreceptor concept seems to be holding up in other neuronal systems—at least until future investigations show otherwise. Nerve terminal functions have been reviewed by Reichardt and Kelly (1983). Beyond this, other presynaptic receptors can respond either to neurotransmitters secreted by other types of neurons or to drugs, all of which regulate neurotransmitter release and biosynthesis. These events are shown in an idealized fashion in Fig. 4.4, and admirably discussed in detail by Cooper, Bloom, and Roth (1986).

## 1.4. Neurotransmitters

The neurotransmitters that have been recognized with reasonable certainty are shown in Table 4.1. Most of them will be discussed in subsequent chapters. Their number has increased rapidly in the past few years as the methodology for their detection has become more sophisticated, and one can look forward to yet more discoveries in this area. At this point it is well to consider that the classical definitions and concepts in this field are undergoing considerable change, and that the

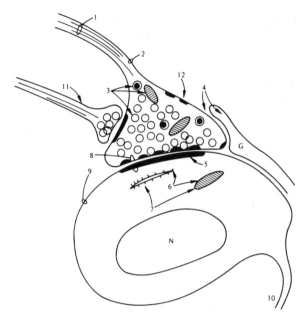

**Fig. 4.4.** Ten steps in synaptic transmission: (1) axonal transport; (2) electrical membrane excitation; (3) synthesis, storage, and release of neurotransmitter; (4) enzymes catabolizing excess transmitter; (5) postsynaptic receptors responding to neurotransmitter; (6) postsynaptic organelles responding to receptor trigger; (7) genetic interaction of nerve cell and cell organelles; (8) variable modifications due to synaptic contact; (9) membrane integration of potential; (11) presynaptic, axoaxonal regulation of neurotransmitter release; (12) autoreceptor. (Reproduced by permission from Cooper, Bloom, and Roth (1986), Oxford University Press, New York)

distinctions between neurotransmitters, neuromodulators, and hormones often become blurred. Many peptide hormones of the hypothalamus and hypophysis, for instance, have been recognized as having neurotransmitter activity at other sites, and neurohormones and the discipline of neuroendocrinology have become increasingly important in the biosciences.

### 1.4.1. Peptide Cotransmitters

In recent years, an explosive development in the discovery of *cotransmitters* has greatly expanded our understanding of neurotransmission, and the homeostatic equilibrium regulated by aminergic and peptidergic coneurotransmitters even in systems as simple as that of *Hydra*. Postsynaptically, cotransmitters can influence the same receptor on the target, bind to two different receptors on the same target, or bind to two different receptors on two different targets. This multipotential reactivity may explain the fact that some drugs and endogenous substances are partial agonists only: they may miss the "help" of a cotransmitter that the full agonist receives. Cross-reactivity of cotransmitter combinations may also explain the many side effects and shortcomings of neuroactive drugs that have been

**Table 4.1.** Structure of Some Neurotransmitters

$CH_3—C(=O)—O—CH_2—CH_2—\overset{\oplus}{N}(CH_3)_3$

Acetylcholine

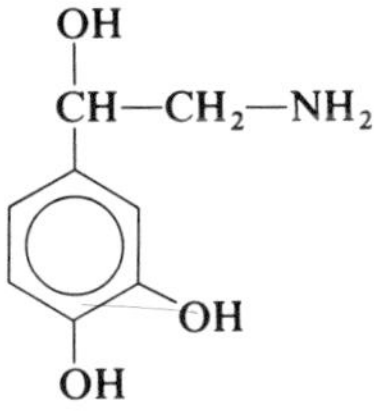

Norepinephrine (noradrenaline)

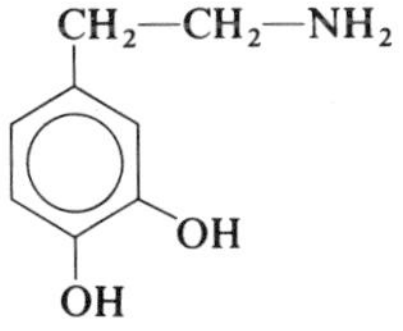

Dopamine

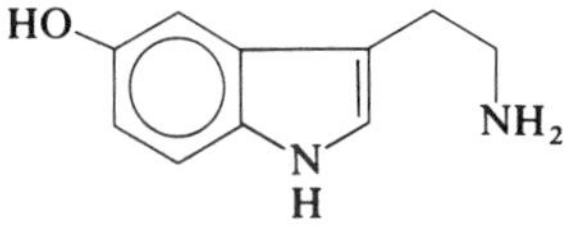

Serotonin 5-HT

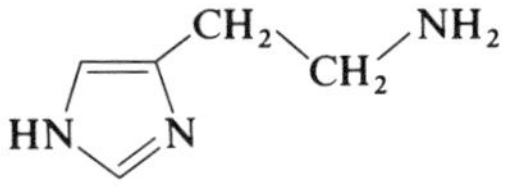

Histamine

$H_2N—(CH_2)_3—COOH$

γ-Aminobutyric acid GABA

$H_2N—CH_2—COOH$

Glycine

$HOOC—CH_2—CH(NH_2)—COOH$

Aspartic acid

$HOOC—(CH_2)_2—CH(NH_2)—COOH$

Glutamic acid

$H_2N—CH_2—CH_2—SO_3H$

Taurine

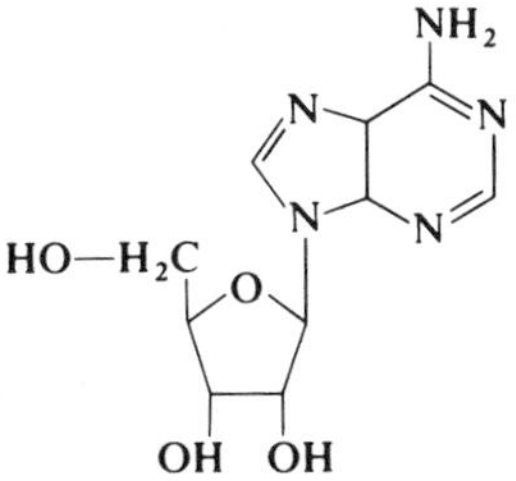

Adenosine

Arg—Pro—Lys—Pro—Gln—Gln—Phe—Phe—Gly—Leu—Met—$NH_2$

Substance P

Tyr—Gly—Gly—Phe—Met—OH
Tyr—Gly—Gly—Phe—Leu—OH

Enkephalins

designed without the benefit of knowing the complete story of *in vivo* processes at the target.

A probably incomplete list of cotransmitter combinations is shown in Table 4.2. It should be kept in mind that a single synapse may operate with as many as four transmitters simultaneously, in any combination of amine and peptide, or even peptide and peptide, within the groupings shown. The peptide neurotransmitters are stored separately, always in large synaptic vesicles; are synthesized in the cell body of the neuron; and are transported to the synapse after posttranslational processing by fast (ATP-driven) transport systems. Amine neurotransmitters are synthesized in the synapse and are stored in small or large vesicles. Low nerve pulses release small

**Table 4.2.** Amine–peptide and peptide–peptide cotransmitter combinations[a]

| Amine–peptide | | Peptide–peptide | |
|---|---|---|---|
| Dopamine | – CCK | | |
| | – Neurotensin | ACTH | – β-Endorphin and β-LPH |
| | – Enkephalin | | – β-Endorphin |
| | | | |
| Norepinephrine | – Somatostatin | β-Endorphin | – β-LPH |
| | – Neurotensin | | – MSH |
| | – Enkephalin | | |
| | – NPY | Dynorphin | – Vasopressin |
| | | | – α-Neoendorphin |
| Epinephrine | – Neurotensin | | |
| | – NPY | Somatostatin | – CCK |
| | – SP | | – APP |
| | – CCK | | – Enkephalin |
| | | | |
| Serotonin | – SP | Substance P | – CCK |
| | – SP and TRH | | – Leu-Enkephalin |
| | – Enkephalin | | – Met-Enkephalin |
| | – CCK | | |
| | | Enkephalin | – APP |
| Acetylcholine | – Somatostatin | | – Neurophysin |
| | – CCK | | – Oxytocin |
| | – VIP | | |
| | – SP | | |
| | – Neurotensin | | |
| | | | |
| GABA | – Somatostatin | | |
| | – Motilin | | |
| | – Motilin and taurine | | |
| | – Serotonin | | |
| | – NTY | | |
| | | | |
| Glycine | – Neurotensin | | |

[a] *Abbreviations*: App, avian pancreatic peptide; CCK, cholecystokinin; LPH, lipotropic hormone; MSH, melanophore-stimulating hormone; NPY, neuropeptide Y; SP, substance P; TRH, thyrotropin-releasing hormone; VIP, vasointestinal peptide.

vesicles only, whereas high pulses cause exocytosis of both small and large vesicles. Different populations of the same type of neurons may differ in their content of cotransmitters. For reviews, see Hökfelt et al. (1986) and Pazoles and Ives (1985).

## 1.5. Neuronal Systems

The neuronal systems of vertebrates are divided into the central nervous system (CNS), comprising the brain and the spinal cord, and the peripheral nervous system (PNS), which serves the rest of the body.

The *brain* is really a collection of highly specialized organs of enormous ana-

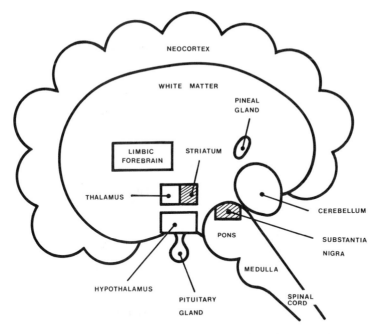

**Fig. 4.5.** Diagrammatic representation of some regions of the human brain.

tomical complexity. The brains of different mammals are very different, and the evolutionary changes in the brain are primarily seen as an increase in relative size and in the complexity of cortical folding, thus increasing the area devoted to *association* (i.e., learning and decision making). Because we will have to discuss the pharmacology of various drugs acting on parts of the CNS, a basic schematic illustration of the human brain is shown in Fig. 4.5.

Commands from the CNS to the periphery and all organs of the body are conveyed by the *autonomic nervous system,* whereas commands to the skeletal muscles are transmitted by the *skeletomotor system.* There is considerable structural difference between the neurons of these two systems (Fig. 4.6).

In the skeletomotor system, a motoneuron may originate from a ventral horn of the spinal cord and continue without interruption, through a myelinated A-fiber, to the muscle. The neuron usually branches in the muscle and forms neuromuscular endplates on each muscle fiber, creating a single motor unit.

The autonomic nervous system differs from this in the interposition of a peripheral *ganglionic synapse* between the CNS and an organ and acts as a kind of switching station. The neurons in the *sympathetic nervous system* orignate in the upper and middle part of the spinal cord and form myelinated B-fibers. Each such fiber makes synaptic connection with the ganglion cell, which continues in a postganglionic, nonmyelinated C-fiber that then synapses on a smooth-muscle cell, a gland, or another neuron. In the sympathetic system, the ganglia are usually in the paravertebral chain, or within some other specialized ganglia. In the *parasympathetic nervous system,* which is the other division of the autonomic nervous system, the

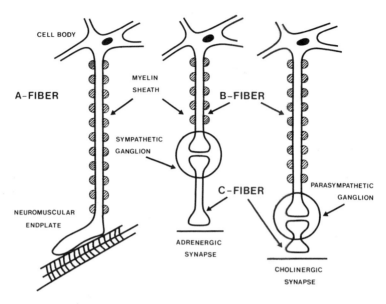

**Fig. 4.6.** Schematic structure of a motor neuron, a sympathetic (adrenergic) neuron, and a parasympathetic (cholinergic) neuron.

**Fig. 4.7.** Interaction of different neurons in a sympathetic ganglion. DA, dopaminergic receptor; M, muscarinic cholinergic receptor; N, nicotinic cholinergic receptor; ACh, acetylcholine; SIFN, small intensely fluorescing interneuron. (After Ryall (1979), *Mechanism of Drug Action on the Nervous System*, Cambridge University Press, Cambridge)

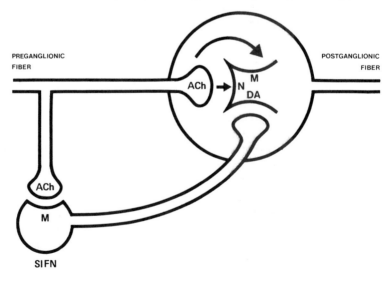

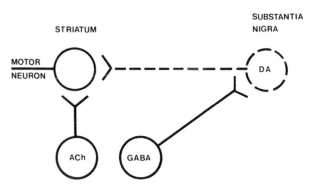

**Fig. 4.8.** Neuronal networks. The motor neuron in the striatum is regulated by an excitatory cholinergic neuron and an inhibitory dopaminergic neuron, which in turn is influenced by a GABAergic fiber. Loss of the dopaminergic nigrostriatal fibers leads to excessive cholinergic stimulation of the motor neuron, resulting in Parkinson's disease.

ganglia are buried in the effector organs and therefore have only short post-ganglionic fibers.

Neuronal systems can be quite complex and can regulate physiological functions through the interaction of sequentially coupled nerve cells that use different neurotransmitters. It should also be kept in mind that a given neurotransmitter can be excitatory in one system but inhibitory in another. For instance, sympathetic ganglia have been shown to contain three kinds of receptors, as illustrated in the schematic diagram in Fig. 4.7. These ganglia, which normally operate on a cholinergic mechanism, also include small, intensely fluorescing neurons (SIFN) which produce dopamine and hyperpolarize the postganglionic neuron, establishing a complex control system.

Only a full understanding of the sequential coupling of neurons permits the treatment of pathological states due to neuronal dysfunction. As an example of this, Fig. 4.8 diagrams the pathophysiology of Parkinson's disease (see Sec. 4.2), a degenerative syndrome characterized by neuromotor disorders such as tremor, rigidity, a stooped posture, and difficulty in initiating and stopping movement. It is due to the disappearance of dopaminergic neurons connecting the *substantia nigra* with the *striatum* (see Fig. 4.5). The loss of dopaminergic inhibition in the striatum permits cholinergic hyperactivity in this brain center, resulting in many of the neuromotor symptoms of the disease. The drug treatment of Parkinson's disease therefore consists of a combination of dopamine replacement and release facilitation, as well as the use of cholinergic blocking agents to control parasympathetic hyperactivity.

### Selected Readings

H. S. Bachelard (1974). *Brain Biochemistry*. Chapman and Hall, London.

J. R. Cooper, F. E. Bloom, and R. H. Roth (1986). *The Biochemical Basis of Neuropharmacology*, 5th ed. Oxford University Press, New York.

M. Göthert (1985). Role of autoreceptors in the function of the peripheral and central nervous system. *Arzneimittelforschung 35*: 1909–1916.

T. Hökfelt, B. Evaritt, B. Meister, T. Melander, M. Schalling, O. Johansson, J. M. Lundberg, A. L. Hulting, S. Werner, C. Cuello, M. Hemming, C. Ouimet, J. Walaas, P. Greengard, and M. Goldstein (1986). Neurons with multiple messengers, with special reference to neuroendocrine systems. *Recent Prog. Hormone Res. 42*: 1–70.

S. W. Kuffler, J. G. Nicholls, and A. R. Martin (1984). *From Neuron to Brain: A Cellular Approach to the Function of the Nervous System*, 2nd ed. Sinauer, Sunderland, MA.

P. M. Laduron (1985). Postsynaptic heteroreceptors in the regulation of neuronal transmission. *Biochem. Pharmacol. 34*: 467–470.

S. Z. Langer (1981). Presynaptic receptors. *Pharmacol. Rev. 32*: 337–363.

C. J. Pazoles and J. L. Ives (1985). Cotransmitters in the CNS. *Annu. Rep. Med. Chem. 20*: 51–60.

L. F. Reichardt and R. B. Kelly (1983). A molecular description of nerve terminal function. *Annu. Rev. Biochem. 52*: 871–926.

J. P. Schadé and D. H. Ford (1973). *Basic Neurology*. Elsevier, New York.

D. J. Triggle and C. R. Triggle (1976). *Chemical Pharmacology of the Synapse*. Academic Press, New York.

E. S. Vizi (1984). *Non-synaptic Interaction Between Neurons: Modulation of Neurochemical Transmission*. Wiley, New York.

## 2. ACETYLCHOLINE AND THE CHOLINERGIC RECEPTORS

The cholinergic neuronal system can be found in the CNS (especially in the *cortex* and *caudate nucleus*), in the autonomic nervous system, and in the skeletomotor system. Acetylcholine (ACh) is the neurotransmitter in all ganglia, the neuromuscular junction, and the postganglionic synapses of the cholinergic (parasympathetic) nervous system. However, the autonomic innervation of most organs utilizes both the parasympathetic (cholinergic) and sympathetic (adrenergic) systems, with the effects of the two usually being opposed. Thus, if one system causes an increase in some physiological action, the other will cause a decrease, and vice versa. Physiology textbooks deal in detail with the innervation of different organs.

As do most neuronal systems, cholinergic receptors show duality, and we distinguish between *nicotinic* and *muscarinic receptors*, which differ in many respects. Whereas **acetylcholine** (4-1) binds to both types of receptors, the plant alkaloids **nicotine** (4-2) and **muscarine** (4-3) trigger a response only from nicotinic or muscarinic cholinergic receptors, respectively. *Nicotinic receptors* are found in all autonomic ganglia (i.e., in the sympathetic system as well as the parasympathetic) and at the neuromuscular endplate of striated muscle. *Muscarinic receptors* occur at postganglionic parasympathetic terminals involved in gastrointestinal and ureteral peristalsis, the promotion of glandular secretion, pupillary constriction, peripheral vasodilation, and reduction in heart rate. Acetylcholine is normally an excitatory neurotransmitter, although it can occasionally show an inhibitory action in cardiac muscle. There, hyperpolarization rather than depolarization occurs because only $K^+$ can cross the muscle membrane. In the CNS, cholinergic inhibition is seen in the thalamus and brainstem.

It is believed that degeneration of cholinergic pathways in the CNS and the

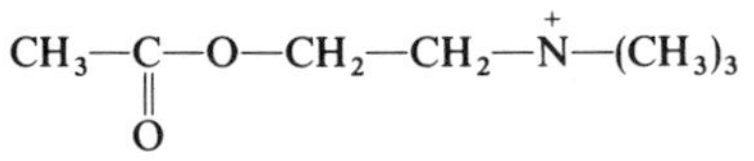

Acetylcholine
4-1

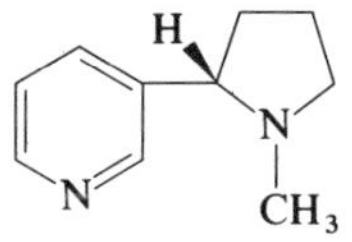

Nicotine
4-2

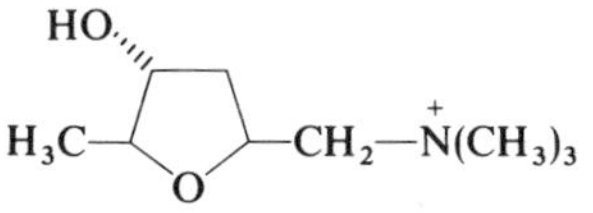

*cis*-L-(+)-Muscarine
4-3

resultant development of "neuronal tangles," histologically apparent irregularities in neuron arrangement, may be the principal, if not only cause of *senile dementia* of the *Alzheimer type*. The disease leads to progressive regression of memory and learned functions, sometimes likened to be the reverse of infant development. Drug treatment with choline replacement, cholinergic agonists, somatostatin, and clonidine have proved to be only marginally successful (Hershenson and Moos, 1986; Hershenson et al., 1986). Since the average age of the population is on the increase, the frequency of Alzheimer's disease is increasing rapidly and requires urgent attention.

### 2.1 Acetylcholine Metabolism

Investigations of ACh metabolism have been helped tremendously by the recent application of such methods as radioimmunoassay (Spector et al., 1978) and pyrolytic gas chromatography. These techniques can detect ACh at the femtomolar $(10^{-15}$ M) level, a sensitivity previously possible only with bioassay.

Acetylcholine is synthesized by the reaction:

$$\text{Choline} + \text{Acetyl-CoA} \rightarrow \text{Acetylcholine} + \text{CoA—SH}$$

which is catalyzed by choline acetyltransferase. Acetyl-coenzyme A (CoA) is ubiquitous; choline is obtained from phosphatidylcholine (lecithin) and free choline. Some of this choline is recycled after ACh is hydrolyzed by acetylcholine esterase (AChE), terminating the neuronal impulse (see Chap. 6, Sec. 3.2). There is a high-affinity transport system ($K_m = 1-5$ M) for choline reuptake in the nerve endings, which can be inhibited by **hemicholinium** (4-4). Unlike most other neurotransmitters, ACh itself is not taken up by active transport into synapses.

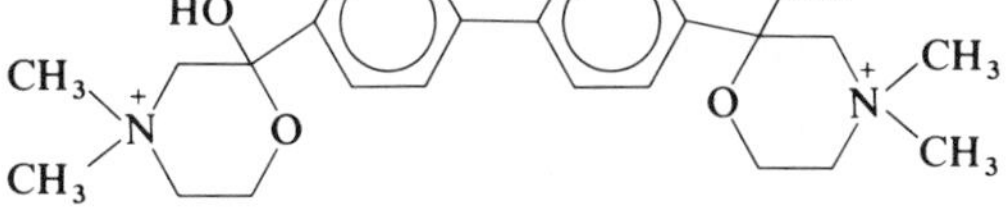

Hemicholinium
4-4

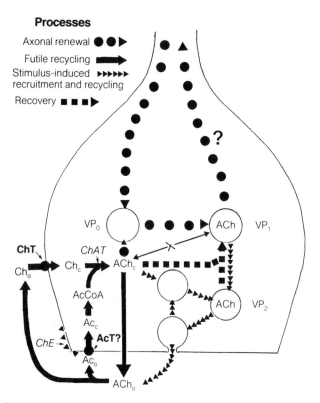

**Fig. 4.9.** The functional organization of the cholinergic terminal. As shown by immuno-cytochemistry and covalent $^{35}$S-labeling, synaptic vesicles are formed in the cell body and transported—largely empty—to the terminal ($VP_0$). Here they fill with cytoplasmic acetyl-choline ($ACh_c$) and enter the pool of reserve vesicles ($VP_1$). On stimulation, a proportion of the $VP_1$ vesicles are recruited into the recycling ($VP_2$) pool. Such vesicles only partially refill from the cytoplasm and undergo partial osmotic dehydration, becoming smaller and denser. At rest, the recycled pool slowly takes up more acetylcholine and reacquires the biophysical properties of the reserve pool. By contrast, there is little or no direct exchange between reserve vesicles and the cytoplasm. Cytoplasmic acetylcholine is subject to "futile recycling" (black arrows). Transmitter leaking out of the terminal ($ACh_0$) is rapidly hydro-lyzed by acetylcholinesterase (AChE) in the cleft and the acetate ($Ac_0$) and choline ($Ch_0$) are salvaged; in both cases, uptake is facilitated by carriers (AcT, ChT). Cytoplasmic cho-line ($Ch_c$) and acetate ($Ac_c$) are resynthesized to cytoplasmic acetylcholine by the soluble enzyme choline acetyltransferase (ChAT); acetate must, however, be first converted to acetyl-coenzyme A (AcCoA). (Not shown in diagram are the endogenous pools of choline and acetyl-CoA which could be called upon to restore cytoplasmic acetylcholine when this is depleted by stimulation.) The cytoplasmic and $VP_2$ pools of transmitter can be specifically labeled by means of false transmitters, and thus cytoplasmic release and that brought about by vesicle recycling can be readily distinguished. A similar cycle exists for synaptic vesicle ATP. There is no evidence yet for antidromic transport of "worn out" vesicles, though by analogy with other systems this is likely to occur. Reproduced by permission from Whittaker, 1986)

As ACh is synthesized, it is *stored* in the neuron or ganglion in at least three different locations. Eighty-five percent of all ACh is stored in a "depot" and can be *released* by neuronal stimulation; it is always the newly synthesized neurotransmitter that is released preferentially. The "surplus" ACh can be released by $K^+$ depolarization only. Finally, there is "stationary" ACh, which cannot be released at all. It was assumed that the neurotransmitter in cholinergic and some other neurons is released through the exocytosis of small transmitter-filled synaptic vesicles, as discussed in Sec. 1.4. However, although the neurotransmitter is undoubtedly at least partly stored in such vesicles, their role in its release is far from settled.

Whittaker (1986) has shown that synaptic vesicles are metabolically inhomogeneous, and those closer to the presynaptic membrane are released preferentially. His ideas are summarized in Fig. 4.9. Several alternative—although not necessarily convincing—hypotheses on ACh release have been proposed. One of these assumes a voltage-dependent $Ca^{2+}$ influx that opens a gate, allowing cytoplasmic ACh release for a timed period; another proposes the containment of ACh in the smooth endoplasmic reticulum, in association with the presynaptic membrane. In conclusion, much is yet to be learned regarding the mechanism of ACh release (Tauc, 1982; Dunant and Israël, 1985).

Acetylcholine release is inhibited by the most potent toxin known, the *botulinus toxin* produced by the anaerobic bacterium *Clostridium botulinum*. The toxin, lethal at 1 ng/kg in humans, enters the synapse by endocytosis at nonmyelinated synaptic membranes and produces muscle paralysis by blocking the active zone of the presynaptic membrane where $Ca^{2+}$-mediated vesicle fusion occurs. The muscle will still respond to direct stimulation by ACh, and Ca entry into the synapse is not inhibited. It has been assumed that the substrates for botulinus toxin may be two proteins, both known as *synapsin-1*. They are thought to participate in vesicle fusion at the presynaptic membrane, and are regulated by cAMP-dependent phosphorylation as well as by Ca/calmodulin-dependent protein kinases (Sellin, 1985).

Acetylcholine is also found in nonneuronal tissues. Ciliary movement in clam gill plates and in the mammalian respiratory tract are both regulated by ACh, and this neurotransmitter also has a direct effect on intestinal smooth muscle and on the heart. It can additionally induce the sporulation of some fungi, and occurs with unknown activity in various locations. It is therefore reasonable to consider ACh a hormone as much as a neurotransmitter, as mentioned previously.

## 2.2. The Nicotinic Acetylcholine Receptor

### 2.2.1. Isolation

The isolation of the nicotinic acetylcholine receptor glycoprotein was achieved almost simultaneously in several laboratories (those of Changeux, O'Brien, Brady, and Eldefrawi) and was helped tremendously by the discovery that the electric organ (*electroplax*) of the electric eel (*Electrophorus electricus*, an inhabitant of the Amazon River) and related species, as well as the electroplax of the electric ray (*Torpedo marmorata*) of the Atlantic Ocean and the Mediterranean Sea, contains acetylcholine receptors (AChR) in a much higher concentration than, for instance, a

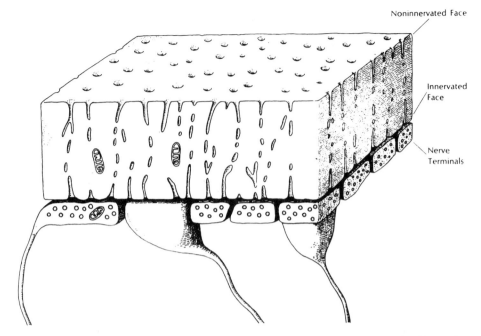

Noninnervated Face

Innervated Face

Nerve Terminals

**Fig. 4.10.** Schematic structure of the electroplax of the electric fish *Torpedo*. The entire innervated face is covered by cholinergic synapses. (Reproduced by permission from Whittaker, in G. Weisman and R. Claiborne (Eds.) (1975), *Cell Membranes: Biochemistry, Cell Biology, and Pathology*, H. P. Publishing, New York)

**Fig. 4.11.** Electron micrograph of an electroplax. NT, nerve terminal; SV, synaptic vesicle; PM, postsynaptic membrane; PRM, presynaptic membrane; CHR, cholinergic receptor area; ACHE, acetylcholinesterase. (Reproduced by permission from Waser et al., in F. Gualtieri, M. Gianella, and C. Melchiorre (Eds.) (1979). *Recent Advances in Receptor Chemistry*, Elsevier Biomedical Press, Amsterdam)

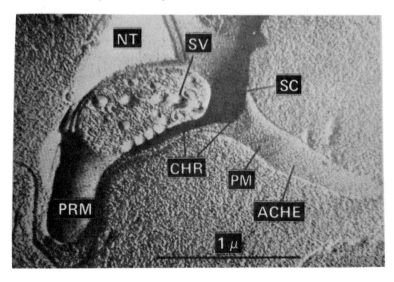

neuromuscular endplate or brain tissue. The relative tissue concentrations are shown in the accompanying table.

| Tissue | AChR |
| --- | --- |
| *Torpedo* | 1000 nmol/kg |
| *Electrophorus* | 50–100 nmol/kg |
| Neuromuscular endplate | 1–50 nmol/kg |
| Brain | 0.1–1 nmol/kg |

The electric organ of the *Torpedo* species (which are related to the sharks and skates) can deliver a shock of 50–60 V; that of the electric eel up to 600 V and about 1 kW energy—sufficient to stun or kill prey or an attacker. The structure of the electroplax is that of modified muscle (Fig. 4.10). The endplate covers one side of the cell completely, hence the enormous concentration of receptors on the postsynaptic membrane. The synaptic area in *Torpedo* can be as high as 50% of the innervated membrane, although in *Electrophorus* it is only 2–3%. Therefore, the receptor protein yields are much higher in *Torpedo*. An electron micrograph of a cross section of the *Torpedo* electric organ reveals a typical synaptic structure (Fig. 4.11).

The discovery that the toxins of *Elapid* snakes bind almost irreversibly to the AChR facilitated the isolation and study of this receptor. The structure of these venoms has been elucidated; the most widely used experimentally are the α-bungarotoxin (BTX) of the Indian cobra and the toxin of the Siamese cobra (Table 4.3). These compounds are peptides containing from 61 to 74 amino acids, five disulfide bridges, and a high proportion of basic arginine and lysine residues,

**Table 4.3.** Amino acid sequence of Siamese cobra neurotoxin

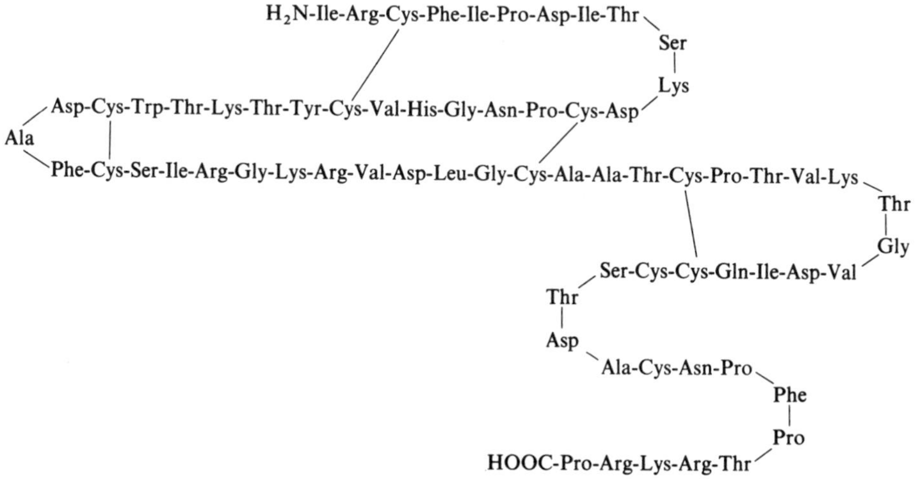

Reproduced by permission from Stenlake (1979), Elsevier/North Holland, New York.

often in close proximity. Venoms are toxic because they block cholinergic neuro-transmission by binding to the receptor.

The AChR is an integral membrane protein, deeply embedded into the postsynaptic membrane. It can be solubilized by nonionic detergents such as Triton X-100, Tween 80, Brij 35, and others, or anionic detergents such as deoxycholate, a bile acid derivative. Functionally, the regulation of ion permeability is lost when the receptor is removed from the membrane; however, the ACh and BTX binding capacity is retained and can be used for following the course of purification. In a typical isolation procedure, the electric organ is homogenized in 1 M NaCl with $Na_2HPO_4$ and EDTA, which solubilizes the acetylcholinesterase. The suspension is then centrifuged and the resulting pellet extracted with detergent, solubilizing the AChR. This receptor "solution" can then be purified further by polyacrylamide gel electrophoresis, by affinity partitioning, or, most efficiently, by affinity chroma-tography (see Chap. 3, Sec. 3.2) either on an immobilized quaternary ligand or on Siamese cobra toxin bound to an agarose bead matrix. The specific activities of the purified preparations range from 8 to 12 $\mu$mol of binding sites per gram of protein, and about 100–150 mg of receptor protein can be obtained from 1 kg of *Torpedo* electric organ. Compared to normal concentration standards, this yield is exception-ally high.

### 2.2.2. Physicochemical Properties and Subunit Structure

The physical and chemical properties of the AChR have been elucidated. Hydrodynamically, a light (L) form with a sedimentation constant of 8.6–9.1 $S$, a heavy form (H) with a constant of 12–13 $S$, and a very heavy (HH) form with a constant of 16.5 $S$ can be isolated, the ratio and exact sedimentation rate depending on the source of the protein. The pI of these forms is 4.9, 5.1, and 5.3, respectively. Optical rotatory dispersion measurements indicate that the receptor consists of about 34% helix and 28–30% $\beta$-sheet structure—a high proportion of ordered secondary structure. Some carbohydrates are part of the molecule. The DNA encoding the receptor has been cloned and sequenced, revealing the complete amino acid sequence of the subunits (Noda et al., 1982).

The *subunit structure* of the AChR varies according to its origin. There are four peptide chains, referred to as $\alpha$ (mass $\sim$40 kD), $\beta$ ($\sim$48 kD), $\gamma$ ($\sim$58 kD), and $\delta$ ($\sim$64 kD), which can be separated by electrophoresis. The receptor of *Torpedo californica* has an $\alpha_2\beta\gamma\delta$ chain composition, giving it a monomeric molecular mass of 250 kD. The receptor is present as a disulfide-linked dimer, joined through the $\delta$ subunit.

The $\alpha$ chain is affinity-labeled specifically by [$^3$H] bromoacetylcholine and by [$^3$H] **MBTA** (4-$N$-maleimidobenzyl-trimethylammonium iodide) (4-5) on Cys-192 and Cys-193 and therefore must be the ACh binding subunit. The other chains have no known specific role, but are integral parts of the receptor and do not dissociate, even in 8 M urea. The different chains have different amino acid sequences but similar compositions. 4-$N$-Maleimidobenzyl-trimethylammonium iodide, a spe-cific affinity reagent developed by Karlin, indicates the important fact that the quaternary-ammonium-ion binding site (the —COO$^-$ of glutamate) and an —SH

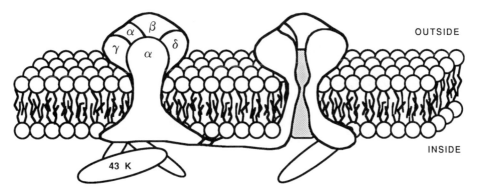

**Fig. 4.12.** This model of the nicotinic acetylcholine receptor shows two pentameric units covalently linked through the δ subunit. One of the units is shown in cross section, indicating the selectivity gate of the ion channel in the closed state. The 43-kD protein is shown associated with the receptor on the cytoplasmic side. (Modified by permission from Kistler et al., 1982)

group binding to maleimide are in close proximity. [³H] Bungarotoxin ([³H] BTX) cross-links the α and β chains and probably also obstructs the ion channel. Additionally, a 43K peptide chain (called *v*) has been found which is not an intrinsic part of the receptor. It binds **procainamide-azide** (4-6), a local anesthetic and photoaffinity label, and is suspected of being an ionophore subunit and local anesthetic binding site. The β and γ chains are preferentially labeled by a nitrene obtained from pyrene-sulfonylazide, a hydrophobic reagent believed to attach itself to proteins within the core of the membrane. The current model of the receptor is shown in Fig. 4.12.

MBTA

4-5

Procainamide-azide

4-6

The *ligand binding sites* of the AChR have been explored by a number of techniques. The ACh binding site was first investigated by Karlin, using the MBTA

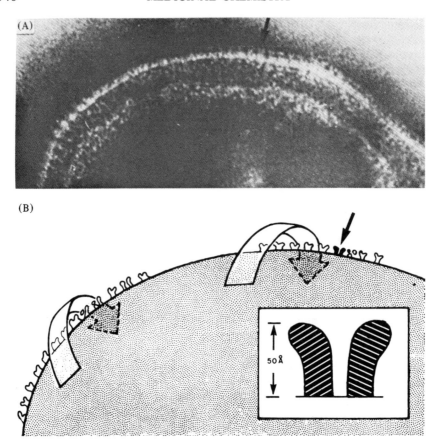

**Fig. 4.13.** (A) Electron micrograph of an electroplax membrane vesicle, showing the receptor protruding on the edge. (B) A schematic of the same vesicle. Magnification 294,000 ×. (Reproduced by permission from Klymowsky and Stroud (1979), *J. Mol. Biol.* 128: 326, Academic Press, London)

affinity label. He worked with membrane-bound receptor, whose binding characteristics differ markedly from those of the isolated glycoprotein. MBTA binds to the native receptor with a $K_D = 8 \times 10^{-5}$ M but binds much more strongly after reduction of the membrane preparation with **dithioerythritol** (4-7), a mild reducing agent. In binding to the AChR, MBTA occupies half the sites that bind $[^{125}I]$BTX.

$$
\begin{array}{c}
CH_2OH \\
| \\
HC-SH \\
| \\
HC-SH \\
| \\
CH_2OH
\end{array}
$$

Dithioerythritol
4-7

*Local anesthetics* (LA) (see Chap. 6. Sec. 1.3) block ion-permeability changes noncompetitively. In the presence of cholinergic agonists, specific and nonspecific local anesthetic sites become evident, perhaps involving the 43K protein mentioned previously. About an equal number of LA and ACh sites have been found on the receptor. Local anesthetics also increase the binding affinity of cholinergic agonists and some antagonists, probably by desensitization. They also change the shape of the binding curve from sigmoidal—a sign of allosteric interaction between the two sites or subunits—to hyperbolic, indicating loss of cooperativity.

Support for the model (Fig. 4.12) has also been provided by electron-microscopic examination of freeze-fractured electroplax specimens, revealing 8-nm-diameter disks surrounding an electron-dense pit of about 2-nm diameter, probably the ionophore. In some cases these disks are packed in a pseudohexagonal array. The receptor protrudes from the lipid membrane about 5 nm on the extracellular side and about 1.5 nm on the intracellular side, showing an overall length of 11 nm. Although the receptor is not axially symmetric, the subunits are arranged in a barrel shape, with a 3-nm channel in the top (outside) unit and a 1-nm channel running through the middle and inner units (Fig. 4.13). The receptor density varies: it is about 50,000/$\mu$m$^2$ in *Electrophorus*, about 30,000/$\mu$m$^2$ in neuromuscular junctions, and 12,000–15,000/$\mu$m$^2$ in *Torpedo* subsynaptic areas.

### 2.2.3. ACh–AChR Site–Ionophore Interactions

These interactions present the central problem in understanding cholinergic neurotransmission: How does agonist binding open the ion channel? What is the duration of the open and closed phases? and How do antagonists or local anesthetics prevent ionophore or AChR site activity? In operational terms, there is a "trigger" (the AChR site), a "gating device," and a "selectivity filter" (the ionophore). A detailed model of an ionophore is shown in Fig. 6.2.

The interaction of ACh with the receptor site can be studied indirectly only. *In vivo*, millimolar concentrations of Ca$^{2+}$ enhance ACh binding. In a complex system, the kinetics of ACh binding can be measured with an "on" rate constant of 2.4 × 10$^7$ M$^{-1}$ sec$^{-1}$. Fluorescence studies have revealed conformational changes in the receptor. The *resting state* (R) has a low affinity for ACh and equilibrates into the *active state* (A), with a medium affinity for the agonist. A *desensitized state* (D) is assumed when the agonist concentration becomes high and the ion channel is closed. This form of the receptor has a high affinity for ACh. The affinity for antagonists is lowest in the A state, whereas that for local anesthetics is lowest in the R state. All of these conformations are present in the membrane *prior to* ACh binding, even though the A-state concentration is very low and the R state is preferred (see also Chap. 2, Sec. 5.2).

The two parts of the ACh regulator, the AChR and the ionophore (the LA site), are, as has been discussed, associated, and their interaction resembles the allosteric interaction of the catalytic and regulatory subunits of the enzyme aspartate transcarbamylase. The ionophore is open only in the A state of the AChR, and shows an all-or-none transition. There are two "open" times, with half-lives of 4 and

0.5 msec, respectively, and the average conductance is 15–30 pmho (pico-ohm$^{-1}$) upon the binding of one to four ACh molecules. During such an open time, about 10,000 Na ions would flow into the cell. The "open" times vary, within one order of magnitude, with the nature of the agonist. Local anesthetics and barbiturates apparently bind to the open channel and inhibit ion transport, whereas general anesthetics seem to influence only the fluidity of the membrane lipid bilayer, exerting a disorganizing effect.

It must be emphasized that this model, although supported by experimental evidence, is still hypothetical in many details. There is, for instance, no good proof that the 43K protein is the ionophore and LA site; reconstruction experiments have not been convincing in this regard. Also, the assumption of only three receptor conformations of the AChR may be an oversimplification.

The physical meaning of *ionophore opening* is also poorly understood. On the basis of structural details of the receptor protein and snake toxins, Stenlake (1979) attempted a molecular explanation of the mechanism of ion-channel opening by agonists and blocking by BTX. He assumed that the agonist–receptor interaction involves the release of an —SH group masked by a $Ca^{2+}$ ion, which then triggers a cascade of disulfide displacements by a series of nucleophilic attacks, resulting in a conformational change that opens the ion channel (Fig. 4.14A).

In snake toxins, cystine disulfide bridges are often close to basic Lys or Arg residues, as shown in Table 4.3, where the 26–30 cystine bridge is close to the Lys-23 and Arg-33. In a mechanism related to the one outlined above, the —S—S bridge of the snake neurotoxin would react with the receptor —SH group, forming a covalent bond, while stabilizing the disulfide bridges of the ionophore and preventing the opening of the ion channel (Fig. 4.14B). Such a mechanism would, incidentally, explain why snake neurotoxins are for all practical purposes irreversible ligands of the AChR, with a $K_D = 10^{-11}$ M.

### Selected Readings

B. M. Conti-Tronconi and M. A. Raftery (1982). The nicotinic cholinergic receptor: correlations of molecular structure with functional properties. *Annu. Rev. Biochem. 51*: 491–530.

Y. Dunant and M. Israël (1985). The release of acetylcholine. *Sci. Am. 252* (4): 58–66.

H. C. Hartzell (1982). Physiological consequences of muscarinic receptor activation. In: *More About Receptors* (J. W. Lamble, Ed.). Elsevier Biomedical Press, Amsterdam.

F. M. Hershenson and W. H. Moos (1986). Drug development in senile cognitive decline. *J. Med. Chem. 29*: 1125–1130.

F. M. Hershenson, J. G. Marriott, and W. H. Moos (1986). Cognitive disorders. *Annu. Rep. Med. Chem. 21*: 31–40.

A. Karlin (1980). Molecular properties of the nicotinic acetylcholine receptor. In: *Cell Surface and Neuronal Function* (G. Poste, G. L. Nicolson, and C. W. Cotman, Eds.). Elsevier Biomedical Press, Amsterdam, pp. 191–260.

A. Karlin, P. N. Kao, and M. DiPaola (1986). Molecular pharmacology of the nicotinic acetylcholine receptor. *Trends Pharmacol. Sci. 7*: 304–308.

R. D. Keynes (1979). Ion channels in the nerve cell membrane. *Sci. Am. 240* (3): 126–135.

J. Kistler, R. M. Stroud, M. W. Klymowski, R. A. Lalancette, and R. H. Fairclough (1982). Structure and function of an acetylcholine receptor. *Biophys. J. 37*: 371–378.

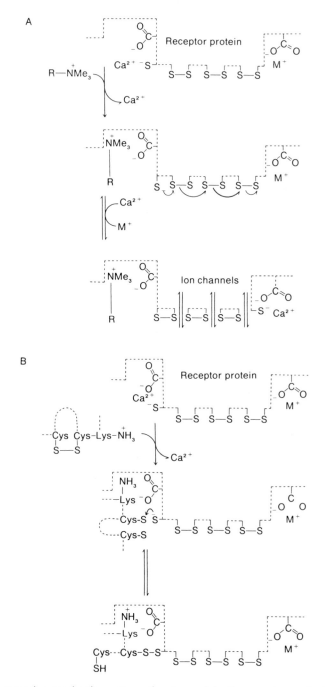

**Fig. 4.14.** A tentative mechanism attempting to explain the opening and closing of ion channels. (**A**) Agonist–receptor interaction opening the channels. (**B**) If snake neurotoxin binds instead of an agonist, the disulfide bonds are stabilized and the channels remain closed. (Reproduced by permission from Stenlake (1979), Elsevier/North Holland, New York)

M. P. McCarthy, J. P. Ernest, E. F. Young, J. Choe, and R. M. Stroud (1986). The molecular neurobiology of the acetylcholine receptor. *Annu. Rev. Neurosci. 9*: 383–413.

M. Noda, H. Takahashi, T. Tanabe, M. Toyosato, Y. Furutani, T. Hirose, M. Asak, S. Inayama, T. Miyata, and S. Numa (1982). Primary structure of α-subunit precursor of *Torpedo californica* acetylcholine receptor from cDNA sequence. *Nature 299*: 793–802.

K. Peper, R. J. Bradley, and F. Dryer (1982). The acetylcholine receptor at the neuromuscular junction. *Physiol. Rev. 62*: 1271–1340.

J. L. Popot and J.-P. Changeux (1984). Nicotinic receptor of acetylcholine: structure of an oligomeric integral membrane protein. *Physiol Rev. 64*: 1162–1239.

L. C. Sellin (1985). The pharmacological mechanism of botulism. *Trends Pharmacol. Sci. 6*: 80–82.

S. H. Snyder (1984). Drug and neurotransmitter receptors in the brain. *Science 224*: 22–31.

S. Spector, A. M. Felix, and J. P. M. Finberg (1978). Radio-immunoassay for acetylcholine. *J. Neurochem. 30*: 685–689.

J. B. Stenlake (1979). Molecular interactions at the cholinergic receptor in neuromuscular blockade. In: *Progress in Medicinal Chemistry* (G. P. Ellis and G. B. West, Eds.), Vol. 16. Elsevier/North Holland, New York, pp. 257–286.

L. Tauc (1982). Nonvesicular release of neurotransmitter. *Physiol. Rev. 62*: 857–893.

D. J. Triggle and C. R. Triggle (1976). *Chemical Pharmacology of the Synapse*. Academic Press, New York.

V. P. Whittaker (1986). The storage and release of acetylcholine. *Trends Pharmacol. Sci. 7*: 312–315.

### 2.3. The Muscarinic Acetylcholine Receptor

Even though the muscarinic receptor, which is present in postganglionic parasympathetic synapses, is much more stereospecific and structure-specific than its nicotinic counterpart, only since the early 1980s have any molecular studies been undertaken to explore similarities and differences between the two classes of AChR. Our knowledge of the structure, biochemical characteristics, and operation of the muscarinic cholinergic receptor is therefore quite vague, since there are no organs rich in this receptor, such as the electroplax, nor are there agents of high selectivity, such as the snake toxins, that can aid in its study. However, it has been labeled with the affinity label [$^3$H] **propyl-benzilylcholine-mustard** (4-8), which is specific for this receptor.

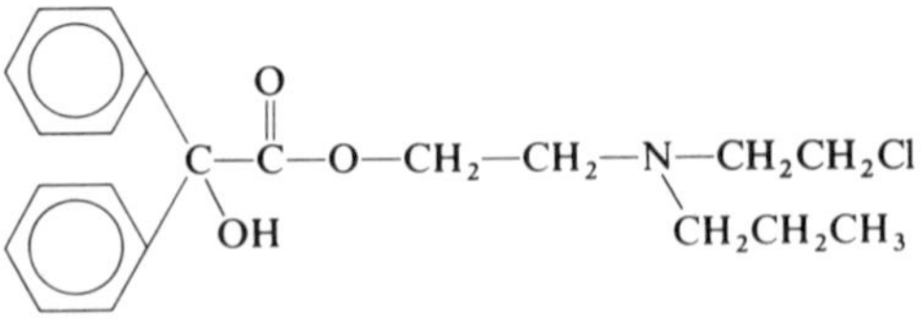

Propyl-benzilylcholine-mustard
4-8

The muscarinic receptor has recently been purified from muscle tissue (pig heart atria). A glycoprotein of mass 78 kD, it has a high- and a low-affinity binding site. In binding experiments, Hammer and Giachetti (1982) have distinguished two sub-types: the $M_1$ and the $M_2$ receptors. The discriminating ligand is **pirenzepine** (4-9) a tricyclic hydrophilic inhibitor of gastric secretion with no central effects, but with very high affinity for the $M_1$ receptor, which closes $K^+$ channels. The $M_2$ receptor is located in heart muscle, the cerebellum, and hindbrain; it is regulated by GTP and inhibits adenylate cyclase. It also seems to regulate a $K^+$ channel that is not directly connected to the recognition site (as is the nicotinic AChR) through an inhibitory GTP-binding protein (Noma, 1986). Recent investigations have shown that muscarinic receptor subtypes $M_1$ and $M_2$ (cardiac) are derived from different mRNAs. It seems, that an $M_2$ (glandular) receptor subtype also exists (Fukuda et al., 1987).

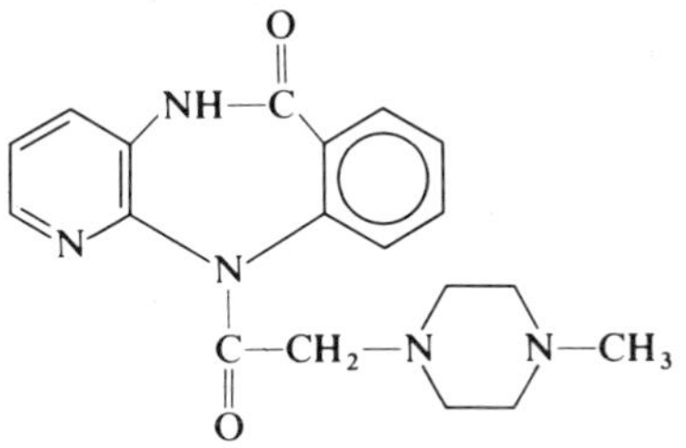

Pirenzepine
4-9

Heart muscle contraction is initiated by an increase in $Ca^{2+}$ ion concentration in the cytosol, where it binds to the protein troponin C, triggering actomyosin contraction. As discussed in the previous section, the cardiac muscle is regulated by muscarinic ACh receptors of the $M_2$ type which control diacylglycerol/inositol triphosphate $(DG/IP_3)$-regulated Ca channels. These channels can be blocked by a new and diverse group of drugs, the *Ca-channel blockers*, which revolutionized the treatment of disorders of heartbeat (arrythmia; see Chap. 6, Sec. 1.4) and vasospasms like angina pectoris as well as hypertension. Because these drugs act specifically on the $Ca^{2+}$ ion channel, they will be discussed in connection with ion channels as targets (Chap. 6, Sec. 1.2).

### Selected Readings

K. Fukuda, T. Kubo, I. Akiba, A. Maeda, M. Mishima, and S. Numa (1987). Molecular distinction between muscarinic acetylcholine receptor subtypes. *Nature 327*: 623–625.

R. Hammer and A. Giachetti (1982). Muscarinic receptor subtypes. *Life Sci. 33*: 2991–2994.

A. Noma (1986). GTP-binding proteins couple cardiac muscarinic receptors to potassium channels. *Trends Neurol. Sci. 10*: 142–143.

D. J. Triggle (1979). The muscarinic receptor: structural, ionic and biochemical implications. In: *Recent Advances in Receptor Chemistry* (F. Gualtieri, M. Gianella, and C. Melchiorre, Eds.). Elsevier/North Holland, New York, pp. 127–146.

J. C. Venter (1983). Muscarinic receptor structure. *J. Biol. Chem. 258*: 4842–4848.

## 2.4. Cholinergic Agonists

Increased stimulation of the AChR can be achieved in two ways: (1) by binding of the *directly acting* cholinergic agonists to the AChR, triggering nicotinic or muscarinic effects, or both; and (2) by binding of the *indirect agonists*, which are drugs that inhibit the hydrolysis of ACh by AChE, thus prolonging the action of available ACh. These drugs act as enzyme inhibitors, and will be discussed in Chap. 6. Sec. 3.2.

**Fig. 4.15.** The principal directly acting cholinergic agonists.

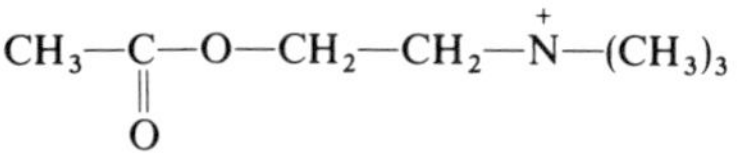

$$CH_3-\underset{\underset{O}{\|}}{C}-O-CH_2-CH_2-\overset{+}{N}-(CH_3)_3$$

4-10   Acetylcholine

$$CH_3-\underset{\underset{O}{\|}}{C}-O-\underset{\underset{CH_3}{|}}{CH}-CH_2-\overset{+}{N}-(CH_3)_3$$

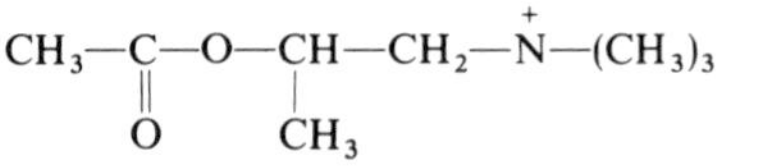

4-11   (+)-Acetyl-β-methylcholine (methacholine)

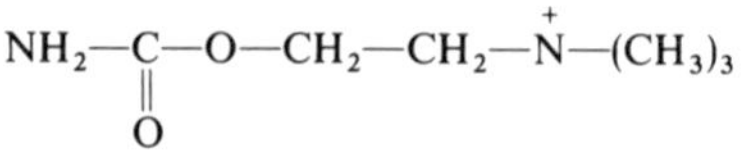

$$NH_2-\underset{\underset{O}{\|}}{C}-O-CH_2-CH_2-\overset{+}{N}-(CH_3)_3$$

4-12   Choline carbamate (carbachol)

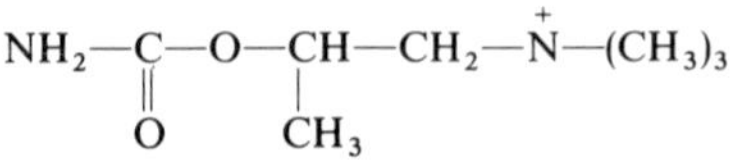

$$NH_2-\underset{\underset{O}{\|}}{C}-O-\underset{\underset{CH_3}{|}}{CH}-CH_2-\overset{+}{N}-(CH_3)_3$$

4-13   Carbamyl-β-methylcholine (betanechol)

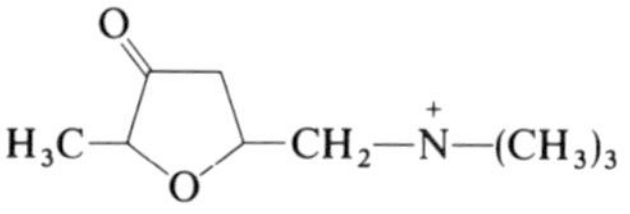

4-14   Muscarone

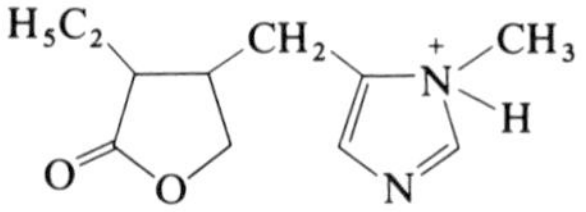

4-15   Pilocarpine

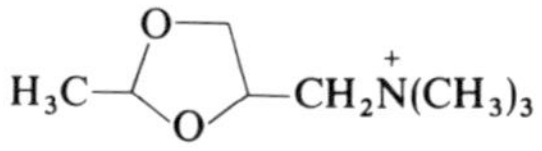

4-16   2-Methyl-4-trimethyl-ammonium-methyl-1,3-dioxolane

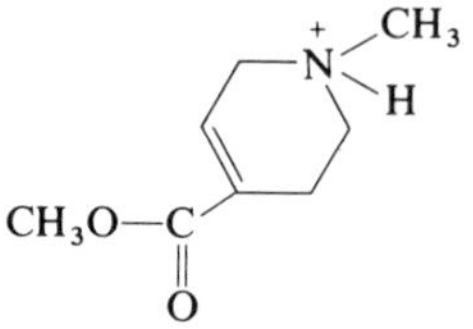

4-17   Arecoline

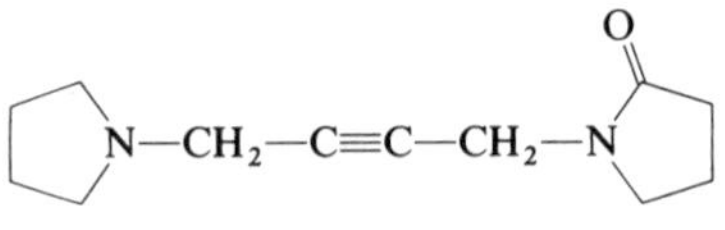

4-18   Oxotremorine

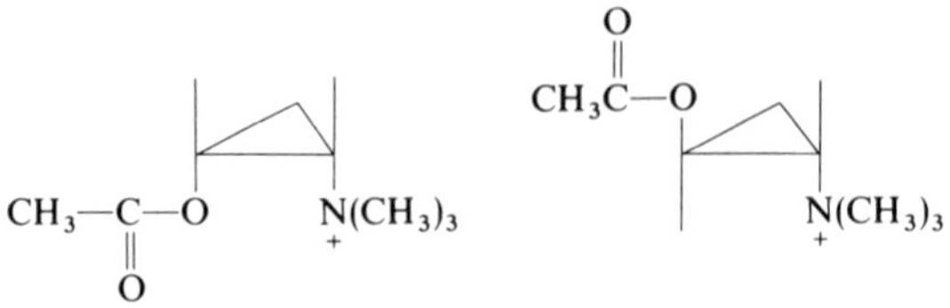

4-19   Acetoxycyclopropyl-trimethylammonium

cis-                    trans-

The principal directly acting cholinergic agonists are shown in Fig. 4.15.

**Acetylcholine** (4-10) has, of course, both nicotinic and muscarinic action. Because it is very rapidly hydrolyzed by AChE and even by aqueous solution, it is not used therapeutically.

**Metacholine** (4-11) is hydrolyzed somewhat more slowly than acetylcholine because of steric hindrance of the ester by the α-methyl group. Its activity is mainly muscarinic, but it is infrequently used.

**Carbachol** (4-12) is a very potent agent because it is not an ester but a carbamate, and is hydrolyzed slowly. It is used in glaucoma to reduce intraocular pressure.

**Betanechol** (4-13) also has a prolonged effect, and finds application in stimulation of the gastrointestinal tract and urinary bladder (both muscarinic effects) to relieve postoperative atony.

The other cholinergic agonists in Fig. 4.15 have no therapeutic use. **Muscarine** (4-3) is an alkaloid of the mushroom *Amanita muscaria*; **muscarone** (4-14) is its semisynthetic analogue. **Pilocarpine** (4-15) is found in the leaves of a shrub and can be used to increase salivation or sweating. **Arecoline** (4-17) is also an alkaloid, and occurs in the betel nut used as a mild euphoriant in India and Southeast Asia. Finally, **oxotremorine** (4-18) is a purely synthetic experimental agent that produces tremors and is helpful in the study of Parkinson's disease and antiparkinsonian drugs whereas the **dioxolane** (4-16) is a muscarine analogue.

### 2.4.1 Structural Modifications of Acetylcholine

Structural modifications of acetylcholine fall into four categories: (1) changes in the quaternary ammonium group; (2) changes in the ethylene chain; (3) changes in the ester group; and (4) the creation of cyclic analogues of the neurotransmitter.

*Ammonium group.* The ammonium group of ACh can be replaced by other "-onium" compounds (phosphonium, arsonium, or sulfonium), but only with the loss of 90% of the activity. One of the methyl groups on the ammonium can be exchanged for larger alkyl residues: for instance, the dimethylethyl derivative is about 25% active. However, the insertion of larger groups or the replacement of more than one methyl leads to an almost complete loss of activity. This finding implies that the size of the quaternary ammonium group and its charge distribution are important to the activity of ACh, since the hydrophobic auxiliary binding site next to the anionic site of the receptor is optimized for two methyl groups. strengthening the ionic interaction. The uncharged carbon analogue **3,3-dimethyl-butyl acetate** has only 0.003% activity, as shown by Burgen, who also calculated the standard free energy ($\Delta G^0$) contribution of the ammonium group to the overall binding of ACh, finding that it amounts to $-20.9$ kJ/mol. However, this energy of binding depends upon overall structure. A detailed discussion of this topic can be found in the monograph of Triggle and Triggle (1976). It is interesting to note that many muscarinic agonists are *tertiary amines*—for example, **pilocarpine** (4-15), **arecoline** (4-17), and **oxotremorine** (4-18). At physiological pH, however, these amines are likely to be protonated and to occur in rigid ring structures. In this way, hydrogen bonding between the protonated amino group and the —COO⁻ group of the anionic site of the receptor is not prevented by the intramolecular interaction of the —C=O and —$NH(CH_3)_2$ groups in the ligand. Such a bond would completely distort the ligand conformation and prevent normal binding to the AChR.

*Ethylene chain.* The ethylene bridge of ACh ensures the proper distance between the ammonium group and the ester group, and is therefore critical in binding to the receptor. Although it is rather dangerous to assign a definite distance between the -onium and ester groups (estimated at about 0.6 nm), the "Rule of Five," proposed by the British pharmacologist Ing, states that there should be no more than four atoms between the $N^+$ and the terminal methyl group. Lengthening of the chain results in rapidly decreasing activity; interestingly, however, the 2-butyne analogue has 50% ACh activity. If the ethylene is branched, only methyl groups are allowed, as shown in the muscarinic agonist **methacholine** (4-11). The α-methyl analogue of ACh has more nicotinic activity.

*Ester Group.* The ester group does not lend itself to much modification either. Large aromatic acid moieties in the ester produce ACh antagonists rather than agonists, some of which are useful as anticholinergic agents (see Sec. 2.5). If ethers and ketones replace the ester, some activity is retained. The only useful replacement for the acetate has been a carbamate group, resulting in **carbachol** (4-12), which is highly active because of its slow hydrolysis.

*Cyclic Analogues of ACh.* Cyclic ACh analogues include the naturally occurring agonists **muscarine** (4-3), **pilocarpine** (4-15), and **arecoline** (4-17), all of which are muscarinic compounds. **Dioxolanes** such as (4-16) are muscarinic analogues of very high potency. The *cis* and *trans* forms of **acetoxycyclopropyl-trimethylammonium iodide** (4-19), mentioned in Chap. 1, Sec. 5, were used in an attempt to determine the conformational requirements of nicotinic and muscarinic activity. The *trans* isomer showed about the same muscarinic effect as ACh itself.

### 2.4.2. Mode of Binding of Acetylcholine

The mode of ACh binding is still somewhat enigmatic despite all of the foregoing findings. The molecule is highly flexible, and its preferred conformation is therefore hard to define. From x-ray crystallographic studies, Chothia and Peter Pauling suggested that ACh acts as the *gauche* conformer (Fig. 4.16), and it is possible to distinguish a methyl side and a carbonyl side, corresponding, respectively, to the muscarinic and nicotinic actions of the molecule. However, this is merely an approximation, for although the muscarinic activity is quite specific, steric parameters are rather irrelevant to the action of nicotinic agonists.

It is generally accepted that the ammonium group of ACh binds ionically to a carboxylate anion of glutamate or aspartate on the receptor, aided by van der Waals interaction of the methyl groups with the adjacent hydrophobic accessory binding site. About 0.59 nm removed from this, the carbonyl group forms a hydrogen bond with an acceptor, perhaps with histidine, as in acetylcholinesterase. In muscarinic agonists a third binding point, involving the methyl group of the acetate, may

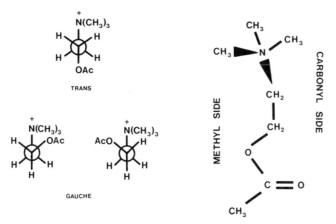

**Fig. 4.16.** Conformation of acetylcholine. According to suggestions by Chothia and Pauling, the methyl side of the gauche conformer would bind preferentially to the muscarinic receptor, whereas the carbonyl side would favor the nicotinic recognition site.

assume increased significance, as shown in Fig. 4.17. Whereas the primary structural requirements for nicotinic agonists are a quaternary ammonium and a carbonyl group, the muscarinic agonists are characterized by an ammonium and a methyl group. Recent calculations (Snyder, 1985) have confirmed that the carbonyl group is the primary hydrogen-binding site in both nicotinic and muscarinic receptors.

**Fig. 4.17.** Binding of acetylcholine to the nicotinic receptor: hydrophobic groups bind two methyl groups, glutamate forms an ionic bond with the ammonium ion, and histidine holds the ester carbonyl through hydrogen bonding. The acetyl methyl will bind preferentially to the muscarinic receptor.

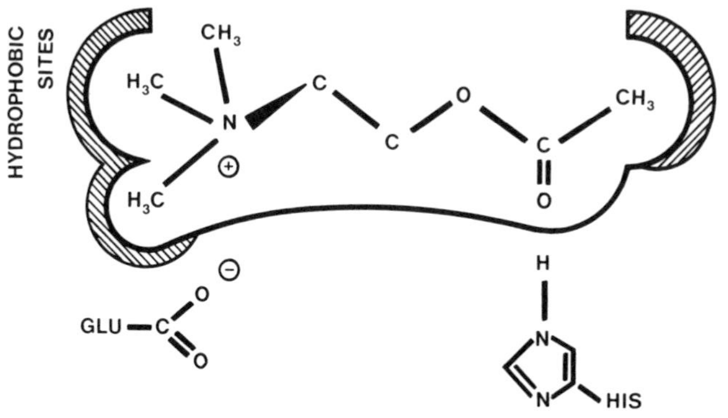

### Selected Readings

J. P. Snyder (1985). Molecular models for muscarinic receptors. *Trends Pharmacol. Sci.* 6: 464–466.

D. J. Triggle and C. R. Triggle (1976). *Chemical Pharmacology of the Synapse.* Academic Press, New York.

## 2.5. Cholinergic Blocking Agents

The peripheral cholinergic synapses (other than neuromuscular endplates) are muscarinic. Drugs that inhibit the interaction of ACh with the AChR are cholinergic blocking agents (or parasympatholytics) and must not be confused with the ganglionic and neuromuscular blocking agents, which act on nicotinic receptors. Anticholinergic agents decrease the secretion of saliva and gastric juice, decrease gastrointestinal and urinary tract peristalsis, and dilate the pupils. Consequently, they are used in treating such gastrointestinal diseases as peptic ulcer, in ophthalmology, and in the treatment of Parkinson's disease (see Sec. 4.2). The last-named use is based on the cholinergic hyperactivity brought about by the loss of central dopaminergic inhibition in parkinsonism, which is controlled by anticholinergic agents.

The oldest anticholinergics are the tropane alkaloids of *Atropa belladonna* (nightshade). **Atropine** (4-20) and **scopolamine** (4-21) are derivatives of tropine, a fused piperidino-pyrrolidine ring system, esterified by atropic acid. Atropine is the

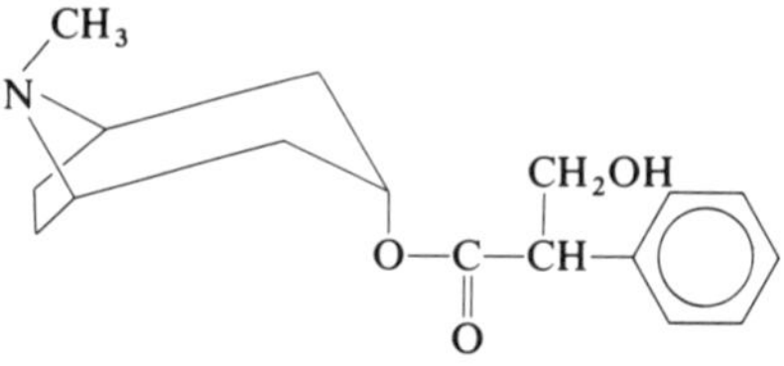

4·21   (±)Atropine
(−)Hyosciamine
4-20

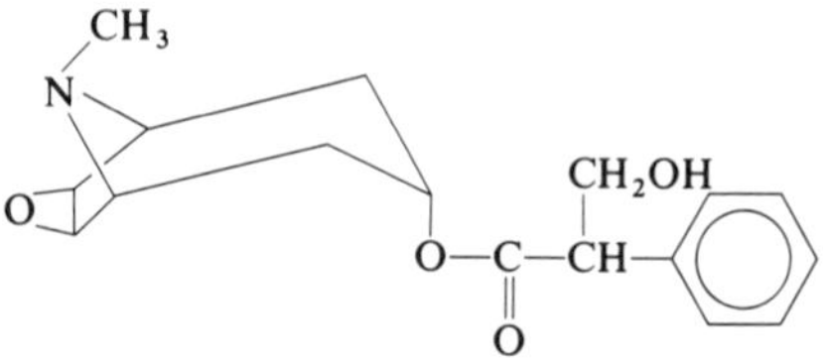

Scopolamine
4-21

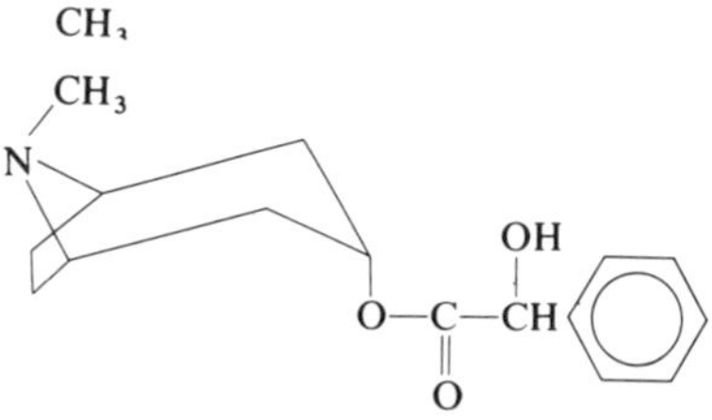

Homatropine
4-22

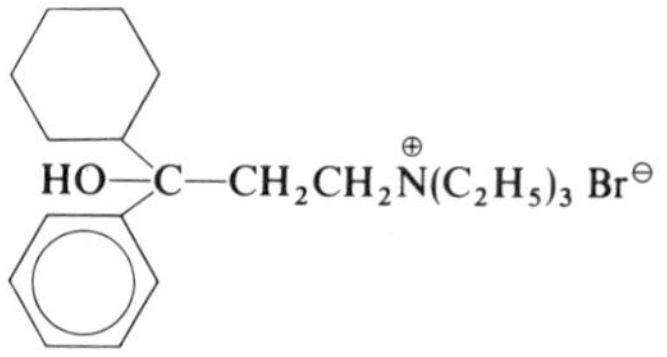

Tridihexethyl bromide
4-23

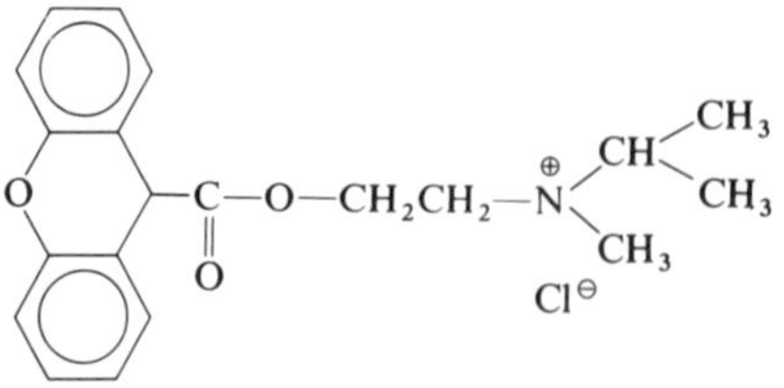

Propantheline chloride
4-24

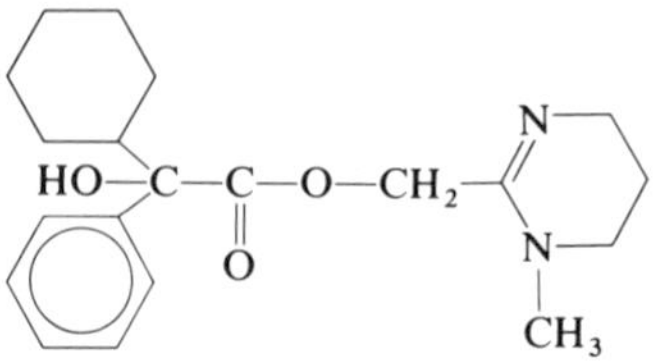

Oxyphencyclimine
4-25

racemate of (−)hyosciamine, whereas scopolamine has an epoxide ring. In large doses, all of these anticholinergic agents have central excitatory and hallucinogenic effects, and were prominent in medieval "witches' brews." A synthetic homologue, **homatropine** (4-22), has a shorter duration of action. All of these agents are mixed $M_1-M_2$ antagonists.

These tropine derivatives are esters of tertiary bases with a bulky acid component (atropic acid, mandelic acid). In general, a number of cholinergic blocking agents have been developed by substituting a larger acid for the acetyl group of ACh and increasing the size of the N-substituents. Among the quaternary compounds, **tridihexethyl bromide** (4-23) and **propantheline bromide** (4-24) are notable; among the tertiary amines **oxyphencyclimine** (4-25) shows high activity. Numerous analogues are known and are used therapeutically (e.g., methylatropine and methylscopolamine). **Pirenzepine** (4-9), as mentioned in Sec. 2.3, is a selective $M_1$ antagonist, ten times more active than histamine $H_2$ antagonists (Sec. 6.3) in the treatment of gastric ulcers. Several derivatives are known.

### 2.5.1. Ganglionic Blocking Agents

Ganglionic blocking agents interfere with the nicotinic ACh receptors in the ganglia. Although ganglia are, functionally, normal receptors, they probably differ structurally from the receptors at the neuromuscular endplate, and show different accessibility. Therefore, ganglionic and neuromuscular blocking agents are two structurally different groups of anticholinergic drugs.

The ganglionic blockade by **tetraethylammonium salts** (4-26) has been known for a long time; however, the prototype blocking agent is **hexamethonium** (4-27), a bisquaternary compound with six methylene groups separating the two cationic groups. Some secondary and tertiary amines such as **trimetaphan** (4-28) and **mecamylamine** (4-29) were in use at one time, since they had longer durations of action in controlling hypertension by decreasing vasoconstriction. However, because none of these compounds can distinguish sympathetic from parasympathetic ganglia, they have numerous side effects. Consequently, they have largely been replaced by the more selective $\beta$-adrenergic blocking agents (Sec. 3.6.4).

$$CH_3-CH_2-\overset{\oplus}{N}(C_2H_5)_3 \qquad (CH_3)_3-\overset{\oplus}{N}-(CH_2)_6-\overset{\oplus}{N}(CH_3)_3$$

Tetraethylammonium salts
4-26

Hexamethonium
4-27

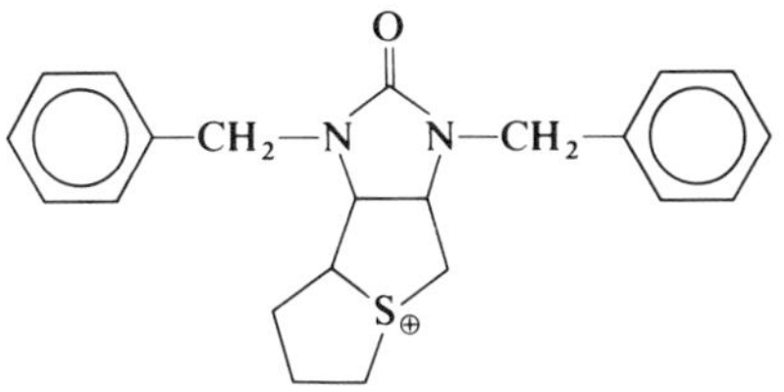

Trimetaphan
4-28

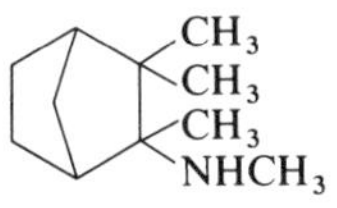

Mecamylamine
4-29

### 2.5.2. Neuromuscular Blocking Agents

Neuromuscular blocking agents are widely used in surgery. They are capable of relaxing the abdominal muscles without the use of deep anesthesia, and make surgery much easier for both the surgeon and patient. There are two major categories of such agents: (1) competitive agents, which occupy the same site as ACh; and (2) depolarizing blocking agents, which mimic the action of ACh but persist at the receptor.

***Competitive Agents.*** Competitive neuromuscular blocking agents were developed through the study of *curare*, the arrow poison of South American Indians. Crude curare contains a number of isoquinoline and indole alkaloids, the best known of which is **tubocurarine** (4-30), a tertiary–quaternary amine in which the distance between the two cations is rigidly fixed at about 1.4 nm (the "curarizing distance"). A similarly rigid, large molecule is the synthetic steroid derivative **pancuronium** (4-31), a bisquaternary derivative with an $N^+-N^+$ distance of 1.1 nm. Two acetylcholine molecules built into a rigid framework are clearly discernible. With curarization, the neuromuscular junction becomes insensitive to ACh and the motor nerve impulse, and the endplate potential falls dramatically. The curarizing agents probably block the ionophore in its closed, resting form by occupying several receptor sites. Numerous bulky analogues are known (Bowman, 1986).

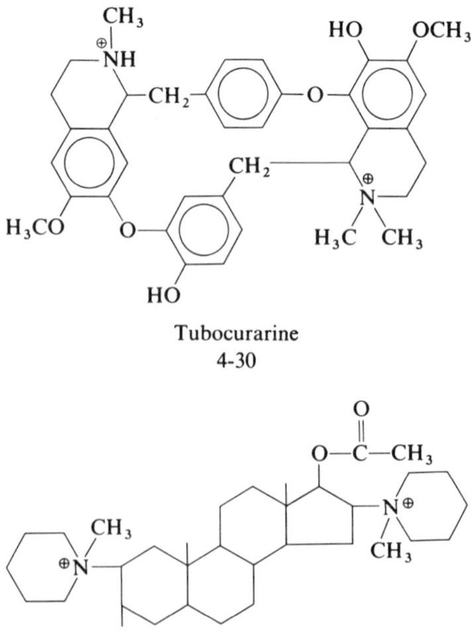

Tubocurarine
4-30

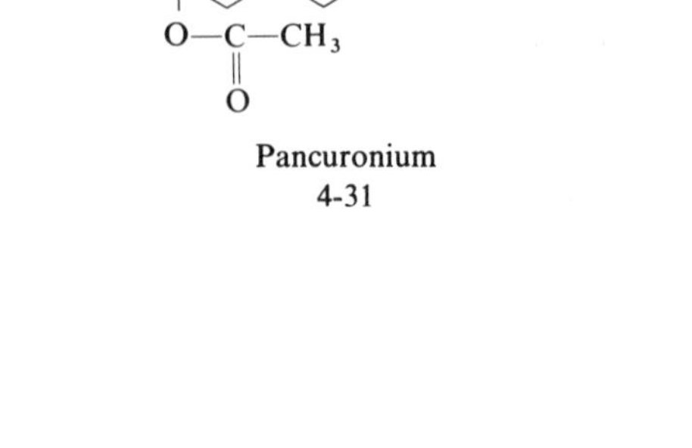

Pancuronium
4-31

*Depolarizing Agents.* The depolarizing neuromuscular blocking agents were discovered through mimicking the $N^+-N^+$ distance described above with aliphatic compounds. **Decamethonium** (4-32), in an extended conformation, approximates this distance, and is the prototype of the depolarizing blocking agents. This drug binds normally to the AChR and triggers the same response as does ACh—a brief contraction of the muscle—which, however, is followed by a prolonged period of transmission blockage accompanied by muscular paralysis. A related compound, **succinylcholine** (succamethonium) (4-33), has the same $N^+-N^+$ distance, even though the 10 intervening atoms are not all carbon. It has a short, self-limiting action since it is easily hydrolyzed by serum cholinesterase (see Chap. 6, Sec. 3.2). Besides depolarizing muscle, both compounds depolarize autonomic ganglia.

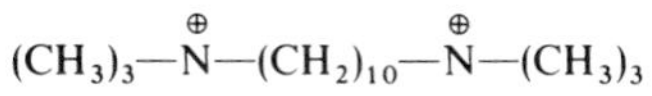

$(CH_3)_3\overset{\oplus}{N}-(CH_2)_{10}-\overset{\oplus}{N}-(CH_3)_3$

Decamethonium
4-32

$(CH_3)_3\overset{\oplus}{N}-CH_2-CH_2-O-\overset{O}{\underset{}{\overset{\|}{C}}}$

$CH_2$

$CH_2$

$(CH_3)_3\overset{\oplus}{N}-CH_2-CH_2-O-\underset{O}{\overset{}{\underset{\|}{C}}}$

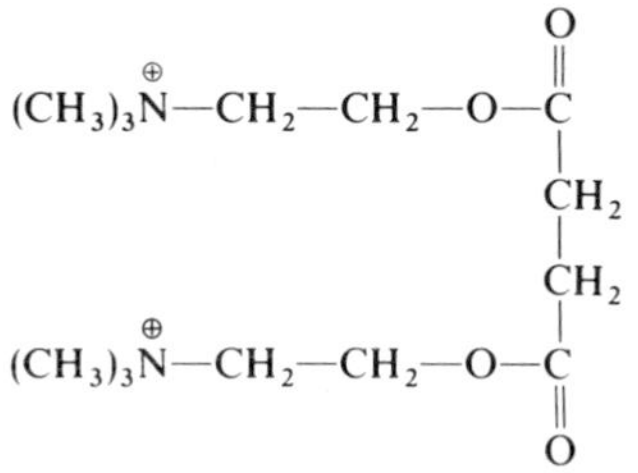

Succinylcholine
4-33

*Structure–Activity Correlations.* The structure–activity correlations of many analogous neuromuscular blocking agents are treated extensively by Triggle and Triggle (1976). The most interesting aspect of these correlations is that between the $N^+-N^+$ distance and the receptor structure. As the number of atoms between the -onium groups is increased beyond 10, the activity decreases until a second peak is reached at around the 16-atom distance [**hexacarbacholine** (4-34) and related compounds], which corresponds to a distance of about 2 nm. It is not necessarily the $N^+-N^+$ distance that is essential; any *induced* positive charge will be appropriate. The *p*-**nitrobenzyl-hexamethonium** chloride derivative (4-35), for instance, carries

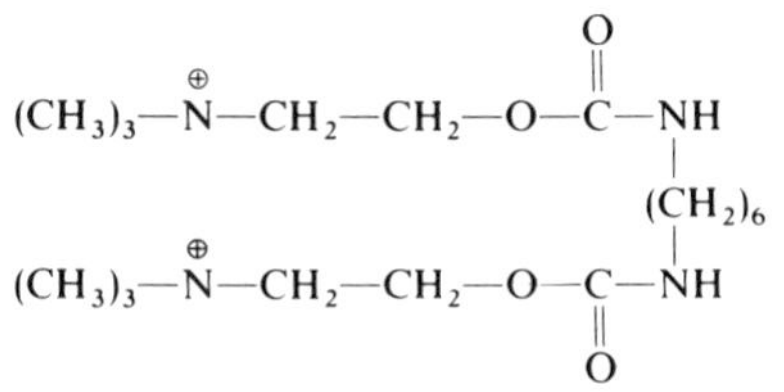

$(CH_3)_3-\overset{\oplus}{N}-CH_2-CH_2-O-\overset{O}{\overset{\|}{C}}-NH$

$(CH_2)_6$

$(CH_3)_3-\overset{\oplus}{N}-CH_2-CH_2-O-\underset{O}{\underset{\|}{C}}-NH$

Hexacarbacholine (Imbretil)
4-34

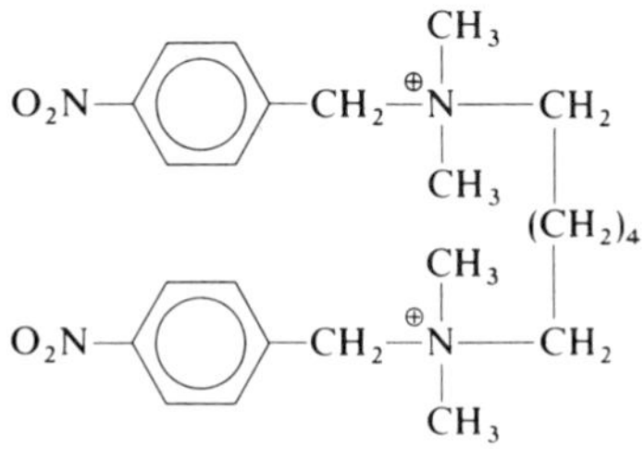

$O_2N-\langle\bigcirc\rangle-CH_2-\overset{CH_3}{\overset{|}{\underset{|}{\overset{\oplus}{N}}}}-CH_2$

$CH_3$

$(CH_2)_4$

$CH_3$

$O_2N-\langle\bigcirc\rangle-CH_2-\overset{|}{\underset{|}{\overset{\oplus}{N}}}-CH_2$

$CH_3$

4-35

the positive charge on its two phenyl rings rather than on formal cationic ammonium ions; this is due to the electron-attracting nitro groups of this compound, which, together with the ammonium ions, dramatically decrease the $\pi$-electron density of the rings. The induced charge distance increases to about 2 nm, and the hexamethonium derivative is therefore inactive as a ganglionic blocker but becomes a very effective curarizing agent. It is interesting to note that lower invertebrates (cladocerans, annelid worms, rotifers), are more sensitive to compounds with an $N^+-N^+$ distance of 16 than those with an $N^+-N^+$ distance of 10, whereas in animals of phylogenetically higher taxa, such as mammals, this sensitivity is reversed. Recent developments in the field of neuromuscular blocking agents are reviewed by Kharkevich (1981) and by Bowman (1986).

At first suggested by Avram Goldstein, the nicotinic AChR in the ganglion cell and the neuromuscular endplate are different. The difference probably consists of dissimilar accessory sites comprising a negative charge. In addition, the neuromuscular site can accommodate not only compounds with an $N^+-N^+$ distance of 10 atoms but also compounds with an $N^+-N^+$ distance of 16 atoms. This may indicate the interaction of such drugs with two instead of just one receptor. Several authors have proposed, on purely hypothetical grounds, receptor lattice arrays accommodating such a multiplicity of drugs.

While these models of the AChR are intellectually pleasing, they are purely hypothetical. Electron-microscopic evidence points to arrangements of 8-nm-diameter receptor rosettes in a pseudohexagonal array, with a 9- to 10-nm distance between the arrays. Assuming an $\alpha_2\beta\gamma\delta$ receptor composition, with AChR sites on the $\alpha$ subunits (see Fig. 4.12), it is hard to see how the 2-nm-long hexadecamethonium could cross-link two or four receptor sites. Clearly the location of accessory sites on the receptor remains unknown, although one indication of this could be the distance of 0.9–1.2 nm between the —SH group and the negative charge.

The Stroud group (Kistler et al., 1982) has suggested more realistic binding sites for bisquaternary compounds based on the amino acid sequences of the four AChR subunits and their arrangement in the membrane (Fig. 4.18). The snake toxins (bungarotoxin, histrionicotoxin, cobra venoms), although all potent neuromuscular blocking agents, are not used therapeutically.

### Selected Readings

B. C. Bowman (1986). Mechanisms of action of neuromuscular blocking drugs. In: *Mechanisms of Drug Action* (G. N. Woodruff, Ed.), Vol. 1. Macmillan, London, pp. 65–96.

D. A. Kharkevich (1981). Main trends in the search for new neuromuscular blocking agents. *Trends Pharmacol. Sci. 2*: 218–220.

J. Kistler, R. M. Stroud, M. W. Klymkowsky, R. A. Lalancette, and R. H. Fairclough (1982). Structure and function of an acetylcholine receptor. *Biophys. J. 37*: 371–378.

D. J. Triggle and C. R. Triggle (1976). *Chemical Pharmacology of the Synapse.* Academic Press, New York.

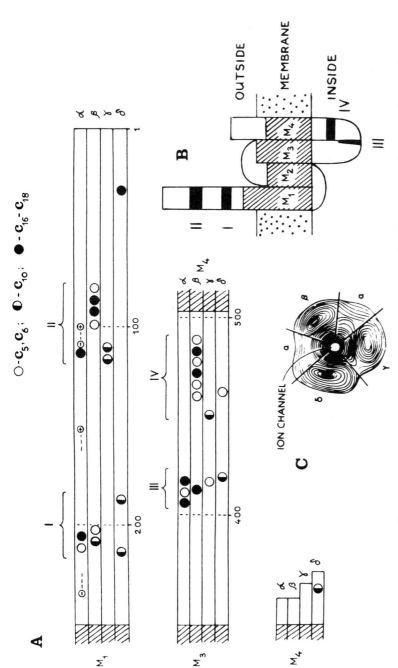

**Fig. 4.18.** Hypothetical binding sites for bisquaternary ammonium compounds (BACs) in the molecule of the electric organ acetylcholine receptor. (**A**) Amino acid sequences of four subunits ($\alpha$, $\beta$, $\gamma$, and $\delta$) aligned for maximal homology. They are concentrated in four areas, indicated I–IV. Two hypothetical sites of ACh binding to the $\alpha$ subunit are indicated as $\ominus$---$\oplus$. (**B**) A model of the electric organ acetylcholine receptor incorporated in the membrane. The intrabilayer portions of the polypeptide chain whose bounds are shown in A ($M_1$–$M_4$) and binding areas shown in A (I–IV) are indicated. (**C**) The electric organ acetylcholine receptor as viewed normal to the membrane, according to the results of electron microscopy and x-ray analysis. The subunits $\alpha$, $\beta$, $\gamma$, and $\delta$ are indicated. (Reproduced by permission from Kistler et al., 1982)

## 3. NOREPINEPHRINE AND THE ADRENERGIC RECEPTORS

### 3.1. The Adrenergic Neuronal System

The adrenergic system, also known as the sympathetic nervous system, is found both peripherally and centrally. As shown in Fig. 4.6, myelinated B-fibers originate in the spinal cord and meet ganglion cells remote from the effector organ. Long, unmyelinated C-fibers then transmit the impulse along the adrenergic axon from the ganglion to the synapses. The synapses, which are thickenings of the axon called *varicosities*, often differ in shape from the synaptic boutons (shown in Fig. 4.2). In the CNS, ganglia may be missing between the adrenergic cell body and the synapse or varicosity.

*Peripherally*, all organs are innervated sympathetically (as well as parasympathetically), and in most cases the adrenergic action of this system is opposite to the cholinergic effects. The neurotransmitter secreted by the nerve endings is norepinephrine and, to a lesser extent, epinephrine.

*Centrally*, two systems can be distinguished:

1. The noradrenergic pathways, primarily situated in the *locus ceruleus*—a deeply pigmented (hence the name, alluding to its blue color) small cell group involved in behavior, mood, and sleep. The cortex, some thalamic and hypothalamic centers, and the cerebellar cortex are innervated from here. The noradrenergic pathways of the *lateral tegmentum* are less well known.
2. The adrenergic pathways that use epinephrine as a neurotransmitter, which have been explored only recently. One of these systems is also tegmental, and mixed with noradrenergic cells. The other is thalamic–hypothalamic, involved with the vagus nerve. Some adrenergic fibers are also found in the fourth ventricle and the spinal cord.

Elucidation of these two sets of pathways, which are usually intermingled with several other neuronal systems, became possible through the application of some ingenious methods. Fluorescence microscopy shows adrenergic nerve endings when histological sections are treated with formaldehyde vapors or glyoxylic acid. Immunocytochemical methods can be used to identify enzymes localized in specific structures that are involved in neurotransmitter metabolism. The functional role of adrenergic and noradrenergic neurons can be assessed by selective destruction of these neurons. A "chemical sympathectomy" (equivalent to the surgical destruction of major sympathetic neurons) can be achieved with **6-hydroxydopamine** (3,4,6-trihydroxy-$\beta$-phenylethylamine; 4-36) and with **xylamine** (4-37) and analogues (Cho and Takimoto, 1985). These compounds deplete norepinephrine in sympathetically innervated organs (except the adrenal medulla) as well as in the CNS, and irreversibly destroy sympathetic terminals and neurons (Jonson, 1980). A similar peripheral "immunosympathectomy" can be achieved with antisera to the nerve growth factor protein. As expected, functional changes in adrenergically innervated structures become evident within a short time after the administration of such antisera. With these methods both the anatomical distribution and the functional role of sympathetic neurons have been elucidated. Histological or electron-microscopic investigations alone could not have achieved this.

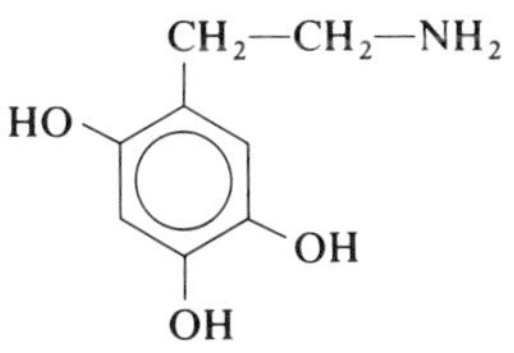

6-Hydroxydopamine          Xylamine
4-36                       4-37

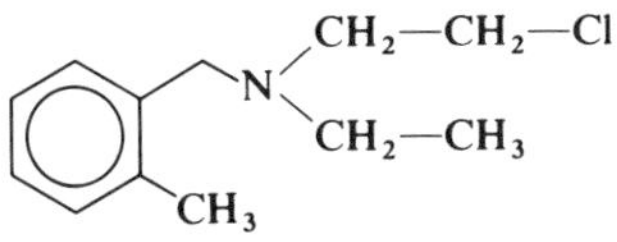

### 3.2. Biosynthesis of Catecholamine Neurotransmitters

The adrenergic system produces neurotransmitters belonging to the class of substances known as *catecholamines*. These are derivatives of **catechol** (4-38) (*o*-dihydroxybenzene), with an $\beta$-aminoethyl side chain. The biogenetically related catecholamines and the pathways leading to their biosynthesis are shown in Fig. 4.19. Starting with **tyrosine** (4-39), the main pathway goes through **dihydroxy-phenylalanine** (4-40) (DOPA), **dopamine** (4-41) (DA), **norepinephrine** (4-42) (NE, also called noradrenaline in the European literature), and finally **epinephrine** (4-43) (E, or adrenaline).

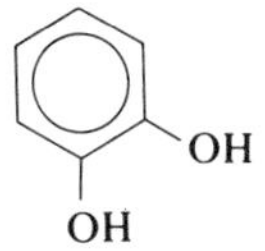

Catechol
4-38

While dopamine is an intermediate for NE and E, it is also a neurotransmitter in its own right. Dopamine and the dopaminergic receptor, as well as drugs that act on it, will be discussed in Sec. 4.

### 3.2.1. Key Enzymes

The enzymes involved in catecholamine biosynthesis have been studied intensively and are the targets of many drugs, as we shall see later. The key enzyme is *tyrosine hydroxylase*, which requires a tetrahydrofolate coenzyme, $O_2$, and $Fe^{2+}$, and is quite specific. As usual for the first enzymes in a biosynthetic pathway, tyrosine hydroxylase is rate limiting, and is therefore the logical point for the inhibition of NE synthesis. *DOPA decarboxylase* acts on all aromatic amino acids and requires pyridoxal phosphate (vitamin $B_6$) as a cofactor. *Dopamine $\beta$-hydroxylase*, located in the membranes of storage vesicles, is a copper-containing protein—a mixed-function oxygenase that uses $O_2$ and ascorbic acid. Finally, *phenylethanolamine*

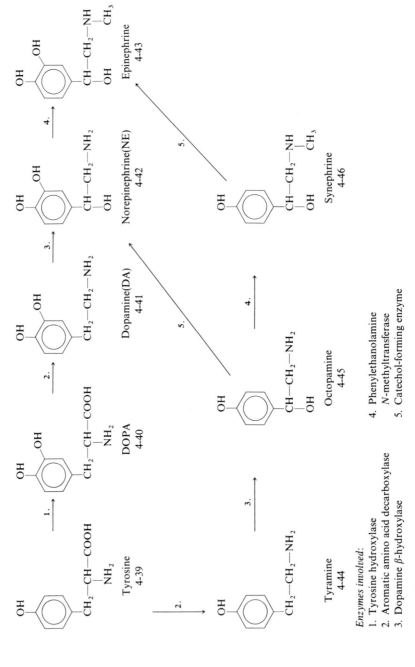

**Fig. 4.19.** Biosynthesis of catecholamines.

*Enzymes involved:*
1. Tyrosine hydroxylase
2. Aromatic amino acid decarboxylase
3. Dopamine β-hydroxylase
4. Phenylethanolamine N-methyltransferase
5. Catechol-forming enzyme

N-*methyltransferase*, located in the adrenal medulla (the main site of epinephrine synthesis) and in the brain, uses S-adenosyl-methionine as a methyl donor.

### 3.2.2. Catecholamine Storage

Catecholamine storage utilizes synaptic vesicles of different sizes in different organs. The largest ones (up to 120 nm) are found in the adrenal medulla and are called chromaffin granules. Catecholamines are stored as their ATP complexes, in a proportion of 4:1, in association with the acidic protein *chromogranin*. This keeps the neurotransmitter in a hypoosmotic form even though its concentration is very high (up to 2.5 M), and also protects it from enzymatic oxidation by monoamine oxidase. The vesicles also contain the enzyme dopamine $\beta$-hydroxylase, proof that NE is synthesized in the vesicle. The vesicles themselves are formed in the cell body and are transported along the axon to the terminal region.

According to recent investigations (Sneddon et al., 1982), ATP acts as a co-transmitter in adrenergic synapses. This corroborates the existence of purinergic receptors (Stone, 1982) and the novel concept of multiple neurotransmitters at a single nerve ending, as outlined in Sec. 1.4. Adenosinergic transmission will be discussed in Sec. 7.

### 3.2.3. Catecholamine Release

The release of catecholamines has been studied mainly in the adrenal medulla, which is analogous to the nerve cell. Our knowledge of events in the peripheral and especially in the central nerve ending is less extensive. In the medulla, the neuronal impulse releases ACh (embryologically the medulla is a modified ganglion and therefore uses ACh as a transmitter). This allows the inflow of $Ca^{2+}$, which triggers fusion of the chromaffin cell membrane with the secretory vesicle, resulting in exocytosis of the entire vesicle contents, including all of the vesicle proteins. Whether neurotransmitter release in noradrenergic varicosities follows the same mechanism is by no means certain. However, exocytosis seems to be a slow and wasteful mechanism in neurons and is being seriously questioned as a concept. Alternative possibilities for the release are shown in Fig. 4.20. As in cholinergic neurons, the freshly synthesized or recycled NE is released first. These facts suggest three different neurotransmitter pools: the vesicular or bound form and two labile pools. Cooper et al. (1986) weigh the pros and cons of these hypotheses.

The release and turnover of catecholamines is subject to complex regulation, the most important type of which is modulation by *presynaptic receptors*. Adrenergic agonists acting on these receptors will decrease—whereas antagonists will increase—neurotransmitter release, and also seem to have an effect on regulating neurotransmitter synthesis. In addition, prostaglandins of the E (PGE) series are potent inhibitors of neural NE release through a feedback loop involving $Ca^{2+}$ ions.

In his review, Langer (1981) described the presence of presynaptic receptors responding to PGE, neuropeptide Y (NPY), enkephalins (Chap. 5, Sec. 3.2), dopamine, muscarinic agonists, and angiotensin (Chap. 5, Sec. 2.8), in addition to adrenergic $\alpha$ and $\beta$ agonists—all of which play a presumably regulatory role.

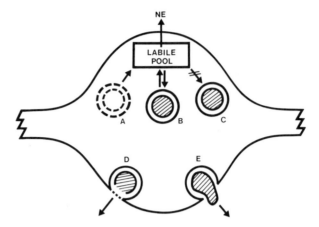

**Fig. 4.20.** Alternative hypotheses of norepinephrine release from an adrenergic varicosity (synapse): (A) Dissolution of vesicle membrane; (B) release into cytosol; and (C) inhibition of NE reuptake into the vesicle could all feed into a labile cytoplamic pool. (D) Release of neurotransmitter through a gap junction; (E) classical exocytosis. (After Trifaro and Cubbedu, 1979)

Acetylcholine and cAMP also seem to regulate catecholamine release. As outlined in Sec. 1.3, these presynaptic heteroreceptors are more likely to have a regulatory role in adrenergic synapses than are the autoreceptors.

### 3.2.4. Catecholamine Metabolism and Reuptake

The metabolism of catecholamines is much slower and more complex than that of ACh. The degradative pathways are shown in Fig. 4.21. The principal, although nonspecific, enzyme in the degradation is *monoamine oxidase* (MAO), which dehydrogenates aliphatic amines in the reaction:

$$R-CH_2-NH_2 \longrightarrow R-CH=NH \xrightarrow{H_2O} R-C{\overset{H}{\underset{O}{\diagup}}} + NH_3$$

The intermediate aldehyde is then oxidized to the corresponding carboxylic acid or, occasionally, is reduced to the alcohol. Monoamine oxidase is found mainly in mitochondrial membranes, and occurs in two isozyme forms. It is a flavoenzyme in that it contains a riboflavin coenzyme. It seems to act only on certain forms of a neurotransmitter. It does not, for example, affect the bound transmitter stored in vesicles, nor, curiously, the transmitter just released. That MAO inhibitors do not increase the intensity of nerve stimulation implies that there is no enzymatic destruction of freshly released transmitter. MAO is discussed further in Chap. 6, Sec. 3.7.

The other enzyme in catecholamine catabolism is *catecholamine O-methyltransferase* (COMT), a cytoplasmic enzyme that uses *S*-adenosyl-methionine to methylate

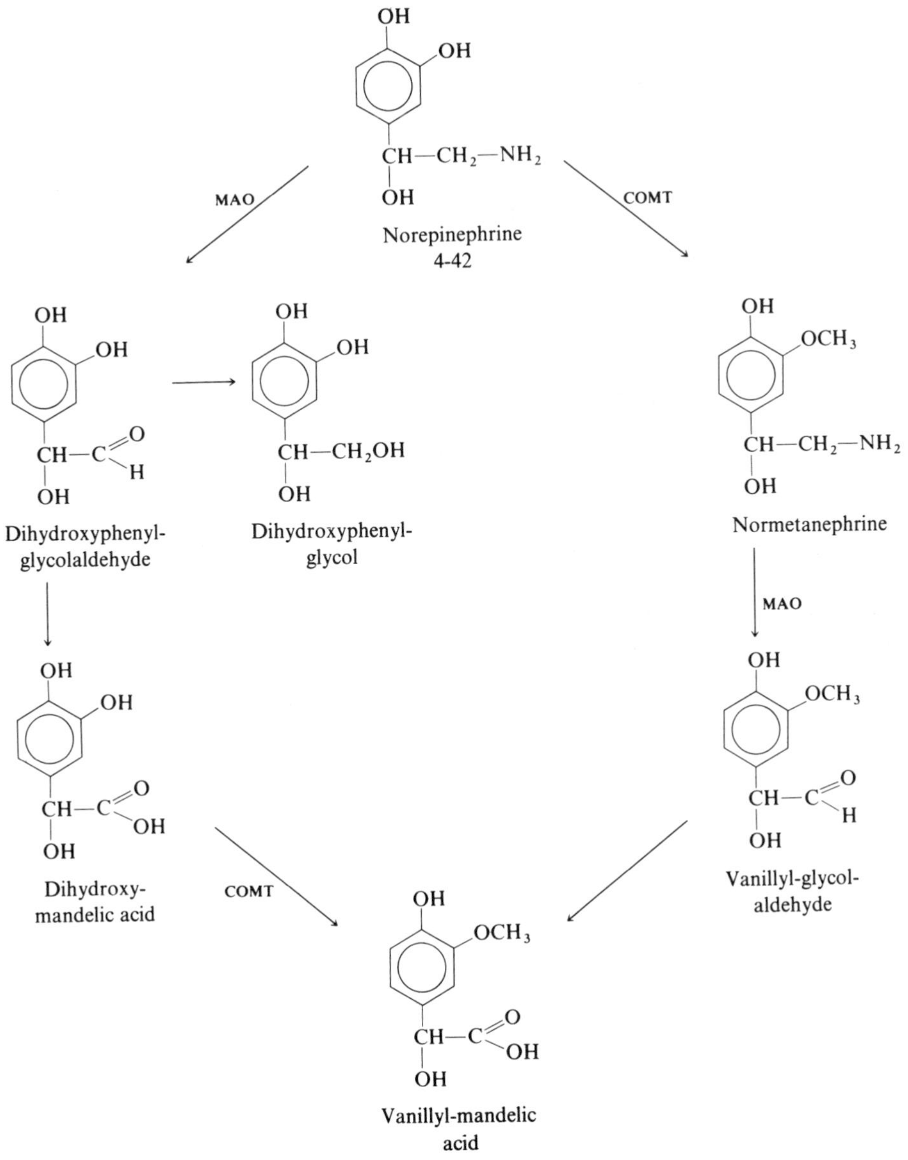

**Fig. 4.21.** Degradative metabolism of norepinephrine.

the 3-OH of catecholamines and render them inactive. The methylated compounds are not taken up into the synapse.

The principal mechanism for the deactivation of released catecholamines is, however, not enzymatic destruction, but *reuptake* into the nerve ending. The presynaptic membrane contains an "amine pump"—a saturable, high-affinity, $Na^+$-dependent active-transport system that requires energy for its function (Knoth et al., 1982). The recycled neurotransmitter is capable of being released again, as experiments with [$^3$H]NE have shown, and can be incorporated into chromaffin granules as well. Besides this "uptake-1," there is a low-affinity postsynaptic "uptake-2." Many drugs interfere with neurotransmitter reuptake and metabolism, as discussed in subsequent sections. Catecholamine metabolism is described in detail in the work of Cooper, Bloom, and Roth (1986).

### 3.3. Adrenergic Receptors

The adrenergic receptors have been studied extensively and thoroughly by pharmacological methods, but much less is known about their biochemistry. There are two major groups of receptors, designated as $\alpha$ and $\beta$, which are in turn subdivided into $\alpha_1$, $\alpha_2$, $\beta_1$, and $\beta_2$ receptors based on their apparent drug sensitivity. The existence of receptor multiplicity was first suggested by Sir Henry Dale, but was formalized and proven by R. P. Ahlquist in 1948. Multiple receptors such as these were termed *isoreceptors*, in analogy to isoenzymes (or isozymes).

The $\alpha$ *receptors* are generally excitatory, as shown in Table 4.4, and mediate a constricting effect on vascular, uterine, and intestinal muscle when stimulated by an $\alpha$ agonist. They respond to different adrenergic agonists in the following order: epinephrine > norepinephrine > isoproterenol (4.74; see Fig. 4.28).

The $\beta$ *receptors* are usually inhibitory on smooth muscle but stimulate the myocardium. Their drug sensitivity is: isoproterenol > epinephrine > norepinephrine. None of these receptors is truly tissue specific, and many organs contain both $\alpha$ and $\beta$ adrenoceptors, although usually one type predominates.

**Table 4.4** Adrenergic receptor differentiation

| Effector organ | Receptor response | | |
|---|---|---|---|
| | $\alpha$ | $\beta_1$ | $\beta_2$ |
| Vascular system | Constriction | — | Dilation |
| Uterus | Constriction | Dilation | Dilation |
| Intestine | Decreased motility | Decreased motility | Decreased motility |
| Heartbeat | | Increase | — |
| Bronchial muscle | Constriction | — | Relaxation |
| Relative agonist sensitivity | | | |
| NE | 100 | 25 | 1–2 |
| Isoproterenol | 5 | 100 | 100 |
| Biochemical effect | | | |
| Regulation of AC | — | + | + |
| Regulation of phosphatidylinositol | + | — | — |
| Lipolysis | — | Increase | — |
| Glycogenolysis | — | — | Increase |

### 3.3.1. Properties of the α Receptor

The properties of the α receptor were recently described by Venter et al. (1984). The $\alpha_1$ receptor is a single 85-kD peptide that exists as a dimer. A 45-kD fragment contains the ligand-binding domain and protrudes into the extracellular space. The development and investigation of polyamine disulfides (4-47 and 4-48)—a novel class of irreversible, selective α antagonists—by Belleau, Melchiorre, and their co-workers (see Melchiorre et al., 1979) opened up new possibilities in studying the α receptor, since these compounds block the receptor completely at a concentration of 20 μM. Moreover, they have no effect on the serotonin and histamine receptors, as do other "irreversible" α-blocking agents. The bound tritiated analogues of the disulfides cannot be washed out and cosediment with the microsomal fractions when the experimental tissue is fractionated. Therefore, these polyamine disulfides are suitable as affinity labels.

$$NH_2-(CH_2)_5-NH-(CH_2)_2-S-S-(CH_2)_2-NH-(CH_2)_5-NH_2$$

APC
4-47

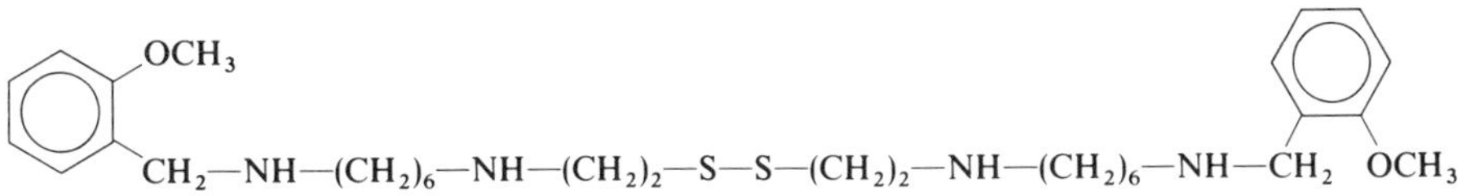

$$CH_2-NH-(CH_2)_6-NH-(CH_2)_2-S-S-(CH_2)_2-NH-(CH_2)_6-NH-CH_2 \quad OCH_3$$

Benextramine
4-48

Investigations with **benextramine** (4-47) analogues have led to the design of a tentative *hypothetical receptor model* (Fig. 4.22). The α receptor is probably a cysteine protein, but the key thiol function seems to be inaccessible in the resting state. The polyamine disulfides can reach this thiol and stabilize the conformation when the —SH group is exposed. The thiol group is surrounded by at least eight anionic sites arranged crosswise and separated by precise distances. In one direction, a $C_6$ chain is the optimal bridging distance; in the perpendicular direction, a $C_8$ chain is necessary. The binding of other α agonists or antagonists to this site has not been explored to the same extent.

Very little about the biochemical role of the α-adrenoceptor is known. It has been suggested that it regulates the phosphatidylinositol cascade and thus is a $Ca^{2+}$ ionophore, which, considering its involvement in smooth muscle contraction, appears to be reasonable. Unlike the β receptor, it is not connected in any way with adenylate cyclase and cAMP formation, except in the CNS.

On the basis of drug selectivity, there is considerable evidence for the existence of two types of α-adrenoceptors, as proposed by Langer (1977) and U'Pritchard et al. (1977). These will be discussed in connection with α-adrenergic agonists in Sec. 3.5.

### 3.4. Properties of the β Receptor

These properties are much better known, as a result of the investigations of the groups of Lefkowitz, Levitzki, Molinoff, and others. As with other receptor studies

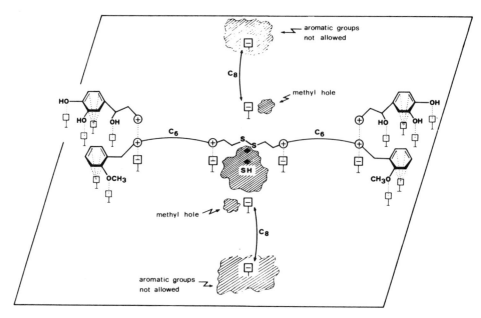

**Fig. 4.22.** Schematic representation of the arrangement of the eight anionic sites on the α-adrenergic receptor, interacting with norepinephrine and benextramine. The common receptor thiol group is shown in the middle, where it can form a covalent bond with the initial addition complex either horizontally or vertically. (Reproduced by permission from C. Melchiorre (1981), *Trends Pharmacol. Sci. 2*: 210, Elsevier Biomedical Press, Amsterdam)

at the molecular level, the key to success in this work was the development of a sufficiently selective radioligand. Two radiolabeled β antagonists—[³H]**dihydroalprenolol** (4-84) (see Fig. 4.29) and [¹²⁵I]**iodohydroxybenzpindolol** (4-49)—were used, the latter at the enormously high specific activity of 2200 Ci/mmol. The experimental tissues were varied and often of nonneuronal origin, such as frog or turkey erythrocyte membranes. These are simple, easily prepared homogeneous model systems, and the information gained from them can be applied to the study of receptors in heart, brain, or muscle tissue (Schorr et al., 1981). The frog erythrocyte receptors have been solubilized and purified by gel chromatography.

Iodohydroxybenzpindolol
4-49

**Table 4.5.** Structural features of $\beta$-adrenoceptor-dependent adenylate cyclase

| Component | Mol. wt. of subunits[a] | Other properties |
|---|---|---|
| $\beta$-Adrenoceptor | 41,000–43,000 | |
| Turkey erythrocyte | 40,000–43,000 | $\beta_1$ subclass |
| Frog erythrocyte | 58,000 | $\beta_2$ subclass |
| Fat cells | 67,000 | $\beta_2$ subclass |
| A43 epidermoid carcinoma | 59,000; 74,000 | $\beta_2$ subclass |
| $G_s$ protein | 45,000 | Possesses GTPase activity |
| $\alpha_s$ | 42,000–45,000 | Binds GTP; undergoes ADP ribosylation by cholera toxin |
| $\beta$ | 35,000 | |
| $\gamma$ | 8,000 | |
| Catalytic unit | | Associated tightly with $G_s$ to forms stoichiometric complex $G_sC$ |
| Rabbit heart | 150,000 | |
| Bovine brain | 120,000 | |
| $G_i$ protein | | Possesses GTPase activity |
| $\alpha_i$ | 41,000 | Binds GTP; undergoes ADP ribosylation by cholera toxin |
| $\beta$ | 35,000 | |
| $\gamma$ | 8,000 | |
| $G_sC$ complex | 215,000 $\pm$ 17,000 | |

[a] In general, $\beta_1$ adrenoceptor subunits have mol. wts. of 41,000–45,000, whereas $\beta_2$ receptors consist of subunits of mol. wt. 63,000–70,000. One report, however, claims that $\beta_1$ and $\beta_2$ receptor subunits are both of mol. wt. $\sim 67,000$. Reproduced by permission from Levitzki (1986).

Only the application of recombinant DNA techniques has solved the structure of the $\beta$-adrenoceptor (Levitzki, 1986; Lefkowitz et al., 1986). Both the $\beta_1$ and the $\beta_2$ receptors are glycoproteins of molecular weight 64K, and their molecular features are shown in Table 4.5. The purified adrenoceptors have been reconstituted in phospholipid vesicles and have functioned properly. The $\beta_2$ receptor gene DNA has been cloned, and the amino acid sequence and tentative tertiary structure have been deduced (Fig. 4.23). This also allowed the elucidation of receptor desensitization: it is accomplished by a specific kinase that phosphorylates the receptor when it is occupied by an agonist. Both $\beta$-adrenoceptor subtypes are coupled to adenylate cyclase, and an inhibitory receptor that activates the inhibitory $G_i$ protein (see Chap. 2, Sec. 5.4) has been postulated (Helmreich and Pfeuffer, 1985; Levitzki, 1986).

The $\beta$ receptor is highly stereospecific, preferentially binding the $(-)$-isomers of drugs. The conformational preference is a phenyl—$NH_3$ *trans* arrangement, meaning that the agonist molecule is extended, with the *m*-OH and $\beta$-OH coincident on the same face of the molecule. The agonist molecule therefore has a polar and a nonpolar side (Fig. 4.24).

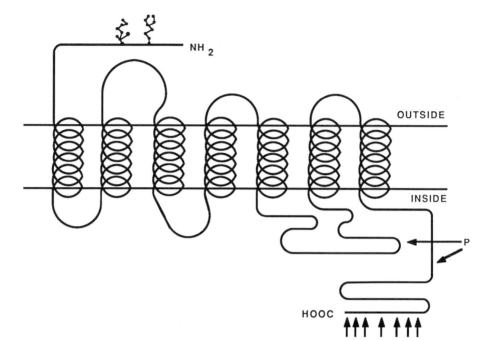

**Fig. 4.23.** Schematic structure of the mammalian $\beta_2$-adrenergic receptor. There are seven membrane-spanning helical regions composed of hydrophobic amino acid sequences, and at least two glutamine-linked glycosylation sites near the N-terminal. P shows potential sites of phosphorylation by cAMP-linked protein kinase, arrows indicate serine and threonine molecules that can be sites of regulatory phosphorylation by receptor kinase. (Modified by permission from Lefkowitz et al. (1986), Elsevier Biomedical Press, Amsterdam)

**Fig. 4.24.** Stereochemistry of norepinephrine.

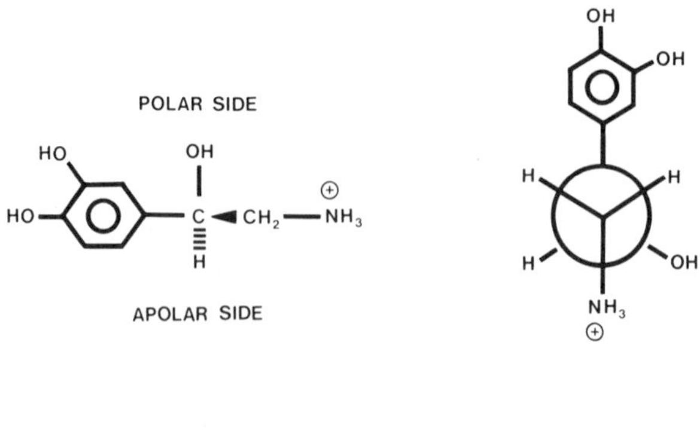

POLAR SIDE

APOLAR SIDE

## Selected Readings

A. K. Cho and G. S. Takimoto (1985). Irreversible inhibitors of adrenergic nerve terminal function. *Trends Pharmacol. Sci. 6*: 443–447.

J. R. Cooper, F. E. Bloom, and R. H. Roth (1986). *The Biochemical Basis of Neuropharmacology*, 5th ed. Oxford University Press, New York.

E. J. M. Helmreich and T. Pfeuffer (1985). Regulation of signal transduction by $\beta$-adrenergic hormone receptors. *Trends Pharmacol. Sci. 6*: 438–443.

G. Jonson (1980). Chemical neurotoxins as denervation tools in neurobiology. *Annu. Rev. Neurobiol. 3*: 169–187.

J. Knoth, M. Zallakia, and D. Njus (1982). Mechanism of proton-linked monoamine transport in chromaffin granule ghosts. *Fed. Proc. 41*: 2742–2745.

S. Z. Langer (1977). Presynaptic receptors and their role in the regulation of neurotransmitter release. *Br. J. Pharmacol. 60*: 481–492.

R. J. Lefkowitz, J. L. Benovic, B. Kobilka, and M. C. Caron (1986). $\beta$-Adrenergic receptors and rhodopsin: shedding new light on an old subject. *Trends Pharmacol. Sci. 7*: 444–448.

A. Levitzki (1986). $\beta$-Adrenergic receptors and their mode of coupling to adenylate cyclase. *Physiol. Rev. 66*: 819–854.

C. Melchiorre, M. S. Yong, B. Benfey, L. Brasili, G. Bolger, and B. Belleau (1979). The catecholamine α-receptor as a polyanionic cysteine protein. Selective covalent occupancy by polyaminedisulfides. In: *Recent Advances in Receptor Chemistry* (F. Gualtieri, M. Gianella, and C. Melchiorre, Eds.). Elsevier/North Holland, New York, pp. 207–220. (See also: *J. Med. Chem. 21*: 1126–1132.)

R. Schorr, R. J. Lefkowitz, and M. G. Caron (1981). Purification of the $\beta$-adrenergic receptor: identification of the hormone binding subunit. *J. Biol. Chem. 256*: 5820–5826.

P. Sneddon, D. P. Westfall, and J. S. Fedan (1982). Cotransmitters in the motor nerves of the guinea-pig vas deferens: electrophysiological evidence. *Science 218*: 693–695.

T. W. Stone (1982). Adenosine symposium. *Trends Pharmacol. Sci. 3*: 423–425.

J. M. Trifaro and L. Cubbedu (1979). Exocytosis as a mechanism of noradrenergic transmitter release. In: *Trends in Autonomic Pharmacology*, Vol. 1 (S. Kalsner, Ed.). Urban and Schwarzenberg, Baltimore.

D. C. U'Pritchard, D. A. Greenberg, and S. H. Snyder (1977). Binding characteristics of a radiolabeled agonist and antagonist at CNS α-NE receptors. *Mol. Pharmacol. 13*: 454–473.

J. C. Venter, P. Horne, B. Eddy, R. Greguski, and C. M. Fraser (1984). $\alpha_1$-Adrenergic receptor structure. *Mol. Pharmacol. 26*: 196–205.

### 3.5. Adrenergic Drugs: Presynaptic Effects

Presynaptic adrenergic drugs can be classified as follows:

1. Drugs acting on catecholamine synthesis
2. Drugs acting on catecholamine metabolism
3. Drugs acting on catecholamine storage
4. Drugs acting on catecholamine reuptake
5. Drugs acting on presynaptic receptors

### 3.5.1. Drugs Interfering with Catecholamine Synthesis

These drugs include some enzyme inhibitors, and therefore are discussed in Chap. 6. However, some of these agents have other, nonenzymatic points of attack. The most widely used of these compounds is **α-methyldopa** (4-50). Like many methyl analogues of enzyme substrates, this drug is a competitive inhibitor of DOPA decarboxylase, and was believed to decrease blood pressure by decreasing available NE through inhibition of its synthesis. Recent findings, however, indicate that α-methyldopa is metabolized to α-methyl-NE, which then stimulates the central presynaptic $α_2$ receptors, thus decreasing NE release (van Zwieten and Timmermans, 1979). The analogous α-methyltyrosine inhibits tyrosine hydroxylase, but is not used as a drug. Other DOPA decarboxylase inhibitors will be discussed in connection with dopamine in Sec. 4.1.

### 3.5.2. Drugs Interfering with Catecholamine Metabolism

This group of drugs consists primarily of compounds that block the enzyme monoamine oxidase (MAO). While useful as hypotensive and antidepressant drugs, their side effects can be serious. We discuss them briefly as enzyme inhibitors in Chap. 6, Sec. 3.7.

### 3.5.3. Drugs Interfering with Catecholamine Storage

These drugs can act in two different ways. The *Rauwolfia* alkaloid **reserpine** (4-51) and related natural or semisynthetic compounds interfere with the membranes of synaptic vesicles and *deplete nerve endings* of NE and dopamine (and, incidentally, of serotonin in serotonergic neurons). The resulting decrease in available neurotransmitter results in hypotension as well as in sedation. It seems that NE reuptake into the vesicles is also impaired. Because more modern and effective drugs are available, reserpine is used only as a hypotensive agent. Interestingly, however, reserpine has been used for centuries in India and is one of the few examples of an "ethnopharmacologic" agent successfully introduced into Western therapeutics.

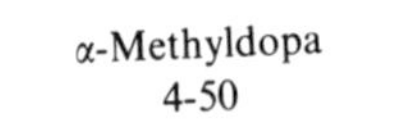

α-Methyldopa
4-50

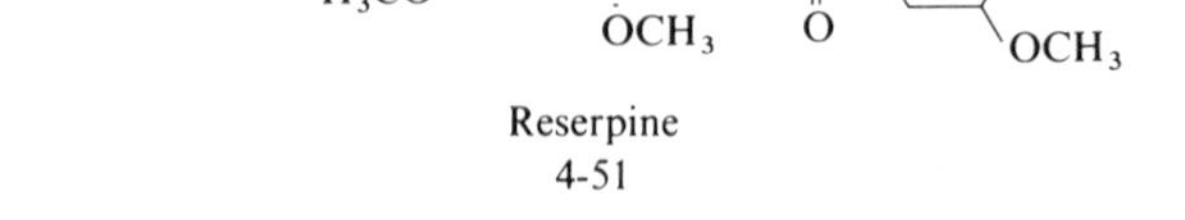

Reserpine
4-51

(+)-**Amphetamine** (phenylisopropylamine) has been used for years as a mood elevator and psychomotor stimulant by persons who must stay awake (truckdrivers, students), and is still used as an appetite suppressant (anorectic). Amphetamines have a multiple neuronal effect: they inhibit neurotransmitter reuptake, increase transmitter release, are direct α agonists, and may also inhibit the enzyme monoamine oxidase. Amphetamine is, however, a dangerous drug. In high doses or when given intravenously (as "speed," the street drug), it can cause symptoms of paranoid schizophrenia by releasing dopamine in the CNS. It also has cardio-vascular effects, and its use is followed by a depressive "letdown" period. Paradoxically, the use of amphetamines is justified only for the treatment of hyperkinetic (hyperactive) children. The cause of this minimal brain dysfunction may be underactivity of central adrenergic pathways due to faulty utilization of catecholamines. Through its neurotransmitter-mobilizing effect, amphetamine al-leviates the disruptive behavior and short attention span of hyperactive children, symptoms that usually disappear spontaneously when the children reach their teens. Recently a specific amphetamine-binding site, related to anorectic activity, has been discovered in the hypothalamus (Paul et al., 1982).

The second mode of interference with neurotransmitter storage is the *prevention of neurotransmitter release* from storage vesicles. Compounds acting in this way are known as adrenergic neuronal blocking agents. Among these are guanidine compounds like **guanethidine** (4-52), and quaternary ammonium compounds like **bretylium** (4-53). Guanethidine, used as a hypotensive drug, also causes some catecholamine depletion, but unlike reserpine it does not cross the blood–brain barrier and thus has no central sedative effects. It acts selectively because it is taken up into the neuron by the same amine pump that transports the neurotransmitter. Since it has no effect on parasympathetic ganglia, guanethidine has replaced nonselective bisquaternary ganglionic blocking agents as a hypotensive drug. Bretylium suffers from poor oral absorption and rapidly induces tolerance, and has therefore been replaced by the guanidines. It is now used as a cardiac antiarrhythmic.

Guanethidine
4-52

Bretylium
4-53

### 3.5.4. Drugs Interfering with Catecholamine Uptake

This group of drugs can also be divided into two categories. The first of these consists of the *false neurotransmitters*. **Tyramine** (4-44) (Fig. 4.19), produced by the decarboxylation of tyrosine (and especially the β-hydroxy derivative of tyramine, **octopamine**; 4-45), can be taken up through the presynaptic membrane by the not very selective uptake-1 mechanism. Tyramine then enters the storage granules to a certain extent (even though the vesicular uptake mechanism is more specific than the presynaptic pump) and displaces NE which, when released, causes post-synaptic effects. In addition, tyramine competes with NE for monoamine oxidase and protects the neurotransmitter from destruction, thus elevating its actual concentration.

**Octopamine** (4-45; Fig. 4.19) which carries a β-hydroxyl group, is taken up even more readily into storage vesicles and is, in turn, released when the neuron fires. As an adrenergic agonist, octopamine is, however, only about one-tenth as active as NE (the $pD_2$ of octopamine is 4.3; that of NE is 5.2); therefore, it acts as a very weak neurotransmitter. Compounds such as this behave like neurotransmitters of low potency, and are called false transmitters. On the other hand, octopamine may be a true transmitter in some invertebrates, with receptors that cannot be occupied either by other catecholamines or serotonin.

The other group of drugs acting on catecholamine recycling are the *true reuptake inhibitors*, which block the amine pump of the reuptake-1 mechanism in central adrenergic, dopaminergic, and serotonergic neurons.

*Tricyclic Antidepressants (Thymoleptics).* According to the classical *amine hypothesis of antidepressant action,* tricyclic antidepressant drugs (TCAs or thymoleptics, some of which are shown in Fig. 4.25) elevate the mood of patients suffering from endogenous depression and decrease the probability of suicide by interfering with the reuptake of NE or serotonin (Sec. 5). Such secondary amines as **desipramine** (4-56) or **nortriptyline** (4-58) are potent inhibitors of NE uptake, whereas the tertiary amines **imipramine** (4-54), **amitriptyline** (4-57), and **doxepin** (4-59) are more effective

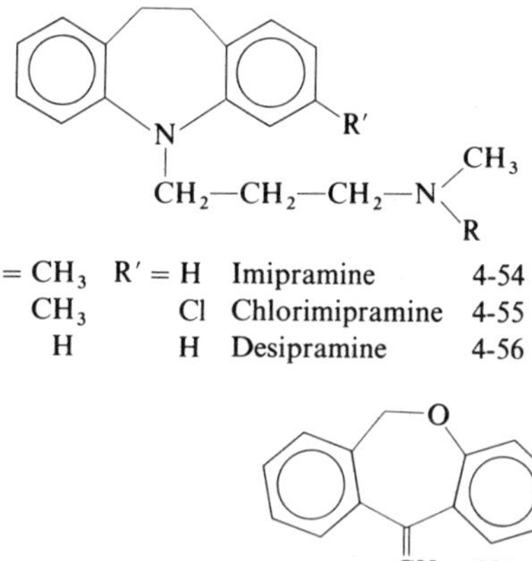

| R = CH₃ | R′ = H | Imipramine | 4-54 |
| CH₃ | Cl | Chlorimipramine | 4-55 |
| H | H | Desipramine | 4-56 |

| R = CH₃ | Amitriptyline | 4-57 |
| H | Nortriptyline | 4-58 |

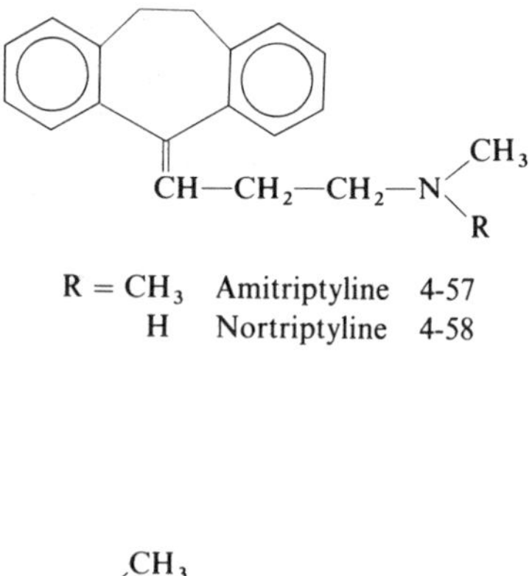

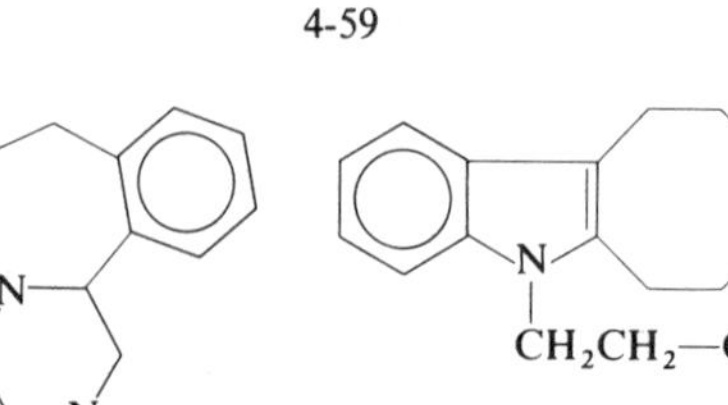

Doxepin
4-59

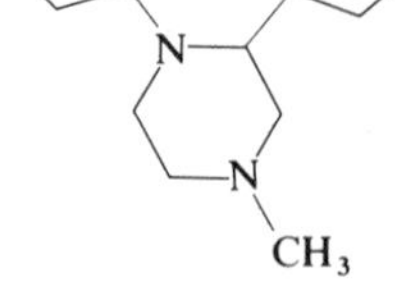

Mianserin
4-60

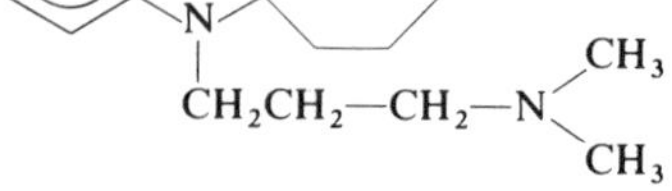

Iprindole
4-61

**Fig. 4.25.** Some widely used thymoleptic (antidepressant) drugs acting by an adrenergic mechanism.

as serotonin uptake inhibitors. According to this hypothesis, reuptake inhibition (i.e., blocking of the amine pump) increases the concentration of the neurotransmitter in the synaptic gap and thus the central adrenergic (or serotonergic) tone, resulting in mood elevation.

Unfortunately, this simple and attractive hypothesis cannot explain a number of facts:

1. The latency period of days or even weeks between the initiation of therapy and the antidepressant effect when tertiary tricyclics are used. Secondary amines act faster, but in both cases, although elevated neurotransmitter levels become rapidly apparent, the clinical improvement lags far behind.
2. **Cocaine** (4-62), the local anesthetic tropane alkaloid of coca leaves, is a potent NE reuptake inhibitor, but has no antidepressant activity.
3. Tricyclics like **mianserin** (4-60) and **iprindole** (4-61) are both excellent, clinically useful antidepressants, but do not block NE or serotonin reuptake.

Cocaine
4-62

Although there are experimental facts supporting the reuptake inhibition hypothesis for endogenous depression, which may indeed be valid in certain cases, Sulser (1979) has called attention to the supersensitivity of central NE-responsive adenylate cyclase, which is unsatisfactorily regulated in depressive patients and is "down-regulated" (desensitized) by tricyclic thymoleptic drugs during chronic treatment. According to this neurotransmitter hypothesis, depression is due not to lack of neurotransmitter but to overstimulation of NE-sensitive structures that are not regulated normally. This would explain the long lag period between the administration of these drugs and the clinical effect, supporting the hypothesis that presynaptic receptors are involved in depression (Kostowski, 1981).

The problem with the current neurotransmitter–receptor hypothesis of antidepressant activity is, that it is based on observations in normal rat brain. Unfortunately, there are very few techniques suitable for *in vivo* work on human CNS receptors of depressed patients. Furthermore, although neuroendocrine investigations are in a very preliminary stage, decreases in melatonin and prolactin secretion (regulated by $\beta_1$-adrenoceptors and 5-HT receptors, respectively) have been reported. In addition, compounds related to dopaminergic and serotonergic functions also act as antidepressants, as will be shown later. Thus, it is likely that endogenous depression is a biologically heterogeneous syndrome, and a single hypothesis explaining the mode of action of all antidepressant drugs is probably not feasible (Hollister, 1986; Leonard, 1986; Stahl and Palazidou, 1986). It is worth noting that a gene assumed to be responsible for endogenous depression has been isolated (Kolata, 1987). This discovery may open up new areas of investigation on the regulatory aspects of affective disorders.

In 1980, Langer et al. discovered specific, *high-affinity receptor sites for TCAs,* with a $K_D$ of 4–6 nM and a $B_{max}$ of 12–16 pmol/g, in the cortex, hypothalamus, and striatum. The binding is saturable, and kinetic studies indicate a noncooperative single population of receptor sites. This suggests the existence of a mood-regulating neuronal system with an unknown endogenous modulator or a presynaptic effect on serotonin uptake. The cerebral cortex of suicides seems to have fewer such receptors (Davis, 1984).

An effect opposite that of tricyclic antidepressants is shown by the simple salt **Li$_2$CO$_3$**, commonly referred to as "lithium." It has been used since about 1970 in the long-term management of *manic-depressive psychosis,* but it is not very useful in acute mania since its calming effects are seen only after 8–10 days. The mode of action of Li salts is not quite clear. $Li^+$ is not handled efficiently by the Na pump, but there are some reports claiming that $Li^+$ (1) accelerates catecholamine reuptake, (2) stimulates NE turnover, and (3) inhibits NE release—all of which are in direct opposition to TCA activities. As described in Chap. 2, Sec. 5.4, Li salts inhibit phosphatidylinositol synthesis, and may therefore interfere with $Ca^{2+}$-mediated intercellular communication. The psychiatric aspects of psychopharmacology are well covered in the book by Lickey and Gordon (1983), written for the educated layperson.

### 3.5.5. Drugs Acting on Presynaptic ($\alpha_2$) Receptors

As mentioned in Sec. 3.5.1 in connection with α-methyldopa, there are two kinds of α receptors in the brain (U'Pritchard et al., 1977) as well as in peripheral tissue (Langer, 1977), as shown by the binding selectivity of various drugs. The crucial experiments have shown that brain tissue prelabeled with [$^3$H]NE will release neurotransmitter upon electrical stimulation or exposure to $K^+$. The release is reduced by the α agonist (Fig. 4.26) **clonidine** (4-63) and stimulated by the α antagonist **yohimbine** (4-68) (Fig. 4.27). Since the adrenoreceptor involved in this

**Fig. 4.26.** α-Adrenergic agonists.

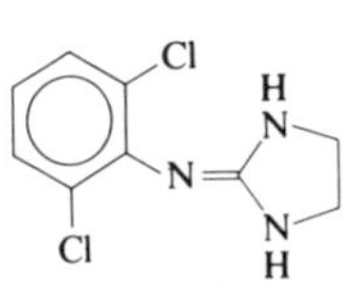

Clonidine ($\alpha_1$ and $\alpha_2$)
4-63

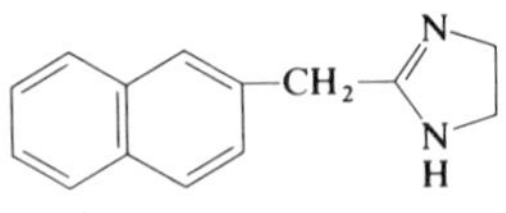

Naphazoline ($\alpha_2$)
4-64

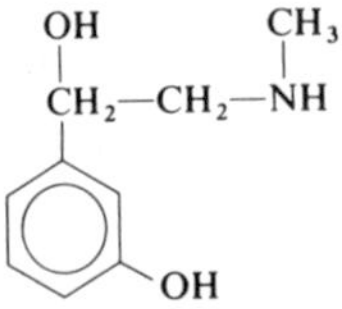

Phenylephrine
4-65

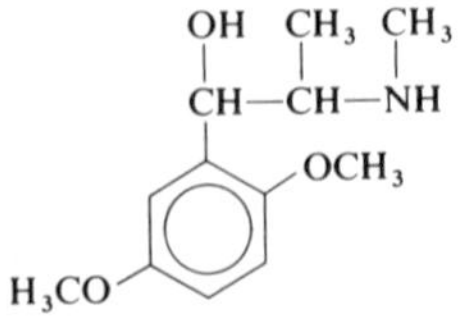

Methoxamine ($\alpha_1$)
4-66

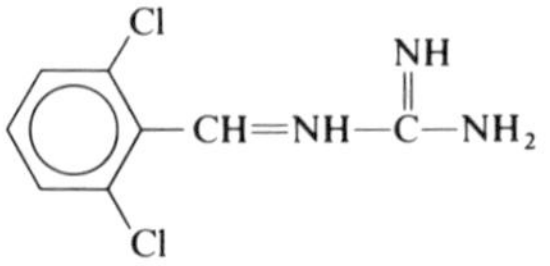

Guanabenz
4-67

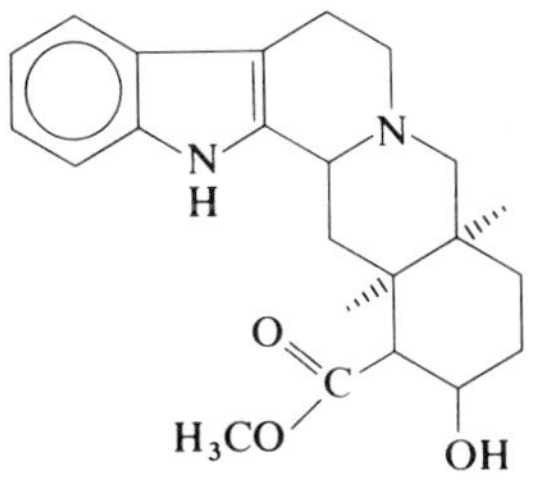


Yohimbine ($\alpha_2$)
4-68

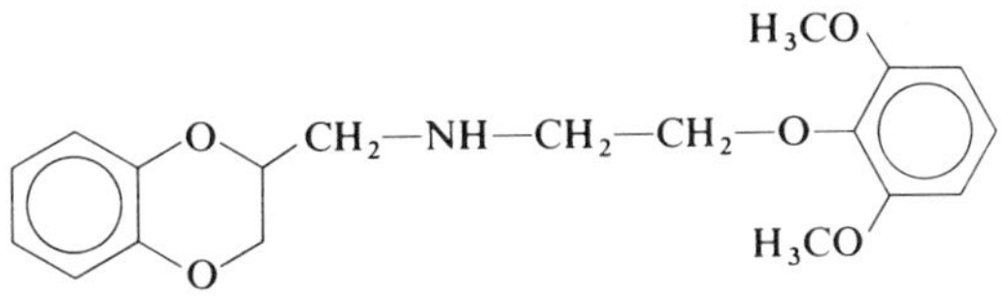


WB 4101 ($\alpha_1$)
4-69

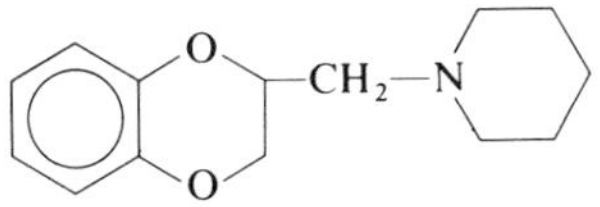


Piperoxan ($\alpha_2$)
4-70

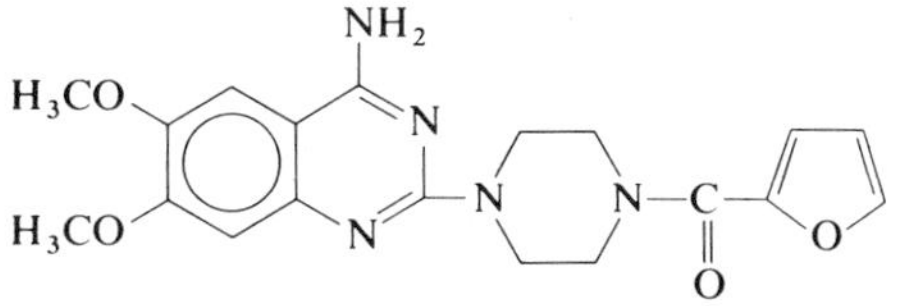


Prazosin ($\alpha_1$)
4-71

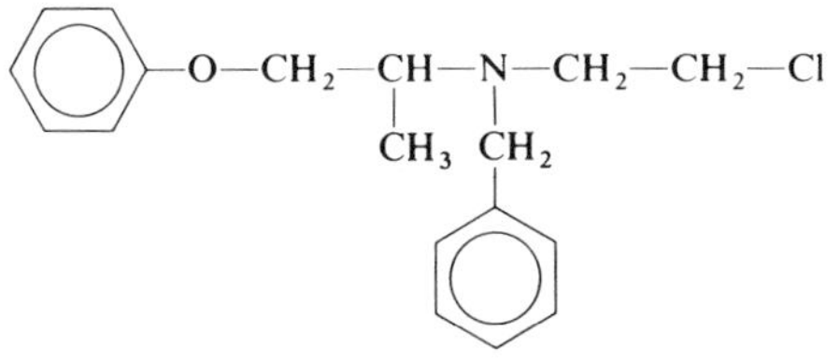


Phenoxybenzamine ($\alpha_1$)
4-72

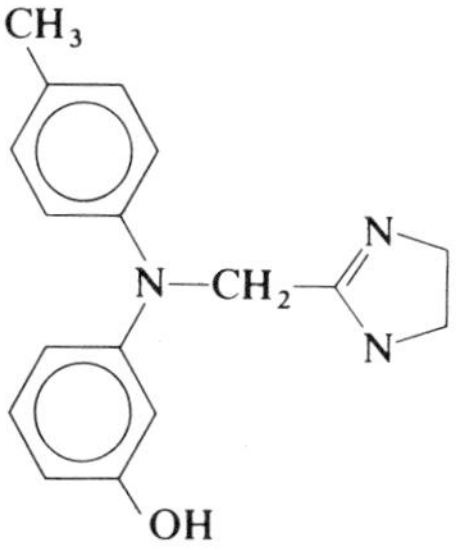


Phentolamine ($\alpha_1$ and $\alpha_2$)
4-73

**Fig. 4.27.** $\alpha$-Adrenergic antagonists.

latter experiment plays a vital role in modulating neurotransmitter release, it must be presynaptic and located on the nerve-ending membrane. A similar selectivity has also been shown by peripheral tissues (heart, uterus), leading to the distinction of $\alpha_1$ (*postsynaptic*) and $\alpha_2$ (*presynaptic*) adrenergic receptors.

There are also *presynaptic $\beta$ receptors*, which show a feedback regulation opposite to that of the $\alpha_2$ receptors; that is, their excitation by a neurotransmitter increases NE release.

Epinephrine and norepinephrine show the same affinity for both $\alpha_1$ and $\alpha_2$ receptors, as do some antagonists such as **phentolamine** (4-73). Sometimes receptor selectivity depends upon the drug concentration: **dihydroergocryptine** (4-107), a partial $\alpha$-blocking agent, binds at a low concentration to $\alpha_1$ receptors; at higher concentrations, however, $\alpha_2$ binding takes over, at the point where the Scatchard plot indicates a positive cooperativity of sites (see Guicheney and Meyer, 1979). This concentration dependence is perfectly logical, considering the NE-release stimulation at a high dose of the blocking agent but not at a low dose, where the blocking action is not severe. It is probable that at higher concentrations all $\alpha$ agonists and antagonists fail to distinguish between $\alpha_1$ and $\alpha_2$ receptors.

Other imidazolines related to clonidine, like **naphazoline** (4-64), are also $\alpha_2$ agonists. In general, $\alpha$-methyl substituents on phenethylamines increase their $\alpha_2$ affinity, as does loss of the 3-OH group. Loss of the 4-OH group of the catechol nucleus promotes $\alpha_1$ activity.

Therapeutically, **clonidine** (4-63) is a newer and very valuable central antihypertensive agent, which may perhaps act on the baroreceptor (blood pressure sensor) reflex pathway, on cardiovascular centers in the medulla, and also peripherally. As evident from the discussion above, clonidine and $\alpha$-methyldopa act in the same way. Clonidine also abolishes most symptoms of opiate withdrawal and stimulates histamine $H_2$ receptors. It seems to have an interesting psychopharmacological activity as well, acting as an antianxiety agent that stimulates $\alpha_2$-adrenoceptors and therefore decreases NE levels.

**Yohimbine** (4-68), an indole alkaloid closely related to reserpine—an $\alpha$-antagonist—has no therapeutic application. Its purported aphrodisiac properties are purely anecdotal (Taberner, 1985).

**Naphazoline** (4-64) and other $\alpha$-agonist imidazoline compounds are nasal decongestants used by inhalation to decrease swelling of the nasal mucosa. Overdependence on and overuse of these drugs can lead to rebound swelling.

### 3.6. Adrenergic Drugs: Postsynaptic Effects

There is a considerable body of classical structure–activity correlation studies in the adrenergic field. It can be summarized as follows:

1. *Phenolic hydroxyls* are important for adrenergic agonist activity. Removal of the 4-OH group leaves intact only $\alpha$-agonist activity, whereas removal of the 3-OH group abolishes *both* $\alpha$- and $\beta$-agonist activity. The 3-OH group can, however, be replaced by a sulfonamide (soterenol) or a hydroxymethyl (salbutamol) group. 3-Amino compounds can be extremely potent. Replacement of the 4-OH group by any such groups leads to an almost total loss of pharmacological action; alternatively, the resultant compound may become an antagonist.
2. The *two-carbon side chain* is essential for activity, although some exceptions are known. The benzylic carbon (next to the ring) must have the *R* absolute configuration.
3. The *alcoholic hydroxyl* can be replaced only by an amino or hydroxymethyl group.
4. Small (—H, —CH$_3$) *N-substituents* produce $\alpha$ activity; larger ones (—CH-(CH$_3$)$_2$, aryl) lead to $\beta$ activity.

More detailed structure–activity relationship studies can be found in the papers of Caron et al. (1978) and Kaiser (1979).

### 3.6.1. α₁-Adrenergic Agonists

These compounds include **NE**, which acts on both $\alpha$ and $\beta$ receptors, and **epinephrine**, which is more active on $\beta$ receptors. As mentioned previously, catecholamines lacking a 4-OH group, such as **phenylephrine** (Neo-Synephrine, 4-65) and **methoxamine** (4-66), show almost pure $\alpha_1$ activity. They are both vasoconstrictors used in treating hypotension (low blood pressure) and nasal congestion. These drugs also inhibit insulin release.

### 3.6.2. α₁-Adrenergic Antagonists

Because of their peripheral vasodilator effect, these drugs are used in the treatment of hypertension. They act beneficially in shock and frost bite by increasing peripheral circulation. Some, like **phenoxybenzamine**, also have cholinergic effects, indicating that these antagonists cross-react with the AChR.

Chemically, adrenergic blocking agents are a varied group, bearing little resemblance to the adrenergic agonists, since they use accessory binding sites of the receptor (Fig. 4.27). Benzodioxanes such as **WB 4101** (4-69) and **piperoxan** (4-70), and quinazolines like **prazosin** (4-71) all carry bulky, basic side chains. **Phentolamine** (4-73) is a rather nonselective older drug of imidazoline structure. There are a few "*irreversible*" alkylating agents such as **phenoxybenzamine** (4-72) and its congeners, which carry a $\beta$-chloroethylamine side chain capable of reacting covalently with nucleophilic —OH or —NH$_2$ groups. Although these compounds are useful drugs and experimental tools, they are slowly removed from the receptor and are therefore not truly irreversible. All of them act through IP$_3$ as second messenger.

### 3.6.3. β-Adrenergic Agonists

These drugs have been thoroughly investigated, but new active compounds continue to emerge as our understanding of structure-activity correlations increases (see Kaiser, 1979 Caron et al., 1978). **Isoproterenol** (4-74) (Fig. 4.28) is a pure $\beta$ agonist, and was the compound that first demonstrated the existence of adrenergic isoreceptors. It acts on both $\beta_1$ and $\beta_2$ receptors, and therefore produces a number of side effects in addition to its primary use as a bronchodilator. Another specific $\beta$ agonist is **methoxyphenamine** (4-75), which seemingly breaks a number of structure-activity correlation "rules."

Modification of the *catechol ring* can dramatically increase $\beta_2$ activity, such as bronchodilation. The $\beta_2/\beta_1$ index (see Kaiser, 1979) increases when a 3-OH group is substituted for a sulfonamide (**soterenol**, 4-76), hydroxymethyl (**albuterol**, 4-77), or methylamino group (4-78). Inclusion of the nitrogen into a carbostyryl ring (an $\alpha$-dihydroquinolone) leads to an experimental compound (4-80) that is 23,000 times more active than isoproterenol and also extremely selective. This compound carries a somewhat different N-substituent, a *t*-butyl group, like albuterol.

Tertiary amines are not active; the $\beta_2$ activity of secondary amines is increased by branched arylalkyl chains.

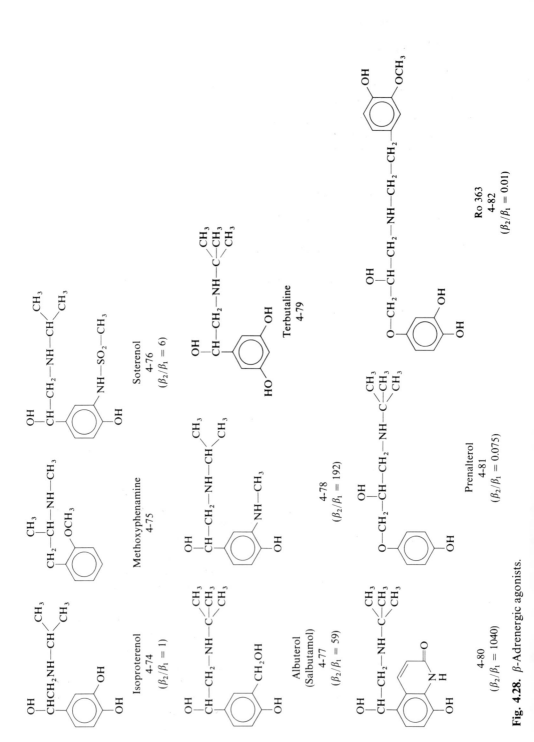

**Fig. 4.28.** β-Adrenergic agonists.

182

Modification of the *aminoethanol side chain* can produce surprising effects, as exemplified by **prenalterol** (4-81), where the insertion of an oxygen and a carbon atom between the alcohol and phenyl groups changes the parent compound ($\beta_2/\beta_1 = 11.5$) into a very selective $\beta_1$ agonist ($\beta_2/\beta_1 = 0.075$). Although many of these $\beta$ agonists are useful in the management of heart failure, their apparently "cardioselective" ($\beta_1$) activity does not necessarily reflect true receptor selectivity. Perhaps only **Ro 363** (4-82) is a true $\beta_1$ agonist (Malta et al., 1985).

### 3.6.4. β-Adrenergic Antagonists

$\beta$ Antagonists (or $\beta$ blockers) are perhaps the most important adrenergic drugs, since they are used extensively in the management of hypertension, a disease very prevalent in the Western world. Structurally, $\beta$ antagonists are much closer to $\beta$ agonists than to either their $\alpha$ counterparts or anticholinergic agents (Fig. 4.29). The first useful $\beta$ antagonist, discovered in 1948, was **dichloroisoproterenol** (4-83) (DCI), obtained by simple replacement of the catechol hydroxyls by chlorine atoms. However, DCI is also a partial $\beta$ agonist, and therefore cannot be used as a hypotensive drug.

Structure–activity studies have led to the development of some rules and regularities:

1. The catechol ring system can be replaced by a great variety of other ring systems, varying from phenylether (**oxprenolol**, 4-85) and sulfonamides (**sotalol**, 4-86) to amides (**labetalol**, 4-88), indoles (**pindolol**, 4-89; **benzpindolol**, 4-90), and naphthalene (**propranolol**, 4-91).
2. The *side chain* is either the unchanged isopropylaminoethanol seen in isoproterenol, or an aryloxy-aminopropanol. The side-chain hydroxyl groups are essential to activity.
3. *N-Substituents* must be bulky to ensure affinity to the $\beta$ receptors; isopropyl is the smallest effective substituent.

It is advantageous to have selective $\beta_1$ or $\beta_2$ blockers, but this goal has been achieved only recently since most organs have both types of $\beta$ receptors in different proportions. While $\beta$ blockers—primarily **propranolol** and the mixed $\alpha$–$\beta$ blocker **labetalol**—are antihypertensive agents, the $\beta_1$ activity of most of these compounds makes them useful in the management of some forms of angina pectoris and in cardiac arrhythmia, and they show great promise in preventing second heart attacks. An extensive tabulation of modern $\beta$ blockers is found in the review of Evans et al. (1979).

Because adrenergic agents find such extensive use as hypotensive drugs, the etiology and drug combination treatment of hypertension are of considerable interest. A discussion in any detail of this complex and confusing field goes beyond the scope of this book, however, and the reader is directed to the reviews of Khosla, Page, and Bumpus (1979), Schier and Marxer (1981), and Smith and Regan (1986) as a point of departure. Other aspects of hypertension will be treated in connection with the

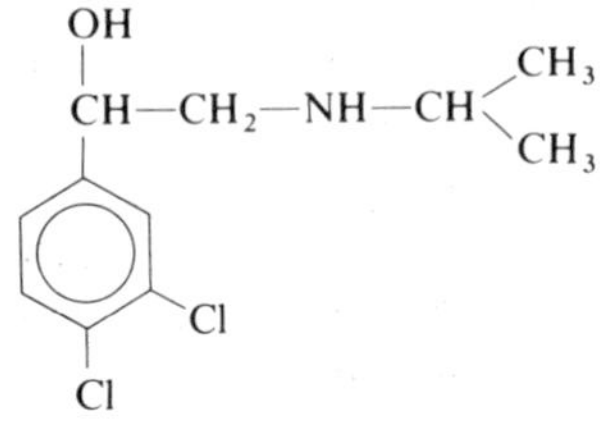

DCl
4-83

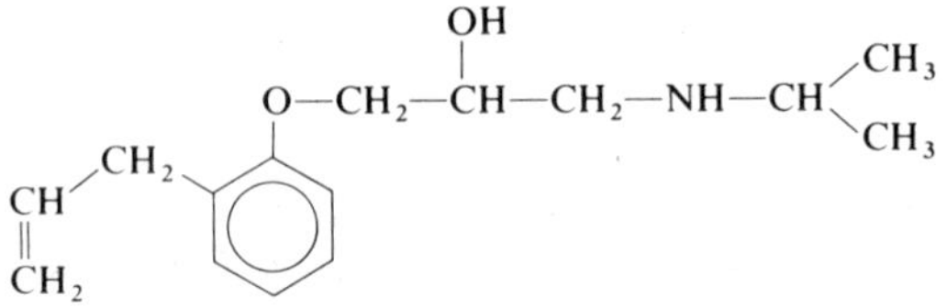

Alprenolol
4-84

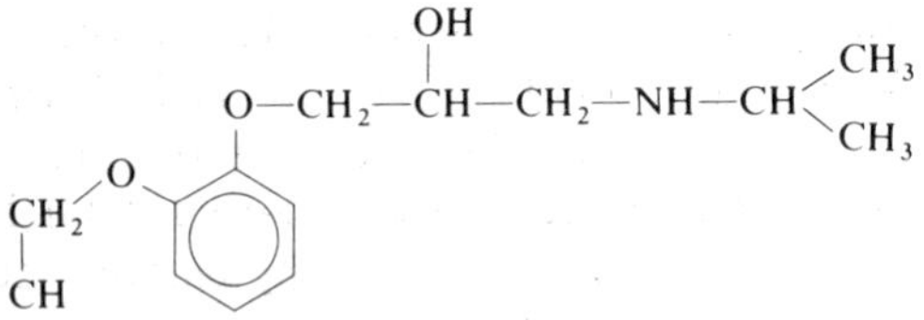

Oxprenolol ($\beta_1$)
4-85

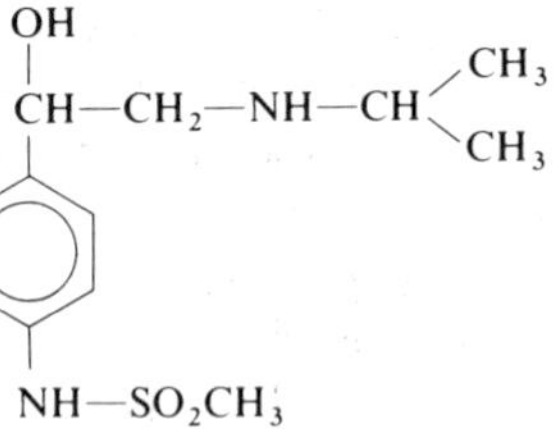

Sotalol
4-86

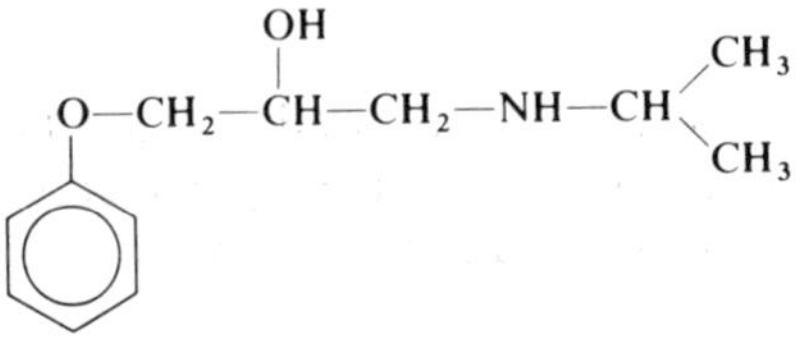

Atenolol
4-87

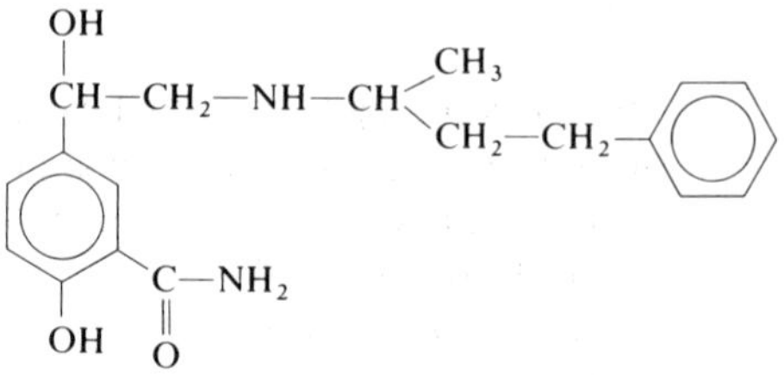

Labetalol (also $\alpha_1$)
4-88

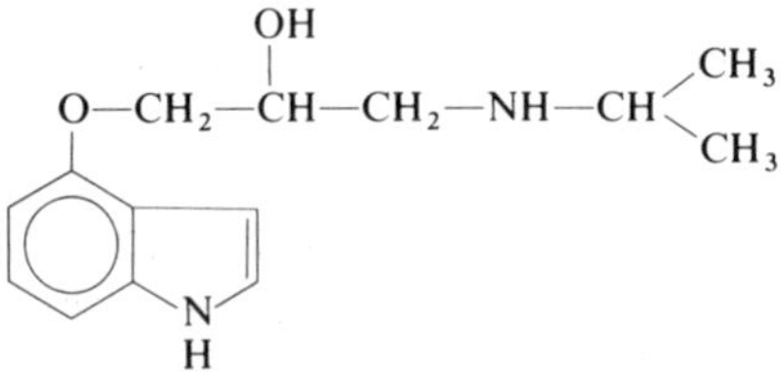

Pindolol
4-89

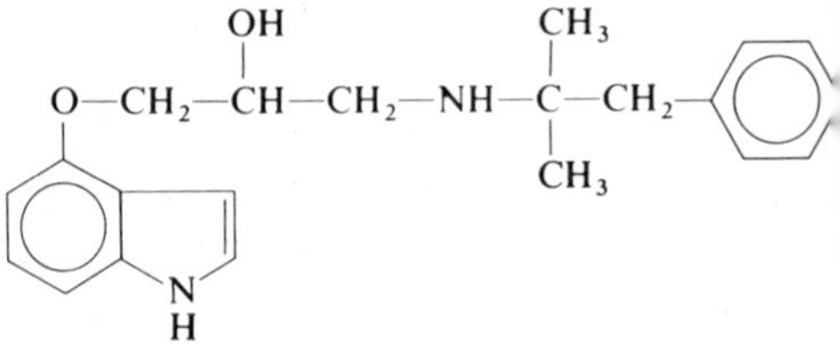

Benzpindolol
4-90

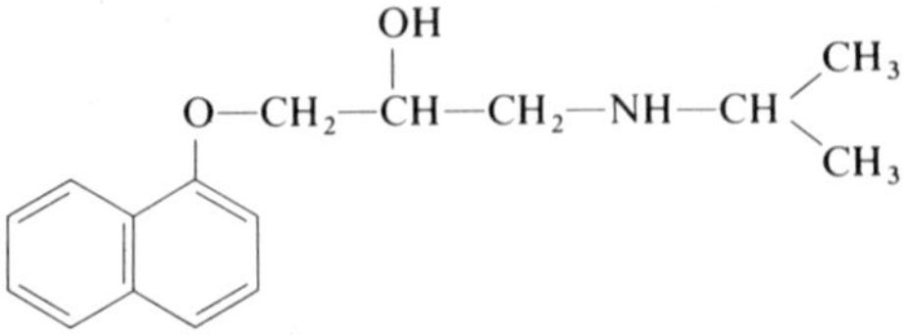

Propranolol
4-91

**Fig. 4.29.** $\beta$-Adrenergic antagonists.

renin and vasopressin systems (Chap. 5, Sec. 2), atriopeptins (Chap. 5, Sec. 5), and calcium channel blockers (Chap. 6, Sec. 1.2).

## Selected Readings

M. C. Caron, C. Mukerjee, and R. J. Lefkowitz (1978). $\beta$-Adrenergic receptors: structure–activity relation determined by direct binding studies. In: *Receptors in Pharmacology* (J. R. Smythies and R. J. Bradley Eds.). Marcel Dekker, New York, pp. 97–121.

A. K. Cho and G. S. Takimoto (1985). Irreversible inhibitors of adrenergic nerve terminal function. *Trends Pharmacol. Sci. 6*: 443–447.

J. R. Cooper, F. E. Bloom, and R. H. Roth (1986). *The Biochemical Basis of Neuropharmacology*, 5th ed. Oxford University Press, New York.

E. Costa and G. Racagni (Eds.) (1982). *Typical and Atypical Antidepressants: Molecular Mechanisms. Advances in Biochemical Psychopharmacology*, Vol. 31. Raven Press, New York.

M. J. Daly and G. P. Levy (1979). The subclassification of $\beta$-adrenoceptors. In: *Trends in Autonomic Pharmacology*, Vol. 1 (S. Kalsner, Ed.). Urban and Schwarzenberg, Baltimore.

A. Davis (1984). Molecular aspects of the imipramine "receptor." *Experientia 40*: 782–794.

D. B. Evans, R. Fox, and I. P. Hauck (1979). $\beta$-Adrenergic receptor blockers as therapeutic agents, *Annu. Rep. Med. Chem. 14*: 81–90.

P. Guicheney and P. Meyer (1979). Biochemical approach to pre- and postsynaptic $\alpha$-adrenoceptors. *Trends Pharmacol. Sci. 1*: 69–71.

B. I. Hoffman and R. J. Lefkowitz (1980). Radioligand binding of adrenergic receptors: new insights into molecular and physiological regulation. *Annu. Rev. Pharmacol. Toxicol. 20*: 581–608.

L. E. Hollister (1986). Current antidepressants. *Annu. Rev. Pharmacol. Toxicol. 26*: 23–37.

G. Jonson (1980). Chemical neurotoxins as denervation tools in neurobiology. *Annu. Rev. Neurobiol. 3*: 169–187.

C. Kaiser (1979). Structure–activity relationships among $\beta$-adrenergic receptor agonists. Implications relating to the steric requirements of adrenoceptors. In: *Recent Advances in Receptor Chemistry* (F. Gualtieri, M. Gianella, and C. Melchiorre, Eds.). Elsevier/North Holland, New York, pp. 189–208.

M. C. Khosla, I. H. Page, and I. M. Bumpus (1979). Interrelations between various blood pressure regulatory systems and the mosaic theory of hypertension. *Biochem. Pharmacol. 28*: 2967–2882.

J. Knoth, M. Zallakia, and D. Njus (1982). Mechanism of proton-linked monoamine transport in chromaffin granule ghosts. *Fed. Proc. 41*: 2742–2745.

G. Kolata (1987). Manic–depression gene tied to chromosome 11. *Science 235*: 1139–1140.

W. Kostowski (1981). Brain noradrenalin, depression and antidepressant drugs: facts and hypotheses. *Trends Pharmacol. Sci. 2*: 314–317.

S. Z. Langer (1977). Presynaptic receptors and their role in the regulation of neurotransmitter release. *Br. J. Pharmacol. 60*: 481–492.

S. Z. Langer (1981). Presynaptic receptors. *Pharmacol. Rev. 32*: 337–363.

S. Z. Langer, C. Moret, R. Raisman, M. L. Dubocovich, and M. Briley (1980). High-affinity [3]H-imipramine binding in rat hypothalamus: association with uptake of serotonin but not norepinephrine. *Science 210*: 1133–1135.

B. E. Leonard (1986). Antidepressant drugs. In: *Mechanism of Drug Action*, Vol. 1 (G. N. Woodruff, Ed.). Macmillan, London.

M. E. Lickey and B. Gordon (1983). *Drugs for Mental Illness.* (W. H. Freeman, San Francisco.

W. L. Matier and W. T. Comer (1979). Antihypertensive agents. *Annu. Rep. Med. Chem. 14*: 61–70.

R. A. Maxwell and H. C. White (1978). Tricyclic and monoamine-oxidase inhibiting antidepressants. Structure–activity correlations. In: *Handbook of Psychopharmacology* (L. L. Iversen, S. D. Iversen, and S. H. Snyder, Eds.), Vol. 14. Plenum Press, New York, pp. 83–156.

E. Malta, G. A. McPherson, and C. Raper (1985). Selective $\beta_1$-adrenoceptor agonists—fact or fiction? *Trends Pharmacol. Sci. 6*: 400–403.

C. J. Ohnmacht, J. B. Malick, and W. J. Frazee (1983). Antidepressants. *Annu. Rep. Med. Chem. 18*: 41–50.

S. M. Paul, B. Hulihan-Giblin, and P. Skolnik (1982). ( + )-Amphetamine binding to rat hypothalamus: relation to anorectic potency of phenylethylamines. *Science 218*: 487–489.

O. Schier and A. Marxer (1981). Antihypertensive agents 1969–1980. In: *Progress in Drug Research* (E. Jucker, Ed.), Vol. 25. Birkhäuser, Basel, pp. 9–32.

R. D. Smith and J. R. Regan (1986). Antihypertensive agents. *Annu. Rev. Med. Chem. 21*: 63–72.

P. Sneddon, D. P. Westfall, and J. S. Fedan (1982). Cotransmitters in the motor nerves of the guinea-pig vas deferens: electrophysiological evidence. *Science 218*: 693–695.

S. M. Stahl and L. Palazidou (1986). The pharmacology of depression: studies of neurotransmitter receptors lead the search for biochemical lesions and new drug therapies. *Trends Pharmacol. Sci. 7*: 349–354.

T. W. Stone (1982). Adenosine symposium. *Trends Pharmacol. Sci. 3*: 423–425.

F. Sulser (1979). New perspectives on the mode of action of antidepressant drugs. *Trends Pharmacol. Sci. 1*: 92–95.

S. Swillens and J. E. Dumont (1980). A unifying model of current concepts and data on adenyl cyclase activation by $\beta$-adrenergic agonist. *Life Sci. 27*: 1013–1028.

P. U. Taberner (1985). Sex and drugs—Aphrodite's legacy. *Trends Pharmacol. Sci. 6*: 49–54.

J. M. Trifaro and L. Cubbedu (1979). Exocytosis as a mechanism of noradrenergic transmitter release. In: *Trends in Autonomic Pharmacology,* Vol. 1 (S. Kalsner, Ed.). Urban and Schwarzenberg, Baltimore.

D. J. Triggle and C. R. Triggle (1976). *Chemical Pharmacology of the Synapse.* Academic Press, New York, Chap. 3.

D. C. U'Pritchard, D. A. Greenberg, and S. H. Snyder (1977). Binding characteristics of a radiolabeled agonist and antagonist at CNS α-NE receptors. *Mol. Pharmacol. 13*: 454–473.

P. A. van Zwieten and P. B. Timmermans (1979). The role of central α-adrenoceptors in the mode of action of hypotensive drugs. *Trends Pharmacol. Sci. 1*: 39–41.

## 4. DOPAMINE AND THE DOPAMINERGIC RECEPTORS

**Dopamine** (3,4-dihydroxyphenyl-$\beta$-ethylamine, DA) (4-41) is a catecholamine intermediate in the biosynthesis of NE and epinephrine (Fig. 4.19). In large doses, DA can act on vascular $\alpha_1$ adrenoceptors and cardiac $\beta_1$ receptors, but it has its own receptors in several vascular (arterial) beds, where its effect is not inhibited by the $\beta$-blocker **propanolol** (4-91). Additionally, there are several very important central structures showing specific DA receptors. Therefore, DA has been accepted as a full-fledged neurotransmitter. Because of their strongly suspected involvement in schizophrenia and proven role in neuromotor disorders such as Parkinson's disease, dopaminergic drugs are the subject of very active research.

*Dopamine metabolism* was dealt with, by implication, in our discussion of general catecholamine biochemistry. Dopamine is stored in synaptic vesicles, and this

**Table 4.6.** Dopamine receptor classification

| Receptor type | $D_1$ (CNS) $DA_1$ (vascular) | $D_2$ (CNS) $DA_2$ (neuronal) |
|---|---|---|
| | high aff. $\underset{Ca^{2+}}{\overset{GTP}{\rightleftarrows}}$ low aff. ($D_3$) | high aff. $\underset{Ca^{2+}}{\overset{GTP + Na^+}{\rightleftarrows}}$ low aff. ($D_4$, auto) |
| Effect on AC | activation | deactivation |
| Selective agonist | SK&F 38393 (4–103) | quinpirole (4–106) bromocriptine (4–108) (–) apomorphine (4–101) |
| Selective antagonist | SCH 23390 (4–104) butaclamol (4–126) | sulpiride (4–127) spiroperidol (4–122) |

storage can be manipulated. Although the reuptake of released DA is the major deactivating mechanism, MAO and COMT act enzymatically on DA in the same way as on NE. However, following the degradative pathway of NE as shown in Fig. 4.21, DA will finally be metabolized to homovanillic acid (3-methoxy-4-hydroxy-phenylacetic acid), since it lacks the $\beta$-hydroxyl group.

The dopamine receptors have been studied extensively by classical pharmacological methods as well as by the technique of receptor labeling. These experiments have revealed that, as with most neurotransmitters, several DA receptor populations exist. There are relatively few peripheral DA receptor sites; the central receptors are of much greater importance. The criteria for classification of the DA receptors are summarized in Table 4.6.

The principal criterion for distinguishing the different receptor types is the post-synaptic effector: $D_1$ receptors activate adenylate cyclase, whereas $D_2$ receptors inhibit it. The latter type seems to constitute the majority, with about 60% in the area of the substantia nigra and striatum. The cyclase-activating receptors are selectively destroyed by **kainic acid** (4-92), a rigid analogue of glutamate (which, incidentally, can also destroy central muscarine receptors), whereas the $D_2$ and $DA_2$ receptors are unaffected.

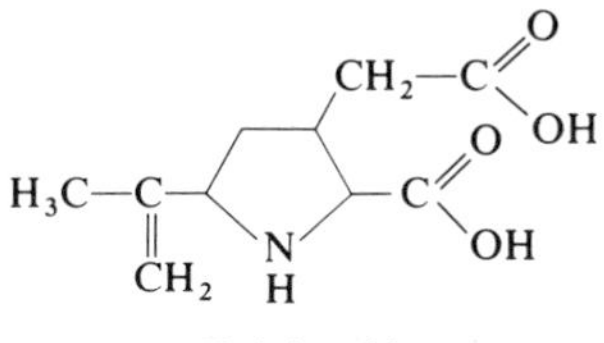

Kainic acid
4-92

Mapping of the DA receptor based on the rigid antipsychotic DA antagonist **butaclamol** (4-126) (see Table 4.7) and a large number of analogues was attempted by Humber and his group (1979). As shown in Fig. 4.30, there is a primary, phenyl binding site of a dimension accommodating two benzene rings, either of which can be occupied (C or C′), an N-binding site which is not ionic but forms a hydrogen bond (site D and G), and an accessory binding site (H) for the t-butyl group.

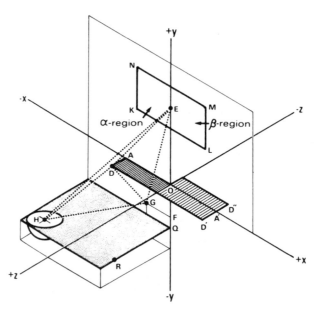

**Fig. 4.30.** A view of the dopaminergic binding site. D and G can hydrogen-bond with nitrogen, and an accessory binding site, H, accommodates the *t*-butyl group of butaclamol $(R_1 = R_2 = H_i, R_3 = —C(CH_3)_3 i; R_4 = OH)$. (Reproduced by permission from Humber et al. (1979), American Chemical Society, Washington, DC)

**Apomorphine** (4-101), a DA agonist, would fit the same site. This model has been criticized for its failure to take into account the effect of halogen substituents on butaclamol; furthermore, the model accommodates both agonists and antagonists.

Isolation of the DA receptors is not as advanced as that of some other receptors. The $D_1$ receptor has been localized in various areas of rat brain (see Kebabian et al., 1986). The $D_2$ receptor has been partially purified (see Strange, 1986) and is a glycoprotein of approximately 110–190 kD. The role of the $D_1$ receptor in the CNS is not well understood, but the $DA_1$ receptor regulates the tone of vascular smooth muscle. The cotransmitter of DA is *neurotensin*, a tridecapeptide, that seems to regulate DA, but is not simply an endogenous antagonist (Nemeroff and Cain, 1985).

### 4.1. Presynaptic Dopaminergic Drug Effects

These can be subdivided in the same way in which adrenergic drugs were classified in Sec 3.5:

1. Dopamine synthesis inhibitors
2. Dopamine metabolism inhibitors
3. Dopamine storage inhibitors
4. Dopamine reuptake inhibitors
5. Presynaptic dopaminergic agonists

### 4.1.1. Dopamine Synthesis Inhibitors

Dopamine synthesis inhibitors interfere with the enzymes involved, and are identical to those discussed in Sec. 3.5 (e.g., **α-methyltyrosine**, a tyrosine hydroxylase inhibitor, and **α-methyldopa**). In this case, $\alpha_2$-adrenergic receptor effects are irrelevant, and only the classical competitive inhibitory effect is of any consequence.

**Carbidopa** (4-93), a hydrazine analogue of α-methyldopa, is an important DOPA decarboxylase inhibitor. It is used to protect the DOPA that is administered in large doses in Parkinson's disease (next section) from peripheral decarboxylation. Therefore, DOPA concentrations in the CNS will increase without requiring the administration of extremely high, toxic doses of DOPA. The exclusive peripheral mode of action of carbidopa is due to its ionic character and inability to cross the blood–brain barrier. **Benserazide** (4-94) has similar activity.

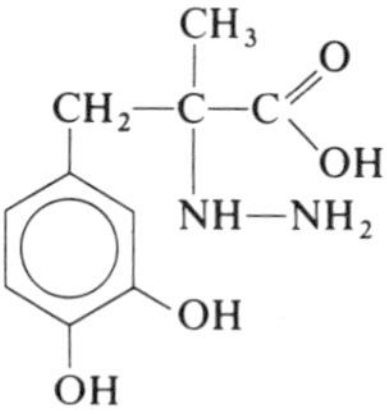

|  |  |
|---|---|
| Carbidopa | Benserazide |
| 4-93 | 4-94 |

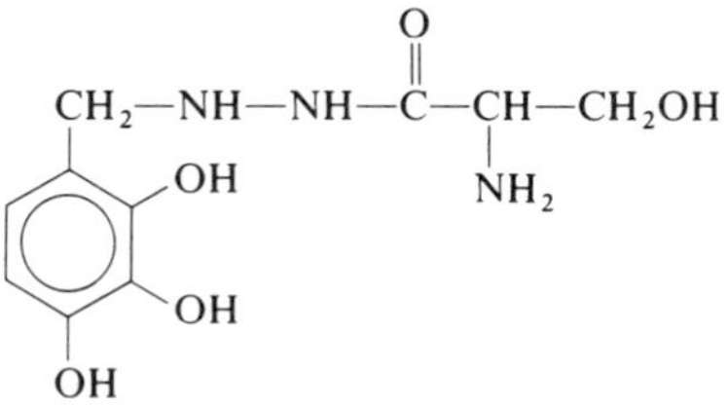

Recently, the specific arylamino acid decarboxylase inhibitory action of DL-α-**fluoromethyldopa** was discovered. By activating the enzyme through covalent binding, this compound completely inhibits both catecholamine and serotonin synthesis. Unlike 6-hydroxydopamine, α-fluoromethyldopa does not destroy the neurons, and unlike reserpine it does not deplete chromaffin tissue in the adrenal gland. Thus, it offers novel approaches to central adrenergic manipulation (Fozard, 1982).

### 4.1.2. Dopamine Metabolism Inhibitors

Dopamine metabolism inhibitors interfere with monoamine oxidase and catecholamine-O-methyltransferase. Monoamine oxidase will be discussed separately in Chap. 6, Sec. 3.7.

### 4.1.3. Dopamine Storage Inhibitors

The storage and release of DA can be modified irreversibly by **reserpine**, just as in vesicles containing other catecholamines and serotonin. Dopamine release can be blocked specifically by **γ-hydroxybutyrate** (4-95) or its precursor, butyrolactone, which can cross the blood–brain barrier. High doses of **amphetamines** do deplete the storage vesicles, but this is not their principal mode of action. Apparently, **amantadine** (4-96), an antiviral drug that is also beneficial in parkinsonism, acts by releasing DA.

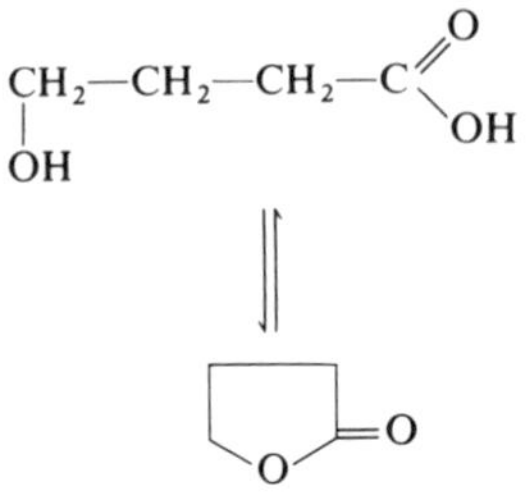

CH₂—CH₂—CH₂—C（=O）OH
|
OH

γ-Hydroxybutyrate
and butyrolactone
4-95

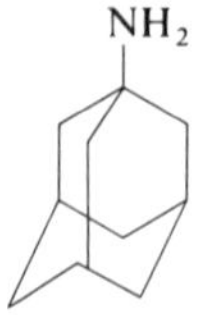

Amantadine
4-96

## 4.1.4. Dopamine Reuptake Inhibitors

Dopamine reuptake can be inhibited specifically by **benztropine** (4-97), an anticholinergic drug, as well as by *amphetamines*. Recently, some specific DA reuptake inhibitors were discovered, such as **tandamine** (4-98), **bupropion** (4-99), and **nomifensine** (4-100), which are all potent antidepressants. Interestingly, a tandamine analogue, **pirandamine** (4-135), is a selective serotonin reuptake inhibitor (see Jirkovsky and Lippman, 1978). Tandamine also inhibits NE uptake.

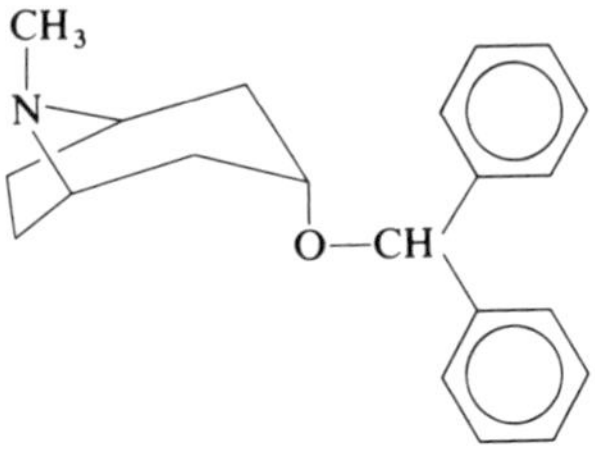

Benztropine
4-97

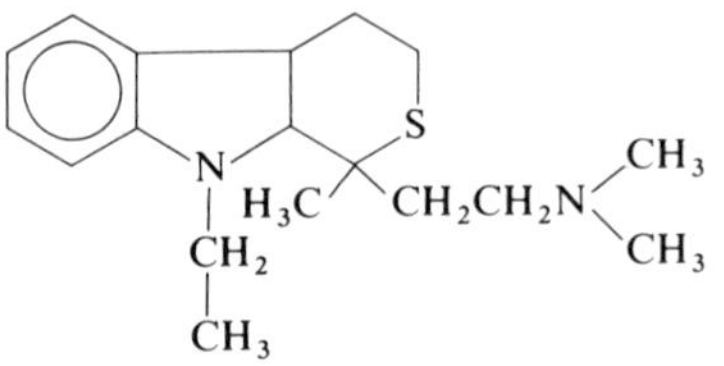

Tandamine
4-98

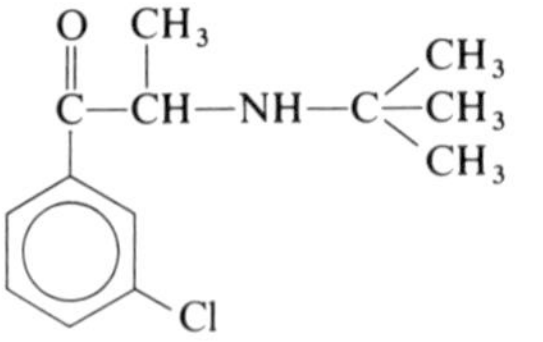

Bupropion
4-99

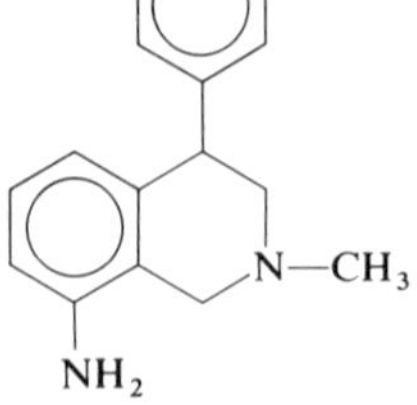

Nomifensine
4-100

## 4.1.5. Presynaptic Dopaminergic Agonists

Presynaptic dopaminergic agonists, like the *N,N*-dimethyl derivative of the tetraline **ADTN** (4-102) and **3-PPP** (4-105) have great potential as antipsychotic agents because they can control DA release. The postsynaptic $DA_2$ antagonists, used in the treatment of schizophrenia but showing extrapyramidal side effects, could thus be replaced by these novel agents.

## 4.2. Postsynaptic Dopaminergic Drug Effects

### 4.2.1. Dopamine Agonists

Besides dopamine itself, several highly active DA agonists are known, all exhibiting the extended $\beta$-phenethylamine structure corresponding to a *trans* conformation. **(−)-Apomorphine** (4-101), known for its emetic effect, is both a pre- and a post-synaptic DA agonist or partial agonist, depending on the system; both hydroxyl groups are necessary for activity. N-Alkylation, $\beta$-hydroxylation (of DA), and $\alpha$-methyl substitution all reduce central DA activity but increase interaction with the peripheral adenylate cyclase.

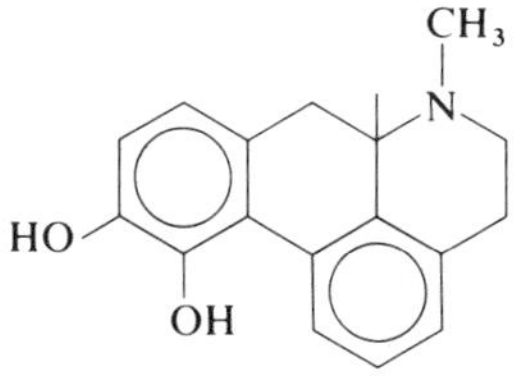

Apomorphine
4-101

Extremely active compounds are found among 2-aminotetralines. **6,7-Dihydroxy-2-aminotetraline (ADTN)** (4-102) and its *N*-(*n*-propyl) derivative are the best known 2-aminotetraline derivatives (see Woodruff et al., 1979). **Nomifensine** (4-100) is related to ADTN but is used as an antidepressant drug. The catechol analogue of nomifensine (with two hydroxyls on the 4-phenyl ring) is also a potent inhibitor of NE and DA uptake. A ring homologue of nomifensine, the experimental drug **SKF 38390** (4-103), seems to be a selective $D_1$ agonist, which dilates the renal vascular bed without having any cardiac effects. It also has central activity, but is not an emetic. The related **SCH 23390** (4-104) is a selective $D_1$ antagonist (Hilditch and Drew, 1985).

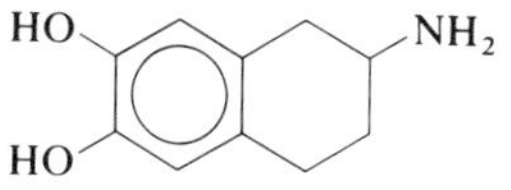

ADTN
4-102

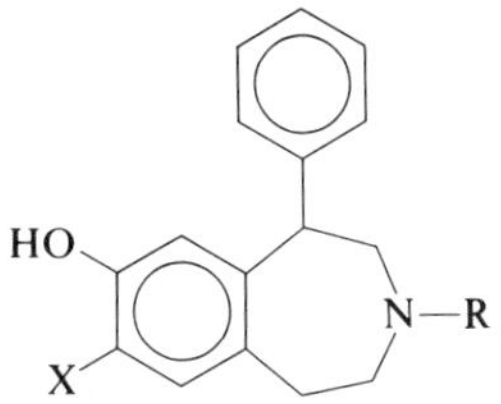

SKF 38393 (X = OH, R = H)
4-103

SCH 23390 (X = Cl, R = $CH_3$)
4-104

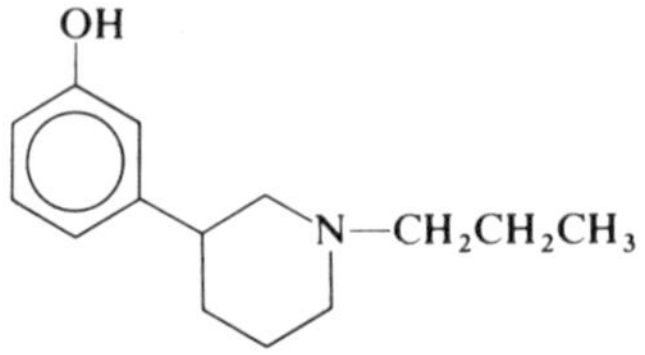

3- (3-Hydroxyphenyl) -
N- (n-propyl)piperidine
4-105

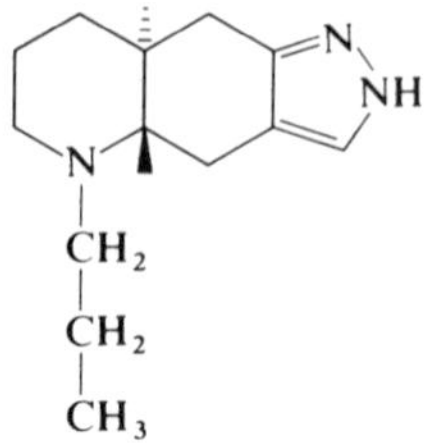

(−)Quinpirole
4-106

The *ergot alkaloids* and their derivatives are a rich source of catecholaminergic drugs (Goldstein et al., 1980). Ergot (*Claviceps purpurea*) is a parasitic fungus found on grasses and cereals (rye). The long black *sclerotium* ("ergot") of the fungus is cultivated. Because the fungus is more valuable than the cereal crop, fields are artificially infected and the mixture of indole alkaloids is extracted from the ripe sclerotia. One of these indole peptide alkaloids, **ergocryptine** (4-107), is an α-adrenergic antagonist, but its **dihydro derivative** (on the double bond of the pyridine ring) is a potent $D_2$ agonist used as a vasodilator with central effects. It is useful as a geriatric performance enhancer, both physical and mental. Other ergot alkaloids have a hypotensive effect, and also cause smooth-muscle contraction, specifically in the uterus. This property is utilized in obstetrics to stop postpartum bleeding. There is a structural correlation between DA and ergot alkaloids: the parent tetracyclic indole acid, lysergic acid, can be considered as containing an extended phenylethylamine moiety. Amides of lysergic acid are hallucinogens, and will be discussed among serotonergic drugs (Sec. 5).

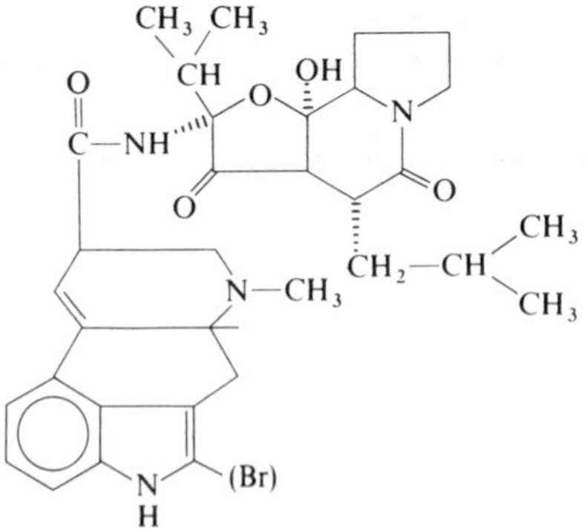

4-107 Ergocryptine
(4-108 Bromocriptine)

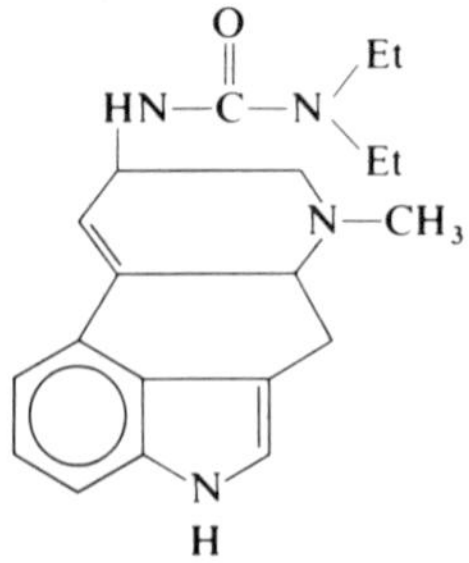

Lisuride
4-109

A novel and very interesting group of compounds is represented by **bromo-criptine** (4-108). This compound seems to be a rather specific $D_2$ agonist at low doses. It is used in parkinsonism and for inhibiting the excessive excretion of *pro-*

*lactin,* a peptide hormone of the pituitary that regulates lactation. It also inhibits the secretion of *growth hormone,* another product of the anterior pituitary, and is being tried in the treatment of acromegaly, a form of giantism (see Chap. 5, Sec. 2.3).

***Parkinson's Disease.*** Deterioration of the dopaminergic neuronal pathways, known under the name of Parkinson's disease, is manifested in a collection of neuromotor syndromes of unknown etiology (Duvoisin, 1976; Ryall, 1979). The symptoms are initial tremor, difficulty in initiating movement (akinesia), rigidity and stooped posture, and speech and swallowing difficulty. Mental functions are not impaired, but the incurable and slowly progressing disease leads to total invalidism.

The mechanism of the neurological symptoms in Parkinson's disease was discovered from the ability of reserpine to cause akinesia in humans by the depletion of central catecholamine stores. The dopamine levels in patients who died from parkinsonism were found to be extremely low because of deterioration of the dopaminergic neuronal cell bodies and the pathways connecting the *substantia nigra* with the *corpus striatum* (nigrostriatal projection; see Fig. 4.5).

Some new light was shed on the possible molecular cause of Parkinson's disease by an accident. In 1982, drug addicts used a "designer" drug (a noncontrolled analogue of a known and illegal narcotic) contaminated with 1-methyl-4-phenyl-1,2,3,6-tetrahydropyridine (**MPTP**, 4-110). Its major quaternary metabolite, **MPP$^+$** (4-111), which seems to be a dopaminergic neurotoxin, produced a severe and tragically permanent parkinsonism. The effect may be age related, and could potentially serve as a model of parkinsonism, although rats do not seem to be sensitive to MPTP (Langston, 1985).

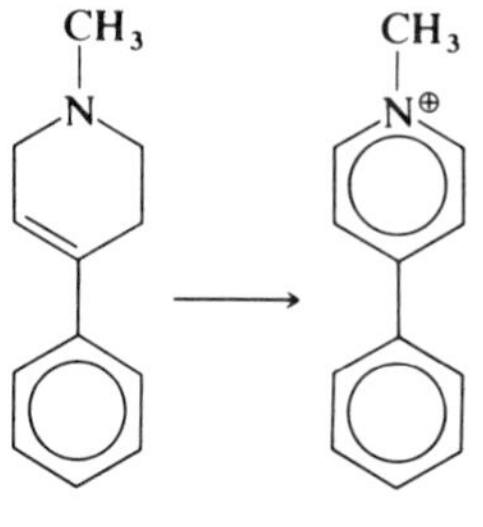

MPTP          MPP$^+$
4-110          4-111

Dopamine replacement therapy cannot be done with DA because it does not cross the blood–brain barrier. However, high doses (3–8 g/day, p.o.) of **L(–)-DOPA-(levodopa),** a prodrug of DA, have a remarkable effect on the akinesia and rigidity. The side effects of such enormous doses over a long time are numerous and unpleasant, consisting mainly of nausea and vomiting. The simultaneous administration of **carbidopa** (4-93) or **benserazide** (4-94)—peripheral DOPA decarboxylase inhibitors—allows the administration of smaller doses, and also prevents the metabolic formation of peripheral DA, which acts as an emetic at the vomiting center in the brainstem. Here the blood–brain barrier is not very effective and can be penetrated by DA.

**Table 4.7.** Examples of neuroleptic (antipsychotic) drugs

Phenothiazines

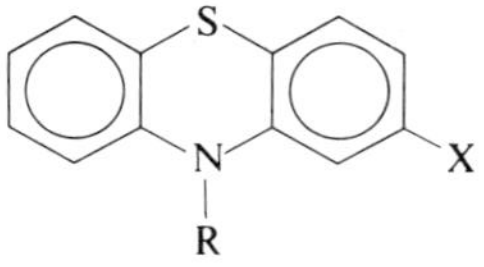

| Number | Drug name | X | R | Average p.o. dose (human) (mg/kg/day) |
|---|---|---|---|---|
| 4-112 | Chlorpromazine | Cl | $CH_2CH_2CH_2N(CH_3)_2$ | 25–50 |
| 4-113 | Triflupromazine | $CF_3$ | $CH_2CH_2CH_2N(CH_3)_2$ | 10–25 |
| 4-114 | Prochlorperazine | Cl | 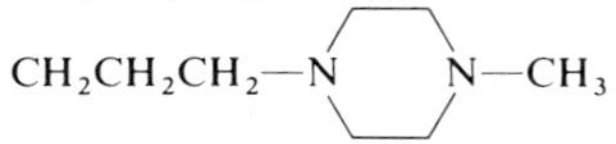 $CH_2CH_2CH_2-N\!\!\diagdown\!\!N-CH_3$ | 5–10 |
| 4-115 | Trifluoperazine | $CF_3$ | 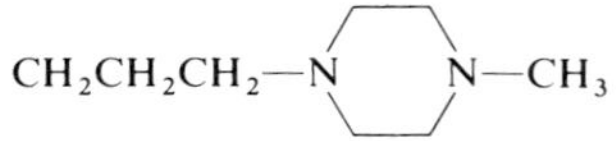 $CH_2CH_2CH_2-N\!\!\diagdown\!\!N-CH_3$ | 2–10 |
| 4-116 | Fluphenazine | $CF_3$ |  $CH_2CH_2CH_2-N\!\!\diagdown\!\!N-CH_2CH_2OH$ | 0.25–0.5 |
| 4-117 | Thioridazine | $SCH_3$ | 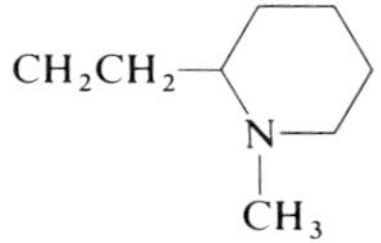 $CH_2CH_2-\!\!\diagdown\!\!N\!-\!CH_3$ | 25–100 |

Thioxanthenes

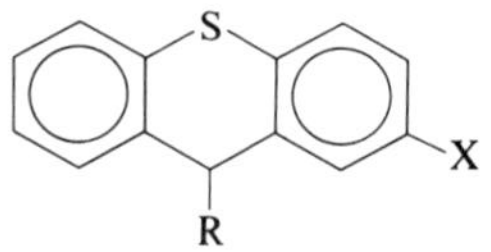

| | | R | X | |
|---|---|---|---|---|
| 4-118 | Chlorprothixene (cis) | =CH—CH₂CH₂N(CH₃)₂ | Cl | 200–300 |
| 4-119 | Flupenthixol | =CH—CH₂CH₂—N〈piperazine〉N—CH₂CH₂OH | 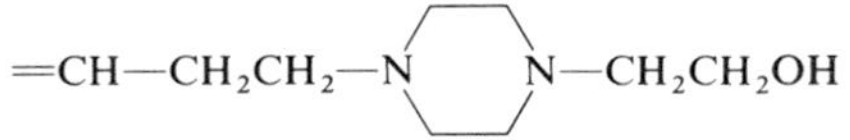 CF₃ | 10–20 |
| 4-120 | Pifluthixol (6-F) | =CH—CH₂CH₂—N〈piperazine〉N—CH₂CH₂OH | 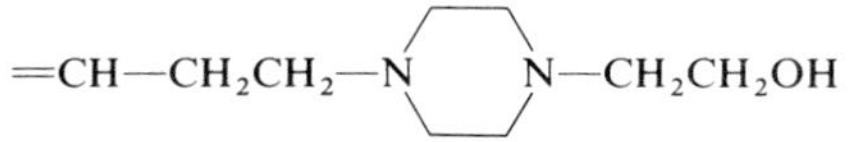 CF₃ | 5–6 |

Butyrophenones

| | | | |
|---|---|---|---|
| 4-121 | Haloperidol | 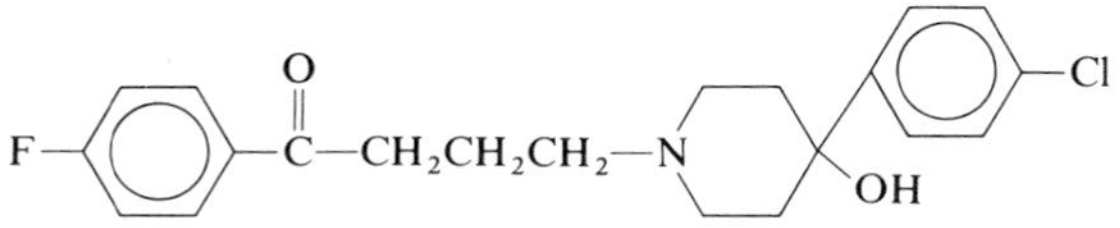 | 0.01–0.05 |
| 4-122 | Spiroperidol | 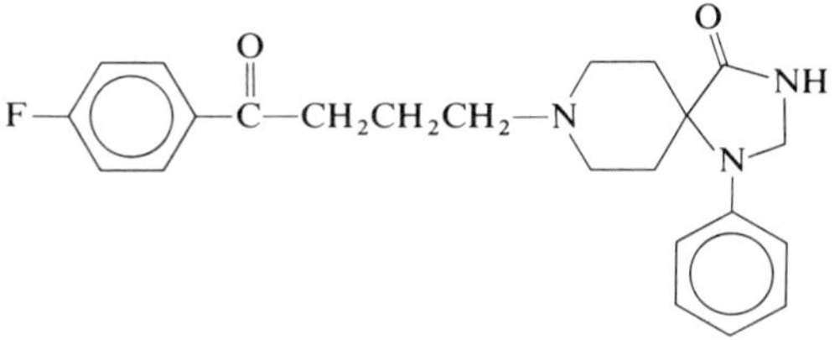 | 0.02–0.05 |
| 4-123 | Benperidol | 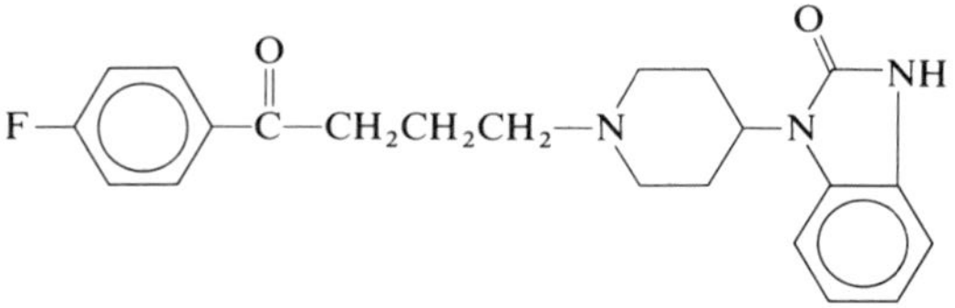 | 0.05–0.1 |

(continues)

195

**Table 4.7.** (*continued*)

Diphenylbutylpiperidines

| 4-124 | Pimozide | 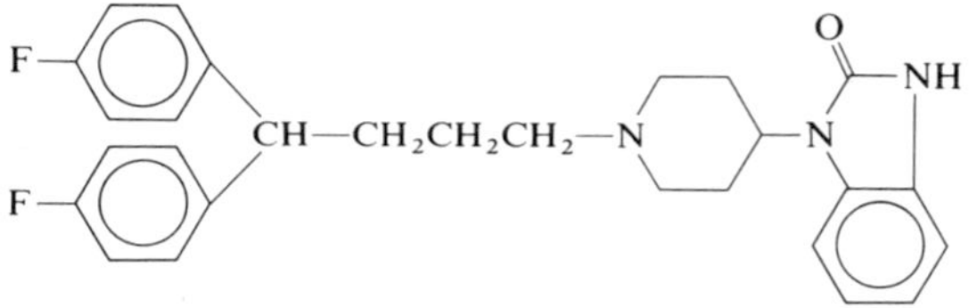 0.05–0.1 |

Other compounds

| 4-125 | Clozapine | 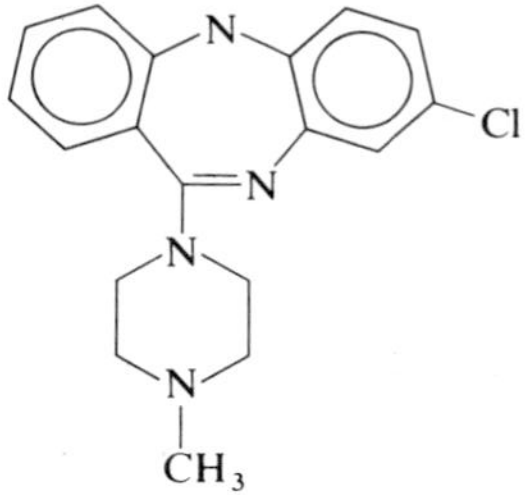 10–20 |

| 4-126 | Butaclamol | 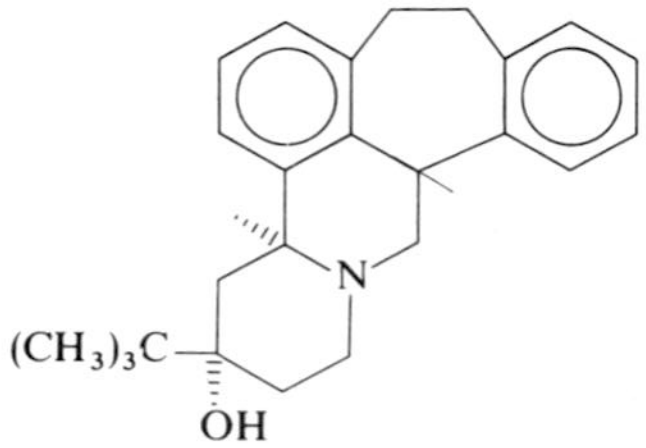 1–2 |

| 4-127 | Sulpiride | 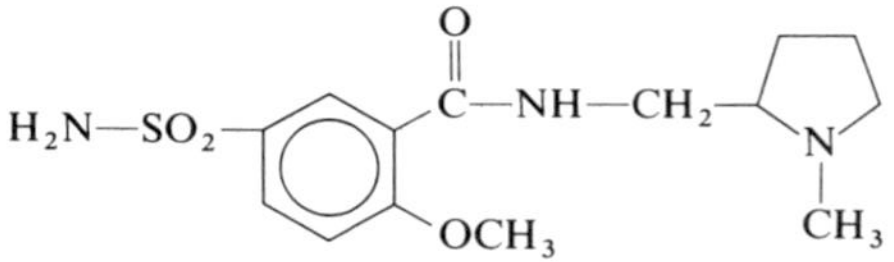 |

(Dosage data from A. Goth, *Medical Pharmacology*, 9th ed. Mosby, St. Louis, 1978, p. 219; L. S. Goodman and A. Gilman, *The Pharmacological Basis of Therapeutics*, 5th ed. Macmillan, 1970.)

In addition to successful DOPA therapy, *antimuscarinic agents* (Sec. 2.5) are also used as antiparkinsonism drugs, because the removal of *inhibiting* dopaminergic effects exaggerates the excitatory cholinergic functions in the striatum. Antimuscarinics thus substitute indirectly for the missing DA.

Drugs that mobilize DA, like **amantadine** (4-96) and **amphetamine**, are of some use in certain forms of parkinsonism. Parkinson's patients tend to become resistant to L-DOPA; moreover, this drug also causes extrapyramidal symptoms (see below) triggering involuntary movements. **Bromocriptine** (4-108) and its analogue **lisuride** (4-109) can be used in such cases. Inhibition of the enzyme monoamine oxidase B (see Chap. 6, Sec. 3.7) by **deprenyl** (6-65) improves the duration of L-DOPA therapy because it inhibits the breakdown of dopamine but not of NE (see Quinn, 1984; Larsen and Calne, 1985).

### 4.2.2. Dopamine Antagonists (Neuroleptics)

Dopamine antagonists are *antipsychotic drugs* (neuroleptics) that are very widely used in the symptomatic management (*not* cure) of all forms of schizophrenia. They were discovered in 1952 by Delay and Daniker, who, when working for the French pharmaceutical company Rhone-Poulenc, became the first to synthesize **chlorpromazine** (4-112, Table 4.7) while searching for a drug with improved antihistaminic properties. Instead, they recognized the major sedative action of the drug in agitated schizophrenics, and a new era in the management of affective disorders began. The tricyclic thymoleptics were derived from chlorpromazine a few years later.

Some examples of the major groups of neuroleptics are shown in Table 4.7. The first and original ring system is that of **phenothiazine**. In order for neuroleptic activity to occur, the distance between the ring nitrogen and side-chain nitrogen must be three carbon atoms. Shorter chains (like promethazine with an ethylamine side chain) are merely antihistamines with a strong sedative action (cf. Sec. 6.3). For optimum activity, the ring substituent in position 2 must be electron-attracting.

*Thioxanthenes* lack the ring nitrogen of phenothiazine, and the side chain is attached by a double bond. In all cases, the *cis* isomer (relative to the substituted phenyl ring) is more active. Electron-attracting substituents seem to have a cumulative effect. For instance, **pifluthixol** (4-120), with a fluorine and a trifluoromethyl substituent, is 5–10 times more potent than its parent **flupenthixol** (4-119), and has an inhibitory effect ($IC_{50} = 9.7 \times 10^{-10}$ M) on the DA-sensitive adenylate cyclase of the striatum.

The *butyrophenones* are chemically unrelated to the phenothiazines, but show a similar antipsychotic action. They were developed by P. A. Jansen, and derived from fentanyl-type analgesics (see Chap. 5, Sec. 3.7). More than 4000 derivatives have been synthesized, of which the three most widely used antipsychotics are shown. **Pimozide** (4-124) is clearly derived from **benperidol** (4-123), even though it is no longer a butyrophenone.

Three novel and nonclassical structures—**clozapine** (4-124), a dibenzodiazepine; **butaclamol** (4-125), an example of a very rigid structure; and the totally different **sulpiride** (4-127)—are also shown. It is remarkable that clozapine shows a very low affinity for the DA receptor, and must therefore act by some other mechanism. The pharmacology of all of these neuroleptics is extremely complex, and the reader is referred either to pharmacology texts or to the reviews of neuroleptics by Neumeyer (1981) and Bradley and Hirsch (1986).

Briefly, phenothiazines and related drugs have a calming effect on psychotic patients, without producing excessive sedation. Other central effects include the important *antiemetic effect* in disease-, drug-, or radiation-induced nausea, but not so much in motion sickness. Butyrophenones are more effective antiemetics than phenothiazines and also potentiate the activity of anesthetics.

***Mode of Action.*** The mode of action of antipsychotic neuroleptics is postsynaptic $D_2$ receptor blockage. The inhibition of [$^3$H] **haloperidol** (4-121) binding by neuroleptics versus the inhibition of apomorphine effects shows an excellent correlation ($r = 0.94$), and even average clinical doses correlate well ($r = 0.87$) with drug binding (Fig. 4.31). Although such a correlation does not prove causality, it is a strong indication of a uniform mechanism of action, especially the correlation with an *in vivo* measure of daily clinical dosage. A very careful critical evaluation of all of the evidence, by Seeman (1977), also favors DA receptor blockage as the mode of action of neuroleptics.

***Side Effects.*** The most common side effects of many antipsychotics are the so-called *extrapyramidal* symptoms: rigidity and tremor (that is, parkinsonian symptoms),

**Fig. 4.31.** Correlation of neuroleptic drug binding with average clinical dose. (Reproduced by permission from Creese et al. (1976), *Science 194*: 546)

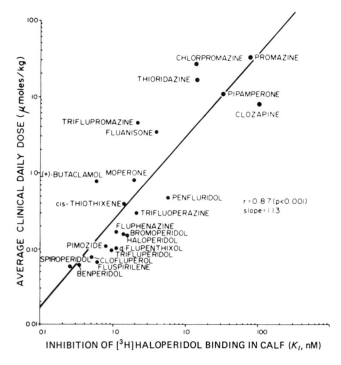

INHIBITION OF [$^3$H]HALOPERIDOL BINDING IN CALF ($K_I$, nM)

**Table 4.8.** Severity of major side effects of antipsychotic drugs at effective clinical doses

| Drug | Extrapyramidal symptoms | Sedative effect | Hypotensive effect |
|---|---|---|---|
| Chlorpromazine | + + | + + + | + +(+) |
| Thioridazine | + | + + + | + + |
| Triflupromazine | + + + | + + | + + |
| Prochlorperazine | + + + | + + | + |
| Haloperidol | + + + | + | + |
| Clozapine | + | + | |
| Pimozide | + | 0 | |

Reproduced by permission from Ryall (1979), Cambridge University Press, Cambridge.

continuous restless walking, and facial grimacing. The final, even more severe side effect of many neuroleptics is *tardive dyskinesia*, which is manifested by stereotypic involuntary movements of the face and extremities. This syndrome, which is more prevalent in older patients after prolonged use of neuroleptics, does not respond well to antiparkinsonian drugs. Tricyclics also have complex cardiovascular side effects, and antimuscarinic activity. Sedation and hypotension are also common problems. Compounds that either act on the presynaptic $D_1$ ($D_3$) receptor (**sulpiride**) or have a low postsynaptic affinity and therefore may act by another mechanism (**clozapine**) are free from extrapyramidal effects (Spano et al., 1979), as shown in Table 4.8. The hypotensive effect is due to $\alpha$-adrenergic activity, but wears off with prolonged administration, just as the sedative activity tends to disappear, even though the latter is quite useful in the management of agitated paranoid schizophrenics.

It should be emphasized that whereas neuroleptic control of schizophrenic symptoms has been spectacularly successful since the 1960s, neuroleptic treatment does not cure the psychotic patient, who will almost certainly relapse if medication is discontinued. Nor does our molecular insight answer any questions about the nature, etiology, or possible biochemistry of mental disorders, and the reader is urged to consult the thoughtful review and criticism by Hornykiewicz (1977) to gain a proper perspective on this field, the book by Lickey and Gordon (1983) on the psychiatric view of psychopharmacology, as well as the books by Bradley and Hirsch (1986) and Iversen (1986).

There are some new and potentially important developments probing into the origin of schizophrenia. The neuropeptide **neurotensin** (NT), a cotransmitter in dopaminergic neurons, may have an antipsychotic effect through modulation of DA release (Nemeroff and Cain, 1985). Thus, drugs acting on NT receptors could be beuroleptics. Similarly, the sulfated octapeptide form of the neuropeptide **chole-cystokinin** (CCK) inhibits DA release by presynaptic depolarization (Phillips et al., 1986). Another report (Tyrer and Mackay, 1986) summarizes findings that indicate reduced CCK and somatostatin concentration in brains of schizophrenics, as well as hitherto unreported anatomical changes in the lateral ventricles and thinner parahippocampal cortices. All these developments are potential avenues for improved control of schizophrenia, especially in patients who do not respond to neuroleptic treatment or show severe tardive dyskinesia.

## Selected Readings

P. B. Bradley and S. H. Hirsch (1986). *The Psychopharmacology and Treatment of Schizophrenia*. Oxford University Press, New York.

I. Creese (1978). Receptor binding as a primary drug screening device. In: *Neurotransmitter Receptor Binding* (H. J. Yamamura, S. J. Enna, and M. J. Kuhar, Eds.). Raven Press, New York, pp. 141–170.

I. Creese, D. R. Sibley, M. W. Hamblin, and S. E. Leff (1983). The classification of dopamine receptors: relationship to radioligand binding. *Annu. Rev. Neurosci. 6*: 43–71.

R. Duvoisin (1976). *Parkinsonism. Clinical Symposia*, Vol. 28, No. 1. Ciba Pharmaceutical Co., Summit, N. J.

J. R. Fozard (1982). Highly potent irreversible inhibitors of aromatic L-amino acid decarboxylase. *Trends Pharmacol. Sci. 3*: 429.

M. Goldstein, D. B. Calne, A. Lieberman, and M. O. Thorner (1980). *Ergot Compounds and Brain Function*. Raven Press, New York.

A. Hilditch and G. M. Drew (1985). Peripheral dopamine receptor subtypes—a closer look. *Trends Pharmacol. Sci. 6*: 396–400.

A. S. Horn (1975). Structure–activity relations for neurotransmitter receptor agonists and antagonists. *Handbook of Psychopharmacology* (L. L. Iversen, S. D. Iversen, and S. H. Snyder, Eds.), Vol. 2. Plenum Press, New York, pp. 179–243.

O. Hornykiewicz (1977). Psychopharmacological implication of dopamine antagonists: a critical evaluation of current evidence. *Annu. Rev. Pharmacol. Toxicol. 17*: 545–549.

L. G. Humber, I. T. Bruderlein, A. H. Philipp, M. Götz, and K. Voith (1979). Mapping the dopamine receptor. I. Features derived from modification in ring E of the neuroleptic butaclamol. *J. Med. Chem. 22*: 761–767.

L. L. Iversen, S. D. Iversen, and S. H. Snyder (Eds) (1978). Neuroleptics and schizophrenia. *Handbook of Psychopharmacology*, Vol. 10: Neuroleptics and Schizophrenia. Plenum Press, New York.

S. D. Iversen (1986). *Psychopharmacology*. Oxford University Press, Oxford.

I. Jirkovsky and W. Lippman (1978). Antidepressants. *Annu. Rep. Med. Chem. 13*: 1–10.

J. W. Kebabian, T. Agui, J. C. van Oene, K. Shigematsu, and J. M. Saavedra (1986). The $D_1$ dopamine receptor: new perspectives. *Trends Pharmacol. Sci. 7*: 96–99.

J. W. Langston (1985). Mechanism of MPTP toxicity: more answers, more questions. *Trends Pharmacol. Sci. 6*: 375–378.

T. A. Larsen and D. B. Calne (1985). Recent advances in the study of Parkinson's disease. In: *Neurotransmitters in Action* (D. Bousfield, Ed.). Elsevier, Amsterdam, pp. 252–256.

M. E. Lickey and B. Gordon (1983). *Drugs for Mental Illness*. W. H. Freeman, San Francisco.

C. B. Nemeroff and S. T. Cain (1985). Neurotensin—dopamine interaction in the CNS. *Trends Pharmacol. Sci. 6*: 201–205.

J. L. Neumeyer (1981). Neuroleptics and anxiolytic agents. In: *Principles of Medicinal Chemistry* (W. O. Foye, Ed.), 2nd ed. Lea and Febiger, Philadelphia, pp. 199–240.

A. H. Philipp, L. G. Humber, and K. Voith (1979). Mapping the dopamine receptor. II. Features derived from modifications in the ring A/B region of the neuroleptic butaclamol. *J. Med. Chem. 22*: 768–773.

A. G. Phillips, R. F. Lane, and C. D. Blaha (1986). Inhibition of dopamine release by cholecystokinin: relevance to schizophrenia. *Trends Pharmacol. 7*: 126–127.

N. P. Quinn (1984). Anti-Parkinson drugs today. *Drugs 28*: 236–262.

R. W. Ryall (1979). *Mechanism of Drug Action on the Nervous System*. Cambridge University Press, Cambridge, Chap 8.

J. M. Schaus and J. A. Clemens (1985). Dopamine receptors and dopaminergic agents. *Annu. Rep. Med. Chem. 20*: 41–50.

P. Seeman (1977). Anti-schizophrenic drugs—membrane receptor sites of action. *Biochem. Pharmacol. 26*: 1741–1748.

P. F. Spano, M. Trabucci, G. V. Corsini, and G. L. Gessa (Eds.) (1979). *Sulpiride and Other Benzamides*. Italian Brain Research Foundation Press/Raven Press, New York.

P. G. Strange (1986). Isolation and characterization of $D_2$ dopamine receptors. *Trends Pharmacol. Sci. 7*: 253–254.

P. Tyrer and A. Mackay (1986). Schizophrenia: no longer a functional psychosis. *Trends Neurosci. 9*: 537–538.

H. M. van Praag (1978). Amine hypothesis of affective disorders. In: *Handbook of Psychopharmacology* (L. L. Iversen, S. D. Iversen, and S. H. Snyder, Eds.)., Vol. 13. Plenum Press, New York, pp. 187–297.

G. N. Woodruff, A. Davis, C. D. Andrews, and J. A. Poat (1979). Dopamine receptors in the mammalian brain. In: *Recent Advances in Receptor Chemistry* (F. Gualtieri, M. Gianella, and C. Melchiorre, Eds.). Elsevier/North Holland, New York, pp. 165–188.

## 5. SEROTONIN AND THE SEROTONERGIC RECEPTORS

**Serotonin** (5-hydroxytryptamine, 5-HT) (4-128) is a central neurotransmitter which is also found peripherally in the intestinal mucosa and in blood platelets, where its role is unknown; it even occurs in plants, such as bananas. Although there is an enormous literature on the biochemistry and pharmacology of serotonin (cf. Essman, 1978), our knowledge of its biological role is very sketchy and fragmented (cf. Peroutka et al., 1981).

Serotonin
4-128

The serotonergic neuronal system in the CNS is rather restricted, and has been localized by fluorescence histochemistry and autoradiography in the raphe region of the pons and brainstem, projecting to the medulla and spinal cord. Because of histochemical difficulties, this neuronal mapping is still not very precise (see Cooper, Bloom, and Roth, 1986). The functional correlations of serotonergic neurons are equally difficult to elucidate, but work in this area has been helped by neurotoxins such as **5,6-** and **5,7-dihydroxytryptamine**, which destroy serotonergic neurons in the same way that 6-hydroxydopamine (4-36) atrophies adrenergic networks (Jacobi and Lytle, 1978).

The characterization and classification of serotonergic receptors is undergoing rapid and controversial development, and has to be viewed with great caution. Table 4.9 attempts a somewhat conservative tabulation of results obtained through

**Table 4.9.** Serotonergic receptor classification

|  | 5-HT$_1$ | 5-HT$_2$ | 5-HT$_3$ |
|---|---|---|---|
| Location | CNS<br>cardiovascular<br>skin<br>ileum | CNS<br>cardiovascular<br>respiratory system<br>urogenital system<br>blood platelets | neuronal? |
| Function | neuronal inhibition<br>vascular contraction | depolarization<br>muscle contraction<br>platelet aggregation | depolarization<br>pain<br>vomiting? |
| Agonist | 8-hydroxy-2-(N-dipropyl)aminotetralin<br>5-carbamoyltryptamine | (+)-S-α-methyl-5-HT | 2-methyl-5-HT |
| Antagonist | methysergide[a] | ketanserin[a] | ICS 205-930[a] |
| Second messenger | cAMP | phosphatidylinositol | ? |

[a] See Fig. 4.34.

the end of 1986. The 5-HT$_1$ receptor has been subdivided into three subcategories (A, B, and C) (see Richardson and Engel, 1986; Middlemiss et al., 1986), but the distinctions are so subtle that they do not belong in a textbook at this stage. This receptor group is widely distributed both centrally and peripherally, and also seems to fulfill the presynaptic autoreceptor role (5-HT$_{1B}$) (Gothert, 1982). It is primarily an inhibitory receptor, and operates through adenylate cyclase. It seems to be involved in the etiology of migraine headaches.

The 5-HT$_2$ receptor, which is better characterized than the 5-HT$_1$ receptor, mediates neuronal depolarization and muscle contraction as well as blood platelet aggregation. It is characterized through binding of its very specific antagonist, **ketanserin** (4-145). Although ketanserin has appreciable affinity for the $\alpha_1$-adrenoceptor and the H$_1$ histamine receptor, it is specific only in terms of the 5-HT receptors. It uses phosphatidylinositol as second messenger.

Although the recently described 5-HT$_3$ receptor is not well known, it has been subdivided by enthusiastic investigators into three subgroups. It seems to stimulate neurotransmitter release in a number of systems (ACh, NE) and may be responsible for pain mediation and drug-induced vomiting (Fozard, 1987). None of the 5-HT receptors has been purified or characterized biochemically; serotonergic synapses can, however, be localized readily by the fluorescence of 5-HT upon treatment with formaldehyde/glutaraldehyde.

### 5.1. The Biosynthesis and Fate of Serotonin

Serotonin metabolism, shown in Fig. 4.32, bears considerable similarities to that of the catecholamines. Serotonin itself is transformed in the *pineal gland* into **melatonin**, a hormone active in lightening skin pigmentation and suppressing the function of the female gonads. The $\beta$-adrenergic innervation of the pineal gland is

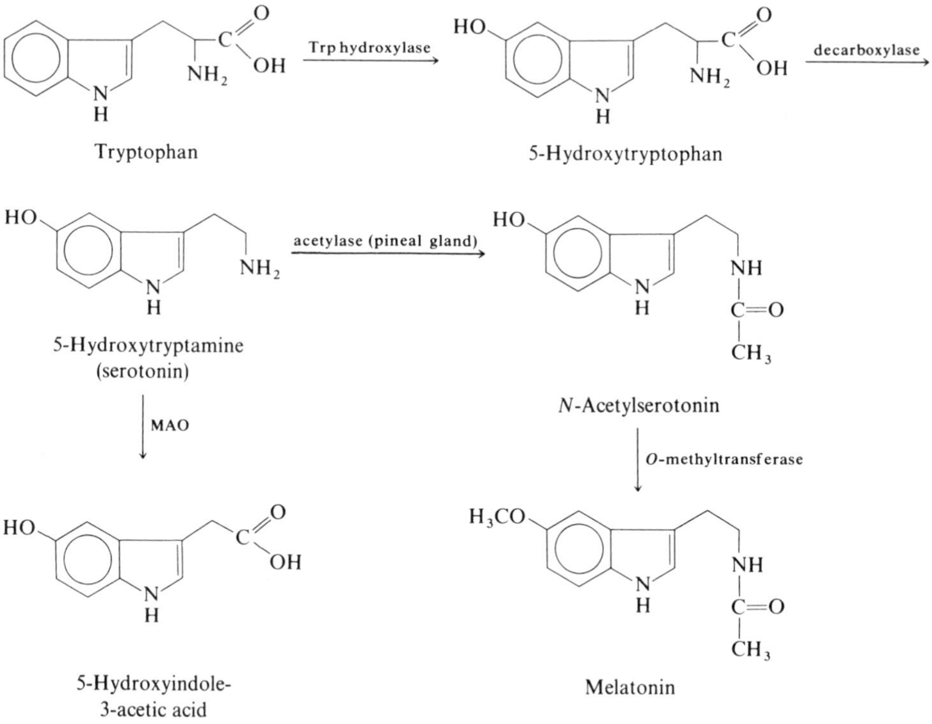

**Fig. 4.32.** Biosynthesis and degradation of serotonin.

governed by light: darkness increases cAMP formation and activation of the acetyltransferase enzyme, resulting in increased melatonin synthesis. Although situated on the thalamus, the pineal gland is not part of the CNS; it is a peripheral organ, as far as the blood–brain barrier is concerned. The role of this gland is still largely unknown.

Serotonin is *stored* in synaptic vesicles and blood platelets in the form of an ATP complex, in the ratio of 2:1 (Nogrady et al., 1972). Very little is known about its release, but exocytosis is the assumed mechanism. The released neurotransmitter is deactivated primarily by *reuptake*, but a significant amount is metabolized by MAO to the corresponding indoleacetic acid.

### 5.2. Presynaptic Serotonergic Drug Effects

#### 5.2.1. Serotonin Synthesis Inhibitors

Synthesis inhibitors block tryptophan hydroxylase, the first rate-determining enzyme in serotonin synthesis. Although **p-chlorophenylalanine** (4-129) can decrease serotonin levels by more than 90%, this treatment does not cause the sedation that is seen after catecholamine depletion with **reserpine**. Therefore, reserpine, although capable of depleting 5-HT vesicles, causes sedation by a catecholaminergic mechanism inhibiting uptake-2. It acts on the membrane of the synaptic vesicle and seems to prevent 5-HT and catecholamine uptake into the granule. **Fenfluramine** (Fig. 4.33, 4-130) is a rather selective 5-HT depletor, used to control appetite. Although structurally an amphetamine, it acts by a serotonergic rather than a catecholaminergic mechanism.

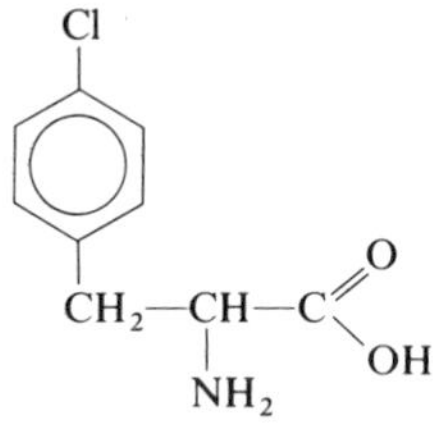

p-Chlorophenyl-
alanine
4-129

### 5.2.2. Serotonin Reuptake Inhibitors

Reuptake through the presynaptic membrane is, again, the major deactivation mechanism for serotonin. It is prevented by the tricyclic thymoleptics (anti-depressants; see Sec. 3.5.4) among which the tertiary amines are more potent at serotonergic terminals than are the secondary bases, whereas the reverse is true for catecholaminergic synapses.

Recently, a number of *selective* 5-HT reuptake inhibitors have been discovered, a few of which are shown in Fig. 4.33 (see Pinder, 1979, and Fuller, 1980). They are dimethylaminoethyl or dimethylaminopropyl derivatives of ring systems usually carrying an electron-attracting substituent ($-CF_3$ or $-CN$). The experimental drug **cyanoimipramine** (4-131) is one of the most potent and selective such drugs, 20 times more active against 5-HT uptake than chlorimipramine. All of these drugs are potent *antidepressants*, with a lower cardiotoxicity than the classical tricyclic agents, strongly suggesting that endogenous depression is a function of the availability of catecholamines as well as of serotonin. This does not contradict the previous discussion of the involvement of NE-dependent adenylate cyclase. In fact, both mechanisms probably contribute to antidepressant action, as shown by the two 5-HT reuptake inhibitors zimelidine and fluoxetin. Whereas **zimelidine** (4-132) elicits subsensitivity of the NE-coupled adenylate cyclase, **fluoxetine** (4-133) does not (see Sulser, 1979). The reuptake inhibitors are usually also active on blood platelets, which—like synapses—also have a high-affinity 5-HT uptake pump.

### 5.3. Postsynaptic Serotonergic Drug Effects

### 5.3.1. Serotonergic Agonists

The serotonergic agonists constitute a relatively small group of compounds, including **serotonin** (4-128) itself, **bufotenin** (4-138), a natural product found in plants as well as toad-skin secretion (hence the name; *Bufo* = toad), bufotenin methyl ether, and two piperazine derivatives: **quipazine** (4-139) and compound 4-140 which acts mainly on the postsynaptic receptors. A large number of analogues have been prepared and tested over the years (see Horn, 1975), allowing the following

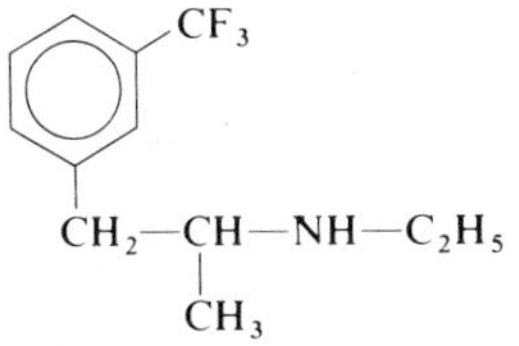

Fenfluramine
4-130

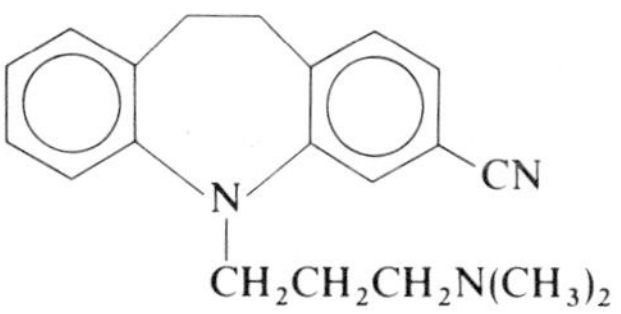

Cyanoimipramine
Ro 11-2465
4-131

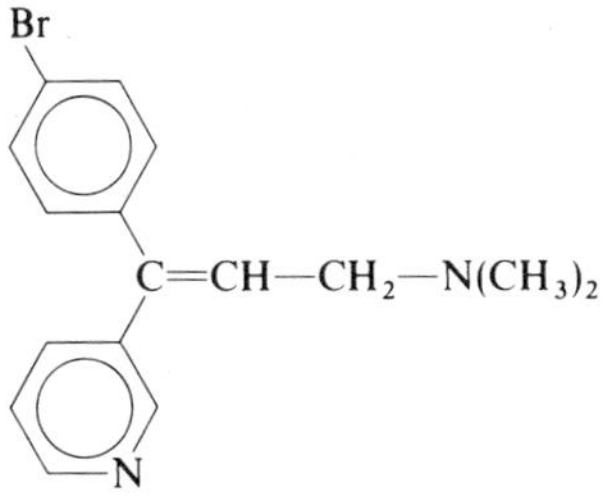

Zimelidine
4-132

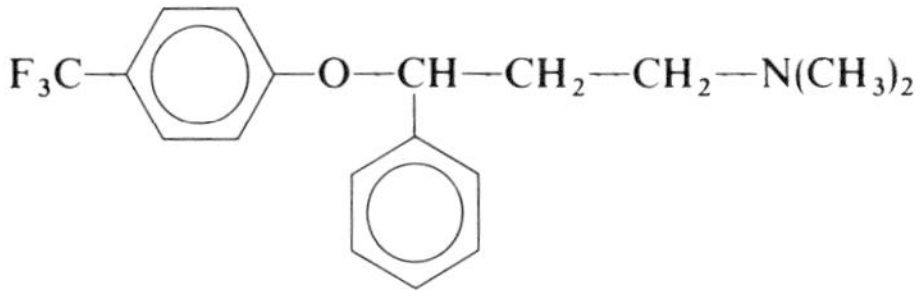

Fluoxetine
4-133

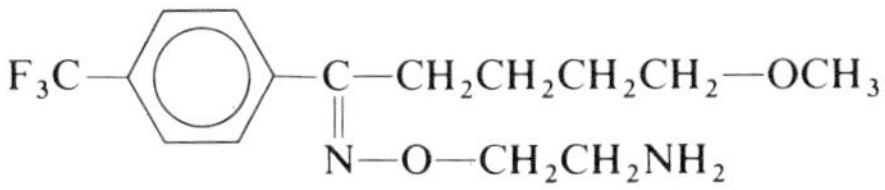

Fluvoxamine
4-134

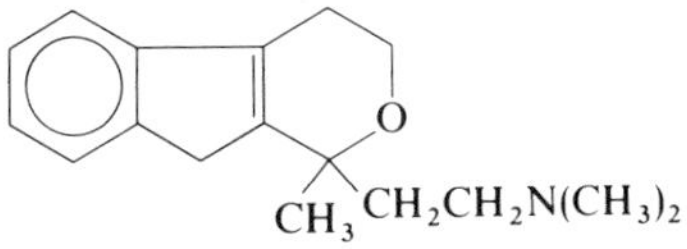

Pirandamine
4-135

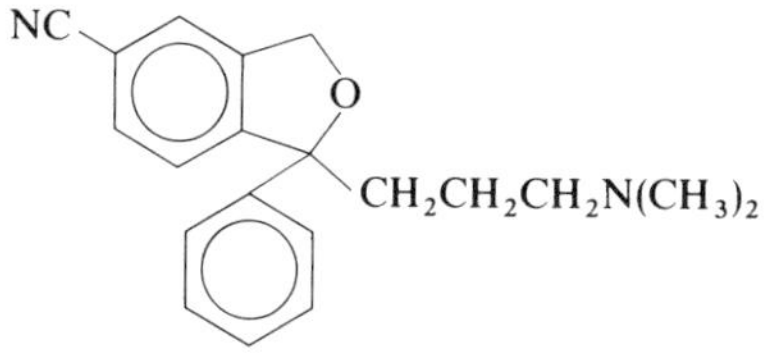

Citalopram
4-136

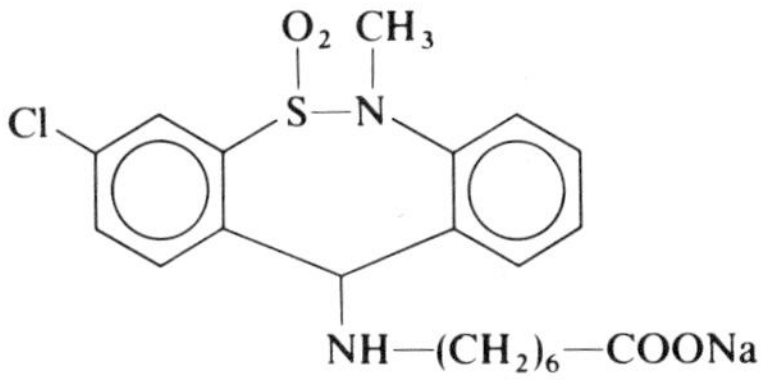

Tianeptine-Na
4-137

**Fig. 4.33.** Selective serotonin uptake inhibitors.

conclusions to be drawn regarding peripheral (uterus, stomach, smooth muscle)
structure–activity correlations:

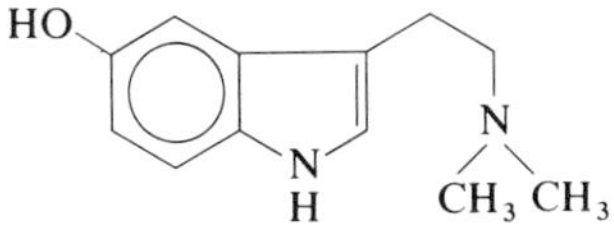

Bufotenin
4-138

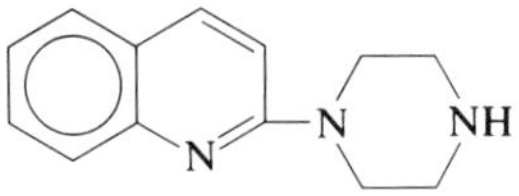

Quipazine
4-139

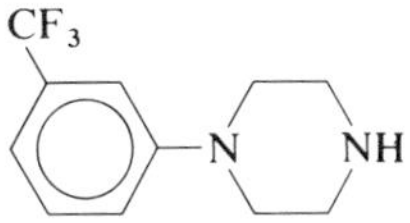

N-(3-Trifluoromethylphenyl)-
piperazine
4-140

1. The free 5-OH group is important, and can be shifted only to position 4 without serious loss of activity. Methylation decreases activity; replacement by halogen or alkyl abolishes it.
2. The position of the side chain on C-3, and its length, are crucial. α-Methyl groups are permissible, but an —OH will abolish smooth-muscle effects. N-Alkylation usually leads to a decrease in activity.

### 5.3.2. Physiological Effects of Serotonin

The physiological effects of serotonin are varied and widespread (Fuller, 1980), some suggesting that they are amenable to modification by serotonergic drugs. Others have led to speculation on the possible involvement of serotonin in mental disease and on the activity of hallucinogenic drugs.

The administration of serotonin leads to powerful *smooth-muscle effects* in the cardiovascular and gastrointestinal systems. Vasodilation and prolonged hypotension result, partly through central effects, if the serotonin concentration in the CNS is increased by administration of the serotonin precursor 5-hydroxytryptophan. Unlike serotonin, this precursor can cross the blood–brain barrier. Reuptake inhibitors like **fenfluramine** (4-130) will increase this activity, but other such drugs are promising hypotensive agents in themselves (e.g., **quipazine**, 4-139). *Intestinal mobility* is also influenced by serotonin.

Serotonin has an effect on the *hypothalamic control of pituitary function* (see Chap. 5, Sec. 2.2), in central *thermoregulation* (attributed to the $5\text{-HT}_{1A}$ receptor), and in *pain* perception (probably the $5\text{-HT}_3$ receptor)—where increased serotonergic function potentiates opiate analgesia. The administration of 5-HT reuptake inhibitors like **fluoxetine** (4-133) or **fenfluramine** (4-130), a 5-HT-depleting amphetamine derivative, increases the anorectic effect of 5-hydroxytryptamine and induces a selective suppression of nonprotein caloric intake in rats. There are obvious favorable implication of this for the control of obesity (Sullivan et al., 1976). The involvement of serotonin in endogenous *mental depression* has been mentioned (see also Glennon, 1987).

**Sleep.** Another very controversial but exciting area of research is the potential role of serotonin in sleep. 5-Hydroxytryptamine may trigger slow-wave sleep (non-REM sleep), whereas the muscarinic AChR and NE are involved in REM sleep (rapid eye movement sleep, *paradoxical* sleep, dream sleep). For a more detailed discussion of this subject, see Vida (1981) and Gillin et al. (1978). Besides the aminergic regulation of sleep, recent research has identified several other presumed sleep factors: delta-sleep-inducing peptide, sleep-promoting substance, interleukin-1, prostaglandin $D_2$ (Inoué and Borbély, 1985), and muramyl peptides (Krueger, 1985). The latter compound resembles the bacterial cell-wall monomer (see Chap. 6, Sec. 2), and although it is not synthesized by mammalian organisms, it may originate from the intestinal bacteria and be modified. In this context, it could be considered a vitamin-like factor.

### 5.3.3. Serotonin Antagonists

Serotonin antagonists are difficult to investigate because the criteria for 5-HT receptor blockage are quite ambiguous. (see Middlemiss et al., 1986). Some

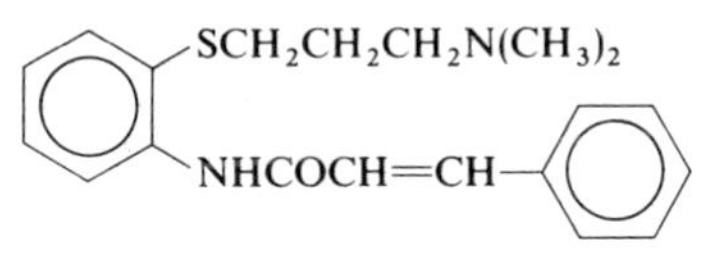

Cinanserin
4-141

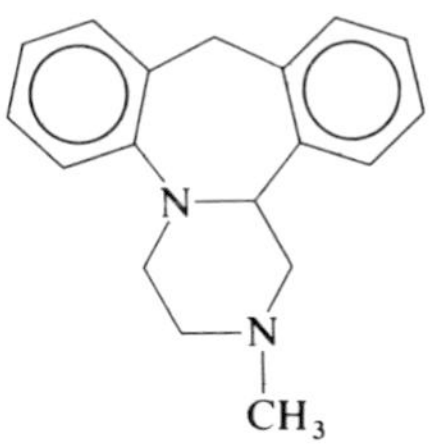

Mianserin
4-142

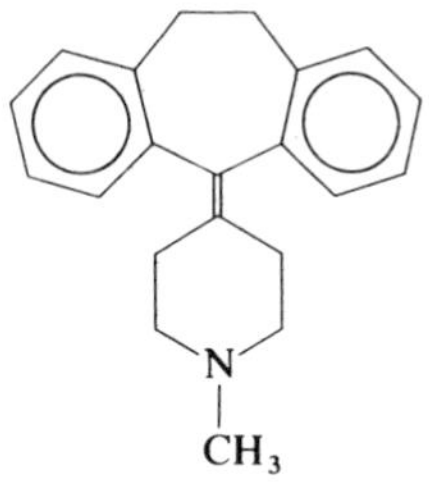

Cyproheptadine
4-143

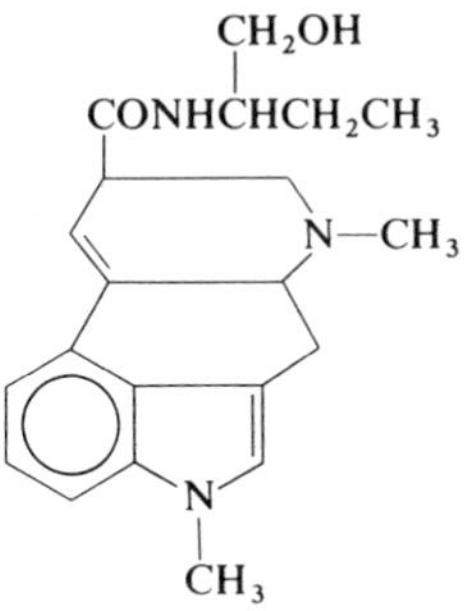

Methysergide
4-144

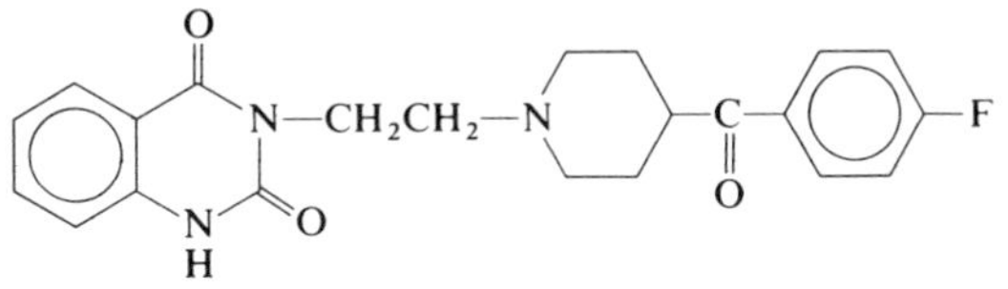

Ketanserin
4-145

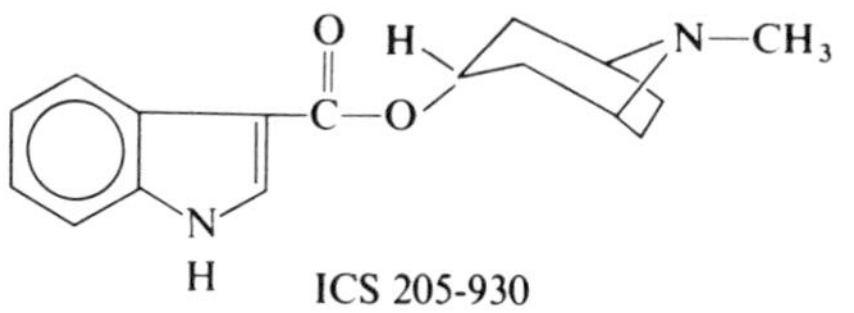

ICS 205-930
4-146

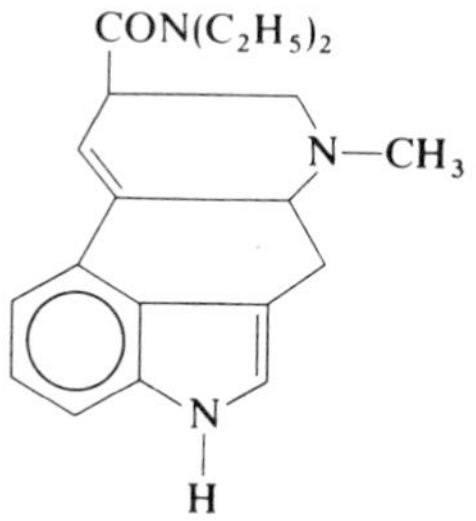

LSD
4-147

**Fig. 4.34.** Some serotonin antagonists.

antagonists that block 5-HT activity peripherally can mimic the neurotransmitter effect if introduced into the CNS by microiontophoresis, indicating multiple receptor sites for 5-HT.

Some serotonin antagonists are shown in Fig. 4.34. **Cinanserin** (4-141) is not only an antiserotonin drug, but also an analgesic and immunosuppresant. **Mianserin** (4-142) is not an antagonist but an agonist in high doses, and is a useful antidepressant. **Cyproheptadine** (4-143) is an antihistamine, in addition to being a 5-HT blocking agent. The other compounds shown are all semisynthetic derivatives of lysergic acid, obtained from *ergot alkaloids* (see Sec. 4.2). **Methysergide** (4-144) is related to the ergot alkaloid ergonovine, and oxytocic (uterus-contracting) drug. It is one of the most potent 5-HT antagonists, and is used for the prevention but not the treatment of migraine headaches; the mechanisms of this action are unknown. **Ergotamine**, another peptide ergot alkaloid, can be used for treating existing migraines, since it probably acts as a vasoconstrictor. **Ketanserin** (4-145), a new $\alpha_1$ and 5-HT$_2$ antagonist, is a clinically effective hypotensive agent acting through a unique mechanism. Related compounds show anxiolytic activity. The diethylamide of lysergic acid, **LSD** (4-147), is a widely studied and abused hallucinogen. (The acronym originates from the German name: Lyserg-säure-diethylamid.)

### 5.3.4. Hallucinogens and Other Psychotomimetic Agents

These drugs seem to act on central 5-HT neurons in a manner that is not clear. They decrease the turnover of serotonin, possibly through a presynaptic receptor in the raphe cells. Since 5-HT is an inhibitory neurotransmitter in many of its actions, the removal of this inhibition could lead to behavioral changes. However, to discredit this simple hypothesis, there are a number of LSD derivatives that are not hallucinogens (e.g., the 2-bromo derivative). The effects of LSD are also seen in animals with raphe lesions that have destroyed the serotonergic neurons. Cooper, Bloom, and Roth (1986) deal with this question, and some aspects are covered by Sankar (1975).

There are indications that other psychotomimetic agents also act through a central 5-HT mechanism. Figure 4.35 shows the structure of some of them. Four groups can be distinguished:

1. Lysergic acid diethylamide and related indolalkylamines.
2. Phenylethylamines (e.g., mescaline).
3. Cannabis derivatives.
4. Anticholinergics; however, since they do not act through a serotonergic mechanism they will not be discussed here.

**Lysergic acid diethylamide** (4-147) is a rather structure-specific compound. Only the (+)-isomer is active, and alkylamides other than the diethyl derivative, including some cyclic analogues (the pyrrolidide and morpholide analogues), have very low activity. Lysergic acid does contain the 3-indolylethylamine moiety, and it is therefore not surprising that other such structures are also hallucinogens. Among natural products, **psilocin** (4-148) and its phosphate ester, psilocybin, occur in a Mexican mushroom (*Psilocybe*). **Harmaline** (4-149), an alkaloid (from *Peganum harmala* and some other plants), and some related compounds are also effective hallucinogens.

After the ingestion of hallucinogens, a great variety of symptoms may occur, including dizziness, tension, perceptual changes of size, time, and distance, visual hallucinations, a crossover from hearing to seeing (colors may be "heard" and sounds "seen"), mood changes, and potential panic. These effects may last for about 12 hours. Tolerance develops quickly, and there is cross-tolerance with phenylethylamines but not with amphetamines. The psychological hazards ("bad trip") of LSD use are very real; the physiologically harmful effects, if any, are not clear-cut.

Phenylethylamines, a fragment of LSD, and a number of alkoxyphenylethylamines have been used by various primitive societies as hallucinogens. The best

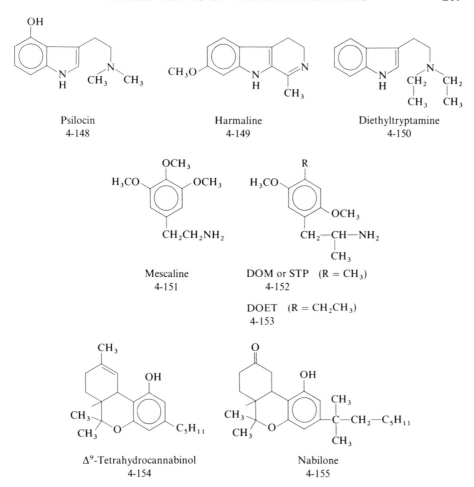

**Fig. 4.35.** Some psychotomimetic compounds.

known is **mescaline** (4-151), which occurs in a number of cacti (*Lophophora,
Trichocereus*) native to Mexico. In large doses (300–500 mg) it causes vivid and
colorful hallucinations, perception of the environment as unusually beautiful, and
increased insight ("mind-expanding experience"). The effect is increased by attaching a methoxy group to the ortho position and using alkyl substituents. **DOM** or
**STP** (4-152), a synthetic compound, is an example of such a drug. One corresponding amphetamine is **DOET** (4-153), which is about 150 times as active as mescaline.

The third group of psychotomimetic compounds are the *cannabinoids*, represented by **Δ⁹-tetrahydrocannabinol** (THC) (4-154), the principal active ingredient of
marijuana and hashish, produced by the hemp *Cannabis sativa*. It is not considered
a hallucinogen, is not habit forming, and seems to have no adverse physiological
effects except in habitual consumers of large quantities. Extremely high doses may
lead to paranoid depersonalization.

The emotional propaganda (both pro and con) surrounding cannabinoids notwithstanding, cannabinoid derivatives—if not THC itself—have interesting therapeutic possibilities (see Lemberger, 1980; Mechoulam, 1986). One of the cannabinoid derivatives in clinical use, **nabilone** (4-155), has a very selective anti-emetic activity in patients suffering from toxic side effects of cancer chemotherapy. Other derivatives show anticonvulsant and analgesic activity, and decrease ocular pressure in glaucoma. The structure–activity correlations of these compounds have been quite thoroughly explored (Razdan, 1986). None of them is psychotomimetic.

Details of psychotomimetic pharmacology are covered in Goodman-Gilman et al. (1985).

### Selected Readings

R. W. Brimblecombe and R. Pinder (1975). *Hallucinogenic Agents.* Wright-Scientechnica, Bristol.

J. R. Cooper, F. E. Bloom, and R. H. Roth (1986). *The Biochemical Basis of Neuropharmacology,* 5th. ed. Oxford University Press, New York.

W. B. Essman (Ed.) (1978). *Serotonin in Health and Disease,* 5 vols. S. P. Medical and Scientific Book Division, Spectrum Publications, New York.

J. R. Fozard (1987). 5-HT$_3$ receptors and cytotoxic drug-induced vomiting. *Trends Pharmacol. Sci. 8*: 44–45.

R. W. Fuller (1980). *Pharmacology of central serotoninergic neurons. Annu. Rev. Pharmacol. Toxicol 20*: 111–127.

J. C. Gillin, W. B. Mendelson, N. Sitaram, and R. J. Wyatt (1978). The neuropharmacology of sleep and wakefulness. *Annu. Rev. Pharmacol. Toxicol. 18*: 563–579.

R. A. Glennon (1987). Central serotonin receptors as targets for drug research. *J. Med. Chem. 30*: 1–12.

A. Goodman-Gilman, L. S. Goodman. T. W. Wall, and F. Murad (Eds.) (1985). *Goodman and Gilman's The Pharmacological Basis of Therapeutics,* 7th ed. Macmillan New York.

M. Gothert (1982). Modulation of serotonin release in the brain via presynaptic receptors. *Trends Pharmacol. Sci. 3*: 437–440.

A. R. Green (1985). *Neuropharmacology of Serotonin.* Oxford University Press, New York.

A. S. Horn (1975). SAR for neurotransmitter receptor agonists and antagonists. In: *Handbook of Psychopharmacology* (L. L. Iversen, S. D. Iversen, and S. H. Snyder, Eds.), Vol. 2. Plenum Press, New York, pp. 179–243.

S. Inoué and A. A. Borbély (1985). *Endogenous Sleep Substances and Sleep Regulation.* VNU Science Press, Utrecht.

J. H. Jacobi and L. D. Lytle (Eds.) (1978). Serotonin neurotoxins. *Ann. N. Y. Acad. Sci. 305.*

J. M. Krueger (1985). Somnogenic activity of muramyl peptides. *Trends Pharmacol. Sci. 6*: 218–221.

L. Lemberger (1980). Potential therapeutic usefulness of marijuana. *Annu. Rev. Pharmacol. Toxicol. 20*: 151–172.

R. Mechoulam (1986). *Cannabinoids as Therapeutic Agents.* CRC Press, Boca Raton.

D. N. Middlemiss, M. Hibert, and J. R. Fozard (1986). Drugs acting at central 5-hydroxytryptamine receptors. *Annu. Rep. Med. Chem. 21*: 41–50.

T. Nogrady, P. D. Hrdina, and G. M. Ling (1972). Investigation of serotonin–ATP association in vitro by NMR and UV spectroscopy. *Mol. Pharmacol. 8*: 565–577.

S. J. Peroutka, R. M. Lebovitz, and S. H. Snyder (1981). Two distinct central serotonin receptors with different physiological functions. *Science 212*: 827–829.

R. M. Pinder (1979). Antidepressants. *Annu. Rep. Med. Chem. 14*: 1–11.

R. K. Razdan (1986). Structure–activity relationships in cannabinoids. *Pharmacol. Rev. 38*: 75–149.

B. P. Richardson and G. Engel (1986). The pharmacology and function of 5-HT$_3$ receptors. *Trends Neurosci. 9*: 424–428.

D. V. S. Sankar (1975). *LSD: A Total Study*. PJD Publishing Co., Westbury, New York.

P. R. Saxena and P. B. Bradley (1986). Functional receptors for 5-hydroxytryptamine. *Trends Pharmacol. Sci. 7*(7): centerfold.

A. C. Sullivan, L. Cheng, and J. G. Hamilton (1976). Agents for the treatment of obesity. *Annu. Rep Med. Chem. 11*: 200–208.

F. Sulser (1979). New perspectives on the mode of action of anti-depressant drugs. *Trends Pharmacol. Sci. 1*: 92–94.

J. A. Vida (1981). Central nervous system depressants: sedative hypnotics. In: *Principles of Medicinal Chemistry*, 2nd ed. (W. O. Foye, Ed). Lea and Febiger, Philadelphia, pp. 155–181.

## 6. HISTAMINE AND THE HISTAMINE RECEPTORS

Since its discovery in 1910, histamine has been considered akin to a "local" hormone (*autacoid*), although lacking an endocrine gland in the classical sense. In the past few years, however, the role of histamine as a central neurotransmitter has been recognized, and a considerable amount of research is now directed toward elucidating its central effects and receptors. The discovery of the duality of the histamine receptor added another dimension to this complex field, leading to new and successful therapeutic as well as theoretical investigations.

### 6.1. Structure, Conformation, and Prototropic Equilibria of Histamine

Protonation has been a important aspect in the design of some histamine antagonists (see Sec. 6.3). Figure 4.36 shows the tautomeric equilibria between different histamine species, and the respective mole percentages of these species. The most important among them is the $N^\tau$—H (tele-) tautomer, which also appears to be the active form of the agonist on both receptors. Tautomerism does not appear to be important in H$_1$-receptor binding (in the intestine); however, it does seem to be important to gastric H$_2$-receptor activity. Histamine may play the role of a proton-transfer agent (Fig. 4.37), in a fashion similar to the charge-relay role of the imidazole ring in serine esterases (e.g., acetylcholinesterase; see Chap. 6, Sec. 3.2). The percentage of the monocation tautomers is greatly influenced by substituents in position 4, which alter the electron density on the $N^\pi$ atom (Table 4.10), an important consideration in modifying the receptor binding properties of histamine.

*Histamine metabolism* differs from that of classical neurotransmitters because histamine is so widely distributed in the body. The highest concentrations in human

| Mole Percentages of Ionic Species | At pH 7.4 | At pH 5.4 |
|---|---|---|
| Dication | 2.4% | 71.5% |
| Monocation | 96.6 | 28.5 |
| Neutral | 1 | $3 \times 10^{-3}$ |
| Anion | $2.5 \times 10^{-7}$ | $1 \times 10^{-11}$ |

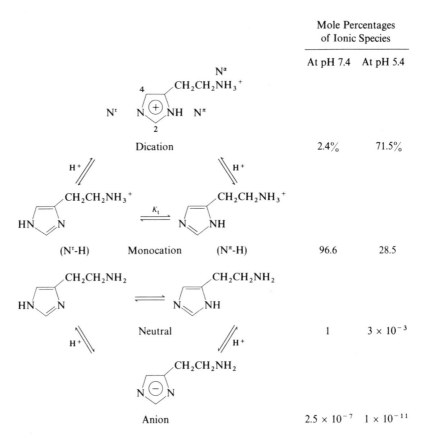

**Fig. 4.36.** Ionic and tautomeric equilibria between histamine species and their respective mole percentages. (Reproduced by permission from Durant (1979), Elsevier/North Holland, New York)

**Fig. 4.37.** 4-Substituted histamine derivatives act as proton-transfer agents in their tele-tautomeric form. This may be important for $H_2$ receptor ligands.

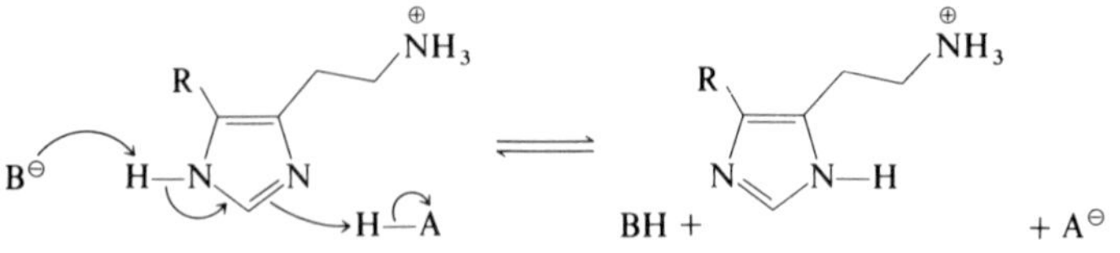

**Table 4.10.** Tautomer concentration ratios, tautomer mole fractions, and relative $H_2$-agonist activities of 4-substituted histamines (histamine = 100)

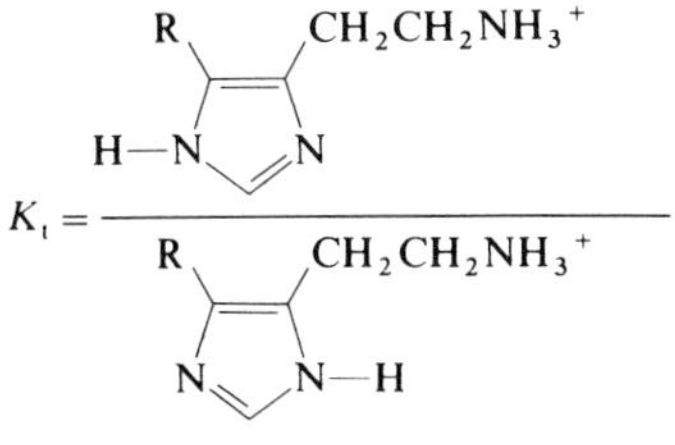

$$K_t = \frac{R \diagdown CH_2CH_2NH_3{}^+ \text{(H—N, N)}}{R \diagdown CH_2CH_2NH_3{}^+ \text{(N, N—H)}}$$

| Compound | R | $K_t{}^a$ | Percent mole fraction[b] $N^\tau$—H | $H_2$ receptor agonist activity (%)[c] |
|---|---|---|---|---|
| Histamine | H | 2.4 | 71 | 100 |
| | $CH_3$ | 4.1 | 80 | 43[d] |
| | Cl | 0.13 | 12 | 11 |
| | Br | 0.11 | 10 | 9[e] |
| | $NO_2$ | 0.009 | 0.9 | 0.6 |

[a] $K_{t,R}$ = antilog $[3.4\sigma_{m,CH_2CH_2NH_3} - \sigma_{m,R}]$; $\sigma_{m,CH_2CH_2NH_3}$ is taken as $+0.11$.

[b] Mole fraction of monocation— not the mole fraction of total species, which would be pH dependent.

[c] Activities determined *in vitro* on guinea pig right atrium, in the presence of propranolol, expressed relative to histamine = 100.

[d] 95% fiducial limits 40–46.

[e] 95% fiducial limits 7.4–10.1.

Reproduced by permission from Ganellin (1977), University Park Press, Baltimore.

tissues are found in the lung, stomach, and skin (33 $\mu g/g$ tissue or less). The very simple histamine metabolic pathways are shown in Fig. 4.38; histamine is produced from histidine in just one step. The principal production takes place in the *mast cells* of the peritoneal cavity and connective tissues; the stored form is granular and can be depleted by special agents like 48/80 [a polymer of *N*-(*p*-methoxyphenylethyl)methylamine with formaldehyde], which destroys mast cells. The gastric mucosa is another major storage tissue, and small amounts of histamine can be found in the brain, but its localization in neurons has not been proven unequivocally.

Histamine can be released from mast cells in antigen–antibody reactions, as in anaphylaxis and allergy, which are the most widely known physiological reactions to histamine. However, these potentially fatal reactions are not caused by histamine alone. Other agents present in mast cells, such as serotonin, acetylcholine, bradykinin (a nonapeptide), and a "slow-reacting substance" or leukotriene (cf. Chap. 5. Sec. 4.1), also contribute. In the stomach, where histamine induces acid secretion, its release seems to be regulated by the peptide hormone *pentagastrin*.

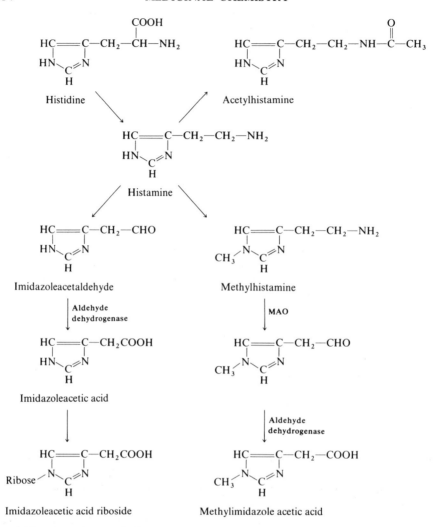

**Fig. 4.38.** Histamine metabolism.

## 6.2. The Histamine Receptors

Classically, these receptors have also been divided into two groups. The first of these, the $H_1$ receptors, were described by Schild in 1966. The $H_2$ receptors were discovered in 1972 by Black et al. (see Ganellin, 1977; Brown and Young, 1985). The $H_1$ *receptor* is found in the smooth muscle of the intestines, bronchi, and blood vessels and is blocked by the "classical" antihistamines (Fig. 4.39). The $H_2$ *receptor*, present in gastric parietal cells, in guinea pig atria, and in the uterus, does not react to $H_1$ blockers, but only to specific $H_2$ antagonists. $H_2$ receptors also appear to be involved in the immunoregulatory system and may be present in T lymphocytes, basophil cells, and mast cells.

Both $H_1$ and $H_2$ receptors are present in the vascular system as well as in the CNS. In the latter, tricyclic antidepressant drugs also seem to interact with histamine receptors.

Histamine receptor subtypes in the CNS and the central neurotransmitter role of histamine have been the subject of many recent investigations. It is currently believed (Schwartz et al., 1986a, 1986b) that there are three central histamine receptors:

1. $H_1$ *receptors* are widely distributed, especially in the cortex. Since histamine does not easily cross the blood–brain barrier (see Chap. 7, Sec. 1.5), central receptors can use only the locally synthesized histamine. Occupation of the $H_1$ receptor by agonist does not activate adenylate cyclase (AC), but seems to potentiate direct activators at $H_2$ and adenosine receptors. It appears to use phosphoinositol as second messenger. $H_1$ receptors are not very specific and are occupied by antidepressants and neuroleptics as well. This explains the sedative effect of all three classes of drugs. The $H_1$ receptors are easily solubilized and have been purified on lectin affinity columns, indicating their glycoprotein nature.

2. $H_2$ *receptors* are coupled to AC; their stimulation has a central disinhibitory effect, due to the decrease in $Ca^{2+}$-activated $K^+$ conductance. Thus the role of the central histamine receptor may not be information transmission, but sensitization of brain areas to excitatory signals from "waking amines."

3. $H_3$ *receptors* have recently been described; these seem to be presynaptic autoreceptors, controlling histamine release and synthesis. They are activated by histamine concentrations that are two orders of magnitude lower than those necessary for triggering postsynaptic receptors. Their blockade may potentially lead to increased blood flow and metabolism combined with a central arousal, whereas their stimulation (or inhibition of central $H_2$ receptors) could have an anticonvulsant or sedative effect. However, no $H_3$-specific drugs have been described to date.

Regarding receptor distribution, direct binding experiments with tritiated **mepyramine** (4-165, Fig. 4.39) show a strikingly uneven distribution of $H_1$ receptors in various brain regions.

Knowledge of the physiological role of histamine in the CNS and evidence for the existence of discrete neuronal networks that could be called histaminergic are currently rather vague (Cooper et al., 1986). Histamine-mediated hypothermia, emesis, and hypertension have been shown to exist, and the well-known sedative effects of $H_1$-antihistamines are centrally mediated.

The histamine receptors have not yet been characterized according to their physicochemical and biochemical properties, mainly because of a lack of sufficiently specific ligands and affinity labels.

### 6.3.  Histamine Agonists

A selection of agonists is shown in Table 4.11, together with an indication of their selectivity for $H_1$ and $H_2$ receptors. This table also gives an idea of the structure–activity effects of variations in the ring. **4-Methylhistamine** (4-156) is highly selective

**Table 4.11.** $H_1$ and $H_2$ agonist activities of some histamine analogues

| | | | Agonist activity rel. to histamine ($=100$) | |
|---|---|---|---|---|
| | Het-CH$_2$CH$_2$X | | $H_1$ receptor[a] | $H_2$ receptor[b] |
| | Het | X | | |
| 4-156 | 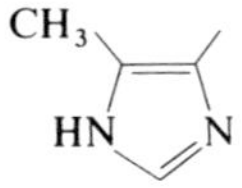 | NH$_2$ | 0.23 | 39 |
| 4-157 | 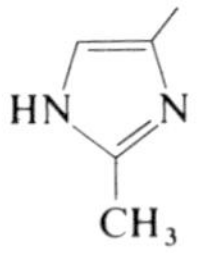 | NH$_2$ | 16.5 | 2.0 |
| 4-158 | 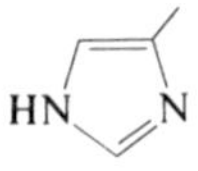 | N(CH$_3$)$_2$ | 44 | 19 |
| 4-159 | 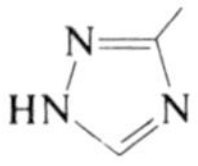 | NH$_2$ | 12.7 | 13.7 |
| 4-160 | 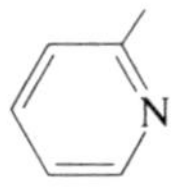 | NH$_2$ | 5.6 | ~0.2 |
| 4-161 | 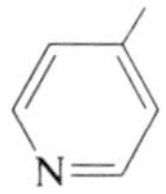 | NH$_2$ | <0.001 | ~0.4 |
| 4-162 |  | NH$_2$ | 26 | ~0.3 |
| 4-163 | 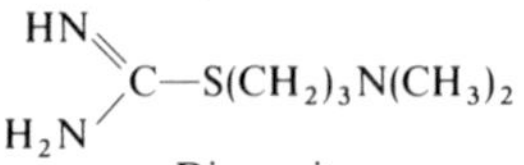 C—S(CH$_2$)$_3$N(CH$_3$)$_2$  Dimaprit | | <0.0001 | 19.5 (70%)[c] |
| 4-164 | 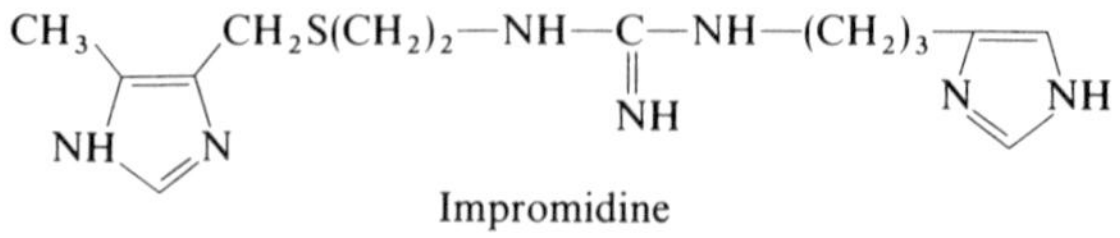 Impromidine | | <0.001 | 1680 |

[a] Tested for contraction on isolated guinea pig ileum in the presence of atropine.

[b] Tested for stimulation of gastric acid secretion in the anesthetized rat.

[c] Guinea pig right atrium.

Reproduced by permission from Durant (1979), Elsevier/North Holland, New York.

for the $H_2$ receptor, whereas the 2-methyl derivative (4-157) is a weak but usable $H_1$ agonist. The fact that the 2-pyridyl (4-160) and the 2-thiazolyl rings (4-162) also lend $H_1$ activity to histamine derivatives shows that tautomerism is not an issue in $H_1$ activity. Large alkyl groups on C-4 decrease activity and lead to partial agonists, whereas side-chain N-substitution enhances the antagonistic properties of the molecule.

The most interesting histamine agonist is **dimaprit** (4-163), which was described a few years ago (Durant, 1979). It is a selective $H_2$ agonist, having between 19% and 70% $H_2$ activity, with no effect on the $H_1$ receptor. The isothiourea system in dimaprit has a planar electron sextet, like that of the imidazole ring in histamine, and is capable of tautomerism as well as of donating and accepting hydrogen. It produces a higher maximum gastric acid secretion in dogs than does histamine. The most active and most selective derivative is **impromidine** (4-164), which acts exclusively on the $H_2$ receptor.

The pharmacological effects of histamine may be summarized as follows:

1. The circulatory effects are manifested as arteriolar dilation and increased capillary permeability, causing plasma loss. The localized redness, edema (hives, wheal), and diffuse redness seen in allergic urticaria (rash) or physical skin injury result from these circulatory changes. Vasodilation also causes a decrease in blood pressure. The involvement of the $Ca^{2+}$-ionophore effect of phosphatidic acid, derived from phosphatidylinositol, has been implied.
2. The effects on the heart ($H_2$ response) are minor, but the heart rate increases.
3. Humans and guinea pigs are very prone to bronchoconstriction by histamine (an $H_1$ effect), and severe asthmatic attacks can be triggered by small doses, provided the person suffers from asthma and is therefore very sensitive to histamine.
4. Stimulation of gastric acid secretion is the most important $H_2$ response; is blocked only by $H_2$ antagonists. As mentioned before, the hormone gastrin may be involved in histamine release, because $H_2$ antagonists block gastrin-induced acid secretion.

## 6.4. Histamine Antagonists

### 6.4.1. $H_1$ Antagonists

Antagonists of the $H_1$ receptor were first discovered by Bovet in 1933. As shown in Fig. 4.39, they do not bear any close resemblance to the agonist, since their binding involves accessory binding sites. Their general structure is

$$\begin{array}{c} Ar \\ \phantom{Ar} \diagdown \\ \phantom{Ar}\phantom{a} X-C-C-\overset{\oplus}{N}H \\ \phantom{Ar} \diagup \\ Ar_1 \end{array} \begin{array}{c} R \\ \diagup \\ \phantom{a} \\ \diagdown \\ R \end{array}$$

**Ethylenediamines** (4-165 and 4-166), aminoalkyl ethers (4-167 and 4-168), and aminopropyl compounds (4-169 and 4-170), for which X is nitrogen, oxygen, and

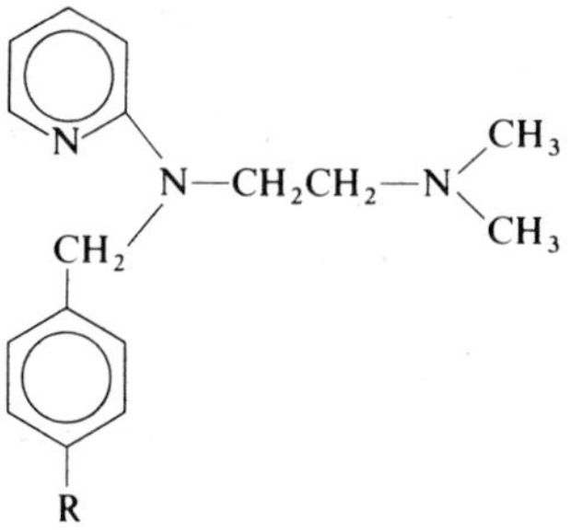

Tripelennamine (R = H)
Mepyramine (R = OCH₃)
4-165

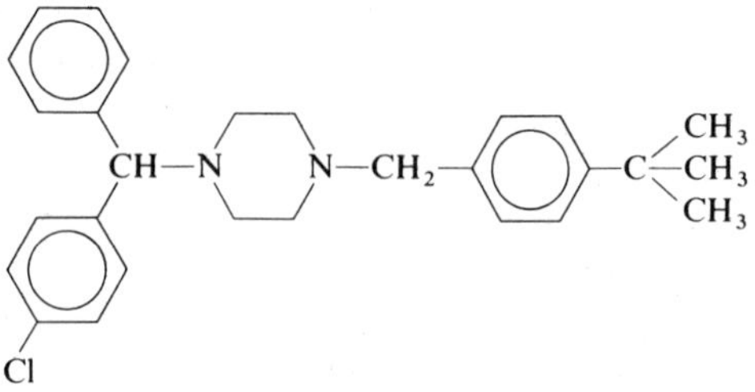

Buclizine
4-166

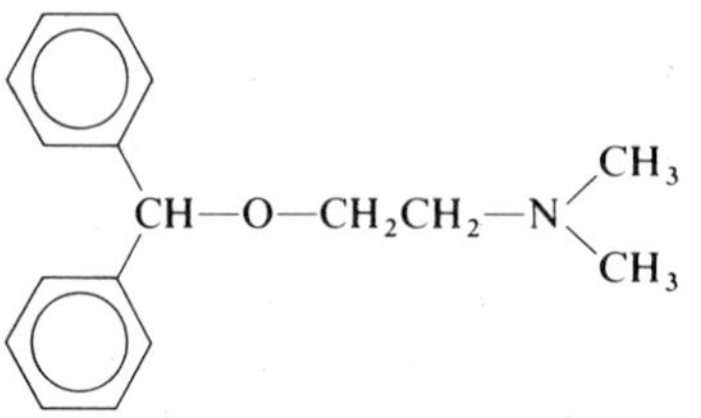

Diphenhydramine
4-167

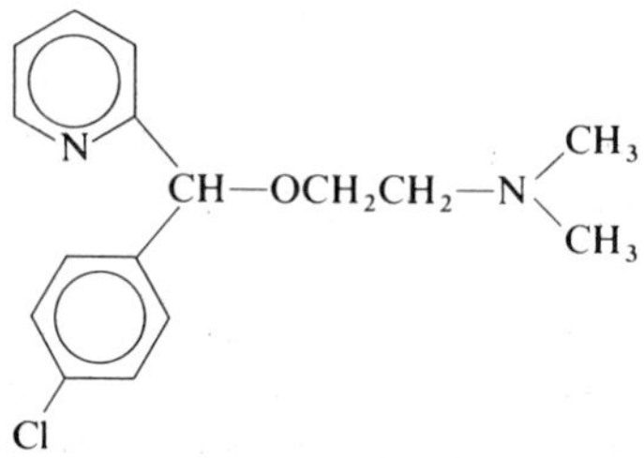

(+)Carbinoxamine
4-168

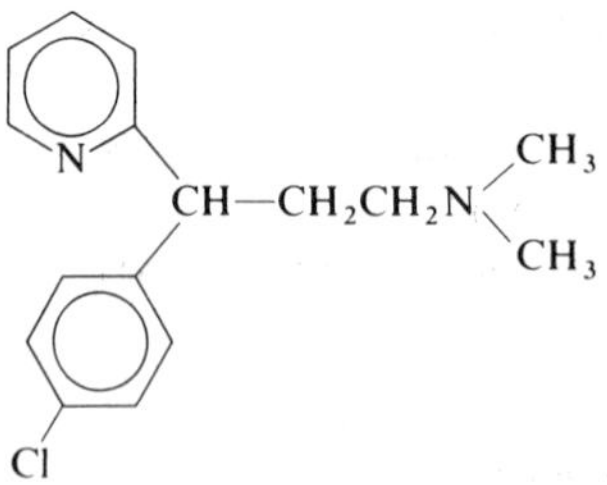

Chlorpheniramine
4-169

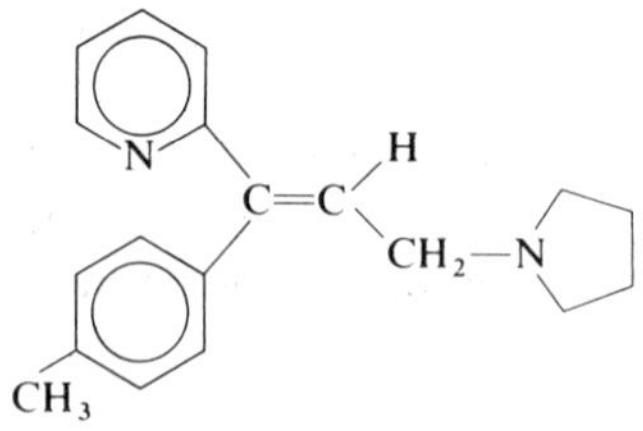

Triprolidine
4-170

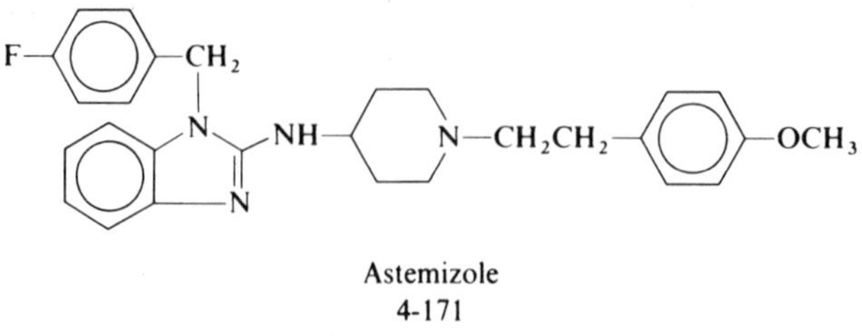

Astemizole
4-171

**Fig. 4.39.** H₁ histamine antagonists.

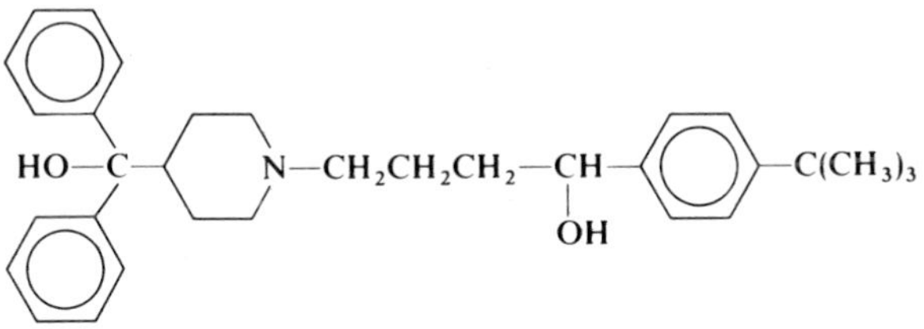

Terfenadine
4-172

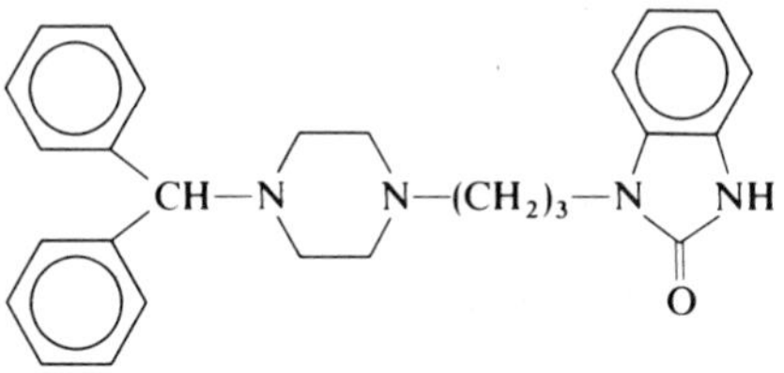

Oxatomide
4-173

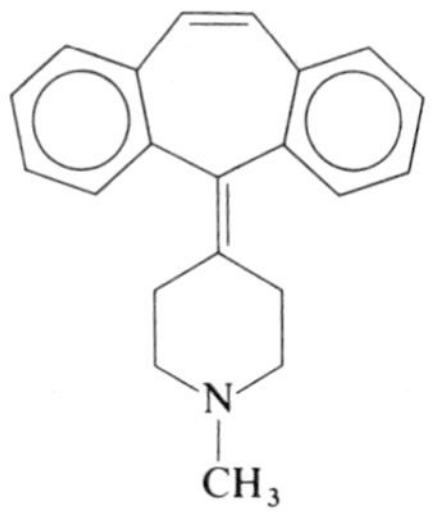

Cyproheptadine
4-174

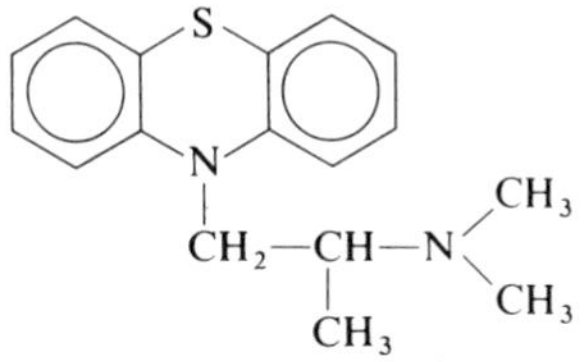

Promethazine
4-175

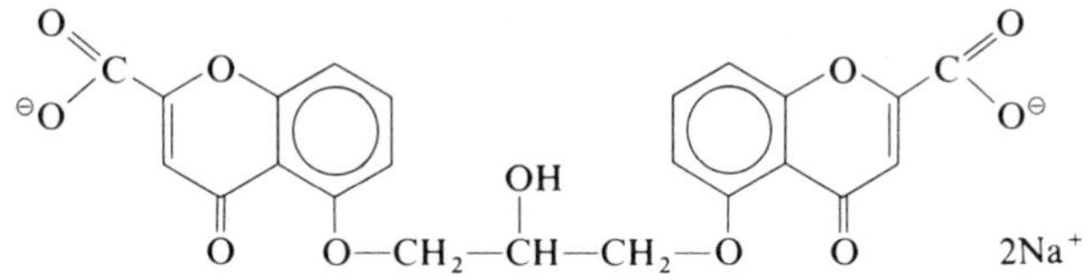

Na-cromoglycate
4-176

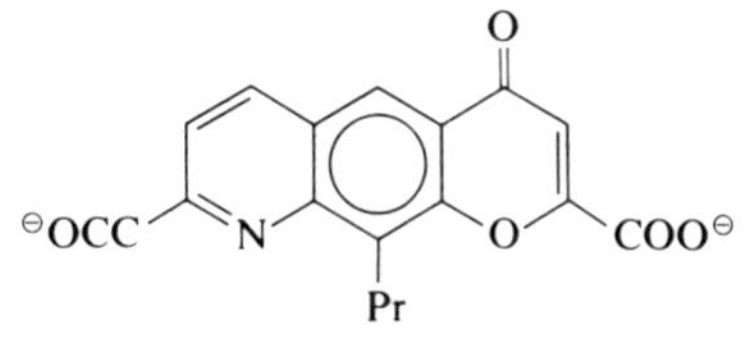

$2Na^+$

Nedocromil-Na
4-177

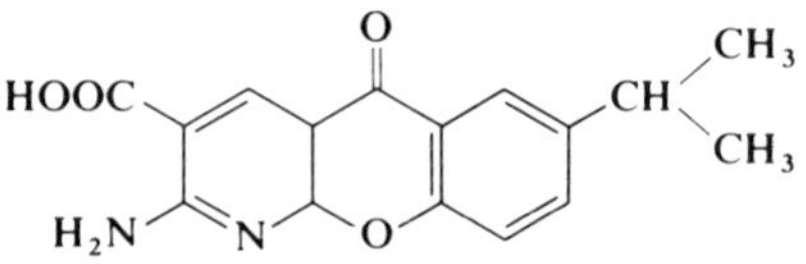

AA-673
4-178

**Fig. 4.39.** (*continued*)

carbon, respectively, show this general structure. **Cyproheptadine** (4-174), a serotonin antagonist, is also a potent antihistamine (about 150 times more active than
diphenhydramine), and so is **promethazine** (4-175) and its derivatives, which
formally at least, can be considered, a result of ring closure connecting the two aryl
rings in a diphenyl-ethylenediamine. A number of related tricyclic antihistamines
are promising experimental drugs (Bell et al., 1979).

The unpleasant sedative CNS effect of most antihistamines, combined with their
slight anticholinergic activity, is exploited for the *prevention of motion sickness.*
**Diphenhydramine** (4-167), in the form of an 8-chlorotheophylline salt (**dimenhydrinate**), is widely used for this purpose. The theophylline derivative was originally
added to counteract the drowsiness produced by diphenhydramine, since it is a
central excitant related to caffeine.

Recently, several nonsedative $H_1$ inhibitors have been marketed—for example,
**astemizole** (4-171) and **terfenadine** (4-172). They are quite polar molecules and
therefore cannot cross the blood–brain barrier to reach central histamine receptors
(Richards et al., 1984; Gleason et al., 1986). Astemizole is especially long acting. An
innovative drug with a combined $H_1$–5-HT–leukotriene antagonism is **oxatomide**
(4-173); it is usable in asthma, where ordinary $H_1$ antagonists are not appropriate
(Ganellin and Schwartz, 1985).

Allergic reactions and especially the bronchospasms in asthma are complex in
origin, and are caused by a number of factors released from mast cells. *Inhibitors
of mediator* (histamine, 5-HT, etc.) *release* are therefore being investigated (see
Bell et al., 1979). The prototype is **Na-cromoglycate** (4-176), a chromone hetero-
cycle useful in the inhalation treatment of bronchial asthma. Similar mediator release inhibitors are the topically active **nedocromil-Na** (4-177) and the orally active
**AA-673** (4-178). They seem to act by phosphorylating a mast cell protein and
thereby stabilize the cell, preventing its disruption (Gleason et al., 1986). In addition, $\beta_2$-adrenergic agonists and $\alpha_1$ blockers can be used in the management of
allergic reactions.

### 6.4.2. $H_2$ Antagonists

Antagonists of the $H_2$ receptor were first reported in 1972 by Black and co-workers,
and work in this area was successively continued by the same group in an elegant
series of investigations based on considerations guided by molecular pharmacological principles. The story of the development of these drugs (Fig. 4.40) has been told
by Durant (1979) and by Brown and Young (1985).

One of the compounds showing weak $H_2$-antagonist activity, guanylhistamine,
was the point of departure in the development of these drugs. Extension of the side
chain was found to increase the $H_2$-antagonist activity, but some agonist effects were
retained. When the very basic guanidino group was replaced by the neutral thiourea,
**burimamide** (4-179) was obtained. Although an effective drug, it cannot be absorbed
orally.

The addition of a 4-methyl group further improved binding to the $H_2$ receptor.
Introduction of the electron-withdrawing sulfur atom into the side chain reduced
the ring $pK_a$. The proportion of the cationic form was also decreased, and the

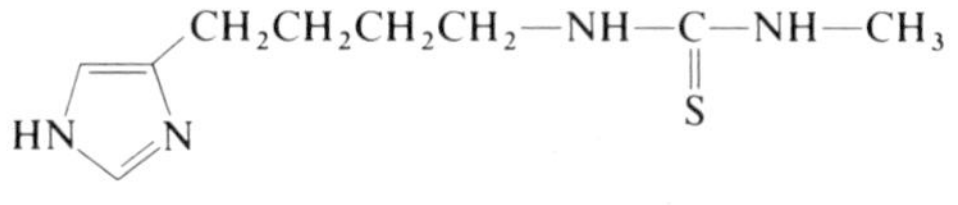

CH$_2$CH$_2$CH$_2$CH$_2$—NH—C—NH—CH$_3$
‖
S

HN⟍⟋N

Burimamide
4-179

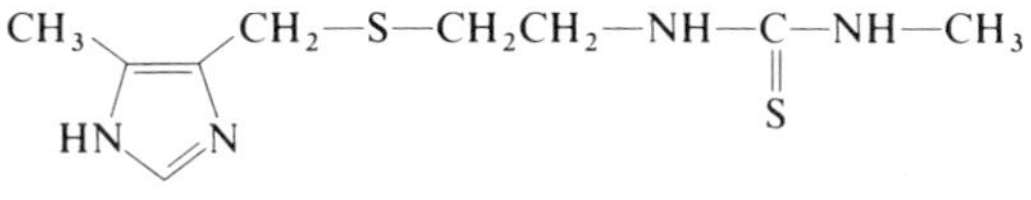

CH$_3$⟍ CH$_2$—S—CH$_2$CH$_2$—NH—C—NH—CH$_3$
‖
S

HN⟍⟋N

Metiamide
4-180

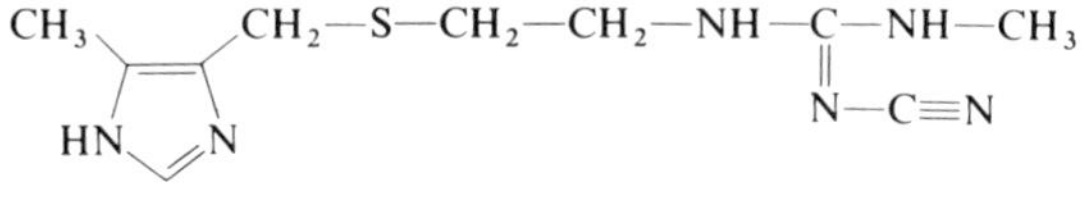

CH$_3$⟍ CH$_2$—S—CH$_2$—CH$_2$—NH—C—NH—CH$_3$
‖
N—C≡N

HN⟍⟋N

Cimetidine
4-181

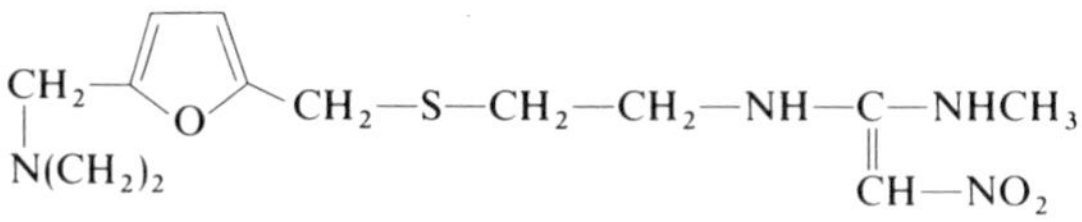

CH$_2$
|
N(CH$_2$)$_2$ ⟨O⟩ CH$_2$—S—CH$_2$—CH$_2$—NH—C—NHCH$_3$
‖
CH—NO$_2$

Ranitidine
4-182

H$_2$N⟍
⟍N—⟨S N⟩—CH$_2$—S—CH$_2$—CH$_2$—C—NH$_2$
H$_2$N⟋
‖
N—SO$_2$NH$_2$

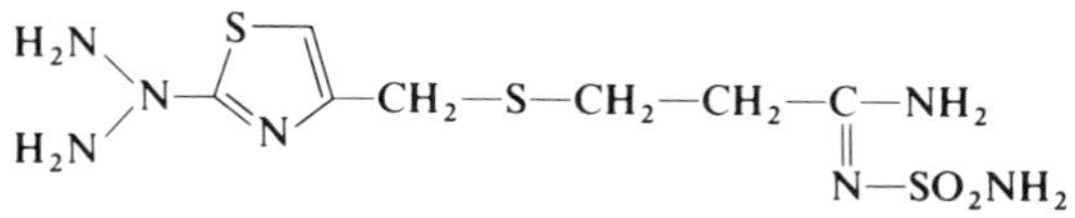

Famotidine
4-183

**Fig. 4.40.** H$_2$ histamine antagonists.

tele tautomer became predominant. Reduced ionization improved the membrane permeability of the molecule; the oral absorption of the resulting compound, **metiamide** (4-180), was excellent, and the compound also had an activity 10 times higher than burimamide. However, metiamide still showed some side effects, in the form of hematological and kidney damage, which were attributed to the thiourea group.

A satisfactory replacement was found by substituting another electron-withdrawing group on guanidine while retaining the appropriate p$K_a$. A cyano group proved suitable, and the safe and effective **cimetidine** (4-181) resulted, which became and still is the drug of choice in treating peptic ulcer.

Lately, it has become clear that an imidazole nucleus is not absolutely necessary for H$_2$-antagonist activity. The furan derivative **ranitidine** (4-182) is even more active than cimetidine, and **famotidine** (4-183) is seven times more active still.

Since none of these compounds is lipid soluble (their average partition coefficient is only 2, compared with coefficients of up to 1000 for typical $H_1$ antagonists), they do not produce any sedative CNS action, since they cannot cross the blood–brain barrier.

Treatment of peptic ulcers is a complicated and multilevel therapy, in which $H_2$ antagonists are very successful and widely used. Other useful drugs are the $H^+$, $K^+$-ATPase inhibitors (Chap. 6, Sec. 3.3) prostaglandins (Chap. 5, Sec. 4.1), the $M_1$ anticholinergic **pirenzepine** (Sec. 2.3), and the neurohormone somatostatin (Chap. 5, Sec. 2.5) (Garay and Muchowski, 1985).

### Selected Readings

S. C. Bell, R. J. Capetola, and D. M. Ritchie (1979). Pulmonary and anti-allergy drugs. *Annu. Rep. Med. Chem. 14*: 51–60.

T. H. Brown and R. C. Young (1985). Antagonists of histamine at its $H_2$-receptor. *Drugs of the Future 10*: 51–69.

K. T. Bunce, D. A. A. Owen, I. R. Smith, and M. R. Vickers (1979). Histamine. In: *Int. Rev. Biochem.*, Vol. 26 (K. F. Tipton, Ed.). University Park Press, Baltimore, pp. 207–256.

J. R. Cooper, F. E. Bloom, and R. H. Roth (1986). *The Biochemical Basis of Neuropharmacology*, 5th ed. Oxford University Press, New York.

G. J. Durant (1979). Chemical aspects of histamine $H_2$-receptor agonists and antagonists. In: *Recent Advances in Receptor Chemistry* (F. Gualtieri, M. Gianella, and C. Melchiorre, Eds.). Elsevier/North Holland, New York, pp. 245–266.

C. R. Ganellin (1977). Chemical constitution and prototropic equilibria in structure–activity analysis. In: *Drug Action at the Molecular Level* (G. C. K. Roberts, Ed.). University Park Press, Baltimore, pp. 1–39.

C. R. Ganellin and M. E. Parsons (Eds.) (1982). *Pharmacology of Histamine Receptors.* Wright PSG, Bristol, England.

C. R. Ganellin and J.-C. Schwartz (Eds.) (1985). *Frontiers in Histamine Research.* Pergamon Press, New York.

G. L. Garay and J. M. Muchowski (1985). Agents for the treatment of peptic ulcer disease. *Annu. Rep. Med, Chem. 20*: 93–105.

J. G. Gleason, C. D. Perchonock, and T. J. Torphy (1986). Pulmonary and antiallergy agents. *Annu. Rep. Med. Chem. 21*: 73–83.

D. M. Richards, R. N. Brogden, R. C. Heel, T. M. Speight, and G. S. Avery (1984). Astemizole: a review of its pharmacological properties and therapeutic efficacy. *Drugs 28*: 38–61.

J.-C. Schwartz, J.-M. Arrang, and M. Garbarg (1986a). Three classes of histamine receptors in the brain. *Trends Pharmacol. Sci. 7*: 24–28.

J.-C. Schwartz, J.-M. Arrang, M. Garbarg, and M. Komer (1986b). Properties and roles of the three subclasses of histamine receptors in the brain. *J. Exp. Biol. 124*: 203–224.

## 7. PURINERGIC NEUROMODULATION AND THE ADENOSINE RECEPTORS

Purines have numerous biochemical functions: they are the building blocks of nucleic acids, the energy transducers ATP and GTP, and the second messenger cAMP. They have also been known to have other direct physiological functions in bronchial constriction, in vasodilation, and in inhibition of platelet aggregation and of central neuronal firing. Adenosine has been found in virtually every synapse and, in view of its bewildering multitude of actions, has been called the "most mysterious neuromodulator" by Cooper et al. (1986). Intensive research in recent years has shed some light on purinergic activity that is (cautiously) defined as neuro-modulation rather than neurotransmission, even though adenosine receptors are reasonably well elucidated. The review of Burnstock (1983) and, more recently, those of Stiles (1986), Katsuragi and Furukawa (1985), and Satchell (1984) provide a useful and concise introduction into this new field.

*Adenosine receptors* are subdivided into two main groups, the $P_1$ and the $P_2$, with a further subdivision of the $P_1$ receptor into the subtypes $A_1$ and $A_2$ (Table 4.12). Because the biochemical properties of the receptors are not known, the classification is based on pharmacological evidence—sensitivity to an agonist series. The $A_1$ receptor inhibits adenylate cyclase, whereas the $A_2$ receptor activates it. In addition, adenosine receptors seem to regulate cardiac $K^+$ conductance without changes in cAMP levels. The transmembrane signaling of the adenosine receptors is apparently the same as that of other cyclase-coupled receptors (Chap. 2, Sec. 5.4).

Purines act both pre- and postsynaptically. Adenosine inhibits the release of NE and ACh in autonomic neuronal terminals, and both adenosine and ATP function as pre- and postjunctional membrane-potential modulators in ganglia. It appears that NE and ATP are cotransmitters in sympathetic synapses, where ATP mediates the fast phasic contraction of smooth muscle through $P_2$ receptors, and NE triggers the slow tonic contraction via $\alpha_1$ adrenoceptors.

**Table 4.12.** Adenosine receptor classification

| | $P_1$ | | $P_2$ |
|---|---|---|---|
| | $A_1$ | $A_2$ | |
| Location | CNS | CNS | intestine |
| | adipocyte | smooth muscle | |
| | heart | liver | |
| | testis | platelets | |
| Function | sedation | transmitter release | contraction |
| | inhibition of lipolysis | inhibition of aggregation | |
| | decrease in contractility | gluconeogenesis | |
| Agonist sensitivity | adenosine > AMP > ADP > ATP | | ATP > ADP > AMP > adenosine |
| Antagonists | clonidine | | arylazido-aminopropionyl-ATP |
| | alkylxanthines | | |
| Second messenger | inhibits AC | stimulates AC | phosphatidylinositol? |

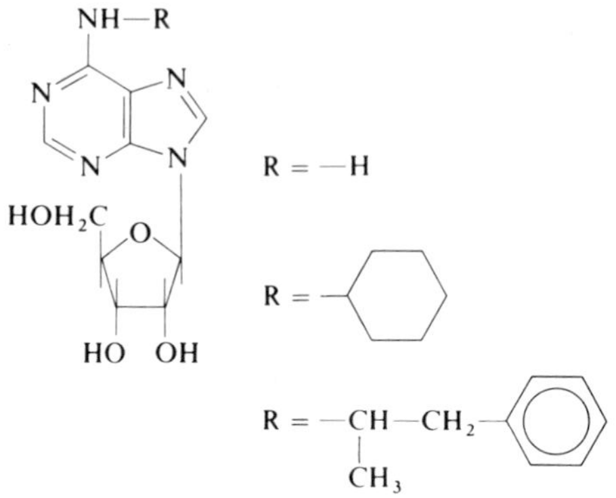

R = —H                          Adenosine
                                4-183

R = —⟨cyclohexyl⟩              Cyclohexyladenosine
                                4-184

R = —CH—CH₂—⟨phenyl⟩           R(−)-phenylisopropyl-adenosine
        |                       4-185
       CH₃

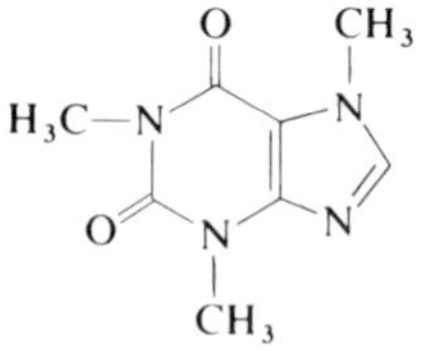

Caffeine
4-186

**Fig. 4.41.** Purinergic ligands.

*Purinergic agonists* (Fig. 4.41) include adenosine and all adenosine phosphates, as well as a number of highly active synthetic $N_6$-substituted adenosine derivatives (cyclohexyl, phenylisopropyl). The $P_1$ receptor is very sensitive to changes in the ribofuranose ring (e.g., epimerization to arabinose), whereas the $P_2$ receptor is not.

*Purinergic antagonists* include **clonidine** (4-63), a potent selective $P_1$ antagonist that is also an $\alpha_2$ and $H_2$ agonist, facilitating purine release. Methylxanthines, especially **caffeine** (4-187), are potent $P_1$ antagonists. When one considers the enormous amount of caffeine consumed in the world, this discovery is significant for understanding the symptoms of caffeine addiction. In view of the stimulating action of caffeine, it is interesting to note that adenosine receptor blockage is directly correlated with locomotor activity in mice.

At the time of writing, there are no drugs based on the adenosine receptor, but coronary vasodilation and perhaps some central effects are potential candidates for such drugs in the future.

### Selected Readings

G. Burnstock (1983). *Purinergic Receptors.* Chapman and Hall, London.

J. R. Cooper, F. E. Bloom, and R. H. Roth (1986). *The Biochemical Basis of Neuropharmacology,* 5th ed. Oxford University Press, New York.

J. W. Daly (1985). Adenosine receptors. *Adv. Cyclic Nucleotide Res.* 18: 29–46.

T. Katsuragi and T. Furukawa (1985). Novel, selective purinoceptor antagonists: investigation of ATP as a neurotransmitter. *Trends Pharmacol. Sci.* 6: 337–339.

D. Satchell (1984). Purine receptors: classification and properties. *Trends Pharmacol. Sci.* 5: 340–343.

S. H. Snyder (1985). Adenosine as a neuromodulator. *Annu. Rev. Neurosci.* 8: 103–124.
G. L. Stiles (1986). Adenosine receptors: structure, function and regulation. *Trends Pharmacol. Sci.* 7: 486–490.

## 8. AMINO ACID NEUROTRANSMITTERS AND THE DRUG EFFECTS THEY MEDIATE

Amino acids occur in the central nervous system as protein building blocks, as sources of energy and, as recently recognized, as potential or proven neurotransmitters or neuromodulators. The elucidation of their biochemical, physiological, and pharmacological roles as neurotransmitters has been exceedingly difficult because of the multitude of nonneural binding sites for amino acids and the lack of specific antagonists to some of them. Without such antagonists, it is impossible to demonstrate their specific action and to prove their role in neurotransmission by rigorous criteria. The field is, however, of great potential importance because a number of neurological disorders can already be connected with amino acid transmitters, and an even greater number show promising correlations. The pharmacological manipulation of these compounds is made especially appealing because they are specific for the CNS, and the lack of any peripheral effect holds the promise of a lowered potential toxicity.

### 8.1. γ-Aminobutyric Acid (GABA)

GABA is the most comprehensively studied inhibitory neurotransmitter, and there are many reviews of its biochemistry and pharmacology (Peck, 1980; Johnston, 1978; and many other sources quoted in the biliography). The reason for this great interest is the discovery that the most "popular" drugs, the benzodiazepine tranquilizers or "anxiolytics," as well as probably the barbiturates, act on the GABAergic neuronal system (Ho and Harris, 1981).

#### 8.1.1. Distribution and Metabolism

There seem to be numerous GABAergic neuronal pathways in the CNS. γ-Aminobutyric acid is found in the highest concentrations in the substantia nigra, is also found in the hypothalamus, and occurs in low concentrations in practically all brain structures as well as in the spinal cord. The amounts present are relatively high—on a $\mu$ moles per gram order of magnitude—rather than the nanomolar quantities seen with most major neurotransmitters. γ-Aminobutyric acid also occurs in glial cells, where its role is unknown.

The biosynthesis of GABA occurs only in the neurons, since it cannot penetrate the blood–brain barrier, and no peripheral precursor is known. As shown in Fig. 4.42, the synthesis is tied to the Krebs cycle through α-ketoglutarate.

γ-Aminobutyric acid is formed by the decarboxylation of L-glutamate, catalyzed by *glutamic acid decarboxylase (GAD)*, an enzyme found only in the mammalian CNS and in the retina. This reaction is irreversible. The cofactor of GAD is pyridoxal phosphate (vitamin $B_6$). Since GAD is the rate-determining enzyme, GABA metabolism can be regulated by the manipulation of this enzyme, the manipulation of pyridoxal, or both.

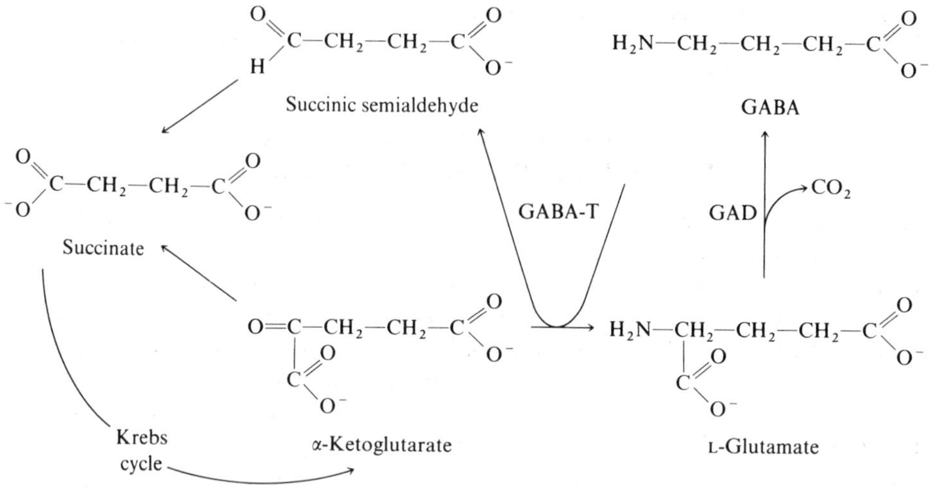

**Fig. 4.42.** γ-Aminobutyric acid (GABA) metabolism. The metabolism is tied into the Krebs cycle through α-ketoglutarate.

γ-Aminobutyric acid can be deactivated and recycled by the transamination reaction with α-ketoglutarate to yield glutamate. This reaction circumvents the usual oxidative route, insofar as glutamate can be decarboxylated to yield GABA once again. This transamination is catalyzed by the enzyme *GABA transaminase* (*GABA-T*), which is widely distributed. Therefore, free GABA cannot be found anywhere except in the brain. The transaminase enzyme also depends on pyridoxal phosphate as a cofactor.

Very little is known about the storage and release of GABA, because it has only recently become possible to localize GABAergic neurons by immunohistochemical methods based on GAD. Although the transmitter is taken up rapidly *in vitro* by brain slices, its *in vivo* uptake is Na dependent and is not uniform (Saelens and Vinick, 1978).

### 8.1.2. Characterization of GABAergic Receptors

GABAergic receptors were thoroughly investigated in the early 1980s. The great increase in research activity in this area is largely due to the recognition that the extremely widely used benzodiazepine tranquilizers (see below) act through the GABA receptor. Costa and his group have conducted a series of ingenious experiments in this area and have suggested the GABA receptor hypothesis presented below (Costa and Guidotti, 1979).

The kinetics of GABA binding suggests that it occurs at more than one site: the *GABA_B receptor* has a low binding affinity $K_D = 200$ nM) and shows great variation in receptor density in various brain areas. The *GABA_A receptor* is a high-affinity binding site ($K_D = 20$ n/M) with a fairly constant density.

When a brain membrane preparation containing the GABA receptors is frozen, thawed, and washed with a 0.5% solution of the detergent Triton X-100, a new, additional population of $GABA_A$ receptors appears. It has been proposed that the detergent treatment removes a thermostable regulatory protein called *GABA-modulin* (Costa and Guidotti, 1979; Toffano et al., 1980), having a molecular weight of 150,000. This protein is an inhibitor of the $Ca^{2+}$-dependent and both the cAMP-dependent and cAMP-independent protein kinases. The removal of GABA-modulin increases protein phosphorylation in the synapses by a factor of 20 fold, and the number of receptors (or their transmitter recognition ability) increases. Only the affinity or number of $GABA_A$ receptors is regulated by GABA-modulin. The receptor assembly thus seems to be composed of the GABA receptor, the protein kinase with attached GABA-modulin, and the ionophore mediating $Cl^-$ ion transport (Fig. 4.43). The GABA-modulin binding site is apparently the same as the benzodiazepine binding site.

The GABA-modulin hypothesis has recently been modified by the Costa group (Costa and Guidotti, 1985), with the isolation of a number of endogenous inhibitors called *GABARINs* (*GABA* Receptor *IN*hibitor). Some of these (the $\beta$-carboline carboxylic acid esters) proved to be artefacts, but the **diazepam-binding inhibitor** (**DBI**), a protein of 105 amino acids isolated from brain, had very high activity. DBI contains two copies of an octadeca-neuropeptide (of 18 amino acids) that can reduce the duration of $Cl^-$ ion channel opening (Racagni and Donoso, 1986). These allosteric inhibitors can be conceptualized as "inverse agonists," and will be discussed in connection with benzodiazepine anxiolytics (Sec. 8.1.5).

The functional role of the $GABA_B$ receptor is unknown, although it may be coupled to adenylate cyclase. Only one selective antagonist, **baclofen** (4-195), is known.

The solubilization of 50% of the benzodiazepine binding sites by detergent (0.5% Lubrol) has been reported. It is a protein of about 57,000 Daltons with a $K_D$ of 11 nM for the brain receptor. In crustacean muscle, where GABA is a peripheral inhibitory neurotransmitter, four types of receptors have been characterized (cf. Andrews and Johnston, 1979). Other methods and results of receptor purification are discussed in detail in the major review of Haefely et al. (1985).

The *neuronal activity of GABA* shows two different inhibitory mechanisms. The first is the partial (presynaptic) depolarization of an excitatory neuron, which causes a decrease in neurotransmitter release when this neuron receives an electrical impulse. Such a mode of action is also suspected in enkephalinergic (opiate-sensitive) neurons (see Chap. 5, Sec. 3). The second mechanism is the conventional hyperpolarization of an excitatory neuron by increased $Cl^-$ ion flux, which makes the neuron unable to fire when it receives a normal impulse.

### 8.1.3. Presynaptic GABAergic Drug Effects

Presynaptic drug effects can interfere with the metabolism, storage, release, and reuptake of GABA, as they can with the functioning of other neurotransmitters.

*GABA Synthesis Inhibitors.* GABA synthesis inhibitors act on the enzymes involved in the decarboxylation and transamination of GABA. Glutamic acid decarboxylase (GAD), the first enzyme in GABA biosynthesis, is inhibited easily by carbonyl reagents such as hydrazines [e.g., **hydrazinopropionic acid** (4-188) or **isonicotinic acid hydrazide** (4-189], which trap pyridoxal, the essential cofactor of the enzyme. A more specific inhibitor is **allylglycine** (4-190). All of these compounds cause seizures and convulsions because they decrease the concentration of GABA.

*GABA Metabolism Inhibitors.* In contrast to GABA synthesis inhibitors, inhibitors of *GABA-T*, the transaminase active in eliminating GABA, increase the concentration of this neurotransmitter. The most potent of these agents are **gabaculine** (4-191) and **4-aminohex-5-enoic acid** (4-192), both of which protect against drug-induced seizures.

$H_2N$—NH—$CH_2$—$CH_2$—COOH

Hydrazinopropionic acid
4-188

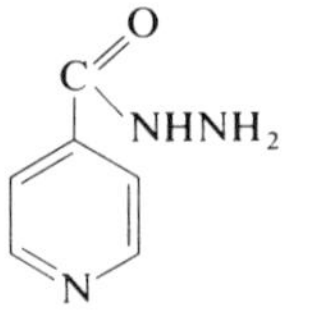

Isonicotinic
acid hydrazide
4-189

$CH_2$=CH—$CH_2$—CH—COOH
|
$NH_2$

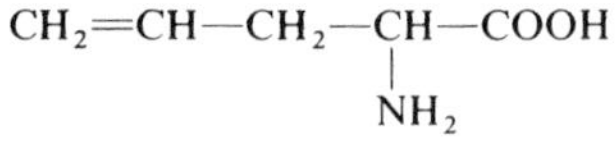

Allylglycine
4-190

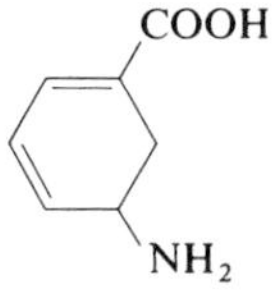

Gabaculine
4-191

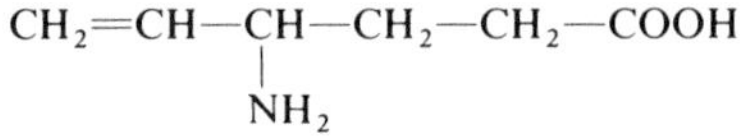

$CH_2$=CH—CH—$CH_2$—$CH_2$—COOH
|
$NH_2$

4-Aminohex-5-enoic acid
4-192

GABA has long been suspected of being involved in epilepsy. Remarkably, only one antiepileptic drug, **sodium valproate** (4-193), has been found to influence GABA metabolism. This drug, used successfully to control epilepsy, blocks *succinic semialdehyde dehydrogenase*, the enzyme oxidizing the semialdehyde (Fig. 4.42). As this metabolite accumulates, GABA-T activity is decreased by end-product inhibition, and the neurotransmitter concentration increases, thus inhibiting seizures.

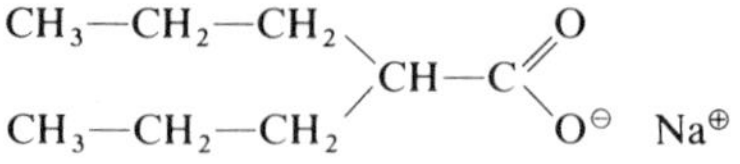

Sodium valproate
4-193

*GABA Reuptake Inhibitors.* Another presynaptic mechanism involves several GABA reuptake inhibitors, such as **nipecotic acid** (4-194) and some other related compounds (see Saelens and Vinick, 1978), which act on neurons and glial cells in a different manner.

*Agents Affecting GABA Release.* High doses of imipramine, haloperidol, and chlorpromazine (at 1 μM concentrations) and pentobarbitone (at 200 μM) are known to inhibit GABA release *in vitro.* **Baclofen,** also known as Lioresal (4-195) [β-(*p*-chlorophenyl)-GABA], is a valuable compound which enhances GABA release and is therefore an indirect agonist. It is an orally active muscle relaxant used in treating the spasticity and muscle rigidity of cerebral palsy, in which it slows the firing rate of dopaminergic neurons.

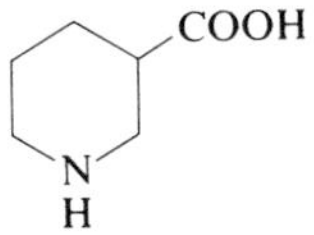

Nipecotic acid
4-194

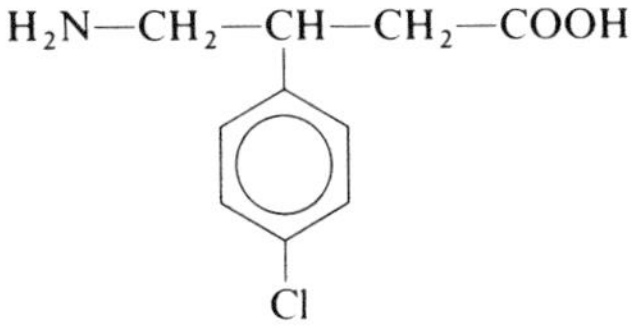

Baclofen
4-195

### 8.1.4. Postsynaptic GABAergic Drug Effects

*GABA Agonists.* Directly acting GABA agonists usually bear some resemblance to the neurotransmitter. **Muscimol** (4-196), an isoxazole isolated from the mushroom *Amanita muscaria* ("deadly fly agaric"), is a hallucinogen with a receptor affinity greater than that of GABA (0.9 nM versus 9.4 nM on Triton-treated membranes) on the GABA$_B$ receptor. A number of related compounds have been synthesized and have also proved to be active (Krogsgaard-Larsen et al., 1982), among them **THIP** **(gaboxadol)** (4-197). **Progabide** (4-198) is another novel GABA agonist and anti-convulsant, which bears a resemblance to the benzodiazepines.

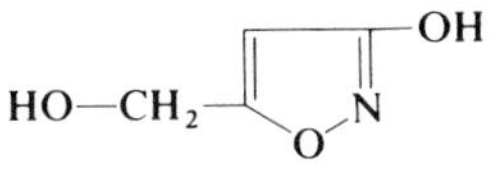

Muscimol
4-196

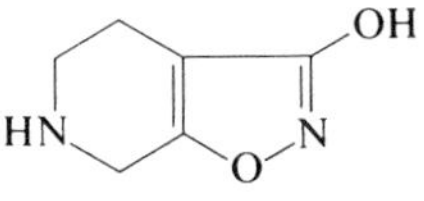

4,5,6,7-Tetrahydro-isoxazolo[5,4-*c*]pyridin-3-ol (THIP)
4-197

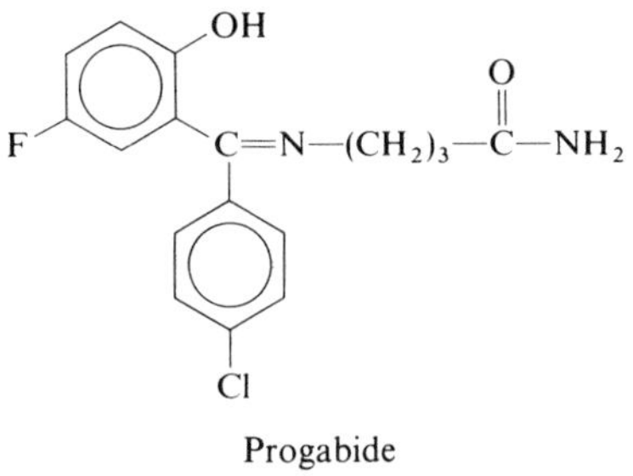

Progabide
4-198

*GABA Antagonists.* The only direct antagonist is the alkaloid **(+)-bicuculline** (4-199), which binds to all synaptic GABA sites. Being a lactone, it is sensitive to hydrolysis. Its binding is influenced in an unknown way by salts; that is, [³H] bicuculline binding in the presence of 50 $\mu$M NaSCN or 200 $\mu$M NaClO$_4$ is more "specific" than in the absence of salts, and only 30–50% of it can be displaced by GABA or muscimol. This is a further indication of GABA receptor multiplicity. Interestingly, **benzylpenicillin** can antagonize GABA in doses below 2 $\mu$M, and can thus be epileptogenic.

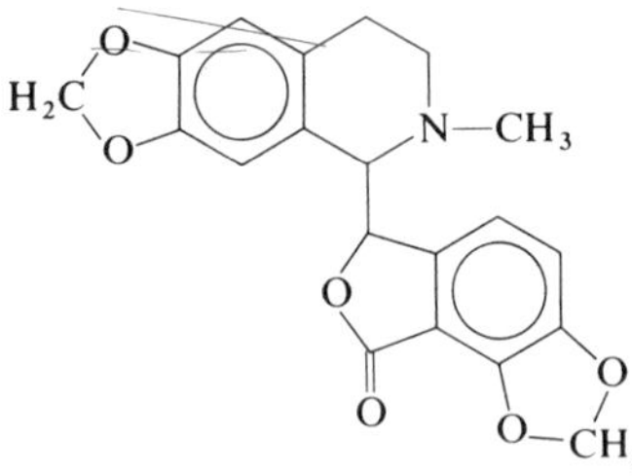

(+)Bicuculline
4-199

**GABA-modulin** or the **diazepam-binding inhibitor (DBI)** be considered an indirect, allosteric antagonist because it inhibits GABA$_A$ receptors, as discussed earlier. Washing with Triton-X-100 removes this protein easily and increases protein phosphorylation at the synapse. GABA-modulin is displaced competitively by the **benzodiazepine** tranquilizers. At present, these drugs are the most widely prescribed and used drugs in attempting to cope with the increasingly stressful way of life in Western civilization, and thus deserve a separate section for their discussion.

### 8.1.5. Benzodiazepines

The benzodiazepines were discovered by Leo Sternbach at the Hoffman-La Roche laboratories, and their pharmacology was elucidated by Randall of the same company. An enormous variety of these compounds exists, a few of which are shown in Table 4.13. Since about 2000 benzodiazepine compounds have been investigated, the structure–activity relationships of these drugs can be generalized as follows:

1. R$_1$ must be an electron-attracting group. No other substituent can be attached to any of the carbons on that ring.
2. R$_2$ and R$_3$ can be varied. Replacement of the lactam oxygen by sulfur decreases activity.
3. The phenyl group is necessary for activity, but only halogen substituents are allowed in the ortho position.

The most widely used benzodiazepine drug is **diazepam** (4-202). It is an anxiolytic, sedative, and muscle relaxant, and also a psychostimulant: the anxious, depressed person becomes more outgoing and active. **Oxazepam** (4-200) and **lorazepam** (4-201) have similar effects. **Temazepam** (4-203), **flunitrazepam** (4-205), and **flurazepam** (4-206) are useful hypnotics and unlike most barbiturates, have no hangover

**Table 4.13.** Some representative benzodiazepine anxiolytics and hypnotics

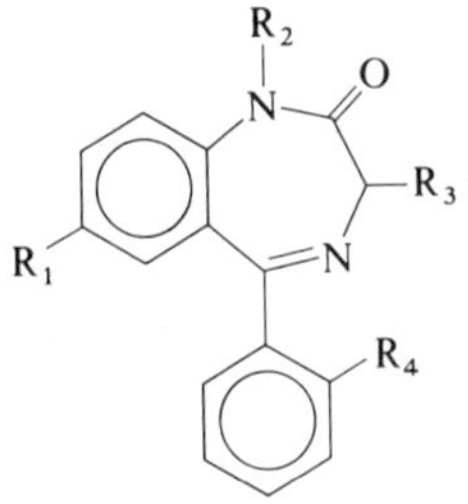

|  | $R_1$ | $R_2$ | $R_3$ | $R_4$ |
|---|---|---|---|---|
| 4-200 Oxazepam | Cl | H | OH | H |
| 4-201 Lorazepam | Cl | H | OH | Cl |
| 4-202 Diazepam (Valium) | Cl | $CH_3$ | H | H |
| 4-203 Temazepam | Cl | $CH_3$ | OH | H |
| 4-204 Clonazepam | $NO_2$ | H | H | Cl |
| 4-205 Flunitrazepam | $NO_2$ | $CH_3$ | H | F |
| 4-206 Flurazepam | Cl | $CH_2CH_2N(Et)_2$ | H | F |

effects. **Clonazepam** (4-204) is a clinically useful anticonvulsant. **Brotizolam** (4-207), a novel benzodiazepine analogue seems to be an effective sedative-hypnotic. **Midazolam** (4-208) is a new imidazolo-benzodiazepine that is water soluble and thus easily injectable. It is a hypnotic sedative with marked amnestic (i.e., memory loss) properties, and is used in dentistry, endoscopies, and induction to anesthetics in the elderly and in cardiac patients.

The benzodiazepine drugs permit normal daytime functioning and do not have a high addiction liability, as do the barbiturates. [$^3$H] Flunitrazepam has also been used successfully in affinity-labeling of the benzodiazepine binding site.

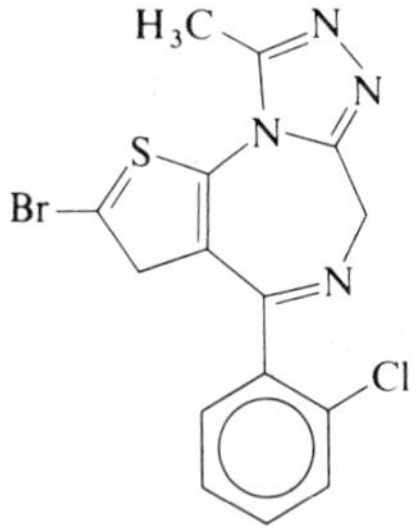

Brotizolam
4-207

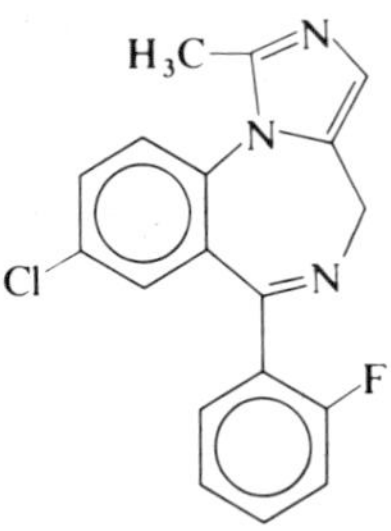

Midazolam
4-208

*Mode of Action.* The mode of action of benzodiazepines is apparently based on

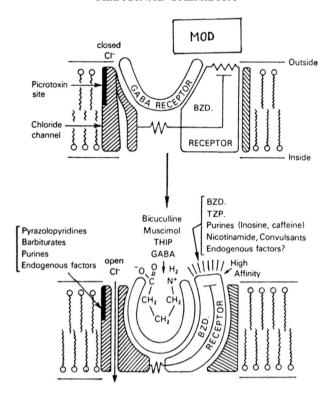

**Fig. 4.43.** Model of the GABA and benzodiazepine receptor complex. In the closed state (top), GABA-modulin (MOD) blocks the benzodiazepine receptor (BZD) as well as the GABA binding site, and the chloride channel is closed. In the open state (bottom), GABA-modulin is removed, the GABA receptor can bind the neurotransmitter, and the associated chloride channel opens up. The benzodiazepine site can now also bind a variety of compounds. (TZP, triazolopyridazines; THIP, 4,5,6,7-tetrahydro-isoxazolo[5,4-c]pyridin-3-ol). (Modified by permission from Skolnick and Paul, 1981)

their displacement of GABA-modulin. They uncover high-affinity GABA$_A$ receptors by allowing phosphorylation of the receptor protein, which is normally suppressed by GABA-modulin. In this way, the opening of the Cl$^-$ ionophore becomes a more efficient process than in the presence of GABA only. This hypothetical model, shown in Fig. 4.43, is unique among all known receptor mechanisms. Benzodiazepines thus could be considered coagonists (Costa, 1982).

***Receptor Characterization and Drug Classification.*** The benzodiazepine receptor seems to consist of three types of recognition sites. The type I receptor site is coupled to the GABA receptor and the ionophore, mediates the anxiolytic effect, and has a high affinity for benzodiazepines and triazolopyridazines (e.g., 4-209), anxiolytics, hypnotics, and anticonvulsants.

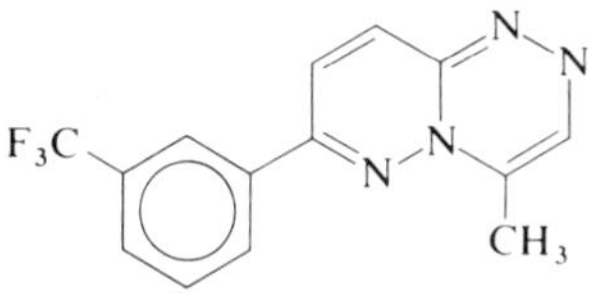

4-209

Type II receptors are *not* coupled to the GABA receptor and the ionophore, mediate effects other than anxiolytic activity, and have a low affinity for triazolopyridazines (Bowling et al., 1982). β-Carbolines antagonize the anxiolytic, anticonvulsant, and sedative effects of benzodiazepines.

*Inverse agonists* like **DMCM** (4-210) are anxiogenic and convulsive: they are called inverse agonists because they bind to agonist sites but have effects opposite to those of GABA. *Competitive antagonists* (e.g., **Ro 15-1788**, 4-211, also bind here; they are inactive by themselves, but prevent agonist and inverse agonist binding (Richards et al., 1986).

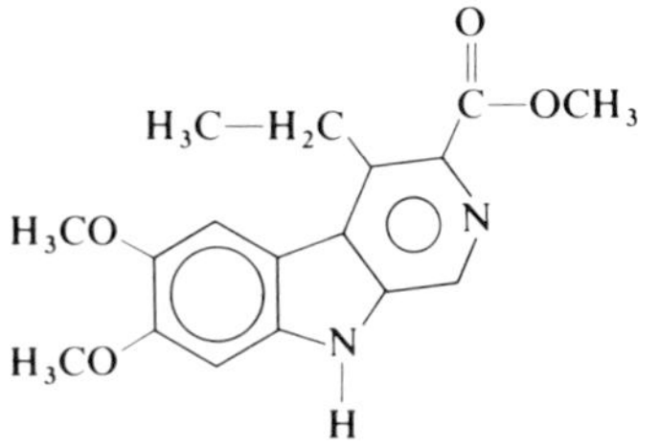

DMCM
4-210

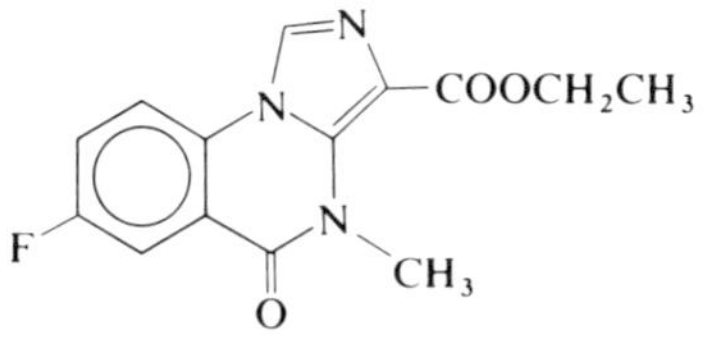

Ro 15-1788
4-211

The recent findings regarding *benzodiazepine antagonists*, which have structures completely unrelated to benzodiazepines, are significant to the future development of the field (see Effland and Forsch, 1982). Quinolines related to compounds (4-212) and (4-213) can differentiate between anticonvulsant and anticonflict effects of BZDs. By "anticonflict" is meant an effect that can decrease anxiety triggered by a situation that an animal cannot resolve, such as when a reward is accompanied by an electric shock. Compound 4-212 is a pure anticonflict drug which might act on a GABA-independent benzodiazepine receptor associated with a Cl⁻ ionophore. It is completely devoid of anticonvulsant and sedative properties. Compound 4-213 has no anticonflict activity, but is a powerful diazepam and barbiturate antagonist.

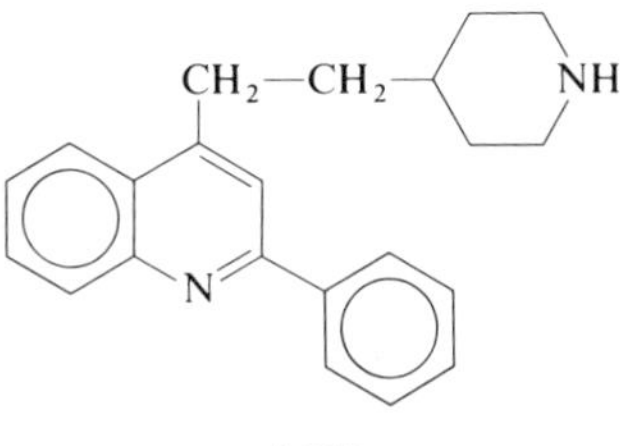

4-212

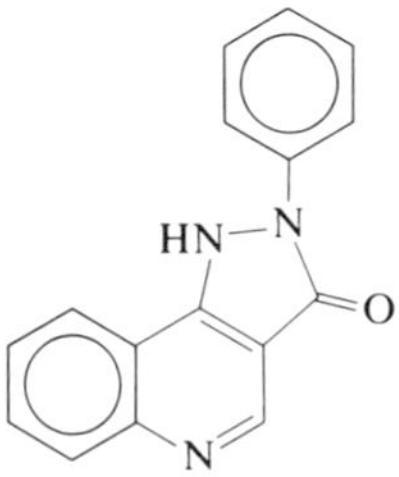

4-213

The type III receptor site cannot be labeled by flunitrazepam photoaffinity labels; it seems to be an allosteric site, inducing a conformational change at the BZD site. Compounds binding here are not benzodiazepines but pyrrolidines, such as **suriclone** (4-214) and **zopiclone** (4-215), which are very active anxiolytic hypnotics. They exhibit no hangover effect or rebound insomnia on drug withdrawal (Trifiletti and Snyder, 1984).

Behavioral testing of atypical anxiolytics revealed a new class of related drugs called *serenics*, which are antiaggressive rather than anxiolytic; they are exemplified by **fluprazine** (4-216) (Olivier et al., 1986).

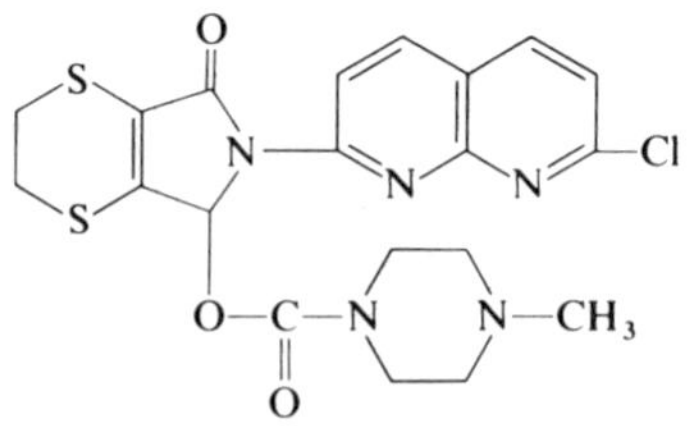

Suriclone
4-214

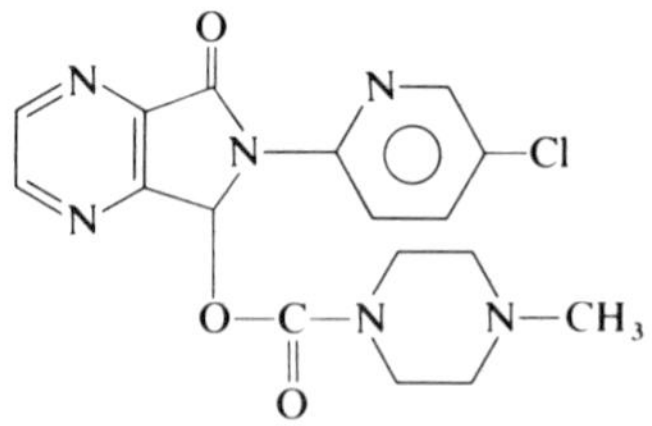

Zopiclone
4-215

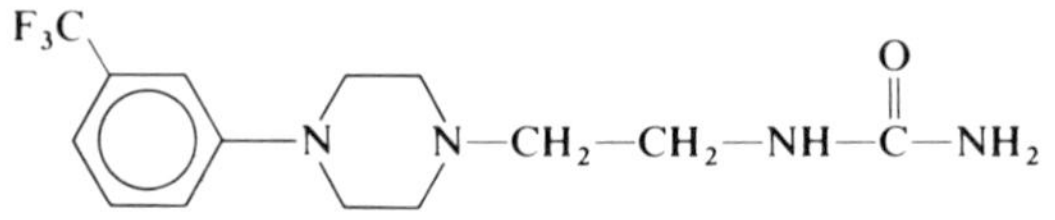

Fluprazine
4-216

In addition to the well-explored central $GABA_A$ receptor, the existence of *peripheral* (type II) GABA receptors has been shown (Erdö, 1985; Anholt, 1986). These receptors seem to be distinct from their CNS counterparts in ontogeny and subcellular localization in just about every organ. Since they seem to be located at the outer mitochondrial membrane, they may be involved in the modulation of intermediary metabolism. In addition, peripheral organs also show GABAergic innervation.

### 8.1.6. Drugs Acting Partially at the GABAergic Ionophore

The GABAergic ionophore (Fig. 4.43) is also a site for drug action. The antagonist **picrotoxinin** (4-217) does not displace GABA from its binding site, but is an effective convulsant, acting by inhibiting $Cl^-$ ion transport. A number of related compounds have a similar action. Using the technique of [$^3$H]dihydropicrotoxinin displacement, it was found that hypnotic, anesthetic, and anticonvulsant drugs displace dihydropicrotoxinin. Thus, they may act, at least partially, as *ionophore agonists*.

**Barbiturates.** A large and still widely used group of these drugs is the barbiturates — sedative/hypnotic compounds which are used in surgical anesthesia and

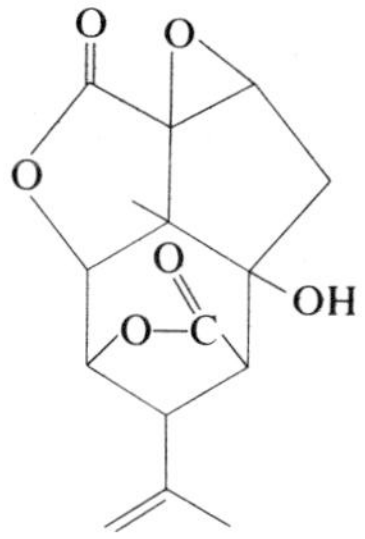

Picrotoxinin
4-217

as "sleeping pills." The principal barbiturates are shown in Table 4.14. Barbiturates are acidic, because of the tautomerism with the enolate. Their $pK_a$ is about 7.3, and therefore even slight changes in body pH will influence their ionization and, consequently, absorption and distribution. The replacement of oxygen by sulfur in position 2 leads to increased lipophilicity and very rapid penetration of the blood–brain barrier. Therefore, compounds like **thiopental** (4-218) are ultrashort-acting intravenous anesthetics, used in surgery for short operations or for inducing anesthesia prior to inhalation anesthetic use. Methylation on the N-1 atom has a

**Table 4.14.** Some representative barbiturate hypnotics

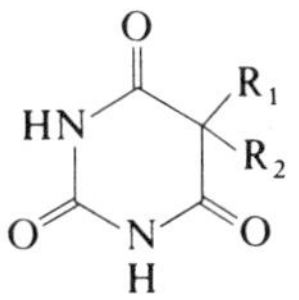

| | | $R_1$ | $R_2$ | |
|---|---|---|---|---|
| 4-218 | Thiopental | $-C_2H_5$ | $-CH-(CH_2)_2CH_3$ / $CH_3$ | (2-S) |
| 4-219 | Hexobarbital | $-CH_3$ | 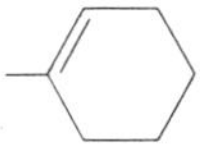 | (1-$CH_3$) |
| 4-220 | Pentobarbital | $-C_2H_5$ | $-CH-(CH_2)_2CH_3$ / $CH_3$ | |
| 4-221 | Amobarbital | $-C_2H_5$ | $-CH_2CH_2CH(CH)_2$ | |
| 4-222 | Phenobarbital | $-C_2H_5$ |  | |
| 4-223 | Barbital | $-C_2H_5$ | $-C_2H_5$ | |

similar effect, as in **hexobarbital** (4-219). Branched side chains on C-5 lead to longer activity (**pentobarbital**, 4-220; and **amobarbital**, 4-221); short side chains, like ethyl, lead to the longest duration of action because the slow entry of the resultant molecule into the CNS affects the onset of action. Aromatic substituents produce anticonvulsant activity. Barbiturate enhancement of GABA binding is proportional to the anesthetic activity of the barbiturate.

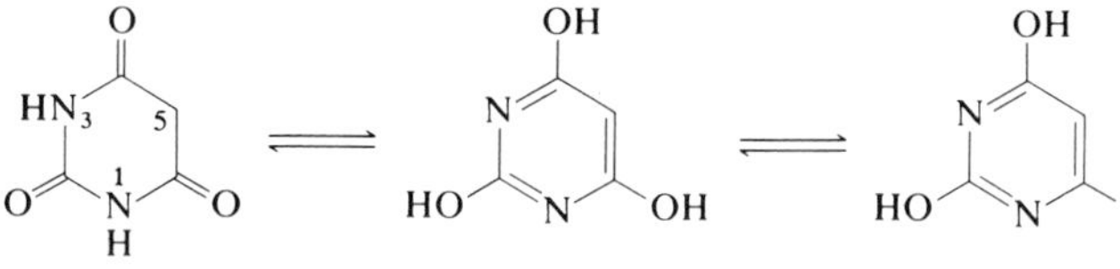

| Lactam | Lactim (Enol) | Enolate |

The principal disadvantages of barbiturates as hypnotics include the development of physical dependence, a relatively low therapeutic index (and the potential of poisoning, as in suicide), suppression of REM sleep, and possible hangover effects.

As mentioned above, benzodiazepines (e.g., flurazepam or brotizolam) are hypnotics as effective as barbiturates, and are much safer in terms of their therapeutic index, addiction potential, and REM sleep-deprivation effects. It is therefore likely that benzodiazepines will displace barbiturates as sedative hypnotics, if not as anesthetics.

There are many *nonbarbiturate hypnotics* among the unsaturated tertiary alcohols and the piperidinediones; some obsolete and dangerous drugs such as chloral hydrate and NaBr are also nonbarbiturate hypnotics.

*Anticonvulsants.* Anticonvulsants are another important group of drugs, some of which act through GABAergic mechanisms. The most severe convulsions are seen in *epilepsy*, which has several different forms. "Grand mal" is the most severe of these, being characterized by generalized and focal seizures; "petit mal" is less severe; and psychomotor "absences" with no convulsions are the mildest form (see Vida, 1981). The cause of epilepsy is unknown; it manifests itself in excessive neuronal firings in the cortex or temporal lobe. When other neurons are exposed to overexcitation, they respond by triggering seizures or convulsions, which can take many forms. The EEG patterns of epilepsy are very characteristic (Eadie, 1984).

Grand mal epilepsy responds well to barbiturates like **phenobarbital** (4-222) and hydantoins such as **phenytoin** (4-224). Petit mal epilepsy, often seen in children, can be treated with drugs having fewer side effects, such as **ethosuximide** (4-225), a succinimide derivative; **carbamazepine** (4-226); or **valproate** (4-193), an enzyme inhibitor (see above). **Clonazepam** (4-204) is also very useful in several varieties of epilepsy. Since different forms of epilepsy respond in different ways to drug treatment, and the wrong drug can be more harmful than beneficial, a proper diagnosis is important.

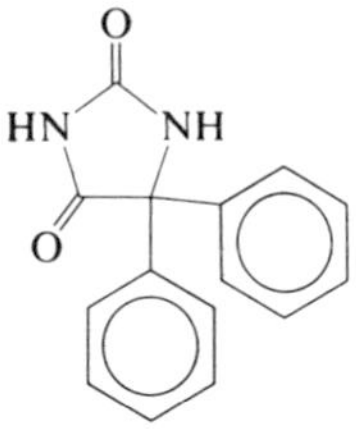

Phenytoin
4-224

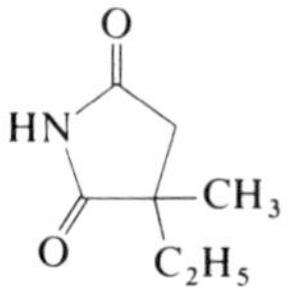

Ethosuximide
4-225

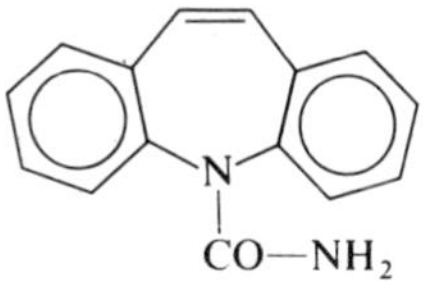

Carbamazepine
4-226

Although some of the foregoing compounds are known to act through the GABAergic system, their mode of action is not clear. Barbiturates and hydantoins also activate Na–K-dependent and Ca-dependent ATPase at high Na:K ratios, and increase Na transport, but whether this is the basis of their action in preventing the spread of seizures is not certain (Johnson and Willow, 1982). Carbamazepine is known to increase available adenosine $A_1$ receptors, and it has been proposed that adenosine is a natural anticonvulsant or convulsion modulator.

### 8.1.7. Defects in GABA Metabolism

*Huntington's chorea*, a neuromotor disorder, has its basis in defective central GABA metabolism. It is a hereditary disease that manifests itself in involuntary movements which disappear only during sleep, and eventually leads to mental deterioration in adults. It has been found that there is a deficiency of glutamic acid decarboxylase (GAD) in Huntington's chorea, and that the GABAergic neurons projecting from the caudate nucleus to the substantia nigra show lesions. Since direct replacement of the lost inhibitory functions on DA neurons is not possible, the usual treatment of Huntington's disease consists of inhibiting excessive dopaminergic activity by the DA-antagonist neuroleptics **haloperidol** (4-121) or **chlorpromazine** (4-112), restoring the balance of GABAergic and dopaminergic functions (Bird, 1980).

$\gamma$-Aminobutyric acid also seems to be involved in a number of other physiological functions, including feeding, sleep, hormonal secretion, cardiovascular functions, and, most importantly, the analgesia that is not mediated by opiate-sensitive neurons. Thus, further study of GABAergic mechanisms may lead to nonaddictive pain-relieving agents (cf. DeFeudis, 1982): The GABA-mimetic **THIP** (tetrahydro-isoxazolo-pyridinol) is reported to be equivalent to morphine. The reason for this correlation may be that enkephalinergic neurons, involved in pain pathways, seem to be regulated by GABAergic neurons.

### Selected Readings

P. R. Andrews and G. A. R. Johnston (1979). GABA agonists and antagonists. *Biochem. Pharmacol. 28*: 2697–2702.

R. R. H. Anholt (1986). Mitochondrial benzodiazepine receptors as potential modulators of intermediary metabolism. *Trends Pharmacol. Sci. 7*: 506–511.

E. D. Bird (1980). Chemical pathology of Huntington's disease. *Annu. Rev. Pharmacol. Toxicol. 20*: 533–551.

A. C. Bowling and R. J. Delorenzo (1982). Micromolar activity BZD receptors: identification and characterization in the CNS. *Science 216*: 1247–1249.

J. R. Cooper, F. E. Bloom, and R. H. Roth (1986). *The Biochemical Basis of Neuropharmacology*, 5th ed. Oxford University Press, New York.

E. Costa (1982). Do benzodiazepines act through a GABAergic mechanism? In: *Chemical Regulation of Biological Mechanisms* (A. M. Creighton and S. Turner, Eds.). The Royal Society of Chemistry, London.

E. Costa, G. di Chiara, and G. L. Gessa (Eds.) (1980). *GABA and Benzodiazepine Receptors. Advances in Biochemical Pharmacology*, Vol. 26. Raven Press, New York.

E. Costa and A. Guidotti (1979). Molecular mechanisms in the receptor action of benzodiazepines. *Annu. Rev. Pharmacol. Toxicol. 19*: 531–545.

E. Costa and A. Guidotti (1985). Endogenous ligands for benzodiazepine recognition sites. *Biochem. Pharmacol. 34*: 3399–3403.

F. V. DeFeudis (1982). $\gamma$-Aminobutyric acid and analgesia. *Trends Pharmacol. Sci. 3*: 444–446.

M. J. Eadie (1984). Anticonvulsant drugs: an update. *Drugs 27*: 328–363.

R. C. Effland and M. F. Forsch (1982). Anti-anxiety agents, anticonvulsants and sedative-hypnotics. *Annu. Rep. Med. Chem. 17*: 11–19.

F. J. Ehlert (1986). "Inverse agonists," cooperativity and drug action at benzodiazepine receptors. *Trends Pharmacol. Sci. 7*: 28–32.

S. L. Erdö (1985). Peripheral GABAergic mechanisms. *Trends Pharmacol. Sci. 6*: 205–208.

G. G. Glaser, J. V. Penry, and D. M. Woodbury (Eds.) (1980). Antiepileptic drugs: mechanism of action. *Advances in Neurobiology*, Vol. 27. Raven Press, New York.

W. Haefely, E. Kyburz, M. Gerecke, and H. Möhler (1985). Recent advances in the molecular pharmacology of benzodiazepine receptors and the structure–activity relationship of their agonists and antagonists. *Adv. Drug Res.* (B. Testa, Ed.), Vol. 14. Academic Press, New York, pp. 165–322.

J. K. Ho and R. A. Harris (1981). Mechanism of action of barbiturates. *Annu. Rev. Pharmacol. Toxicol. 21*: 83–111.

G. A. R. Johnston (1978). Amino acid receptors. In: *Receptors in Pharmacology* (J. R. Smythies and R. J. Bradley, Eds.). Marcel Dekker, New York, pp. 295–333.

G. A. R. Johnson and M. Willow (1982). GABA and barbiturate receptors. In: *More About Receptors* (J. W. Lamble, Ed.). Elsevier Biochemical Press, Amsterdam.

F. Kofod, P. Krogsgaard-Larsen, and J. Scheel-Kruger (Eds.) (1979). *GABA Neurotransmitters: Pharmacology, Biochemistry and Pharmacochemical Aspects*. Munksgaard, Copenhagen.

P. Krogsgaard-Larsen, P. Jacobsen, E. Falch, and H. Hjeds (1982). GABA-agonists: chemical, molecular, pharmacologic and therapeutic aspects. In: *The Chemical Regulation of Biological Mechanisms* (A, M. Creighton and S. Turner, Eds.). The Royal Society of Chemistry, London.

B. Olivier, D. van Dahlen, and J. Hartog (1986). A new class of psychotropic drugs: serenics. *Drugs of the Future 11*: 473–489.

E. J. Peck, Jr. (1980). Receptors for amino acids. *Annu. Rev. Physiol. 42*: 615–627.

G. Racagni and A. O. Donoso (Eds.) (1986). *GABA and Endocrine Functions*. Raven Press, New York.

J. G. Richards, P. Schoch, H. Möhler, and W. Haefely (1986). Benzodiazepine receptors resolved. *Experientia 42*: 121–126.

J. U. Saelens and F. J. Vinick (1978). Agents affecting GABA in the CNS. *Annu. Rep. Med. Chem. 13*: 31–40.

P. Skolnick and S. M. Paul (1981). Benzodiazepine receptors. *Annu. Rep. Med. Chem. 16*: 21–29.

S. H. Snyder and R. H. Junis (1979). Peptide neurotransmitters. *Annu. Rev. Biochem. 48*: 755–782.

G. Toffano, A. Leon, M. Massotti, A. Guidotti, and E. Costa (1980). GABA-modulin: a regulating protein for GABA receptors. In: *Receptors for Neurotransmitters and Peptide Hormones* (G. Pepeu, M. J. Kuhar, and S. J. Enna, Eds.). Raven Press, New York, pp. 132–142.

R. R. Trifiletti and S. H. Snyder (1984). Anxiolytic cylopyrrolones zopiclone and suriclone bind to a novel site linked allosterically to BZD receptors. *Mol. Pharmacol. 26*: 458–469.

J. A. Vida (1981). Central nervous system depressants: sedative-hypnotics. In: *Principles of Medicinal Chemistry*, 2nd ed. (W. O. Foye, Ed.). Lea and Febiger, Philadelphia, pp. 155–181.

M. Williams and N. Yokoyama (1986). Anxiolytics, anticonvulsants and sedative hypnotics. *Annu. Rep. Med. Chem. 21*: 11–20.

## 8.2. Glycine

The simplest amino acid is a neurotransmitter unique to vertebrates, found in the brainstem, spinal cord, and probably the retina. Like GABA, it is also a predominantly inhibitory transmitter. There is little evidence linking disorders of glycine neurotransmission to any neurological or psychiatric syndromes, except perhaps spasticity.

Whereas the peripheral biochemistry of glycine is well explored, central glycine metabolism is not well understood. Therefore, the uptake and metabolism of this substance cannot be influenced by drugs. It is probable that in the spinal cord, there are two glycine binding sites, in addition to an ionophore. The indole alkaloid **strychnine** seems to bind to the ionophore with a $K_D = 35$ nM, whereas a second receptor site binds glycine with a low affinity (in the $\mu$M range). *Tetanus toxin* also binds to the glycine site, and blocks glycine release as well (Peck, 1980). The highly lethal convulsant action of strychnine is thus a result of glycine antagonism in the spinal cord.

At one time it was thought that the benzodiazepine tranquilizers act on the glycine receptors. It has been found, however, that although they can displace [$^3$H]strychnine, glycine binding remains unaffected. Glycine involvement in anesthesia has been proposed (Chap. 1, Sec. 3.3).

The first tentative hypothesis on the structure of the *glycine receptor* has emerged (Betz, 1987). The polymeric glycoprotein apparently consists of two to three copies of a 48-kD protein, one to two copies of a 58-kD protein, and one copy of a 93-kD subunit; the total mass of the receptor is about 350 kD. The strychnine-binding site is primarily on the 48-kD subunits, but the role of the 93-kD protomer is unknown. The receptor subunits probably form the Cl⁻ ion channel (Fig. 4.44).

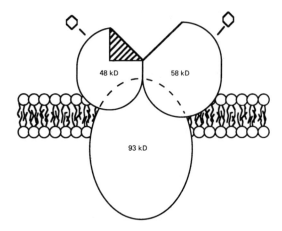

**Fig. 4.44.** Cross section of the glycine receptor. The two outer subunits are glycoproteins, as indicated by the hexopyranose rings. The hatched area is strychnine binding site. (Modified by permission from Betz, 1987)

### Selected Readings

H. Betz (1987). Biology and structure of the mammalian glycine receptor. *Trends Neurosci.* *10*: 113–117.

E. J. Peck (1980). Receptors for amino acids. *Annu. Rev. Physiol. 42*: 615–627.

### 8.3. Taurine

Taurine (2-aminoethanesulfonic acid) is another recently recognized inhibitory neurotransmitter in the brain. It is formed from cysteine, and its accumulation can be prevented by the cardiac glycoside **ouabain** (see Chap. 6, Sec. 3.3). The only known antagonist is strychnine and the effects of taurine therefore cannot be differentiated from those of glycine. Although receptor sites and specific actions cannot be elucidated without an antagonist, taurine has been implicated in epilepsy, mongolism (Down's syndrome), and, potentially, in heart disease.

There are a large number of physiological effects attributed to taurine (Huxtable and Sebring, 1986), among them cardiovascular (antiarrythmic), central (anticonvulsant, excitability modulation), muscle (membrane stabilizer), and reproductive (sperm motility factor) activity. The general mode of action is probably membrane based, involving the modulation of $Ca^{2+}$ and $Cl^-$ ion effects, and osmoregulation;

$$\text{(structure)} \quad \text{N—CH}_2\text{—CH}_2\text{—SO}_2\text{—NH—R}$$

4-227

the neurotransmitter nature of taurine is therefore questionable. The **phthalimino-taurinamide** (4-227) and its *N*-alkyl derivatives are less polar than taurine, and are more effective antiepileptics than valproate.

### *Selected Readings*

R. J. Huxtable and L. A. Sebring (1986). Towards a unifying theory for the action of taurine. *Trends Pharmacol. Sci.* 7: 481–485.

C. E. Wright, H. H. Tallan, Y. Y. Lin, and G. E. Gaull (1986). Taurine: biological update. *Annu. Rev. Biochem.* 55: 427–453.

## 8.4. Excitatory Amino Acid Transmitters

**Glutamate** and **aspartate** have long been known as excitatory transmitters, first in crustacean muscle and later in the vertebrate CNS. As amino acids, they have many other important biochemical roles; thus their concentration is uniformly high. Certain areas in the spinal cord, interneurons of the reflex arc, and a pathway from the *cortex* to the *striatum* are presumed sites of activity. Because of the uniform glutamate and aspartate distribution, mapping of the receptors was accomplished only recently (Monaghan et al., 1983). For the same reason, specific nonendogenous natural-product agonists had to be used for receptor characterization, in the same manner as in the differentiation of the nicotinic and muscarinic cholinoceptors.

Table 4.15 summarizes the properties of these receptors. Figure 4.45 shows the structure of the agonists and antagonists active at the receptors (see Sharif, 1985; Cooper et al., 1986). It is noteworthy that **kainate** (4-229), a rigid analogue of glutamate, is an irreversible neurotoxin, causing destruction of glutaminergic as well as cholinergic and dopaminergic neurons.

*N*-**Methyl-D-aspartate** (**NMDA**) ($A_1$) receptors have been implicated in the mechanism of information processing, memory, and learning, through long-term potentiation of neuronal pathways, involving NMDA receptors (Dingledine, 1986) as well as **kainate/quisqualate** receptors (Collingridge, 1985).

**Table 4.15.** Excitatory amino acid receptor classification

|  | *N*-Methyl-D-aspartate ($A_1$) | Quisqualate ($A_2$) | Kainate ($A_3$) |
|---|---|---|---|
| Location | cortex, hippocampus | hippocampus caudate nucleus | hippocampus caudate nucleus |
| Function | convulsions neuropotentiation degenerative effects | unknown | neurotoxic |
| Agonists | NMDA ibotenate | glutamate > aspartate quisqualate | kainate domoate |
| Antagonists | 2-amino-5-phosphonovalerate | glutamyltaurine | glutamyltaurine |

**Selective agonists**

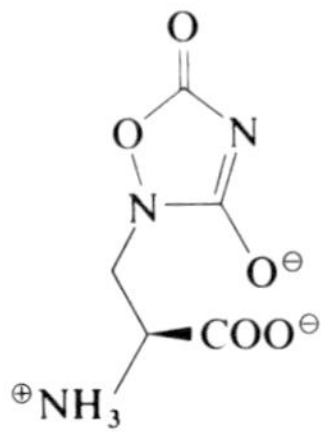

Quisqualate
4-228

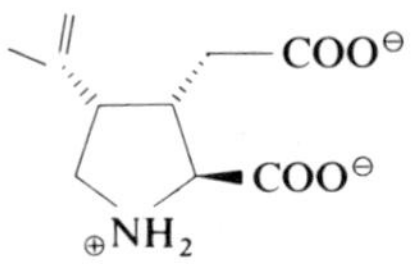

Kainate
4-229

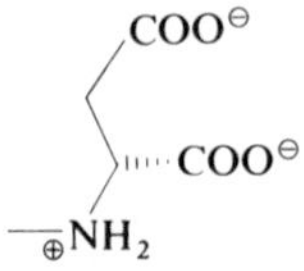

*N*-Methyl-D-aspartate
(NMDA)
4-230

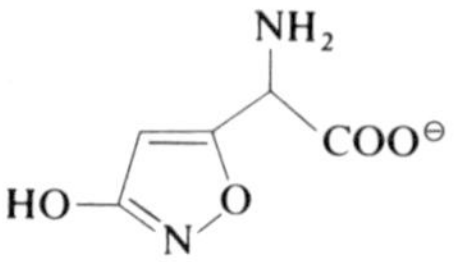

Ibotenate
4-231

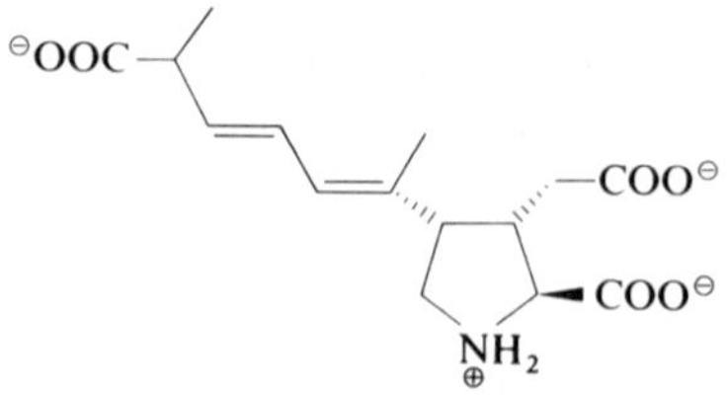

Domoate
4-232

**Endogenous agonists**

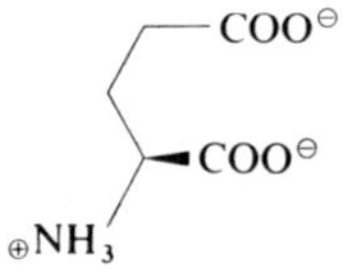

L-Glutamate
4-233

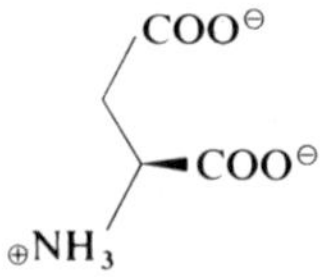

L-Aspartate
4-234

**Antagonists**

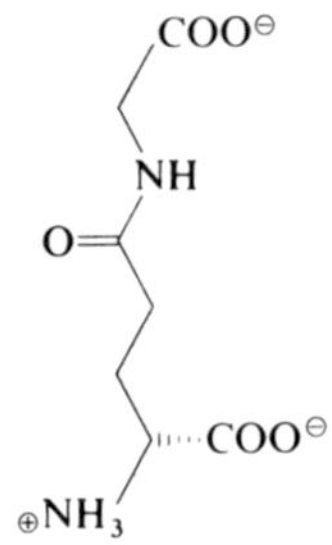

γ-D-Glutamylglycine
(DGG)
4-235

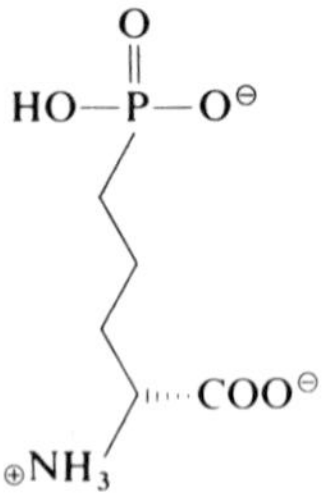

D-2-Amino-5-phosphonovalerate
D-2-Amino-5-phosphonopentanoate
(APV, AP5)
4-236

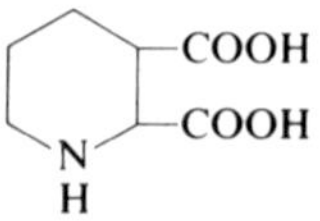

*cis*-2,3-Piperidine
dicarboxylic acid
4-237

**Fig. 4.45.** Excitatory amino acid agonists and antagonists.

The most recent investigations indicate (see Mayer, 1987) that excitatory amino acids may utilize a single common channel with multiple conductance (i.e., not just an open and closed state) for kainate, quisqalate, and NMDA. To complicate matters further, quisqalate may not operate on an ion channel at all, but may instead cause the synthesis of inositol-1,4,5-triphosphate resulting in an increase in $Ca^{2+}$. In this, it joins ACh and GABA, which are also known to activate both an ion channel as well as a G-protein-mediated second messenger.

### Selected Readings

G. L. Collingridge (1985). Long term potentiation in the hippocampus: mechanisms of initiation and modulation by neurotransmitters. *Trends Pharmacol. Sci. 6*: 407–411.

J. R. Cooper, F. E. Bloom, and R. H. Roth (1986). *The Biochemical Basis of Neuropharmacology*, 5th ed. Oxford University Press, New York.

R. Dingledine (1986). NMDA receptors: what do they do? *Trends Neurosci. 9*: 47–49.

P. Krogsgaard-Larsen and T. Honore (1983). Glutamate receptors and new glutamate agonists. *Trends Pharmacol Sci. 4*: 31–33.

M. Mayer (1987). Two channels reduced to one. *Nature 325*: 480–481.

D. T. Monaghan, V. R. Holets, D. W. Toy, and C. W. Cotman (1983). Anatomical distribution of four pharmacologically distinct $^3$H-L-glutamate binding sites. *Nature 306*: 176–179.

N. A. Sharif (1985). Multiple synaptic receptors for neuroactive aminoacid transmitters—new vistas. *Int. Rev. Neurobiol.*, Vol. 26. Academic Press, New York, pp. 85–150.

### 8.5. Substance P

Substance P (for "powder") (SP), an undecapeptide of the sequence Arg-Pro-Lys-Pro-Gln-Gln-Phe-Phe-Gly-Leu-Met-$NH_2$, is an extremely active excitatory peptide neurotransmitter. The mechanisms of its biosynthesis and inactivation are unknown. Moreover, there are few antagonists, but the localized neuronal presence of SP and its release in the salivary gland and in several brain regions suggest that it has a neurotransmitter or neurohormonal role. Its coexistence in some serotonergic neurons in the CNS is an interesting phenomenon. Substance P may be involved in pain mediation, as suggested by the fact that its injection into the brain has produced analgesia (Sandberg and Iversen, 1982) and it may regulate catecholamine turnover. The second messenger for SP seems to be phosphatidylinositol, which mobilizes $Ca^{2+}$.

SP is involved in mediating pain responses peripherally as well as centrally. Small fibers of the peripheral pain-sensitive neurons use SP as excitatory transmitter, and neurotransmitter release is inhibited by opiates at the dorsal horn of the spinal cord (see Chap. 5, Sec. 3.1, on pain physiology). In the CNS, the effect of SP is inhibitory, analgesic, and is stimulated by the endogenous opiate neuropeptide met-enkephalin (Chap. 5, Sec. 3.5). In stressed animals, SP can block the analgesic effect of endogenous opiates. The peripheral sensory effects of SP are mediated by the C-terminal fragment (6–8 amino acids) of SP, whereas the central analgesic effects are due to the N-terminal fragment, which also stimulates learning and memory. Thus posttranslational peptide-modifying enzymes may decide which effect will prevail, in addition to the receptors at the site of action.

Since SP and related peptides (**eledoisin** and **physalaemin**, collectively called *tachykinins*) react differently at different sites, three receptors—SP-E, SP-K, and SP-P—have been proposed (see Buck and Burcher, 1986). Research has been hampered by lack of sufficiently active antagonists, although the (D-Pro$^2$, D-Trp$^{7,9}$) analogue (where the superscript denotes the position of the amino acid in the peptide chain) blocks the antiinflammatory responses of SP. Unfortunately, some of these antagonists are also neurotoxins on motor nerves (Hanley, 1985).

### Selected Readings

S. H. Buck and E. Burcher (1986). The tachykinins: a family of peptides with a brood of 'receptors.' *Trends Pharmacol. Sci. 7*: 65–68.

M. R. Hanley (1985). Substance P antagonists. In: *Neurotransmitters in Action* (D. Bousfield, Ed.). Elsevier, Amsterdam, pp. 170–172.

R. A. Nicoll, C. Schenker, and S. E. Leeman (1980). Substance P as a transmitter candidate. *Annu. Rev. Neurosci. 3*: 227–268.

B. E. B. Sandberg and L. L. Iversen (1982). Substance P. *J. Med. Chem. 25*: 1009–1015.

# 5

# Drugs Acting on Hormones, Neurohormones, and Their Receptors

## 1. STEROID HORMONES AND THEIR RECEPTORS

The steroid hormones include the estrogens, gestagens, androgens, and adrenocorticoids, as well as their precursor, cholesterol. They are all found preponderantly in animals, and are based on the four-ring **steran** carbon skeleton (5-1). Other pharmacologically interesting steroids, like the heart-active cardenolides, are compounds of plant origin. Since they act in an entirely different fashion, they are discussed separately in Chap. 6, Sec. 3.3.

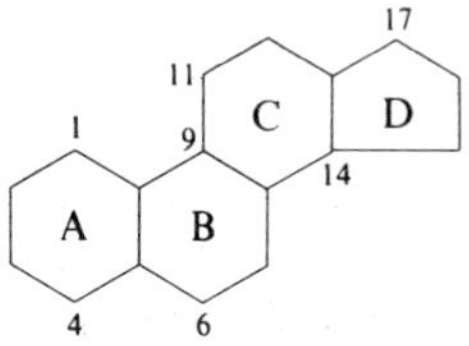

Steran skeleton
5-1

Seemingly minor changes in the stereochemistry and substitution pattern of the steran skeleton result in vastly different, specific physiological and pharmacological effects, which in turn influence morphological, developmental, metabolic, and behavioral phenomena. Lately we have begun to understand the molecular basis of this finely tuned distinction between related compounds. This chapter attempts to compare and contrast the structure and mode of action of various steroids, their role in regulating hormonal secretion, and the timing of this regulatory action; in other words, the "biological message" that these small hormone molecules convey to many target tissues.

The organic chemistry and biochemistry of steroids is the subject of many excellent books and an enormous amount of research and patent literature. The reader is referred primarily to the reviews by Jones (1973 and 1976), the textbooks by Lehninger and Foye, and the monograph series edited by Litwack (cf. Liao, 1977) to name just a few. Nevertheless, a brief summary of steroid structure, classification, and biochemistry follows, as well as a discussion of our present understanding of the general steroid-receptor mechanism.

## 1.1. The Structure and Conformation of Steroids

All steroids are based on the steran skeleton (5-1), a fully hydrogenated **cyclo-pentano-phenanthrene**. Traditionally, the rings of this skeleton are labeled A, B, C, and D. The numbering of the carbon atoms is as shown. All four rings are in the chair conformation in naturally occurring steroids; additionally, rings B, C, and D are always *trans* with respect to each other, whereas rings A and B can be *trans* (as in cholestanol) or *cis* (the coprostanol conformation). It is simple to conceptualize this ring anellation (fusion) if one observes the relation of substituents (including hydrogen) on the carbon atoms common to the rings in question. For rings A and B, the relative positions of the 19-methyl group (attached to C-10) and the hydrogen on C-5 are determining, and their *trans* or *cis* configuration is easily visualized. In general, neighboring substituents are *trans* if they are diaxial or diequatorial, and *cis* if they are axial–equatorial. The two methyl groups on C-10 and C-13 are always axial relative to rings B and D, with the C-10 substituent (which is not necessarily methyl) being the conformational reference point.

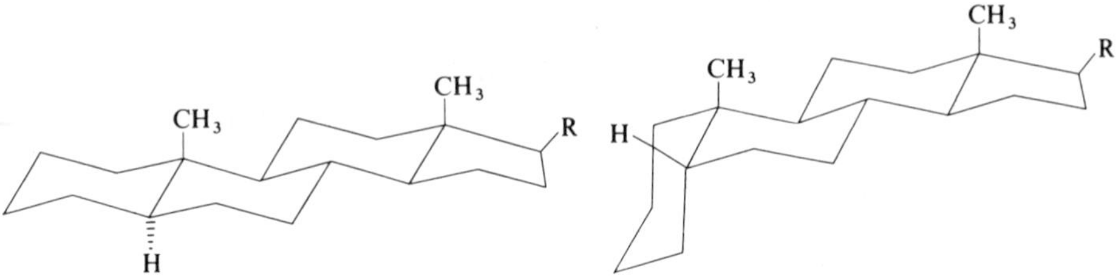

Cholestane, A–B *trans*                    Coprostane, A–B *cis*

A somewhat obsolete but still valid nomenclature determines substituent con-formations relative to the plane of a cyclohexane-type ring: thus, the 19-methyl group in the steroid ring system is designated $\beta$ and is *above* the plane of the molecule, while H-5 in cholestane is $\alpha$ and *below* the plane. The $\alpha$–$\beta$ convention for steroids must not be compared or mixed with the usual axial–equatorial convention (Chap. 1, Sec. 5.3) since with the latter convention, the flipping of a cyclohexane ring (for example) from one chair form to another changes the position of a $\beta$ substituent from axial to equatorial, and vice versa:

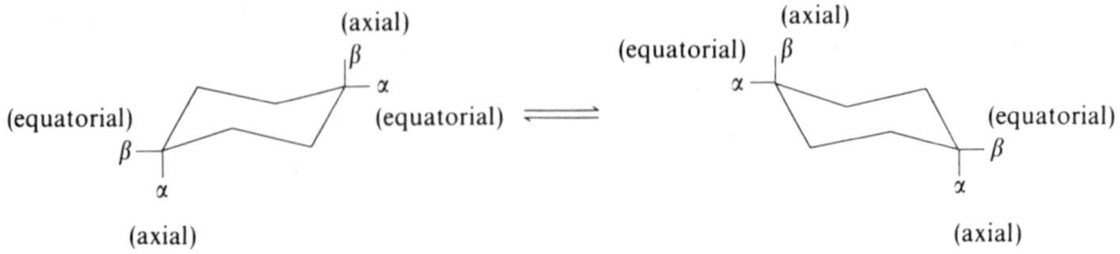

Since a substituent designated $\alpha$ (or $\beta$) will remain $\alpha$ (or $\beta$) but can be either axial or equatorial, confusion can arise. The stability, reactivity, and spectroscopy of a substituent will, however, change depending on its axial or equatorial position. Equatorial substituents are normally more reactive and less stable than their epimers and show slightly different absorption spectra. The physiological and pharmacological properties of the different molecules are also different, as might be expected.

These considerations apply to all cycloalkane derivatives, including steroids. However, the chair form of a ring is inherently more stable than the boat form. Moreover, the fused-ring nature of the system lends it a very considerable rigidity, and *cis–trans* isomerization would necessitate the breaking and formation of covalent bonds. Therefore, steroid substituents maintain their conformation at room temperature, whereas cyclohexane substituents usually do not.

Steroids are classified according to their substituents (in addition to their occurrence). Representatives of each group are shown in Fig. 5.1, with salient features accented by bold lines.

**Fig. 5.1.** Representative examples of the main steroid groups. Characteristic features are shown in bold print.

Sterols

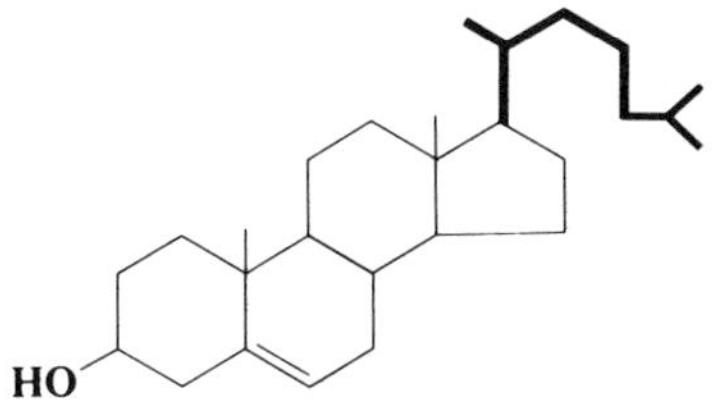

Cholesterol

Bile acids
(fat detergents)

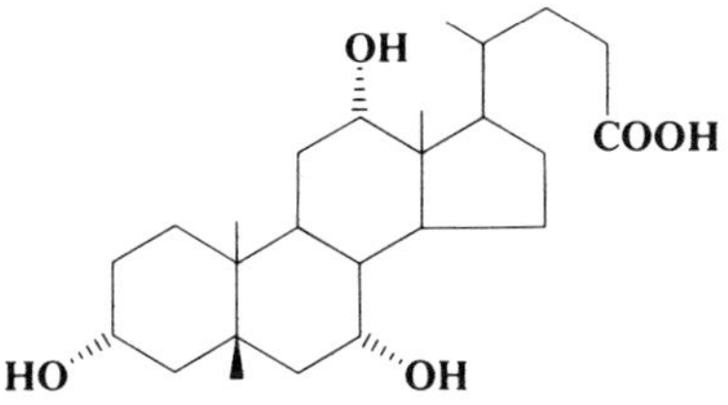

Cholic acid

Estrogens
(female sex hormones)

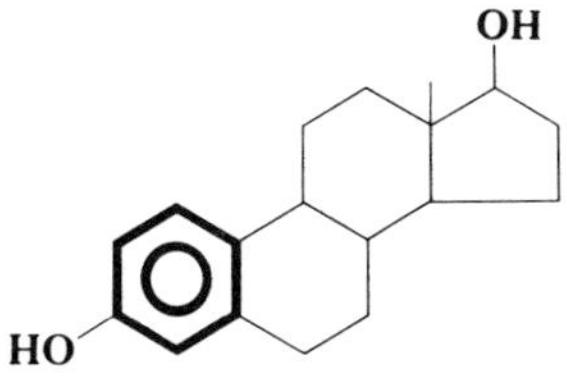

Estradiol

Gestagens
(female sex hormones)

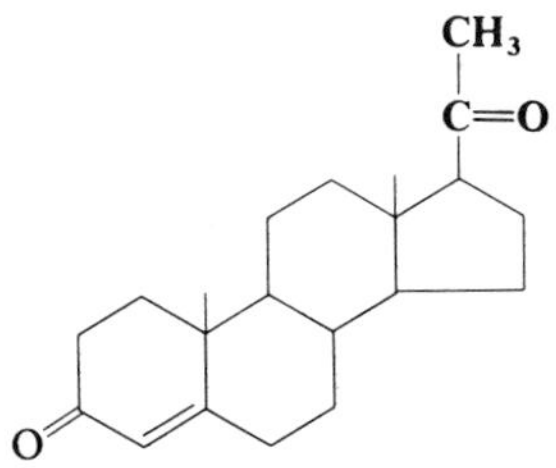

Progesterone

Androgens
(male sex hormones)

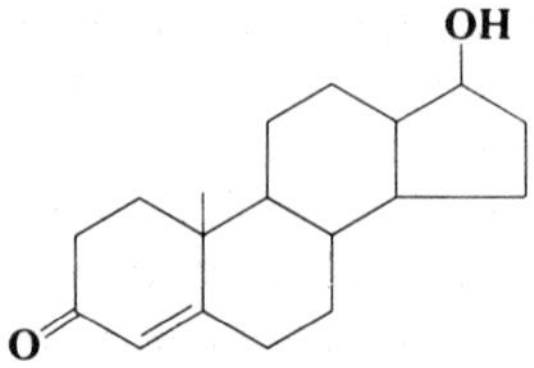

Testosterone

Corticoids
(metabolic regulators)

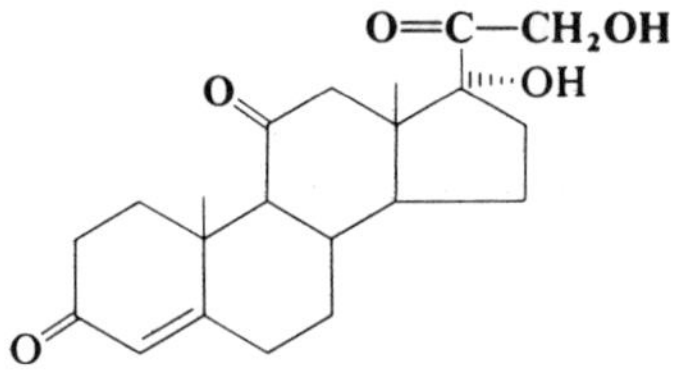

Cortisone

Cardenolids
(cardioactive drugs)

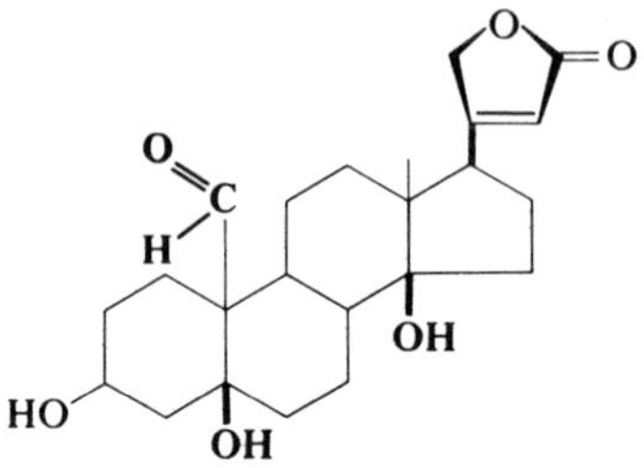

Strophanthidine

Sapogenins

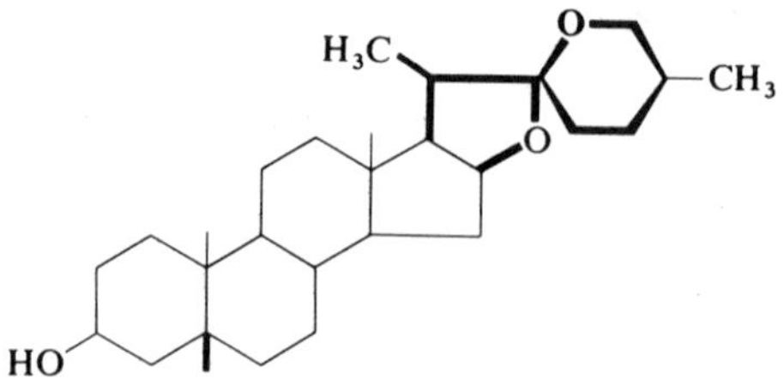

Sarsasapogenin

Steroid alkaloids

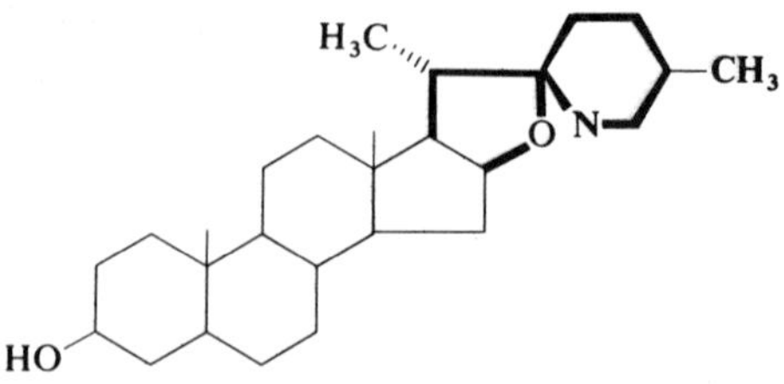

Tomatidine

**Fig. 5.1.** (*continued*)

### 1.1.1. Steroid Biosynthesis

The biosynthetic correlations of steroids are very complex, as one would expect when all of the compounds in a group have to be derived from a single precursor (cholesterol). A very abbreviated summary of the biogenesis of animal steroids is shown in Fig. 5.2, with many steps and intermediates left out. The ultimate source of all the compounds involved in steroid synthesis is acetate, in the form of acetyl-coenzyme A. Cholesterol, besides being ingested in food, is synthesized in large amounts, and an adult human contains about 250 g of cholesterol. In contrast, the steroid hormones are produced at the milligram level or lower.

*Regulation.* The regulation of steroid biosynthesis is achieved by an intricate network of peptide hormones, some of which are discussed in Sec. 2.2. The production of these hormones is under neuroendocrine influence, starting with the *hypothalamus* in the CNS, which itself is under dopaminergic control. The hypothalamus produces a number of small peptides hormones that act as the releasing factors for a second series of peptide hormones synthesized in the *anterior pituitary gland*. Among these hormones adrenocorticotropic hormone (ACTH) regulates corticosteroid synthesis in the adrenal cortex, whereas luteinizing hormone (LH) and follicle stimulating hormone (FSH) (the gonadotropins) act on the ovaries and testes. Gonadotropins induce the production of estrogens and gestagens in the female, which, in turn, produce appropriate changes in the reproductive tract. In the male, LH and FSH regulate androgen formation. At the same time, the steroids have a feedback regulatory effect on the hypothalamo-pituitary axis, setting up an exquisitely tuned regulatory loop. In addition, gonads produce the protein *inhibin*, which suppresses production of follitropin (FSH) and gonadoliberin in the pituitary and hypothalamus, respectively (see Secs. 2.5 and 2.3), and therefore, indirectly, steroid hormone production. These correlations are summarized in Fig. 5.3 and Table 5.5. There are many opportunities to exercise direct or indirect drug control over a large number of metabolic and reproductive phenomena in such a sensitive multicomponent feedback system. Understanding the physiology, regulation, and molecular mechanism by which the steroids act is therefore the basis of rational drug design and therapy.

## 1.2. Steroid Receptors

These receptors are highly specific macromolecules found in the target tissues (uterus, vagina, prostate), in regulatory organs (pituitary, hypothalamus), and in low concentrations in the brain, liver, kidney, ovary, and many other tissues. In all of their target organs, steroids exert their influence directly on protein synthesis, at the level of transcription of the genetic message. They also influence many enzyme systems through cAMP-dependent protein kinases.

While the macromolecules and target tissues involved show extreme specificity for the appropriate steroid hormones and their congeners, the general scheme of steroid-receptor mechanism seems to be remarkably uniform. We can therefore deal with this receptor model in a general way, mentioning specific details as appropriate in the subsections of this chapter.

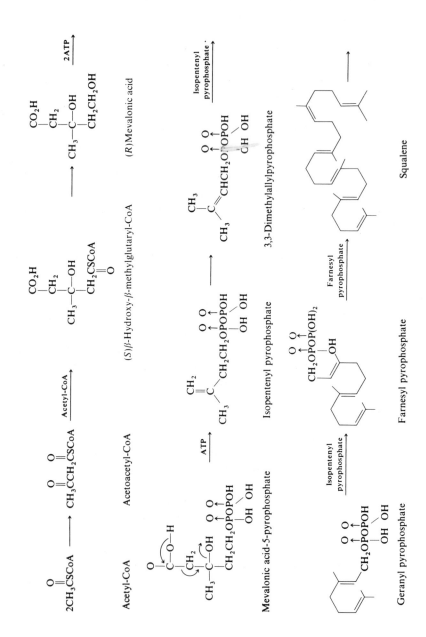

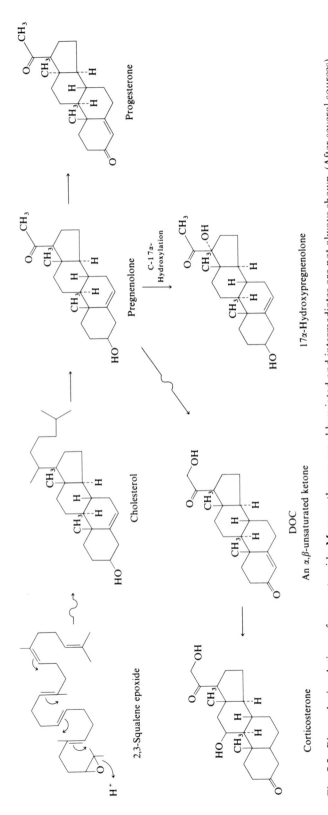

**Fig. 5.2.** Biosynthetic relations of some steroids. Many pathways are abbreviated and intermediates are not always shown. (After several sources)

251

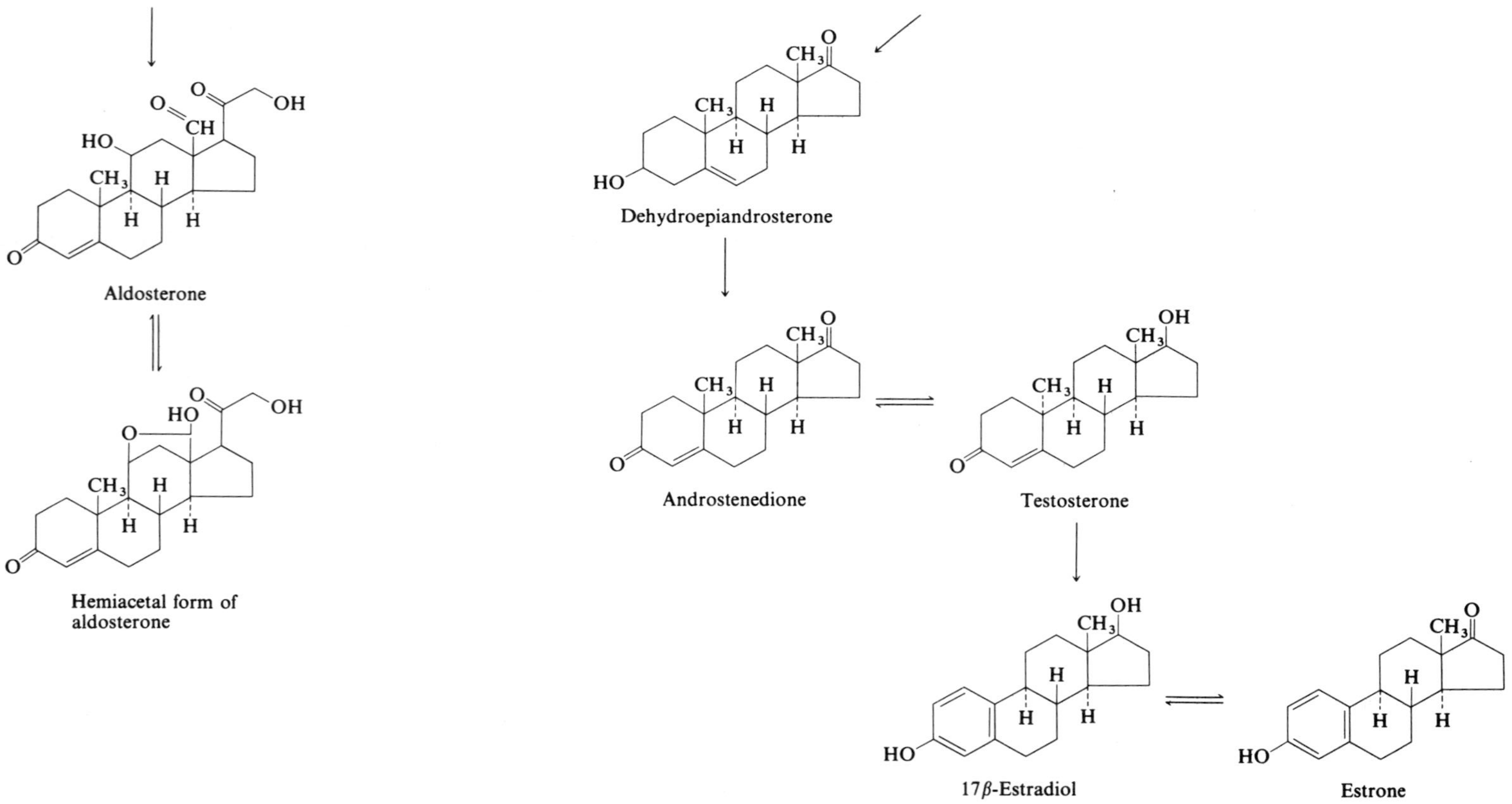

Dehydropiandrosterone

Androstenedione

Testosterone

17-β-Estradiol

Estrone

Aldosterone

Hemiacetal form of
aldosterone

**Fig. 5.2.** (*continued*)

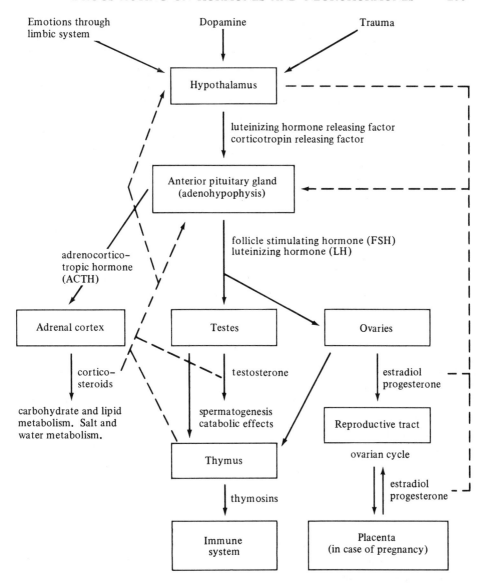

**Fig. 5.3.** Regulation of steroid hormone production and some physiological effects of these hormones. Dashed lines show feedback regulation.

The general steroid-receptor hypothesis is based mainly on estrogen and progesterone receptors. The currently accepted mechanism is unique, and consists of several steps at different subcellular structures:

1. Cytoplasmic receptor activation
2. Translocation of the hormone–receptor complex to the nucleus
3. Binding of the complex to chromatin acceptor sites
4. Activation of transcription

The steroid hormones are transported to their target cells via the bloodstream in a protein-bound form, but diffuse into the cell as free steroids. Here, they encounter a *cytoplasmic receptor protein*, which in the case of estrogens (the best-studied hormones) has a 4 $S$ sedimentation constant, corresponding to a molecular weight of about 75,000. This asymmetric, ellipsoidal protein undergoes estrogen-catalyzed activation by forming a dimer with another, different protein, resulting in a 7–9 $S$ (200,000-Dalton) complex. This additional protein is found in nontarget as well as target tissues. It also seems that in the absence of the steroid, the 4 $S$ receptor is associated with an inhibitory macromolecule from which it must dissociate before activation. The same sequence of activation is found in the case of glucocorticoid receptors (Sec. 1.7). This activation process is temperature- and ionic strength-dependent. Selective proteolysis can be used to distinguish the steroid-binding portion of the cytosol receptor from the site necessary to bind to nuclear chromatin. Antibodies raised against the receptor from calf uterus cross-react with receptors from other mammals, but do not show any immunologic reaction with androgen or progesterone receptors.

Both of the cytosol receptor subunits described above bind steroid hormones before they can be translocated to the nucleus. The cytosol receptors have been isolated and purified to homogeneity.

### 1.2.1. Translocation

Translocation of the *steroid–receptor complex* to the cell nucleus occurs by an unknown mechanism, perhaps by diffusion only.

The dimeric steroid–receptor complex enters the nucleus of the target cell and binds through its B subunit (the "specifier") to the appropriate *nuclear acceptor site*. This is an acidic fraction of nonhistone protein designated AP$_3$ protein. It can be isolated from any cell and can replace the identical protein in the target cell without any loss of binding affinity. The B subunit of the steroid–receptor dimer thus transports the A subunit to the vicinity of the cistron specifying the appropriate protein to be synthesized. This A subunit cannot bind to chromatin, but only to "naked" DNA. After the initial binding through its B protomer, as described above, the A–B dimer must therefore dissociate.

The A subunit will expose the appropriate steroid target gene, which was previously hidden in the supercoiled chromatin, by destabilizing the DNA duplex, allowing RNA polymerase to bind and thus transcribe the genetic message into mRNA. Unfortunately, how such precise recognition of the appropriate DNA segment occurs is unknown. About 3,000 to 4,000 initiation sites per picogram of DNA have been found, with the number in most cases being directly related to the concentration of the steroid–receptor complex. The synthetic response in the cell is very rapid: within 15 minutes a considerable increase in the concentration of RNA polymerase can be detected, and within 30 minutes induced protein synthesis is measurable. These early responses, occurring in the first 4–6 hours, can be triggered by hormones that have a lower than optimum affinity for the receptor (such as estriol in uterus, which binds much more weakly than estradiol). However, the late effects of steroid action (e.g., uterine weight increase after 24 hours, or DNA synthesis) are

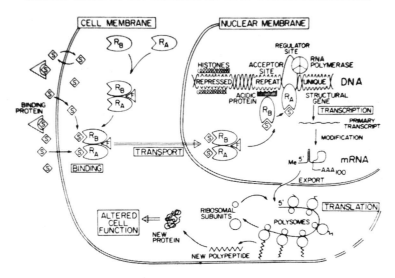

**Fig. 5.4.** A general model for steroid-hormone action, patterned mainly on estrogens. For details, see text. (Reproduced by permission from Schrader et al. (1978), Marcel Dekker, New York)

stimulated only by high-affinity steroids (see Katzenellenbogen, 1980, and Ringold, 1985).

The steroid–receptor complex remains in the nucleus for a limited time only, and eventually dissociates from chromatin. About 40% of the receptors released in this dissociation are recycled and used again; the rest are probably destroyed and resynthesized. Steroid hormones can apparently regulate the level of synthesis of their own receptors, and sometimes the synthesis of other steroid receptors as well; however, other regulatory factors are also involved, such as some of the peptide hormones. This model, developed by O'Malley and Schrader (1976) and his group, is shown in Fig. 5.4.

This model has been tentatively modified by Gorski (see Gorski et al., 1984; Gorski, 1986) and Walters (1985). These authors claim that the "cytoplasmic receptors" are in fact nuclear all the time and are immobilized; or, they migrate in the unoccupied state to the nucleus. The steroid hormones present in both the cytoplasmic and nuclear compartment bind to the nuclear receptor, triggering a conformational change to the "active" form, which has a high affinity for the nuclear acceptor site and thus initiates transcription.

### 1.3. Cholesterol

#### *1.3.1. Role in Atherosclerosis*

Cholesterol metabolism and its involvement in atherosclerosis and heart disease play a central role in research efforts directed at decreasing the incidence of cardiovascular disease, the number-one killer in Western civilization. There is a statistical correlation between serum cholesterol levels and the incidence of atherosclerosis. In the latter condition, lesions (atheromas) containing cholesterol and other lipids form in the arterial walls, followed by the fibrosis of these plaques, and narrowing of the arterial diameter, causing a decrease in blood flow, changes in the blood-clotting mechanism, and eventual total arterial occlusion. This then leads to a heart attack or stroke. Nevertheless, the entire field of cholesterol metabolism, dietary involvement (e.g., excessive consumption of saturated fatty acids), exercise, and life-style in the etiology of cardiovascular disease is a highly controversial topic, fraught with half-truths and misinformation, especially in popular reports on this topic.

There is increasing evidence that atherosclerosis can be arrested and even decreased (see Cayen, 1979). However, the evaluation of drug therapy for this condition is complicated by the fact that the regulation of lipoprotein levels and cholesterol metabolism does not necessarily have an effect on atherosclerosis.

Besides high levels of serum cholesterol and cholesterol esters, a combination of other factors appears to be necessary for the development of atherosclerosis. From a molecular viewpoint, the ratio of the various serum lipoproteins that carry cholesterol is of major importance. There are three groups of these: very-low-density ($d = 0.95 - 1.006$), low-density ($d = 1.006 - 1.063$), and high-density ($d = 1.063 - 1.21$) lipoproteins. The first two groups seem to increase, whereas the high-density lipoproteins (HDL) seem to decrease the incidence of atheromas. High-density lipoproteins may even facilitate the removal of cholesterol from the arterial wall. This is probably accomplished in two ways: (1) by the increased esterification of cholesterol, and (2) by inhibition of the LDL–cholesterol complex uptake by the cells of the arterial wall (Brown and Goldstein, 1984).

***Drug Therapy.*** According to our present knowledge there are two potential ways to decrease cholesterol-containing atheroma formation: (1) by decreasing circulating cholesterol levels, and (2) by increasing the HDL concentrations. Their combined use is preferable to the use of either method alone.

The most widely used drug for inhibiting cholesterol biosynthesis is **clofibrate** (5-2), an isobutyrate derivative. It seems to block cholesterol synthesis at the point at which cholesterol exerts feedback inhibition on its own synthesis: at the formation of mevalonate (Fig. 5.2). It also blocks acetyl-CoA carboxylase, the enzyme producing malonyl-CoA. However, clofibrate is far from being an ideal drug, having quite a few side effects.

Clofibrate
5-2

**Nicotinic acid** (5-3) in large doses influences the lipoprotein ratio, decreasing the concentrations of very low and low-density lipoprotein, but has no effect on HDL–

cholesterol complexes. **Acipimox** (5-4), a new pyrazine derivative, is 20 times more active than nicotinic acid. Similarly, **metformin** (5-5), an $N,N$-dimethyl-biguanidine, has no effect on cholesterol biosynthesis, but only on lipoprotein composition. It produces a 50% reduction in the serum triglyceride level, which is usually also very high in persons prone to hyperlipidemia. The drug is also a hypoglycemic agent and lowers blood glucose levels. Unfortunately, nothing is known about the mode of action of these compounds.

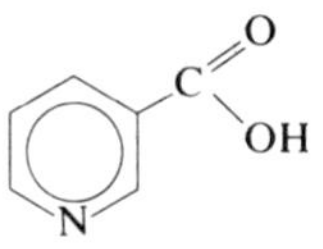

Nicotinic acid
5-3

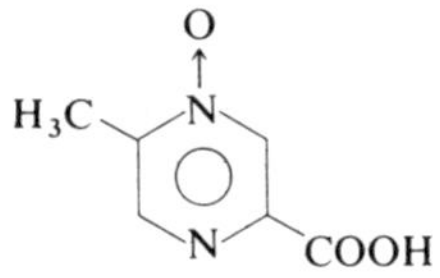

Acipimox
5-4

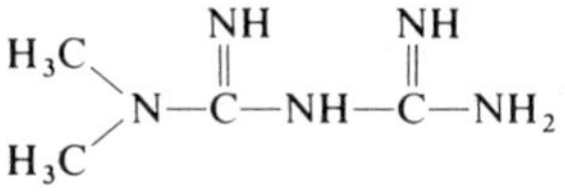

Metformin
5-5

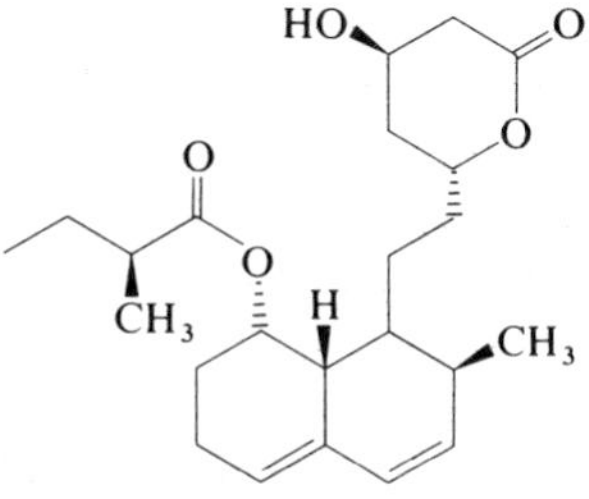

Compactin
5-6

Among recent advances in treating hypercholesterolemia is the use of anion-exchange resins like **cholestyramine** and **colestipol** which sequester bile acid in the intestine, excrete them, and thus increase their synthesis in the liver by a feedback mechanism. Increased bile acid synthesis increases cholesterol metabolism and also decreases LDL concentration. The resins work very well in combination with cholesterol biosynthesis inhibitors like the fungal metabolites **compactin** (5-6) and its methyl derivative **mevinolin**. Kane and Havel (1986) and Newton and Krause (1986) summarize recent results in this important area. The beneficial effect of prostacyclin (also called prostaglandin $I_2$) on atherosclerosis will be discussed in Sec. 4.

### 1.3.2. Bile Acids in Digestion

Cholesterol is metabolized in the liver to bile acids (Fig. 5.1), which are necessary for digestion, since they act as natural detergents and solubilize dietary fats. All of the hydroxyl groups of the various bile acids (a maximum of three OH groups) are axial, the A–B ring anellation is *cis*, and the polarity of the side chain is increased by conjugation with glycine or taurine. Consequently, there are both hydrophilic and hydrophobic portions of the molecule, which can thus act as a detergent and form inclusion compounds with fatty acids, promoting their absorption through the intestinal wall. The excessive excretion of cholesterol can lead to its crystallization and the formation of *gallstones*. There has been some success in dissolving these stones through the chronic (1–2 years) administration of bile acid derivatives.

### 1.3.3. Introduction to Reproductive Steroid Hormones

The reproductive steroid hormones are divided into three classes:

1. *Estrogens*, which regulate ovulation and the development of the secondary female sex characteristics
2. *Gestagens*, or progestins, which maintain pregnancy
3. *Androgens*, the male sex hormones

## 1.4. Estrogens

These steroids are produced mainly in the ovaries when the latter are stimulated by follicle-stimulating hormone. Under such stimulation, the estrogen levels rise until the middle of the menstrual cycle (when ovulation takes place), remain at a fairly constant concentration, and then decline if fertilization does not take place. The final result is menses, the shedding of the uterine endometrium (lining). Estrogens also regulate uterine growth in immature animals, and, as noted above, are responsible for all female secondary sex characteristics.

Although estrogens are easily isolated from the urine of pregnant women, their most abundant source is the urine of stallions, in which they appear as metabolites of androgens. The primary estrogenic hormone from this source is **17β-estradiol** (5-7, Fig. 5.5), which is metabolized to the ketone **estrone** (5-8), only one-tenth as active. Horses also excrete **equilenin** (5-9), a steroid containing the napthalene ring system. **Estriol** (5-10), the 16α-ol derivative of estradiol, is also less active than estradiol. All of these steroids have an aromatic A-ring and therefore lack a 19-methyl group.

Two semisynthetic, orally active estrogens are **ethinyl estradiol** (5-11) and its 3-methyl ether (**mestranol**). Both of these are used in oral contraceptives (see Sec. 1.5) **Moxestrol**, the 12-methoxy derivative, is used in receptor labeling (Ojasoo and Raynaud, 1978).

### 1.4.1. Nonsteroidal Estrogens

Nonsteroidal estrogens have also been synthesized. The first such compounds were ***trans*-diethylstilbestrol** (5-12) and its reduced derivative **hexestrol** (5-13). The way in which these stilbene (diphenylethylene) derivatives are usually drawn suggests a resemblance to the steroid skeleton. The resemblance is purely incidental, since the two ethyl groups are not indispensable for estrogenic activity. For example, four methyl groups will give comparable pharmacological activity. However, it seems that the thickness of the molecule is important to its activity: the planes of the two phenyl groups must have a 60° torsion angle, and four alkyl methyl groups will provide the same steric hindrance as two ethyls. The dimethyl derivative is therefore inactive. Because their use may lead to cancer to the vulva in the daughters of their

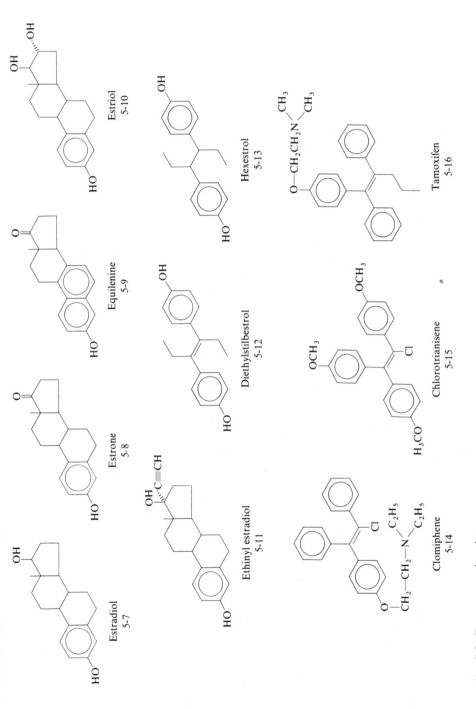

**Fig. 5.5.** Estrogens and antiestrogens.

259

users, stilbestrol derivatives are obsolete and dangerous drugs. This delayed action was discovered only recently.

**Chlorotrianisene** (5-15) is vaguely related to the stilbestrols. Since it is highly nonpolar, it is stored in body fat and liberated slowly, probably as an active metabolite. It is therefore considered a pro-drug.

Estrogens are used therapeutically to replace or augment hormones whose natural production is insufficient during menopause, in menstrual disorders, or as a result of insufficient development of the female reproductive tract. In androgen-dependent prostate carcinoma, estrogens are used therapeutically to suppress androgen formation and thus tumor growth.

### 1.4.2. Antiestrogens

Antiestrogens are used for two purposes: as fertility drugs and as antitumor agents. The first application is based on the fact that estradiol, the natural hormone, inhibits the secretion of the gonadotrophic hormones LH and FSH by feedback inhibition. The result of this inhibition is the production of a single ovum in every menstrual period, thus preventing overlapping pregnancies. Antiestrogens block this inhibition in women who are infertile because of anovulation resulting from excessive estradiol production. Therefore, multiple pregnancies are rather common in women treated with antiestrogens. This incidence of multiple births can rise up to 10%, much higher than the normal rate.

Antiestrogens are also active as antitumor agents in estrogen-dependent mammary carcinoma (breast cancer), a neoplasm which has estrogen receptors. These are found in about two thirds of all breast tumors. Such tumors can sometimes be treated with androgens, preferably the nonvirilizing derivatives (see Sec. 1.6), in addition to removal of the ovaries. The two antiestrogens in use are **clomiphene** (5-14) and **tamoxifen** (5-16), both of which are aminoether derivatives of stilbene. Tamoxifen inhibits synthesis of estrogen-regulated proteins by blocking the $G_1$ phase of the cell cycle, and is also an inhibitor of calmodulin-mediated enzyme systems. Its molecular mechanism of action is depicted in Fig. 5.6, where the estrogen-induced conformational change in the receptor is prevented by the side chain of tamoxifen (Jordan, 1984). The *cis* isomer of tamoxifen is estrogenic rather than being an estrogen antagonist.

### 1.5. Gestagens (Progestins)

These hormones are essential for the maintenance of pregnancy. The only natural progestin hormone, **progesterone** (5-17, Fig. 5.7), is produced by the *corpus luteum*, an endocrine gland formed in the ovary by the ruptured ovarian follicle after the level of luteinizing hormone peaks. If pregnancy occurs, the *corpus luteum* persists for the first 3 months of the pregnancy; after that, its role is taken over by the placenta as the major source of progesterone as well as estrogens. Through its feedback effect on the hypothalamus, progesterone prevents ovulation and also

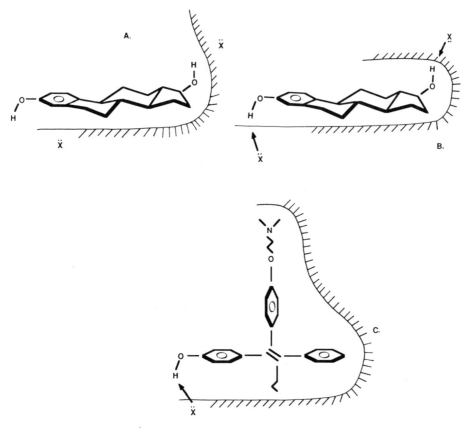

**Fig. 5.6.** Receptor effect of antiestrogens shows binding of flat estradiol molecule through hydrogen bonding sites (**X**) and folding of receptor protein induced fit (**A** and **B**). The tamoxifen molecule (**C**) can also bind but does not allow conformational change of the receptor. (Modified by permission from Jordan, 1984)

stops uterine movement, thus avoiding dislodgement of the fertilized egg or embryo. In the absence of pregnancy, the progesterone level, together with the estrogen concentration, declines, resulting in estrus or menses—the shedding of uterine endometrium along with the unfertilized egg. At the same time, low steroid levels disinhibit hypothalamo–pituitary endocrine secretion, the peptide hormone levels rise, and the cycle starts again.

Progesterone itself is used in threatened spontaneous abortion and the treatment of menstrual disorders. The main use of gestagens however, is based on their antifertility effect. Since progesterone is poorly absorbed, it cannot be given orally; furthermore, it is not particularly potent. **Ethisterone** (5-18), an acetylenic compound prepared from androsterone, is an orally active progestin, but also has some male hormone action. A large number of semisynthetic progesterone derivatives (progestogens) have been synthesized, and have found their principal use as oral contraceptives.

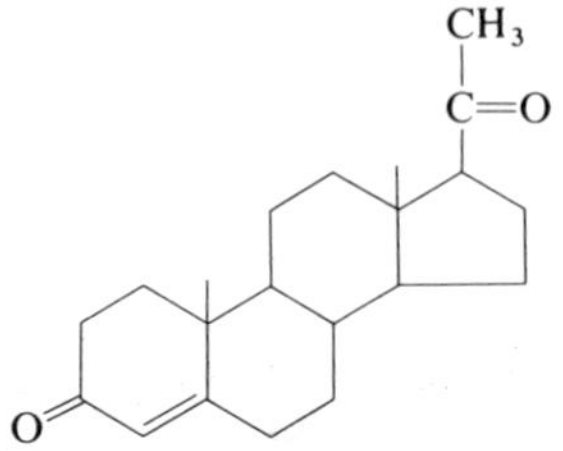

Progesterone
5-17

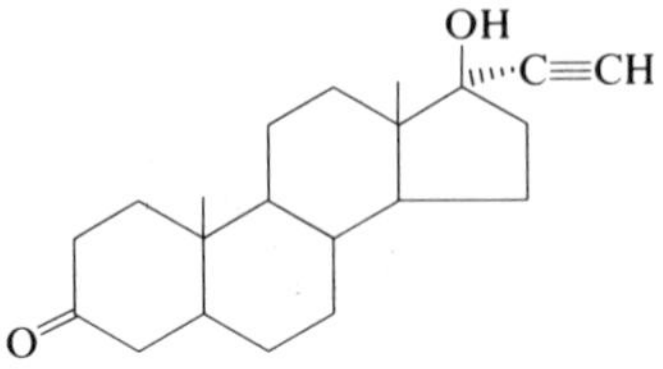

Ethisterone
5-18

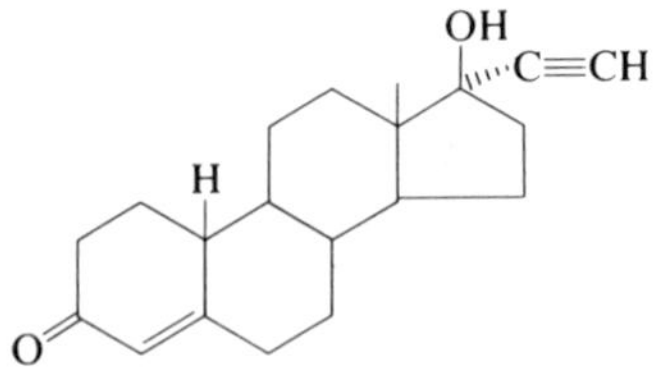

Norethindrone
5-19

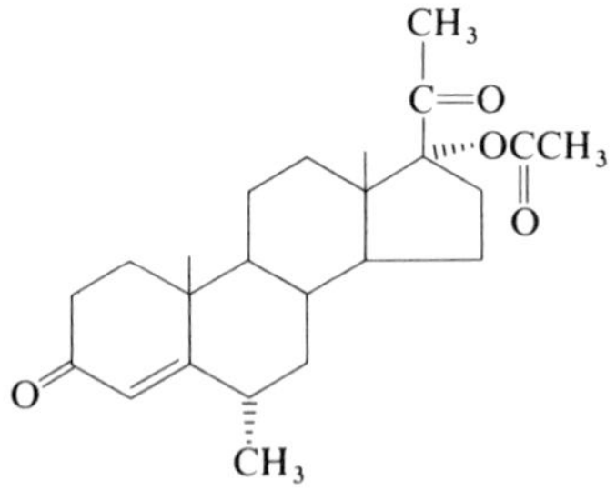

Medroxyprogesterone-acetate
5-20

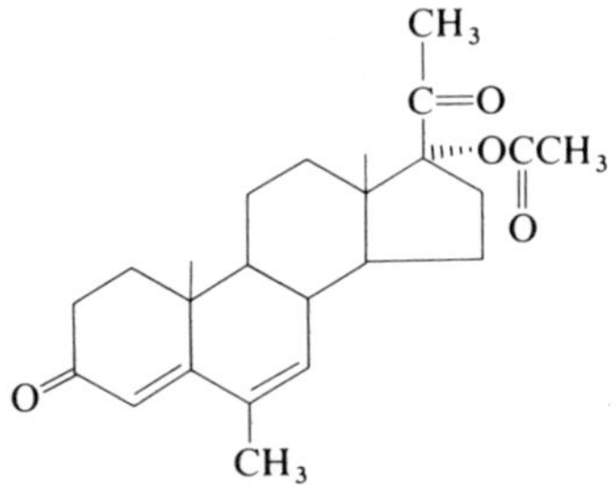

Megesterol acetate
5-21

**Fig. 5.7.** Gestagens or progestins.

### 1.5.1. Oral Contraceptives

The concept of oral contraception was pioneered by Gregory Pincus in the early 1950s. Modifications of the ethisterone molecule—removal of the 19-methyl group (5-19) or the introduction of additional methyl groups in position C-6 of 17α-acetoxyprogesterone—led to a series of highly active gestagens, shown in Fig. 5.7. Acetylation increases the lipid solubility and therefore extends the duration of activity of such derivatives, whereas the introduction of a methyl group on C-6 (and C-11) interferes with metabolic destruction of the drug and increases its oral potency.

Oral contraceptives are used in three ways:

1. A progestogen and an oral estrogen, such as mestranol, are combined and taken from the fifth to the twenty-fifth day of the cycle. Withdrawal then results in menstruation. The effect is due to inhibition of ovulation through the hypothalamo–pituitary mechanism, and probably by alteration of the viscosity of the cervical mucus. Phasic contraceptives vary the gestagen dose during the cycle, mimicking its variation under physiological conditions.
2. The sequential method, now in disrepute, used estrogen alone between days 5 and 20 of the cycle, followed by a progestogen for 4–5 days. The estrogen almost completely prevents ovulation; the gestagen fosters eventual maturation of the egg but induces shedding of the endometrium immediately during withdrawal bleeding.
3. A low dosage of gestagen ("mini-pill") is preferred nowadays, in the form of **medroxyprogesterone acetate** (5-20) or its 6-chloro analogue (**chlormadinone acetate**), both of which are active at very low doses. These compounds do not seem to inhibit ovulation, but rather interfere with the endometrium, cervical mucus, and perhaps the physiology of the fallopian tubes. Their use prevents most of the side effects of oral contraception, specifically nausea, water retention, and in some cases thrombophlebitis. They cause menstrual irregularities and therefore have not gained widespread acceptance.

It is to be expected that these extremely widely used oral methods of contraception will eventually be replaced by luteinizing hormone analogues as antiovulatory agents (see Sec. 2.2), or possibly by immunological methods directed against human chorionic gonadotrophin (hCG), which has both LH and FSH activity. Prostaglandins also show promise (see Sec. 4), and reliable male contraception is also a future possibility (Bell et al., 1986).

### 1.6. Androgens (Male Sex Hormones)

These steroids control the development of male characteristics: sperm production and growth of the sex organs, prostate, and seminal vesicle, as well as controlling metabolic effects during growth in adolescence. They cause nitrogen retention by stimulating protein synthesis (called the anabolic effect).

**Testosterone** (5-22, Fig. 5.8), the hormone synthesized by the testes, must be reduced to **dihydrotesterone** (DHT) before binding to the receptor can take place. Among highly active synthetic testosterone analogues, the **7α-methyl-19-nortestosterone** (5-23) and **oxandrolone** (5-24) have about a hundred times greater activity than testosterone as androgens. **17α-Methyltestosterone** (5-25) is orally active.

The distinction of anabolic from androgenic action is important in the anabolic therapy of such wasting conditions as cancer, trauma, osteoporosis, and the effects of immobilization. These conditions necessitate nitrogen and mineral retention; however, the masculinizing effects of androgenic agents would be undesirable in female patients. **Fluoxymesterone** (5-26), the 9α-fluoro-11-hydroxy-17α-methyl derivative of methyltestosterone, has 10 times the androgenic and 20 times the

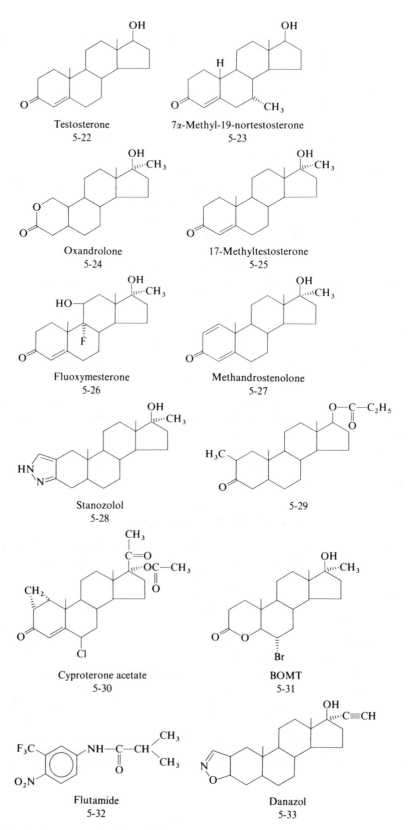

**Fig. 5.8.** Androgens, anabolic steroids, and antiandrogens.

anabolic activity of the parent compound, but is used mainly as an androgen in hormone-replacement therapy, hypogonadism, and some forms of breast cancer. Some other anabolic agents are **methandrostenolone** (5-27) and **stanozolol** (5-28), a pyrazole derivative. An interesting nonvirilizing androgen that is useful in the treatment of breast cancer in young women is the **2-methyl derivative** of **testosterone propionate** (5-29). Use of anabolic steroids by athletes is a widespread but dangerous, unproven, and banned misuse of these drugs (cf. Doerge, 1982).

### 1.6.1. Molecular Action

The molecular action of androgens differs slightly from the general model described previously, which is principally based on estrogens and progesterone. In the prostate, testosterone is reduced enzymatically to DHT, which then binds with very high affinity ($K_D = 10^{-11}$ M) to the so-called $\beta$-protein, forming complex II. A low-affinity ($10^{-7}$ M), high-capacity protein is also present. However, this is a much less selective protein, since it binds estradiol in addition to DHT. The cytosol complex II sediments at 7.5–8.5 S, and is transformed to the 3–4 S nuclear form.

The binding of androgens to the cytosol receptor is a process requiring several hours, since the steroid apparently becomes "enveloped" into the protein. This could also be responsible for the very high binding constants ($10^{-11}$–$10^{-12}$ M) observed for androgens. Molecular flatness, especially in the A–B ring area, is a prerequisite for effective binding, and A–B *cis* compounds are therefore inactive (see Liao, 1977). Accessory binding sites for a 7α-methyl group seem to be present, which strengthens the binding, and the 17β-hydroxyl group is essential for activity.

Following binding, the cytosol complex II migrates to the target-cell nuclei and initiates transcription and protein synthesis in a manner identical to that described in the general steroid–receptor model. A heat-labile acceptor factor seems to be responsible for chromatin binding of the complex. As many as 10,000 receptor molecules may bind to any nucleus in the target tissues. The rat ventral prostate contains about $100 \pm 40$ fmol ($= 10^{-15}$ moles) of DHT-binding receptors per milligram of protein. Latest results indicate that the cytosol receptor may be an artifact, perhaps in analogy to the estrogen receptor (Rowley and Tindall, 1986).

### 1.6.2. Antiandrogens

Antiandrogens, such as **cyproterone acetate** (5-30), **BOMT** (5-31), or the nonsteroidal **flutamide** (5-32), are competitive antagonists on the cytosol receptor. They do not prevent DHT formation, but inhibit the nuclear retention of DHT in the prostate. They cause feminization in male fetuses and decrease libido in males. Cyproterone is also an extremely active progestogen (Rasmusson, 1986).

### 1.6.3. Male Contraception

Male contraception, involving the suppression of spermatogenesis, is an intensely investigated area, but there have been few positive results that could be used on a practical scale. **Danazol** (5-33), a gonadotropin inhibitor in females, is also active in males if it is taken together with long-acting testosterone derivatives. No complete spermatozoan elimination has been achieved with male contraceptive agents (Bell et al., 1986), although a total absence of sperm is probably not necessary for sterility. **Gossypol** (5-34), a phenolic compound isolated from cottonseed oil has direct antispermatogenic activity. It is used in China, but it has considerable and partly irreversible side effects (Bell et al., 1986).

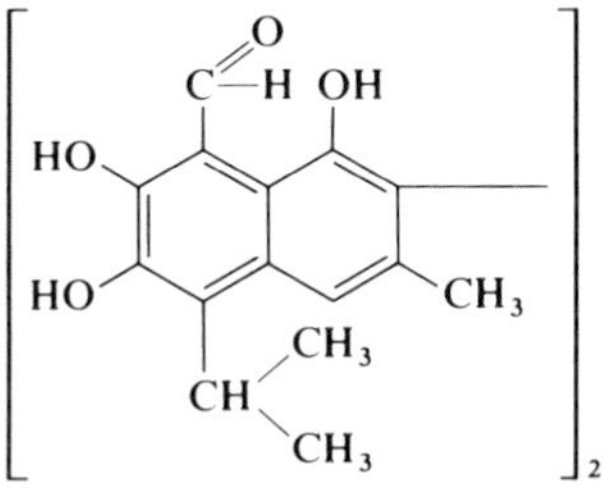

Gossypol
5-34

## 1.7. Adrenal Steroids

The adrenal gland is a caplike organ sitting on top of the kidney. Histologically the gland consists of the inner *medulla*, the site of catecholamine synthesis; and the outer *cortex* (shell), where steroid synthesis take place. Like most endocrine glands, the adrenal cortex is regulated by the hypothalamo–pituitary peptides.

The hypothalamus secretes **corticotropin releasing factor** (CRF) (see Sec. 2.2), which controls the release of **adrenocorticotropin** (ACTH), a peptide consisting of 39 amino acids. Corticotropin secretion is under feedback regulation by the adrenal steroids, as well as being under the control of higher CNS centers; stress or epinephrine can increase corticosteroid production.

On the basis of biochemical effects, two groups of corticosteroids can be distinguished: *glucocorticoids*, which act on carbohydrate, fat, and protein metabolism; and *mineralocorticoids*, which regulate electrolyte balance through $Na^+$ retention.

### 1.7.1. Glucocorticoids

Glucocorticoids promote gluconeogenesis by increasing the pyruvate carboxylase concentration, and thus increase the concentration of oxaloacetate in the mitochondrial pathway of pyruvate–phosphoenolpyruvate synthesis. Another effect appears to be the inhibition of pyruvate oxidation. Lipolysis is stimulated primarily through the activation of adenylate cyclase. As do all steroids, glucocorticoids increase the rate of enzyme synthesis in the nucleus of target cells and achieve their effect on overall protein synthesis in this manner. The principal target of glucocorticoids is the liver, although other organs—notably the muscles and brain—are also rich in glucocorticoid receptors. The overall metabolic result is increased glycogen storage and hyperglycemia.

**Table 5.1.** Relative glucocorticoid and mineralocorticoid activity of some cortico-steroid derivatives.

|  | Relative glucocorticoid potency | Relative antiinflammatory activity | Relative mineralocorticoid activity |
|---|---|---|---|
| Cortisol | 1 | 1 | 1 |
| Fluorocortisone | 10 | 0 | 300 |
| Prednisone | 3 | 5 | 0.8 |
| Methylprednisolone | 5 | 6 | 0.5 |
| Triamcinolone | 2 | 5 | 0.1 |
| Dexamethasone | 10 | 30 | 0.05 |
| 11-Desoxycorticosterone | 0 | 0 | 30 |
| Aldosterone | 0.2 | 0 | 600 |

Most glucocorticoids have some mineralocorticoid effect, usually considered an undesirable activity. Molecular modifications can separate the two activities, as shown in Table 5.1.

***Pharmacological Activity.*** The most important pharmacological action of glucocorticoids and their semisynthetic analogues is their *antiinflammatory and antirheumatic activity*, discovered in 1949 by Hench and co-workers. While revolutionizing the treatment of osteoarthritis and rheumatoid arthritis, these drugs did not provide a cure for these crippling, painful, and widespread diseases. Although the mode of action of glucocorticoids is unknown, there are certain indications that they interfere with prostaglandin and collagenase synthesis, and the circulatory distribution of leukocytes in inflamed tissue, as well as inhibiting edema by decreasing capillary permeability (see Fauci, 1978). These steroids are also effective in some allergic diseases, such as bronchial asthma and in urticaria and other types of dermatitis. In the case of skin symptoms, the external application of steroid drugs is possible, eliminating many systemic side effects. Patients suffering from acute leukemia and lymphosarcoma may also benefit temporarily from glucocorticoids. However, the symptomatic nature of the treatment must be emphasized in all of these conditions.

Glucocorticoids are used in replacement therapy in cases in which the adrenal cortex is destroyed for some reason. The resulting syndrome, called *Addison's disease* and described more than 120 years ago, is lethal if not treated. The symptoms are weakness, anemia, nausea, hypotension, depression, and an abnormal skin pigmentation. Hyperadrenalism, or *Cushing's syndrome*, on the other hand, may result from adrenal tumors or increased ACTH levels. Its main symptoms (also seen after prolonged corticosteroid treatment) are facial "mooning," acne, osteoporosis, hypertension, weight gain, and a decreased resistance to infection.

***Glucocorticoid Receptor.*** Although the glucocorticoid receptor differs from the estrogen or progesterone receptors, the basic principles of its action seem to be the same. Circulating hormones, bound to a serum protein (transcortin), pass into

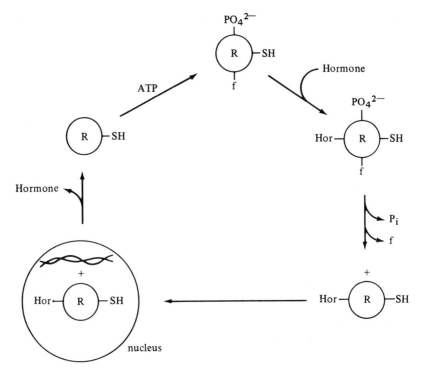

**Fig. 5.9.** A proposed glucocorticoid receptor mechanism. The receptor (R. on top) carries an —SH group, a phosphate ($P_i$), and an inactivating, heat-stable factor (f); the receptor becomes activated by losing both the phosphate and the f factor. After the hormone (Hor) binds, the charged receptor–hormone complex can translocate into the nucleus and bind to DNA. After initiating transcription, the complex leaves the nucleus and dissociates; then the receptor is reactivated through phosphorylation by a protein kinase. (Modified after Schmidt and Litwack, 1982)

the target cell and are taken up by the cytoplasmic receptor, which seems to differ in structure from the estrogen receptor (Fig. 5.9).

Binding to the glucocorticoid receptor is somewhat dependent on temperature, and is optimal at 37°C. Activation of the receptor is catalyzed by a cytoplasmic transforming factor. The steroid–receptor complex is then translocated to the nucleus, binds to chromatin, and induces mRNA production.

*Structure–Activity Correlations.* The structure–activity relationships of gluco-corticoids (Fig. 5.10) are based on two natural hormones **cortisol** (5-35) and **corti-costerone** (5-36). The characteristic structural features of these hormones are the conjugated 3-ketone, the 11-OH group, and the 17β-ketol side chain. Molecular modifications have been aimed at deriving compounds with glucocorticoid and antiinflammatory actions but lacking in mineralocorticoid effects and side effects. Substituents added to the cortisol molecule alter is activity independently of other functional groups present on the molecule, and can be quantified by an "enhance-ment factor" that gives an indication of the binding capability of the derivative to

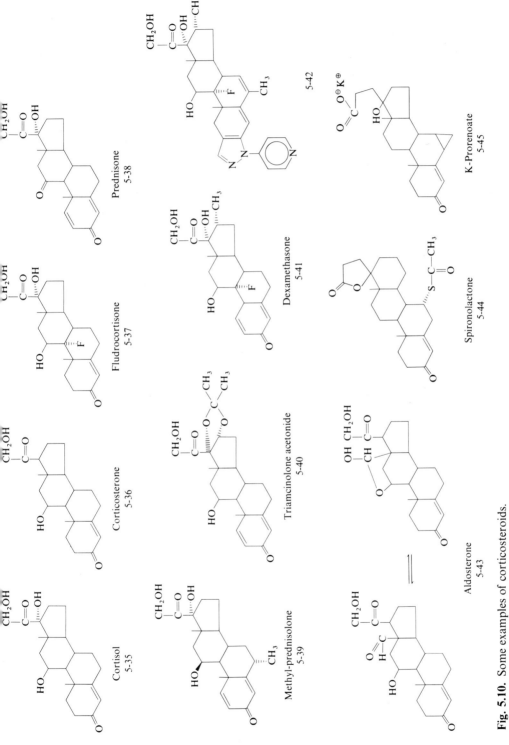

**Fig. 5.10.** Some examples of corticosteroids.

Cortisol
5-35

Corticosterone
5-36

Fludrocortisone
5-37

Prednisone
5-38

Methyl-prednisolone
5-39

Triamcinolone acetonide
5-40

Dexamethasone
5-41

5-42

Aldosterone
5-43

Spironolactone
5-44

K-Prorenoate
5-45

269

the cytosol receptor. The contribution of various substituents to binding was measured by Wolff and co-workers (Table 5.2). While a high receptor binding ability does not necessarily reflect overall pharmacological activity, it is an important factor in glucocorticoid drug design. Table 5.1 shows modifications that increase activity:

1. *Halogenation.* 9α-Fluorocortisone is only about 10 times more active than its parent compound, but its mineralocorticoid activity is 300–600 times greater. This is undesirable since it leads to edema; thus, the compound **fludrocortisone** (5-37) is used only topically, in ointments. Interestingly, transcortin, the glucocorticoid transport protein, does not bind fludrocortisone, indicating that the α-side of the molecule is important in transcortin binding but not in receptor affinity.
2. *Additional double bonds.* $\Delta^1$-compounds (where $\Delta^1$ indicates the position of a double bond) were introduced, like **prednisone** (5-38), a $\Delta^1$-11-ketone and **prednisolone**, its 11-hydroxy analogue. Changing the geometry of the A ring increased the potency without augmenting mineralocorticoid activity.

The introduction of methyl group in **methylprednisolone** (5-39) resulted in a slight increase in activity; however, the greatest improvements in activity came from the combination of a double bond, halogen, and methyl substituents.

**Triamcinolone** (5-40), in the form of its acetonide (an acetone ketal), shows a 9α-fluoro group in addition to $\Delta^1$ unsaturation, and a 16α-OH. It is used in treating

**Table 5.2.** Contribution to glucocorticoid-receptor binding of progesterone substituents

| Substituent | Free energy (kcal/mol) | Fried glycogen-deposition-enhancement factor (rat) |
|---|---|---|
| $\Delta^1$ | −0.29 | 3–4 |
| 6α-F | −0.36 | — |
| 6α-CH | −1.09 | 2–3 |
| 9α-F | −0.57 | 10 |
| 9α-Cl | −0.71 | 3–5 |
| 9α-Br | +1.89 | 0.4 |
| 9α-OCH$_3$ | +2.21 | — |
| 11β-OH | −0.89 | — |
| 11β-OH, 11-keto | +2.23 | — |
| 11-keto | +1.67 | — |
| 16α-CH$_3$ | −0.11 | — |
| 16β-CH$_3$ | −0.21 | — |
| 16α-OH + acetonide | −0.35 | — |
| 17α-OH | +0.49 | 1–2 |
| 21-OH | −0.56 | 4–7 |

Reproduced by permission from Wolff et al., (1977), *Biochemistry* 7: 3201, American Chemical Society, Washington, DC.

psoriasis and other dermatological problems. **Dexamethasone** (5-41), perhaps the most active and highly stable glucocorticoid, is the 16α-methyl analogue of triamcinolone.

Among the newer glucocorticoid derivatives, the [3, 2-c]-(2'-arylpyrazolo) derivatives such as 5-42, show a high activity (2000 times that of cortisol). Auxiliary binding sites are obviously involved in the receptor affinity of these compounds. The optimum glucocorticoid structure shows a 1α,2β-half-chair conformation for ring A, with ring D a 13-envelope (C-13 is bent up) or a half-chair. Halogenation is most effective in positions 6, 7, 9, and 12. The compounds bind on their β face by hydrophobic binding forces.

It is interesting to note that progesterone binds well to the glucocorticoid receptor despite a missing 11-oxygen functional group, but fails to elicit gene activation in glucocorticoid target cells, shedding light on the role of the 11-OH group.

### 1.7.2. Mineralocorticoids

Mineralocorticoids like the natural hormone **aldosterone** (5-43) regulate electrolyte concentration by stimulating $Na^+$ retention in kidney cells. **11-Desoxycorticosterone** and **fludrocortisone** (5-37) are much less active, but are used for maintaining electrolyte balance in adrenal insufficiency. Aldosterone synthesis is probably regulated by ACTH and angiotensin (see Sec. 2.8), a peptide hormone. This hormone has its own receptors in kidney cells. Hyperaldosteronism can play a role in high blood pressure and muscular tetany. *Aldosterone antagonists* like **spironolactone** (5-44) and the safer **K-prorenoate** (5-45) are therefore useful as hypotensive diuretic agents, because the increase in $Na^+$ excretion that they promote is always paralleled by an increased urine volume.

Structurally, aldosterone differs from glucocorticoids in having an 18-aldehyde group and lacking the 17α-OH. The aldehyde participates in a tautomeric equilibrium, forming a cyclic hemiacetal ring. **Spironolactone**, a synthetic compound and antiandrogen, has a lactone ring attached to C-17 through one common carbon (a "spiro" compound) and the 7-thiolester group, while **K-prorenoate** has a 6,7-cyclopropyl ring. The side chain is the hydrolyzed lactone ring of spironolactone. K-prorenoate is 3–5 times more active than spironolactome, does not cause potassium loss, and is not a potential carcinogen.

Arriza et al. (1987) has elucidated the structure of the mineralocorticoid receptor through cloning of its cDNA. It shows considerable structural and functional similarity to the glucocorticoid receptor.

### Selected Readings

H. B. Anstall (1974). Steroid hormonal analogues. In: *Biochemistry of Hormones* (H. V. Rickenberg, Ed.), Int. Rev. Sci., Biochemistry Series 1, Vol. 8. Butterworths, London, pp. 243–282.

J. L. Arriza, C. Weinberger, G. Cerelli, T. M. Glaser, B. L. Handelin, D. E. Housman, and R. E. Evans (1987). Cloning of human mineralocorticoid receptor c-DNA: structural and functional kinship with the glucocorticoid receptor. *Science* 227: 268–275.

M. R. Bell, F. H. Batzold, and R. C. Winneker (1986). Chemical control of fertility. *Annu. Rep. Med. Chem. 21*: 169–177.

M. S. Brown and J. L. Goldstein (1984). How LDL receptors influence cholesterol and atherosclerosis. *Sci. Am. 255*(5): 58–66.

M. H. Cake and G. Litvack (1975). The glucocorticoid receptor. In: *Biochemical Actions of Hormones* (G. Litvack, Ed.), Vol. III. Academic Press, New York, pp. 317–390.

M. N. Cayen (1979) Disorders of lipid metabolism. *Annu. Rep. Med. Chem. 14*: 198–208.

H. Danielson and J. Sjövall (Eds.) (1985). *Sterols and Bile Acids.* New Comprehensive Biochemistry, Vol. 12. Elsevier, Amsterdam.

C. E. Day (1978). Pharmacologic regulation of serum lipoproteins. *Annu. Rep. Med. Chem. 13*: 184–195.

R. F. Doerge (Ed.) (1982). *Wilson and Giswold's Textbook of Organic Medicinal and Pharmaceutical Chemistry*, 8th Ed. J. B. Lippincott, Philadelphia.

A. S. Fauci (1978). Mechanism of action of glucocorticosteroids. *Annu. Rep. Med. Chem. 13*: 179–183.

W. O. Foye (Ed.) (1981). *Principles of Medicinal Chemistry*, 2nd ed. Lea and Febiger, Philadelphia.

J. Gorski (1986). The nature and development of steroid hormone receptors. *Experientia 42*: 744–750.

J. Gorski, W. Welshons, and D. Sakai (1984). Remodeling the estrogen receptor model. *Mol. Cell. Endocrinol. 36*: 11–13.

M. J. Green and B. N. Lutsky (1976). Steroids. *Annu. Rep. Med. Chem. 11*: 149–157.

W. W. Grody, W. T. Schrader, and B. W. O'Mally (1982). Activation, transformation and subunit structure of steroid hormone receptors. *Endocrinol. Rev. 3*: 141–163.

O. Jenne and C. W. Bardin (1984). Androgen and antiandrogen receptor binding. *Annu. Rev. Physiol. 46*: 107–118.

W. F. Jones (1973). *Steroids.* Int. Rev. Sci., Organic Chemistry Series I. Vol. 8. Butterworths, London.

W. F. Jones (1976). *Steroids.* Int. Rev. Sci., Organic Chemistry Series II. Vol. 8. Butterworths, London.

V. C. Jordan (1984). Biochemical pharmacology of antiestrogen action. *Pharmacol. Rev. 36*: 245–276.

J. P. Kane and R. J. Havel (1986). Treatment of hypercholesterolemia. *Annu. Rev. Medicine 37*: 427–435.

B. S. Katzenellenbogen (1980). Dynamics of steroid hormone receptor action. *Annu. Rev. Physiol. 42*: 17–35.

A. L. Lehninger (1982). *Principles of Biochemistry.* Worth, New York.

S. Liao (1977). Molecular action of androgens. In: *Biochemical Action of Hormones* (G. Litwack, Ed.), Vol. 4. Academic Press, New York, pp. 351–406.

W. R. Nes and M. L. McKean (1977). *Biochemistry of Steroids and Other Isopentenoids.* University Park Press, Baltimore.

R. S. Newton and B. R. Krause (1986). Approaches to drug intervention in atherosclerotic disease. *Annu. Rep. Med. Chem. 21*: 189–200.

T. Ojasoo and J. P. Raynaud (1978). Unique steroid congeners for receptor studies. *Cancer Res. 38*: 4186–4198.

B. W. O'Mally and W. T. Schrader (1976). Receptors of steroid hormones. *Sci. Am. 234*(2): 32–40.

G. H. Rasmusson (1986). Chemical control of androgen action. *Annu. Rep. Med. Chem. 21*: 179–188.

G. M. Ringold (1985). Steroid hormone regulation of gene expression. *Annu. Rev. Pharmacol. Toxicol. 25*: 529–566.

D. R. Rowley and D. J. Tindall (1986). Androgen receptor protein: purification and molecular properties. In: *Biochemical Action of Hormones* (G. Litwack, Ed.), Vol. 13. Academic Press, New York, pp. 305–324.

T. J. Schmidt and G. Litwack (1982). Glucocorticoid receptor activation. *Physiol. Rev. 62*: 1131–1192.

W. T. Schrader, R. W. Kuhn, R. E. Buller, R. J. Schwartz, and B. W. O'Mally (1978). Target cell gene regulatory processes: control by progesterone receptor complexes in vitro. In: *Receptors in Pharmacology* (J. R. Smythies and R. J. Bradley, Eds.), Marcel Dekker, New York, pp. 67–95.

M. R. Walters (1985). Steroid hormone receptors and the nucleus. *Endocrinol. Rev. 6*: 512–543.

P. Y. G. Wong (1980). Pharmacology of male contraception. *Trends Pharmacol. Sci. 1*: 254–257.

## 2. PEPTIDE AND PROTEIN HORMONES AND NEUROHORMONES

### 2.1. An Introduction to Neuropeptides

Since the mid-1970s, a major revolution has occurred in our understanding of neurotransmission and endocrinology, combining these two previously distinct disciplines into the single unified field of *neuroendocrinology*. The conceptual breakthrough was achieved by the finding that hypothalamic releasing factors were also produced in the CNS outside of the hypothalamus, and these "hormones" proved to have neuroexcitatory activity. These neuropeptides can, therefore, mediate communication between neurons either directly or indirectly. Consequently, secretory processes can be combined into a continuum of secretory cell–target cell combinations (Table 5.3).

Thus neurons are capable of synthesizing peptides that were previously classified as hormones, using them for communication either over the short-range (synaptically) or over the long range (via the circulation); conversely, endocrine glands produce compounds that can act as neurotransmitters elsewhere. About 50 neuropeptides have been described thus far (see Table 5.4), and the end is not in sight. Amino acid sequences have been compiled by van Nispen and Pinder (1986). Neuropeptides share a number of functional characteristics with small amine neurotransmitters: their release is normally Ca dependent, and they operate through

**Table 5.3.** Neuroendocrine secretory processes

| Type | Secretory cell | Target cell |
|------|----------------|-------------|
| Autocrine | tissue cell, unicellular organism | same as secretory cell |
| Paracrine | tissue cell, e.g., pituitary | neighboring cell of same organ |
| Endocrine | tissue cell of "gland" | any organ via circulation |
| Neurotransmission | neuron | neuron or other organ via synapse |
| Neuroendocrine | neuron | any organ via circulation |

**Table 5.4.** Categories of mammalian brain peptides cited as to original tissues in which they were shown to be localized, or according to some functional aspects

| | |
|---|---|
| Hypothalamic releasing hormones | Gastrointestinal peptides |
| Thyrotropin-releasing hormone | Vasoactive intestinal polypeptide |
| Gonadotropin-releasing hormone | Cholecystokinin |
| Somatostatin | Gastrin |
| Corticotropin-releasing hormone | Substance P |
| Growth hormone-releasing hormone | Neurotensin |
| Neurohypophyseal hormones | Met-enkephalin |
| Vasopressin | Leu-enkephalin |
| Oxytocin | Insulin |
| Neurophysin(s) | Glucagon |
| Pituitary peptides | Bombesin |
| Adrenocorticotropic hormone | Secretin |
| $\beta$-Endorphin ($\beta$-LPH) | Somatostatin |
| $\alpha$-Melanocyte-stimulating hormone | TRH |
| Prolactin | Motilin |
| Luteinizing hormone | Pancreatic polypeptide |
| Growth hormone | Growth factors |
| Thyrotropin | IGF II |
| Opioid peptides | EGF |
| Dynorphin | FGF |
| $\beta$-Endorphin | Endothelial cell growth factor |
| Met-enkephalin | Others |
| Leu-enkephalin | Angiotensin II |
| Invertebrate peptides | Bombesin |
| FMRF amide | Bradykinin |
| Hydra head activator | Carnosine |
| | Sleep peptide(s) |
| | Calcitonin |
| | CGRP (calcitonin gene-related peptide) |
| | Neuropeptide Yy |
| | Thymosin |
| | Cardionatriuretic peptide |

After Krieger (1986).

ion channels or second messengers. There are, however, a number of complicating factors peculiar to this all-important group of "first messengers."

First, we have already discussed the co-occurrence of small amine and peptide neurotransmitters: their release is normally Ca dependent, and they operate through signal transmission. They are also capable of regulating each other's release and even the synthesis, clustering, and affinity of receptors (see Krieger, 1986). Neuroendocrine cells are capable of producing more than one peptide, and thus an amine/peptide as well as a peptide/peptide combination is possible. It is known, for instance, that the vagus nerve contains substance P, vasointestinal peptide, enkephalin, cholecystokinin, and somatostatin.

Our ideas about the selectivity of peptide neurohormones have also undergone profound development. Because most of these neurohormones act both centrally and peripherally, one has to surmise that the receptors in different organs are *isoreceptors*, in the sense of, say, the $\beta_1$- and $\beta_2$-adrenoceptors. Although the neuropeptide binds to both, the "command" executed will be appropriate to the receptor and the organ. In addition, different parts of the peptide may carry a different message, as in the "sychnologic" functions of ACTH (see Sec. 2.4), or the example already discussed: the peripheral pain mediation of substance P by the C-terminus, and the central analgesia by the N-terminus (Chap. 4, Sec. 8.5). This hierarchy of selectivities is necessary in a neurohormone that is distributed by the blood circulation because all cells are equally exposed to its message. There is also a difference in the onset and duration of action between the ultrafast, small amine neurotransmitters and the slow but durable and persistent peptides. It seems that the two classes also differ in the average amount present in tissue $(1-10 \ \mu mol/mg$ tissue for amines; fmol to pmol/mg tissue for peptides); this difference, however, need not involve the rate of turnover. Peptide concentrations can also vary by several orders of magnitude in different organs.

The *functions* of CNS neuropeptide/amine combinations seem to be multiple, because the possible variations and fine-tuning of signals offer nearly inexhaustible possibilities that may even satisfy the requirements of the enormous complexity of the homeostatic integration and control of feeding, temperature regulation, circulation, pain, learning, and many other behavioral and developmental challenges confronting organisms continuously. Our modest initial recognition of at least the immense complexity of neuroendocrine functions raises the hope that herein lies the key to future understanding of higher mental function like learning and memory. The various functions and their neuroendocrinology have been admirably organized in the book of Krieger, Brownstein, and Martin (1983).

The metabolism of neuropeptides is not noticeably different from that of other proteins. Neuropeptides are unusual, however, in that the majority are synthesized in the form of *prohormones* that may contain several copies of several smaller individual peptides, sometimes even of unrelated activity. These have to be modified (by glycosylation, disulfide bridge formation, methylation, etc.), cleaved by exo- and endopeptidases, and the fragments further modified (by C-terminal amidation, pyroglutamate formation) (van Nispen and Pinder, 1986). These processes vary from organ to organ; thus the same prohormone can undergo alternative posttranslational modification, appropriate to the specific needs of the organ or species. Neuropeptides do not undergo reuptake like the majority of amine transmitters, but rather are eliminated by proteolysis. Protein synthesis must occur in the cell body of the neuron, and the protein must be packaged in the Golgi apparatus and transported by a fast transport system $(3-5 \ \mu m/sec)$ to the synapse; thus we need to learn more about neuropeptide economies in terms of production and use. It is obvious that the complex nature of neuropeptide and the variety of factors involved in neuroendocrine physiology provide drug designers with ample opportunities for interference with these processes.

### Selected Readings

J. R. Cooper, F. E. Bloom, and R. H. Roth (1986). *The Biochemical Basis of Neuropharmacology*, 5th ed. Oxford University Press, New York.

D. T. Krieger (1986). An overview of neuropeptides. In: *Neuropeptides in Neurologic and Psychiatric Disease* (J. B. Martin and J. D. Barchas, Eds.), Raven Press, New York.

D. T. Krieger. M. J. Brownstein, and J. B. Martin (1983). *Brain Peptides*. Wiley-Interscience, New York.

D. R. Lynch and S. H. Snyder (1986). Neuropeptides: multiple molecular forms, metabolic pathways, and receptors. *Annu. Rev. Biochem. 55*: 773–799.

J. van Nispen and R. Pinder (1986). Formation and degradation of neuropeptides. *Annu. Rep. Med. Chem. 21*: 51–62.

## 2.2. The Hypothalamic and Pituitary Neurohormones

The ventromedial part of the hypothalamus—the floor of the third ventricle in the brain—connects with the cerebrospinal fluid and contains capillaries of the portal vein. Various parts of the median eminence of the hypothalamus produce a number of peptides that enter the portal vein and reach the pituitary gland (or hypophysis), situated immediately below the hypothalamus. The pituitary, however, is not part of the CNS because it lies outside the blood–brain barrier. The hypothalamic peptides are hormones, but since they are secreted by neurons they can therefore also be considered neurotransmitters, and some seem to fulfill the role of true neurotransmitters. Additionally, these hormones regulate the synthesis of other peptide hormones produced by the pituitary, and are called "*releasing hormones*," "releasing factors," or "inhibitory factors" as the case may be. The release of these hypothalamic neurohormones is regulated by higher brain centers through cholinergic, dopaminergic, and GABAergic intervention; their synthesis is adjusted by feedback mechanisms from the target organs.

The correlation of the hypothalamus and its hormones with the hormones of the anterior pituitary gland is summarized in Table 5.5. An anatomical diagram of the pituitary is shown in Fig. 5.11. As this diagram indicates, there is no direct vascular connection between the hypothalamus and the posterior lobe of the pituitary that would correspond to the portal vein system for the anterior lobe of the gland. Only two major neurons, which originate in the supraoptic and paraventricular nuclei, descend to the posterior lobe. These neurons produce two further small peptide hormones, **oxytocin** and **vasopressin** in the hypothalamus, which then bind to the transport proteins called *neurophysins*. The complex is then packaged into neurosecretory granules (equivalent to synaptic vesicles) and transported to the posterior lobe of the pituitary gland. From here they are released directly by neural impulses. Oxytocin and vasopressin are therefore true neurohormones, since they are produced and transported directly by neurons without the help of a gland or blood circulation.

Not all hypothalamo–pituitary hormones will be discussed in the subsequent sections. Only those that are well-defined chemical entities or have a direct connection with drug action are considered.

## 2.3. Hypothalamic Releasing Factors

These peptides are also known as hypophysiotropic hormones; they were isolated and their structure was elucidated primarily in the laboratories of Andrew Schally

**Table 5.5.** Hypothalamic hormones, their target hormones in the anterior pituitary (adenohypophysis), and the glands controlled by these pituitary hormones. Their ultimate physiological effects are also shown.

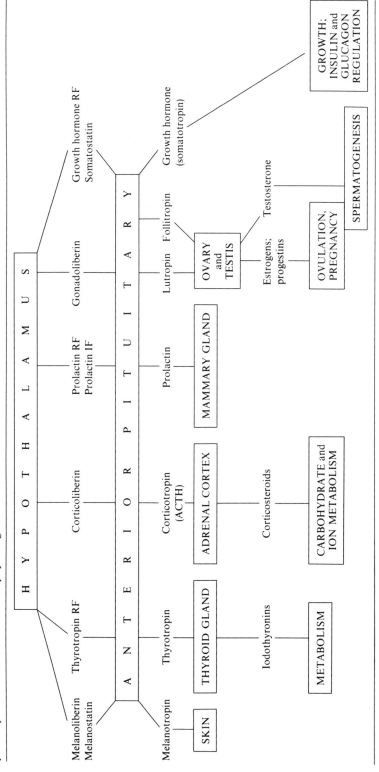

RF = releasing factor; IF = inhibiting factor.

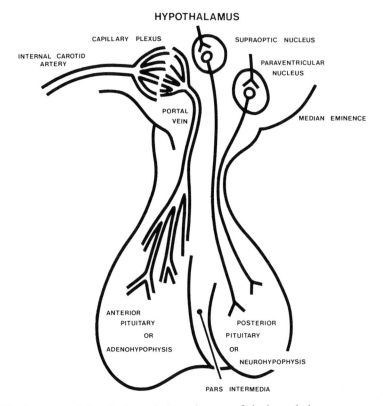

**Fig. 5.11.** Anatomy of the pituitary gland and a part of the hypothalamus.

and of Roger Guillemin, who shared the Nobel Prize in medicine for this work in 1977. The extremely tedious and difficult isolation of minute amounts of labile hormones from thousands of organs, their separation, and their peptide sequence determination were in large measure successful; and since the early 1970s neuroendocrinology has undergone an explosive development as a result. These hormones have an extremely high binding affinity, with a $K_D$ ranging from 2 to $20 \times 10^{-10}$ M. They act on the plasma membrane receptors of pituitary cells, triggering an energy-requiring process that probably involves $Ca^{2+}$ and cAMP (see Vale et al., 1977). Brain cell membranes also show the presence of receptors for these hormones, as would be expected in a system regulated by feedback.

### 2.3.1. Thyroliberin

Thyroliberin (thyrotropin releasing hormone, TRH) was the first releasing factor to be isolated and synthesized. It has the simple tripeptide structure of pyroGlu-His-Pro-NH$_2$ (5-46). The three rings on the peptide decrease the accessibility of the peptide bonds to hydrolysis by proteolytic enzymes and account for some oral activity, which is rare among peptide hormones. Nevertheless, the hormone is quickly inactivated *in vivo*. It is active in picogram amounts and liberates 200–2000 times its own amount of thyrotropin (thyroid stimulating hormone, TSH). Interestingly, it also promotes the release of prolactin, another pituitary hormone, even though there is a specific prolactin-releasing factor (PRF) in the hypothalamus; however, prolactin can also be released by numerous other substances (see Schally et al., 1978, p. 100).

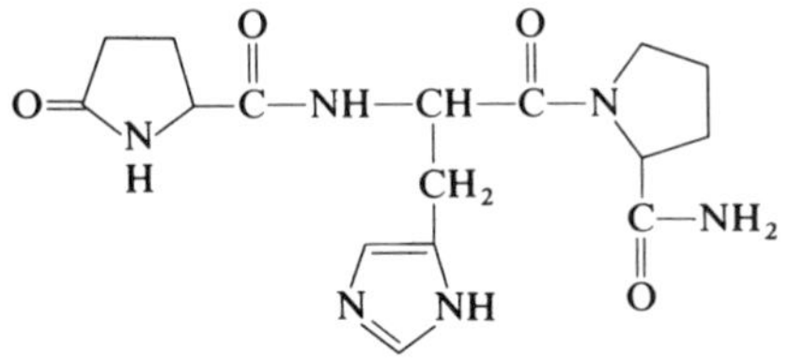

Thyroliberin
5-46

There are indications that thyroliberin is neurotropic, acting as a neuromodulator or transmitter in the brain and spinal cord, and that it exhibits antidepressant activity.

**TRH Analogues.** Analogues of TRH have been synthesized, showing that the $\pi$ electron system and the basic imidazole ring are necessary for activity: the N-formyl-Pro-Met-His-Pro-NH$_2$ (5-47) has 40% of the full activity of TRH but is resistant to serum inactivation. Some analogues can differentiate between pituitary and CNS receptors: homo-pyroGlu-His-Pro-NH$_2$ (5-48) is equipotent to TRH in the pituitary but about 10 times more active in the CNS.

**TRH Antagonists.** Antagonists of TRH have also been synthesized. For example, cyclopentylcarbonyl-thienylalanyl-pyrrolidine amide (5-49) inhibits TSH release at high doses. Thyroliberin is used diagnostically only, to distinguish between hypothalamic and pituitary hypothyroidism.

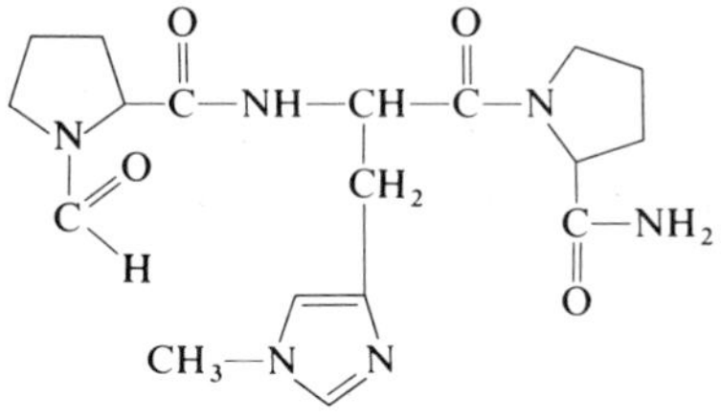

5-47

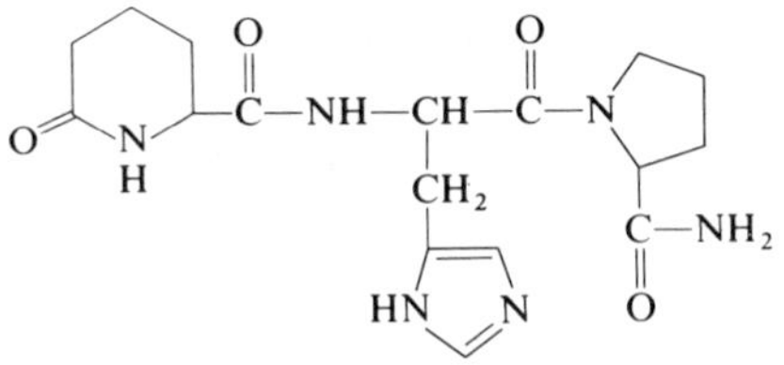

5-48

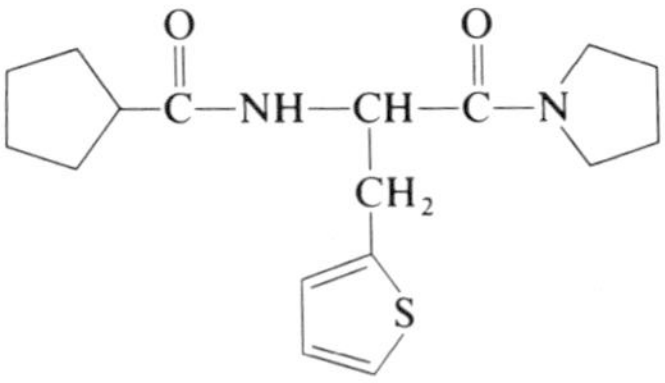

5-49

### 2.3.2. Gonadoliberin

Gonadoliberin (gonadotropin releasing hormone, GnRH; luteinizing hormone/ follicle stimulating hormone-releasing factor, LH-RH) releases both gonadotropins, lutropin (LH) and follitropin (FSH). It has been isolated and synthesized by the Schally group and has an identical structure in all vertebrates:

$$\text{pyroGlu—His—Trp—Ser—Tyr—Gly—Leu—Arg—Pro—Gly-NH}_2$$

Gonadoliberin does not release equal amounts of LH and FSH, because *inhibin*, the release-inhibiting protein produced by the gonads (Sec. 1.2) suppresses the production of follitropin but not lutropin, whereas steroid feedback regulation affects both LH and FSH. In any case, no FSH-RF has been found. Apparently, GnRH synthesis is under GABAergic control and acts through adenylate cyclase.

***GnRH Agonists.*** Some "superagonist" analogues of GnRH have been prepared among the more than 1000 compounds investigated. The [D-Trp$^6$, Pro$^9$-*N*-Et], or [D-Ala$^6$, *N*-Me-Leu$^7$] analogues (where the superscript indicates the amino acid position in the original peptide) have a 150-fold greater activity than GnRH. They induce ovulation and spermatogenesis, increasing gonadotropin and sex-steroid levels at a dose of about 10 $\mu$g. In large doses (250 $\mu$g) they block implantation of the fertilized egg, causing luteolysis or dissolution of the *corpus luteum*, and act as postcoital contraceptive agents. This means that GnRH analogues can be used either to increase fertility in men or women or to inhibit ovulation, as well as to terminate pregnancy. They also inhibit spermatogenesis in rats after the repeated administration of 100-ng doses over a prolonged period, thus allowing the potential for decreasing male fertility without decreasing libido. Many of these highly active analogues have a long-lasting action because the D-amino acids in their structure prevent attack by proteolytic enzymes. Additionally, they are active not only parenterally (by injection) but as nasal sprays or intravaginal suppositories.

It is thus understandable that there is great activity in research into these compounds, since they may become the new "natural" contraceptive agents for both males and females. Several of these compounds have reached the clinical application stage. **Buserelin**, a D-Ser (*t*-Bu)$^6$-Pro-NHEt$^9$ derivative, is one of the compounds used in the treatment of prostate cancer as a chemical castrating agent, replacing estradiol which causes feminization. It is also useful in treating precocious puberty. **Nafarelin**, [3-(2-naphthyl)-alanyl]$^6$-LHRH, is a long-acting antiovulatory agonist that has a half-life of about three hours, as compared to 10 minutes for gonadoliberin, and has about a 200-fold greater activity than the native hormone (Labrie et al., 1984; Dutta and Furr, 1985).

All of these agonists act through Ca$^{2+}$, which triggers the release of gonadotropin secretory granules to the extracellular space. Calmodulin antagonists (e.g., **pimozide**) inhibit lutropin release. The natural gonadoliberin is released in a pulsatile fashion; continuous administration down-regulates the receptors, and the tissue becomes refractory to stimulation, resulting in inhibition of ovulation (Buckingham, 1984; Conn, 1984).

Gonadoliberin also occurs outside the hypothalamus, and regulates the activity of testes, ovary, and placenta. In addition, it seems to be involved in tumor growth, and, as a central neurotransmitter, induces lordosis, the copulating posture in female rats; thus it has an effect on behavior.

*GnRH Antagonists.* Gonadoliberin antagonists can be obtained by modifying the first three amino acids of the natural peptide. The N-terminal is considered the active center, whereas the rest of the molecule serves only in the binding process. The [D-Phe$^2$, Pro$^3$, D-Trp$^6$], [D-pyro-Glu$^1$, D-Phe$^2$, D-Trp$^3$, D-Trp$^6$] and similar compounds can block ovulation at 200- to 750-$\mu$g doses, but there are very few data on the activity of GnRH antagonists in humans.

### 2.3.3. Somatocrinin

The *growth hormone releasing factor* (somatocrinin) was isolated only in 1982. It consists of 44 amino acids, and in addition to stimulating growth hormone production it also stimulates release of **somatomedins**, which are responsible for the many anabolic effects of growth hormone (Felix et al., 1985; Guillemin et al., 1984). Its production by recombinant DNA techniques have made it possible to undertake physiological studies on its activity, but results are still preliminary.

### 2.3.4. Somatostatin

Somatostatin (SS, growth hormone release-inhibiting hormone, GH-RIH) is perhaps the best investigated and most important of the inhibitory factors produced by the hypothalamus. It is a cyclic tetradecapeptide isolated by the Guillemin group in 1973, and has the following structure:

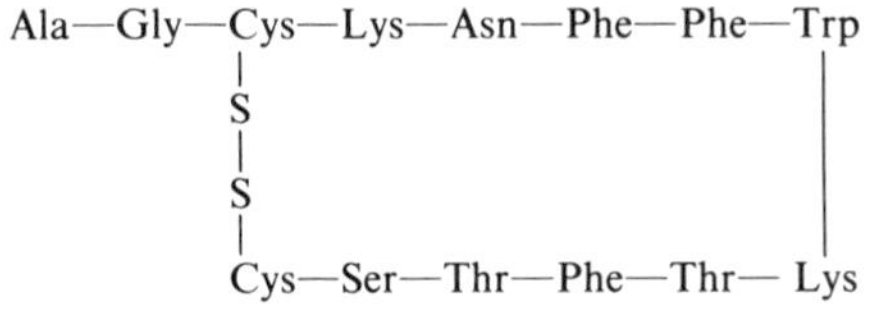

$$
\begin{array}{c}
\text{Ala—Gly—Cys—Lys—Asn—Phe—Phe—Trp} \\
\qquad\quad | \qquad\qquad\qquad\qquad\qquad\quad | \\
\qquad\quad \text{S} \qquad\qquad\qquad\qquad\qquad\quad | \\
\qquad\quad | \qquad\qquad\qquad\qquad\qquad\quad | \\
\qquad\quad \text{S} \qquad\qquad\qquad\qquad\qquad\quad | \\
\qquad\quad | \qquad\qquad\qquad\qquad\qquad\quad | \\
\text{Cys—Ser—Thr—Phe—Thr— Lys}
\end{array}
$$

It is equally active in the reduced linear or the cyclic form, and has recently been produced by methods of "genetic engineering" that incorporate the human somatostatin gene into the genome of *Escherichia coli*, providing a cheap and efficient source of the hormone.

Somatostatin is very active at nanomolar levels, but is also very labile, and shows a half-life of only a few minutes when injected. It is hydrolyzed by endopeptidases between the Trp$^8$—Lys$^9$ residues, and the therapeutic applications of the native hormone are therefore rather restricted.

*Activity.* The principal activity of somatostatin is inhibition of the release of growth hormone (somatotropin) from the pituitary. Excessive growth hormone production leads to acromegaly, a form of giantism, whereas its lack results in dwarfism. Since acromegaly is a relatively rare endocrine disease, other actions of somatostatin have received more attention—primarily its action on the pancreas. Somatostatin suppresses the production of both pancreatic hormones, insulin and glucagon (see Sec. 2.7). According to present views, *diabetes mellitus* is caused by lack of insulin and a loss of insulin receptors as well as an excess of glucagon, causing hyperglycemia (an excessive blood glucose concentration), faulty glucose metabolism, lipolysis, amino acid mobilization from proteins, and a number of chronic effects including blindness. The contemporary treatment of diabetes mellitus concentrates on the replacement of insulin or the administration of nonpeptide hypoglycemic drugs. Supplementary somatostatin administration, to decrease the release of glucagon, has dramatic effects on glucose metabolism in the disease and can reduce the insulin requirement of diabetics to less than half of their usual dose.

Somatostatin also reduces gastric acid secretion and has potential use in treating gastric ulcers. Additionally, being distributed throughout the CNS, mainly in the spinal cord, and found in nerve endings, it is therefore assumed to be a neurotransmitter. It potentiates some of the effects of L-DOPA, induces sedation and hypothermia, and affects sleep patterns by inhibiting central epinephrine secretion. Two types of somatostatin (SS) receptors have been found (Tran et al., 1985): the pituitary and the pancreas contain $SS_A$ receptors, whereas the brain contains both $SS_A$ and $SS_B$ receptors. The suspected CNS targets include the extrapyramidal motor system and perhaps cognition.

***Synthetic Analogues.*** In view of the many actions of somatostatin that potentially have great therapeutic importance, a very large number of synthetic analogues have been prepared (see Verber and Saperstein, 1979), with particular attention being given to overcoming the impracticably short half-life of the native hormone.

Modifications designed to enhance the enzyme resistance and prolong the activity of SS derivatives have been quite successful. The use of D-amino acids instead of the normal L-enantiomers (e.g., in $Trp^8$), or replacement of the disulfide link by a nonreducible ethylene bridge leads to an increased duration of activity, approaching 3 hours. Several analogues, like the [D-Ala$^2$, D-Trp$^8$]somatostatin, which was 20 times the activity of SS on growth hormone release, show a greatly increased effect. The NH-terminal outside the cyclic dodecapeptide is not essential for activity.

Selectivity of action results from manipulation of the cysteines in SS analogues. When $Cys^3$ is replaced by its D-enantiomer, insulin release is preferentially inhibited; the L-$Cys^{14}$ is replaced, an increase in the inhibition of glucagon secretion occurs. When the C-terminal cysteine forms a lactam with the N-terminal of $Ala^1$, resulting in cyclo-SS, growth-hormone repression becomes enhanced. None of these analogues has yet been introduced into therapeutic use. However, the conformationally restricted cyclic analogue 5-50 is only half the size of SS, but has the same potency, is biologically stable, and is orally active (see Hruby, 1985). The compound 5-51 is about 50–100 times more potent than SS in inhibiting insulin, glucagon, and growth hormone release. Other analogues (DeFeudis and Moreau, 1986) have been designed as antitumor agents with activity against some endocrine tumors.

$$(CH_3—N)-Ala—Phe—D-Trp \qquad\qquad (CH_3—N)-Ala—Tyr—D-Trp$$
$$Phe—Thr—Lys \qquad\qquad\qquad Phe—Val—Lys$$
$$5\text{-}50 \qquad\qquad\qquad\qquad 5\text{-}51$$

## 2.4. Oxytocin and Vasopressin

The supraoptic and paraventricular nuclei of the hypothalamus produce a number of peptide hormones which are bound to the transport proteins known as neurophysins and are transported along axons to the neurohypophysis (Fig. 5.11) (see Walker, 1975). The two most important hormones in this group are oxytocin and vasopressin.

**Oxytocin** (OT) (5-52) is a nonapeptide in which six amino acids form a ring closed by a disulfide bridge, while the ring itself forms an antiparallel "pleated sheet" (Fig. 5.12A). The "tail" portion of the peptide, composed of Pro-Leu-Gly-$NH_2$, is also rigidly held in a folded conformation. It is suggested that the so-called cooperative conformation of the hormone on the receptor is different, with the aromatic ring of the Tyr residue folded back (Fig. 5.12B) over the large ring of the peptide.

**Fig. 5.12.** Conformation of oxytocin dissolved in dimethylsulfoxide (DMSO) (**A**) and at the receptor site (**B**). (Reproduced by permission from Meienhofer, in Wolff (Ed.) (1979), *Burger's Medicinal Chemistry*, Part II, p. 797, Wiley-Interscience, New York)

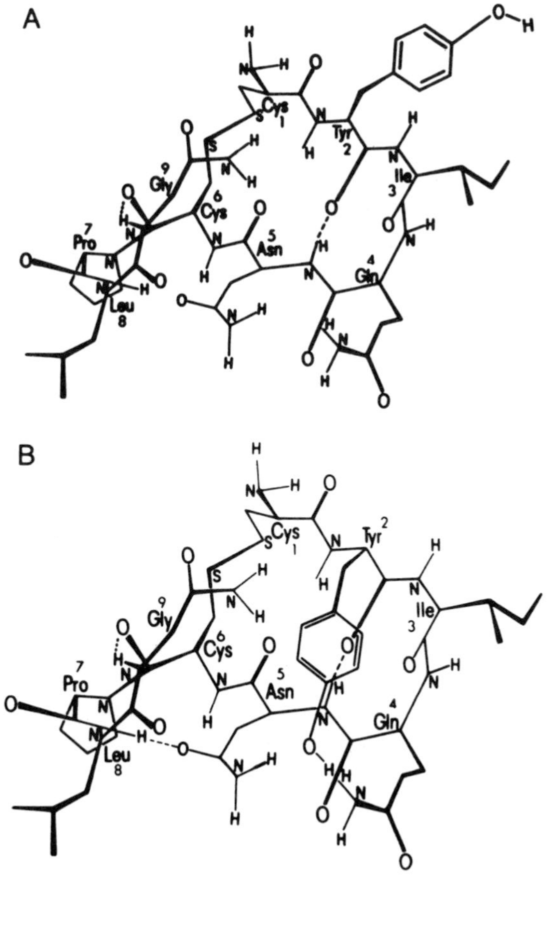

Cys—Tyr—Ile—Glu—Asn—Cys—Pro—Leu—Gly—NH$_2$

Oxytocin
5-52

Oxytocin causes the powerful contraction of some smooth muscles and plays a vital role in milk ejection (not to be confused with milk secretion, which is regulated by prolactin). It also has uterotonic action, contracting the muscles of the uterus, and is therefore used clinically to induce childbirth. The natural role of OT in childbirth is, however, a controversial subject. In birds, OT triggers an analogous effect: contraction of the oviduct and ovipositor.

**Vasopressin** occurs in two variations: arginine-vasopressin (AVP, 5-53) and lysine-vasopressin (LVP), in which Arg$^8$ is replaced by Lys. The conformation of these hormones is almost identical to that of oxytocin, except that the terminal "tail" is conformationally free and not held by the ring.

Cys—Tyr—Phe—Glu—Asn—Cys—Pro—Arg—Gly—NH$_2$

Arginine-vasopressin
5-53

The physiological role of the vasopressins is the regulation of water reabsorption in the renal tubules (i.e., an antidiuretic action). In high doses, they promote the contraction of arterioles and capillaries and an increase in blood pressure; hence the name of these hormones. Because of their very similar structures, OT and VP overlap in a number of effects.

### 2.4.1. Structures–Activity Correlations

The elucidation of the conformation–activity relationships of OT by Walter (1977) was of the utmost importance in the design of highly active analogues of these hormones. Amino acids 3, 4, 7, and 8 are not involved in the hydrogen bonding that determines the ring conformation, and can therefore act as sites of binding to the oxytocin receptor. The Tyr$^2$-hydroxyl group, the intact hexapeptide ring, and the amide of Asn$^5$ are essential for the biological activity of this hormone. Therefore, the "corner" amino acids 3, 4, and 8 can be varied, yielding more selective compounds: the [Thr$^4$, Gly$^7$] OT has an oxytocin/antidiuretic activity ratio of 135,000:1, compared to the 200:1 ratio of the natural hormone. On the other hand, the [1-deamino, D-Val$^4$, D-Arg$^8$] VP has a 125,000:1 antidiuretic/pressor activity ratio (AVP has a 1:1 ratio). A few oxytocin inhibitors have also been prepared, such as the [3, 5-dibromo-Tyr$^2$] OT, which inhibits the uterotonic effect of the hormone.

### 2.4.2. Clinical Applications

The clinical applications of OT and vasopressin are widespread. Oxytocin is used to induce labor in childbirth as well as to promote the expulsion of the placenta, although the antidiuretic activity of the native hormone is a disadvantage. *Diabetes insipidus*, or diuresis due to low levels of vasopressin, is treated with the highly active synthetic analogues mentioned above. It is significant that vasopressin levels rise dramatically after the smoking of cigarettes, from 3–4 ng/liter to 50–150 ng/liter. Vasopressin has also been implicated in learning and memory (although this idea has been challenged recently: see Gash and Thomas, 1985) and is a local neurotransmitter used by the hypothalamus (Bloom, 1981; Snyder, 1980).

Other drugs with uterotonic activity include the *ergot alkaloids*. Ergot is the sclerotium of the fungus *Claviceps purpurea*, which infects cereals, mainly rye (see Chap. 4, Sec 4.2). A number of indole alkaloids have been isolated from this source, in which the indole moiety is lysergic acid. The latter forms amides with both cyclic tripeptides (e.g., **ergocristine**, 5-54) and with the amino-alcohol L-alaninol in **ergonovine** (5-55). The peptide alkaloids have a slow and cumulative action, whereas the water-soluble ergonovine and its derivatives are fast acting. The latter is used to prevent postpartum hemorrhage by the compression of uterine blood vessels through uterine muscle contraction. Some of these alkaloids are α-adrenergic blocking agents and have been used with moderate success in the treatment of migraine headaches.

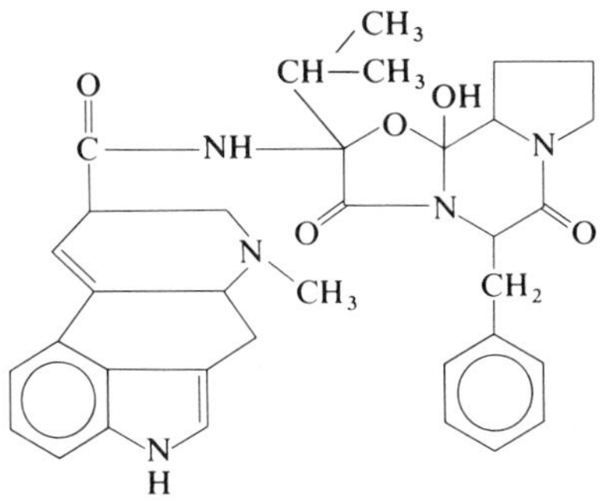

Ergocristine
5-54

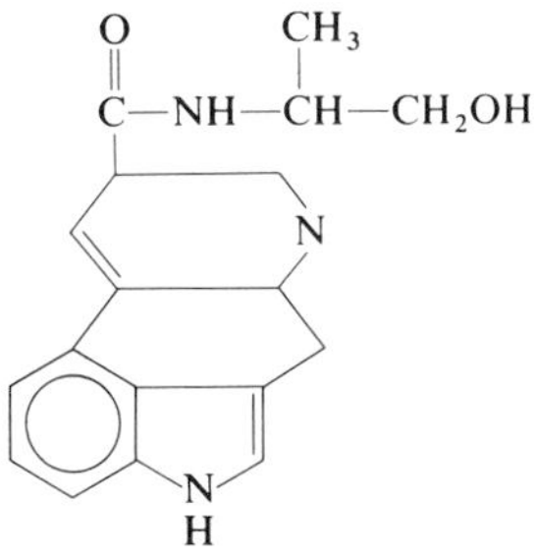

Ergonovine
5-55

## 2.5. Pituitary Hormones

### 2.5.1. Gonadotropins

The gonadotropins are produced by the anterior pituitary (adenohypophysis) and the placenta. This group of glycoproteins (carbohydrate-containing proteins) includes the following hormones:

1. Lutropin (LH, luteinizing hormone, which in the male is called interstitial cell stimulating hormone, ICSH)
2. Follitropin (FSH, follicle stimulating hormone)
3. Human chorionic gonadotropin (hCG)
4. Human menopausal gonadotropin (hMG)

These are very complex hormones, having molecular weights of around 28,000. The α subunit consists of 89 amino acids, the β subunit consists of 115 amino acids in

LH and FSH, and of 145 in hCG. The complete amino acid sequences are described by Butt (1975). While the α subunits of all gonadotropins are very similar in their amino acid sequence, the β subunits of the various hormones are quite different. The carbohydrate portions of both the α and β subunits contain oligosaccharides attached at specific amino acids (Asn), which branch at a mannose group and contain galactose, glucosamine, galactosamine, and acetylneuraminic acid residues. The carbohydrate portion of the hormones influences their biological and immunological properties as well as their stability. Removal of the terminal sialic acid from FSH reduces its half-life from 90 minutes to 2–3 minutes.

The gonadotropins are released in a pulsating manner, every few hours. As discussed in Sec 1.4, lutropin and follitropin act together to regulate ovarian functions, egg maturation, and follicular transformation to the *corpus luteum* in females. In the male, spermatogenesis depends on these hormones. Ovarian and testicular steroids are also produced as a result of gonadotropin action, and these in turn have a feedback regulatory effect on the hypothalamus and pituitary. Human chorionic gonadotropin, produced by the placenta, shows LH activity and is more stable than the other gonadotropins.

Therapeutically, gonadotropins are used to induce ovulation in infertile women. The antiestrogens **clomiphene** (5-14) and **tamoxifen** (5-16) (see Fig. 5.5) are also used for this purpose, since they counteract the ovulation-inhibitory effect of estrogens. Pregnancy tests depend on the presence of an increased hCG concentration in the urine after fertilization, which is monitored by radioimmunoassay.

### 2.5.2. Corticotropin

Corticotropin (adrenocorticotropic hormone, ACTH) regulates the function of the adrenal cortex and has many other effects on metabolism. It contains 39 amino acids in the form of a random coil, owing to the presence of several proline residues which prevent helix formation. Species differences are seen in amino acids 31 and 33 only; the rest of the ACTH molecule is identical in all animals and humans. The first 24 amino acids are responsible for all of the biological action of ACTH, whereas the remaining amino acids provide stability. Since a large number of ACTH analogues have been synthesized, the information contained in the molecule has been analyzed in detail, as shown in Fig. 5.13. The receptor binding core seems to reside in positions 15–18, whereas steroid synthesis is regulated by the sequence of amino acids 6–13. The term "sychnologic" was proposed by Schwyzer (1980) to denote peptides having defined sequences as regulating parts of their physiological function.

Adrenocorticotropic hormone has the following biological activities:

1. A direct effect on the adrenal cortex: the regulation of steroid synthesis and depletion of vitamin C from the adrenal gland
2. Indirect effects mediated by the adrenal gland: thymus involution and an increase in glucose utilization
3. Extraadrenal effects: melanotropic hormone and growth hormone release; lipolytic action; and such neurological effects as stretching and yawning

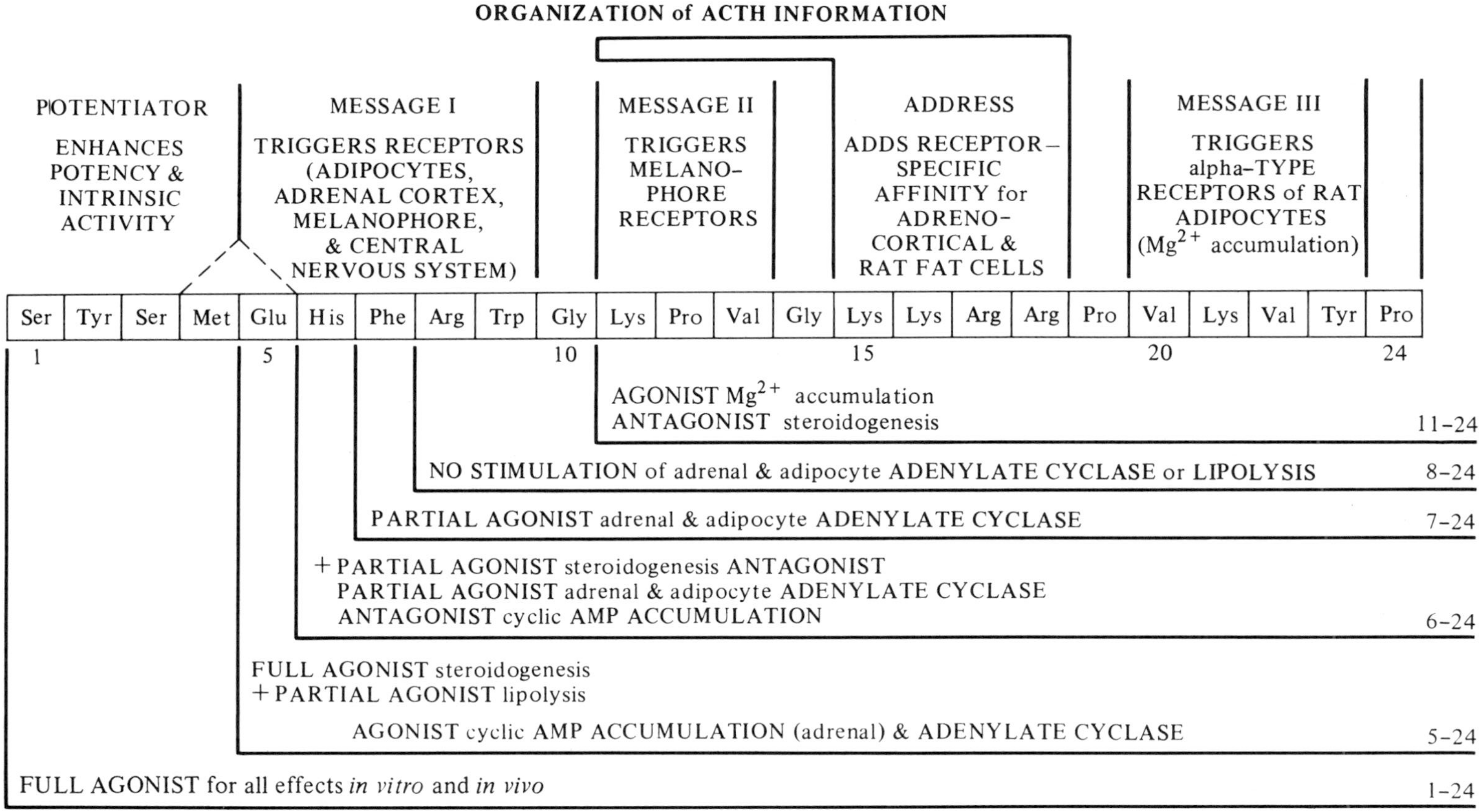

**Fig. 5.13.** Information organization in the adrenocorticotropin (ACTH) 1–24 fragment. The C-terminus of the hormone, comprising amino acids 25–39, contains information about transport properties, species labels, and stability, but does not trigger any physiological activity. (Reproduced by permission from Schwyzer (1980), Elsevier/North Holland Biomedical Press, Amsterdam)

The primary effect of ACTH seems to be cAMP production by interaction of the hormone with two populations of receptors: one with a $K_D$ of $9 \times 10^{-11}$, which is present only in 60 sites per cell, and a second population with a $K_D$ of $3 \times 10^{-7}$ but numbering 600,000 sites per cell. The difference between the extent of steroidogenesis and cAMP formation indicates different receptor affinities in different organs and perhaps the existence of spare receptors for adenylate cyclase activation.

The $ACTH_{4-10}$ fragment showed remarkable behavioral effects in humans: it acted as a stimulant, and restored optimal performance during long, monotonous task. In rats it was a positive reinforcer in self-administration experiments (Branconnier, 1986).

## Selected Readings

M. R. Bell, F. H. Batzold, and R. C. Winekker (1986). Chemical control of fertility. *Annu. Rep. Med. Chem. 21*: 169–177.

F. E. Bloom (1981). Neuropeptides. *Sci. Am. 245*(4): 148–168.

R. J. Branconnier (1986). The human behavioral pharmacology of $ACTH_{4-10}$. In: *Neuropeptides and Behavior* (D. de Wied, W. H. Gispen, and T. B. van Winersma Greidanus, Eds.), Int. Encyclopedia of Pharm. Therapy, Vol. 114. Pergamon, Oxford, pp. 421–434.

J. C. Buckingham (1978). The hypophysiotrophic hormones. In: *Progress in Med. Chem.* (G. P. Ellis and G. B. West, Eds.), Vol. 15. Elsevier/North Holland, New York, pp. 165–210.

J. C. Buckingham (1984). LHRH: fertility and antifertility drugs. *Trends Pharmacol. Sci. 5*: 136–137.

W. R. Butt (1975). *Hormone Chemistry*, Vol. 1, 2nd ed. Wiley, New York.

P. M. Conn (1984). Molecular mechanism of gonadotropin releasing hormone action. *Biochem. Action of Hormones* (G. Litwack, Ed.), Vol. 11. Academic Press, New York, pp. 67–92.

F. V. DeFeudis and J.-P. Moreau (1986). Studies on somatostatin analogs might lead to new therapies for certain types of cancer. *Trends Pharmacol. Sci. 7*: 384–386.

A. S. Dutta and B. J. A. Furr (1985). Luteinizing hormone releasing hormone (LHRH) analogs. *Annu. Rep. Med. Chem. 20*: 203–214.

A. M. Felix, E. P. Heimer, and T. F. Mowles (1985). Growth hormone releasing factors. *Annu. Rep. Med. Chem. 20*: 185–192.

D. M. Gash and G. J. Thomas (1985). What is the importance of vasopressin in memory processes? In: *Neurotransmitters in Action* (D. Bousfield, Ed.). Elsevier, Amsterdam, pp. 305–308.

R. Guillemin, P. Brazeau, P. Böhlen, F. Esch, N. Ling, W. B. Wehrenberg, B. Bloch, C. Mougin, F. Zeytin, and A. Baird (1984). Somatocrinin, the growth hormone releasing factor. *Recent Prog. Hormone Res.* (R. O. Greep, Ed.), Vol. 40. Academic Press, New York, pp. 233–299.

V. J. Hruby (1985). Design of peptide hormone and neurotransmitter analogs. *Trends Pharmacol. Sci. 6*: 259–262.

L. L. Iversen, S. D. Iversen, and S. H. Snyder (Eds.) (1982). *Neuropeptides, Handbook of Psychopharmacology*, Vol. 16. Plenum Press, New York.

F. Labrie, A. Bélanger, and A. Dupont (Eds.) (1984). *Luteinizing Hormone Releasing Hormone and Its Analogs*. Elsevier, Amsterdam.

S. L. Lightman and B. J. Everitt (1986). *Neuroendocrinology*. Blackwell Scientific, Oxford.

A. V. Schally, D. H. Coy, and C. A. Meyers (1978). Hypothalamic regulatory hormones. *Annu. Rev. Biochem. 47*: 89–118.

R. Schwyzer (1980). Organization and transduction of peptide information. *Trends Pharmacol. Sci. 1*: 327–331.

S. H. Snyder (1980). Brain peptides as neurotransmitters. *Science 209*: 976–983.

J. A. Thomas and E. J. Keenan (1986). *Principles of Endocrine Pharmacology*. Plenum Press, New York.

V. T. Tran, M. F. Beal, and J. B. Martin (1985). Two types of somatostatin receptors differentiated by cyclic somatostatin analogs. *Science 228*: 492–495.

R. H. Unger, R. E. Dobbs, and L. Orci (1978). Insulin, glucagon and somatostatin secretion in the regulation of metabolism. *Annu. Rev. Physiol. 40*: 307–343.

W. Vale, C. Rivier and M. Brown (1977). Regulatory peptides of the hypothalamus. *Annu. Rev. Physiol. 39*: 473–527.

D. F. Verber and R. Saperstein (1979). Somatostatin. *Annu. Rep. Med. Chem. 14*: 209–218.

R. Walker (Ed.) (1975). Neurophysins, carriers of peptide hormones. *Ann. N.Y. Acad. Sci. 248*: 1–512.

R. Walter (1977). Identification of sites in oxytocin involved in uterine receptor recognition and activation. *Fed. Proc. 36*: 1872–1878.

## 2.6. Hormones of the Thyroid Gland

The thyroid gland, which surrounds the larynx, has an enormous variety of metabolic functions. It is itself regulated by **thyroliberin**, which in turn regulates production of **thyrotropin** (thyroid stimulating hormone, TSH). The latter is a pituitary glycoprotein whose $\alpha$ subunit is identical with that of lutropin (LH), and whose specific $\beta$ moiety is composed of 112 amino acids carrying a single carbohydrate side chain.

### 2.6.1. Calcitonin

Calcitonin (CT) is a peptide containing 32 amino acids. The hormone isolated from fish exhibits a potency about 20–25 times higher than that of mammalian CT because of its higher receptor affinity in bone and the kidney. Calcitonin decreases the $Ca^{2+}$ content of plasma by increasing $Ca^{2+}$ and $PO_4^{2-}$ excretion in the urine, as well as inhibiting the absorption of these ions in the intestine. It also decreases the activation (hydroxylation) of vitamin $D_3$, another regulator of $Ca^{2+}$ and phosphorus metabolism. In addition, CT prevents the mobilization of Ca from bone.

### 2.6.2. Parathyrin

Parathyrin (parathyroid hormone, parathormone, PTH) is secreted by the parathyroid glands, which are located on the four poles of the thyroid gland. A peptide of 84 amino acids, it has two precursors during its biosynthesis, a pre-pro-PTH with an additional 31 amino acids and a pro-PTH, with an additional 6 amino acids only. Parathyroid hormone increases the serum $Ca^{2+}$ content and decreases the inorganic phosphate content of the body by inhibiting renal phosphate absorption and mobilizing $Ca^{2+}$ from bone. It therefore has an action opposite to that of calcitonin. Both hormones act through the production of cAMP, but at different receptors in different zones of the bones and kidneys.

### 2.6.3. Iodothyronines

L-**Thyroxine (tetraiodothyronine, $T_4$)** (5-56) is 3,5,3′,5′-tetraiodo-*p*-hydroxyphen-oxy-phenylalanine. The **3,5,3′-triiodo analogue ($T_3$)** is also a naturally occurring and active hormone. In most of its physiological effects, $T_3$ is more active than $T_4$. Both hormones are amino acid derivatives, discovered more than 60 years ago and synthesized in the thyroid gland from tyrosine by iodination and transfer of the iodophenol portion of a second iodotyrosine molecule. The synthesis takes place in the follicles (acini) of the thyroid gland, a large endocrine organ weighing about 20 g in adults. The lumen of the follicle is filled with a colloidal, viscous solution of *thyroglobulin*. This protein binds $T_3$ and $T_4$ with high affinity, and is also efficient in binding circulating iodide, which enters the thyroid gland by active transport coupled to an ATPase. Iodide uptake by the gland is enhanced by thyrotropin (TSH) and inhibited by large anions such as $ClO_4^-$. The iodide is then oxidized and attached to tyrosine by peroxidase and a flavoprotein monooxygenase in a process that involves NADPH. Thyroglobulin (MW, 650,000; 19 *S*) binds the thyronines very stongly, and the hormones can enter the circulation and reach other cells only after the binding protein is lysed. This proteolysis is stimulated by TSH.

Thyroxine ($T_4$)
5-56

The iodothyronines are very insoluble molecules and are kept in solution by *transport proteins*. The most important of these is **thyroxine binding globulin (TBG)**, which carried about 65% $T_4$ and 70% $T_3$. It is a small (MW, 60,000–65,000) glyco-protein consisting of four subunits. It has a single, high-affinity binding site for $T_4$, with an estimated $K_D$ of $1.2 \times 10^{-10}$ M.

*Thyroxine binding prealbumin (TBPA)* carries about 30% $T_4$ but no $T_3$. Its affinity is only of the order of $10^{-8}$ M, but it is much more abundant in serum than TBG. The amino acid sequence as well as the structure of this protein is known. Four identical subunits (127 amino acids each) form a prolate ellipsoid. Noncovalent interactions between the subunits form a channel of 1-nm diameter along the long axis, which has a funnel-shaped opening of 2.5 nm. The $T_4$ molecule is held in one arm of this channel, binding to Lys$^9$ and Lys$^{15}$. The 4′-OH is a requirement for binding (Fig. 5.14), which shows negative cooperativity. It is interesting that in addition to $T_4$, four molecules of the small retinol-binding protein, the vitamin A carrier, are also bound to TBPA.

*Serum albumin* also transports $T_4$, but with a $K_D$ of only $10^{-6}$ M. In addition, there is a small amount of free $T_3$ and $T_4$ in the serum (about 2 μg/liter).

**Biological Effects.** The biological effects of iodothyronines are numerous, and are outlined below.

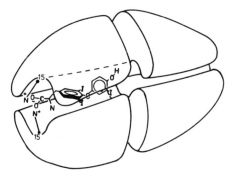

**Fig. 5.14.** Schematic diagram of the thyroxine-binding prealbumin. $T_3$ is shown in the central channel between the four subunits. The carboxylate ion of the alanine side chain forms an ion pair with $Lys^{15}$ of a protein subunit. (Reproduced by permission from Jorgensen (1978), Academic Press, New York)

1. These hormones induce *amphibian metamorphosis*, the change of a tadpole into a frog—an obviously very complex series of biochemical and developmental reactions. The jellyfish *Aurelia* is also sensitive to thyroxine: its polyp stage will segment and produce young mobile medusae. Human fetuses will show skeletal abnormalities as well as neuromuscular and mental retardation if born with an inadequately functioning thyroid gland.

2. Thyroid deficiency has been connected to cretinism and myxedema. Cretins are born mentally defective, are small, and have coarse hair and thick skin. Myxedema, seen in older people. is characterized by subcutaneous semifluid deposits, causing puffiness of the hands and face. The basal metabolism of these patients is depressed to 30–40% below normal, and their body temperature and pulse rate are also reduced. Women suffering from myxedema often give birth to cretins.

3. In hyperthyroidism the metabolic rate is increased, resulting in vasodilation and sweating, weight loss, and nervous excitability. When the thyroid gland enlarges as a result of increased activity, the so-called goiter develops. In certain geographical locations, this can be caused by a chronic lack of iodide, and all developed nations therefore use iodized table salt. In addition, hyperthyroidism results in exophthalmos, in which the eyeballs protrude markedly. This probably occurs through an autoimmune process, and can be very persistent even after hyperthyroidism is cured. The increased metabolic rate in the disease manifests itself as an increase in the oxygen demand of all tissue (except those of the brain), and an increased sensitivity to $\beta$-adrenergic agonists.

***Mode of Action.*** The mode of action of the iodothyronines is the regulation of protein synthesis in the nuclei of cells sensitive to these hormones. Their nuclear action is, however, very different in its details from that of steroid hormones.

   While high-affinity ($K_D = 4 \times 8^{-10}$ M for $T_3$) and saturable cytosol-binding proteins have been isolated from thyroid cells, they do not seem to play a role in transport of the iodothyronines to cell nuclei; nor are they precursors of nuclear acceptors for these hormones, as opposed to the case with steroid hormones.

A *nuclear receptor* for thyroid hormones has been isolated, showing a $K_D$ for $T_3$ on the order of $10^{-9}$–$10^{-11}$ M in many cell types. The molecular weight of the receptor is 60,000–65,000, and its binding affinity for $T_4$ is only about 10% of that for $T_3$. It is a nonbasic, nonhistone protein associated with nuclear DNA and involved in regulating the transcription of mRNA for a presumably large number of enzymes. The hypothetical binding of $T_3$ to the nuclear receptor is shown in Fig. 5.15.

***Structure–Activity Correlations.*** The structure–activity relationships of thyroid hormones have been studied with both qualitative and quantitative methods, including the Hansch correlation, and are summarized by Jorgensen (1978). The structural requirements for receptor binding, and therefore hormone activity, are:

1. Two aromatic rings perpendicular to each other and separated by a spacer atom (O, S, or C) which holds the rings at an angle of about 120°.
2. Halogen or methyl groups on the 3 and 5 positions of the ring that bears the alanine side chain. These substituents keep the rings perpendicular to each other and participate in hydrophobic bonding to the receptor.
3. An anionic side chain of two or three carbons long, para to the bridging atom, forming an ion pair with the nuclear receptor. The —NH$_2$ group decreases receptor affinity but plays a role in transport of the hormones, and delays their metabolic degradation.
4. A phenolic 4'-OH group, which may be generated metabolically (e.g., by oxidation *in vivo*) if originally absent.
5. A lipophilic halogen, alkyl, or aryl substituent in the 3' position. An isopropyl group has the optimal effect.

**Fig. 5.15.** Schematic diagram of the binding of trisubstituted thyronine ($T_3$) to the nuclear receptor. The diagram shows why a 5'-iodo substituent (absent here) would interfere with the binding of the 4'-OH group, which is essential to binding. (Reproduced by permission from Jorgensen (1978), Academic Press, New York)

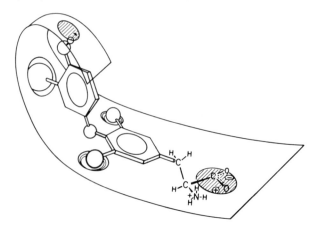

6. A 5′ substituent reduces activity in direct proportion to its size. It interferes with the binding of the 4′-OH group, increases the binding to transport proteins, and therefore reduces the concentration of available free hormone.

***Thyroid Antagonists.*** *Hypothyroidism* can be treated with a regular intake of iodide, but in case of destruction of the gland, thyroxine must be used. The treatement of *hyperthyroidism* is less straightforward. Surgery, in the form of thyroidectomy, is employed, although inadvertent removal of the parathryoid glands may complicate the subsequent hormone treatments. Irradiation with ingested $^{125}I_2$, primarily a $\gamma$-emitting isotope, or $^{131}I_2$, a $\beta$-emitting isotope, is useful because it destroys thyroid follicles selectively. While these methods offer advantages over thyroidectomy, all of the dangers inherent in radioisotope treatment must be carefully considered prior to their use.

Goitrogens or compounds that produce goiter, were recognized a long time ago and are intrathyroidal antagonists. Some are naturally occurring, like the **progoitrin** (5-57) that is transformed into the cyclic **goitrin** (5-58) in rutabaga plants. Large anions such as thiocyanate or perchlorate were once used as competitive inhibitors of iodide uptake by the thyroid gland, but have severe side effects. The drugs of choice for controlling hyperthyroidism are **6-propylthiouracil** (5-59) and **methimazole** (5-60), both of which are cyclic derivatives of thiourea, inhibiting the thyroid peroxidase that is essential to biosynthesis of the thyroid hormones.

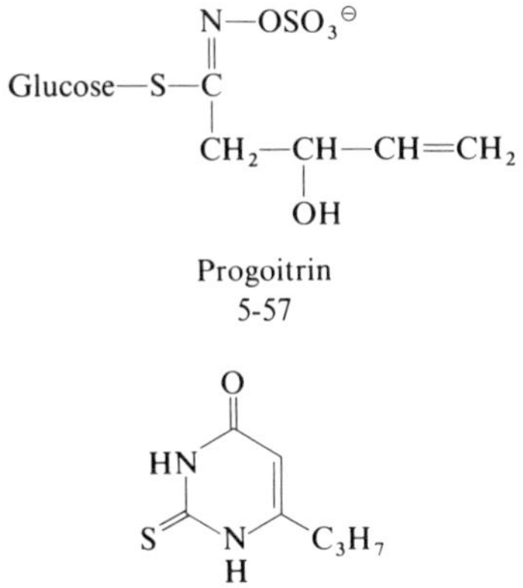

Progoitrin
5-57

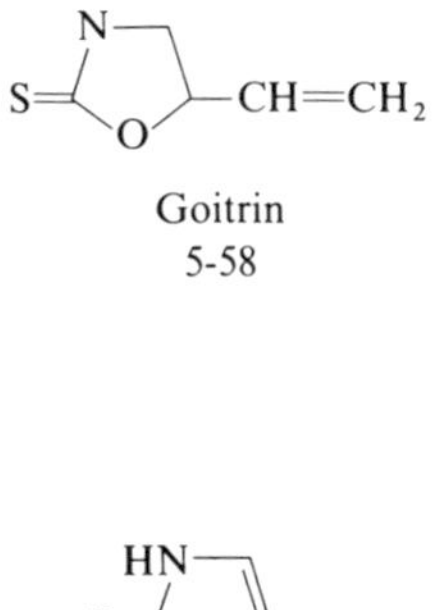

Goitrin
5-58

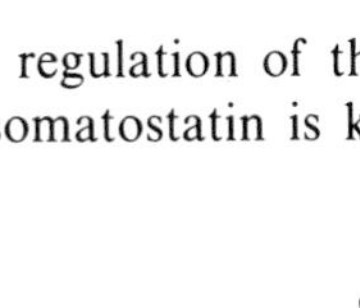

6-Propylthiouracil
5-59

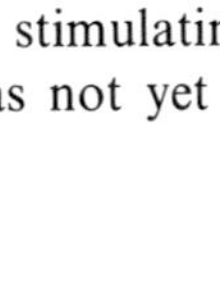

Methimazole
5-60

The feedback regulation of thyroliberin and thyroid stimulating hormone by thyroxine and somatostatin is known to occur, but has not yet been exploited therapeutically.

### Selected Readings

W. R. Butt (1976). *Hormone Chemistry*, 2nd ed., Vol. 2: *Thyroid Hormones*. Wiley, New York, Chap. 6.

E. N. Cheung (1985). Thyroid hormone action: determination of hormone–receptor interaction using structural analogs and molecular modeling. *Trends Pharmacol. Sci. 6*: 31–34.

V. Cody (1979). Thyroid hormones–receptor interactions: binding models from molecular conformation and binding affinity data. In: *Computer Assisted Drug Design* (E. C. Olson and R. E. Christoffersen, Eds.). American Chemical Society Symposium Series Vol. 112, Washington, DC, pp. 281–299.

D. S. Cooper (1984). Antithyroid drugs. *N. Engl. J. Med. 311*: 1353–1357.

E. C. Jorgensen (1976). Structure–activity relationships of thyroxine analogs. *Pharmacol. Ther. [B] 2*: 661–682.

E. C. Jorgensen (1978). Thyroid and hormones and analogues. II. Structure–activity relationships. In: *Hormonal Proteins and Peptides* (C. H. Li, Ed.), Vol. 6. Academic Press, New York, pp. 107–204.

J. H. Oppenheimer and M. I. Surks (1975). Biochemical basis of thyroid hormone action. In: *Biochemical Action of Hormones* (G. Litwack, Ed.), Vol. 3. Academic Press, New York, pp. 119–157.

K. Sterling and J. H. Lazarus (1977). The thyroid and its control. *Annu. Rev. Physiol. 39*: 349–371.

## 2.7. The Hormones of the Pancreas: Insulin and Glucagon

The homeostatic regulation of nutrient levels is a basic biochemical task, not currently well understood. Three peptide hormones occupy a central role in this regulation of carbohydrate, lipid, and amino acid metabolism: insulin, glucagon, and somatostatin. A lack of insulin leads to *diabetes mellitus,* characterized by high blood glucose levels, excretion of glucose in the urine (hence "mellitus": honeyed), and failure to utilize carbohydrates and lipids. The disease is invariably fatal if untreated. Recently, hypotheses of immunological and even viral origin of diabetes have been proposed.

Whereas insulin causes hypoglycemia, the other pancreatic hormone, glucagon, mobilizes glucose from its stores and causes hyperglycemia. Somatostatin, originally discovered as a hypothalamic hormone (Sec 2.3.4) that inhibits growth-hormone release, is also found in the pancreas, where it inhibits the secretion of both insulin and glucagon. This effect gives it great importance, since according to present views, diabetes is due as much to the overproduction of glucagon as to the lack of insulin. Future treatment of the disease may therefore consist of the administration of a combination of insulin (or an analogue) plus somatostatin. The interrelationship of these three hormones is discussed by Unger et al. (1978).

### 2.7.1. Insulin

**Structure.** In 1916, Schafer discovered that the $\beta$ cells of the "islets of Langerhans" in the pancreas secrete an antidiabetic substance which he named insulin ( = "of an islet"). This was followed by the isolation of the peptide by Banting and Best in 1922, its sequencing by Sanger in 1955, and its total synthesis by several groups in the early 1970s. Recently, the cloning of the human insulin gene and its transfer into bacteria has been achieved, and human insulin is now produced by fermentation technology.

Insulin is synthesized by the pancreatic $\beta$ cells in the form of **proinsulin** (Fig. 5.16), in which a connecting "C-peptide" consisting of amino acids 31–60 joins the A

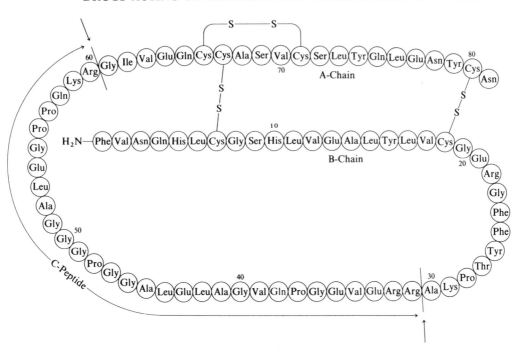

**Fig. 5.16.** Structure of bovine proinsulin. On activation to insulin, the C-peptide (31–60) is excised and leaves behind insulin, in which the two peptide chains are connected only by disulfide bonds.

chain at $A^1$ through an Arg—Lys, to $B^{30}$ through an Arg—Arg fragment. In this way the two connecting disulfide bonds can meet properly. Proinsulin is more soluble than insulin, but *in vitro* has only about 35% insulin-like activity. Recently, a pre-proinsulin was discovered with an additional 23 amino acids attached to the N-terminus of the B chain; however, it has a lifetime of only a few seconds before being converted to proinsulin; its structure was actually deduced from the nucleotide sequence of its mRNA. Proinsulin is transformed slowly into insulin in the storage granules of the Golgi apparatus of pancreatic cells.

Insulin therefore consists of two peptide chains which are connected by two disulfide bonds, since the C-peptide is cleaved off. There are some species-specific differences in the amino acid sequence of the hormone. X-ray diffraction studies have shown that insulin occurs as a hexameric protein containing two Zn atoms. The dimers are first held by four hydrogen bonds and a hydrophobic bond along the $\beta^{21-30}$ sequence, in the form of an antiparallel $\beta$ sheet. The dimers then bind by interaction of the $B^{14}$-Ala, $B^{17}$-Leu, and $B^{18}$-Val residues. The core of the hexamer contains water.

***Structure–Activity Correlations.*** This detailed knowledge of the three-dimensional structure of insulin led to the recognition that its biological activity resides in an *area* of the molecule rather than in specific amino acid residues, just as dimerization and further association of the molecule also depend on an intact spatial structure (Fig. 5.17).

The foregoing concept is corroborated by structural modifications of the hormone. The last three amino acids of the B chain can be removed without a loss of activity, but cleavage of the C-terminal of the A chain (Asn²¹) results in a total loss of activity. Amino acids can be replaced inside the chains only if such substitution does not change the overall geometry of the molecule. The structure–activity relationships of insulin derivatives are inconsistent and not always comparable.

***Characterization of Binding.*** Different equilibrium binding constants for insulin have been reported by different authors. As measured with $^{125}$I-labeled insulin in adipocytes (fat cells), the $K_D$ varies from $5 \times 10^{-11}$ to $3 \times 10^{-9}$ M. There are about 10,000–12,000 binding sites in every fat cell. Although Scatchard plots indicate a single population of receptors in fat cells, other cell populations (liver cells, lymphocytes) show a Scatchard plot that is not linear. There are two possible explanations for this. The first suggests that two binding sites for insulin may exist. This possibility is indicated by the finding that some anti-insulin antibodies inhibit the activation of glucose oxidation but do not hinder the receptor effect on amino acid incorporation into proteins. The second explanation is based on the negative cooperativity of the receptor, which implies that binding is inhibited and dissociation accelerated in the presence of more insulin. Negative cooperativity is functionally "useful," since it means greater responses at lower hormone concentrations. With regard to insulin, negative cooperativity can be explained by the nature of the sites involved in dimerization of the hormone, as outlined in the

**Fig. 5.17.** Schematic view of the crystal structure of insulin, indicating areas of interaction. (Reproduced by permission from Meienhofer, in Wolff (Ed.) (1979), *Burger's Medicinal Chemistry*, 4th ed., Part II, Wiley-Interscience, New York)

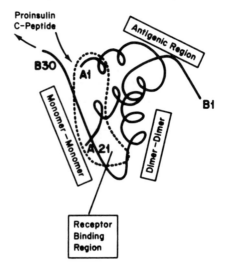

previous section: amino acids 28–30 of the B chain are expendable with no loss of activity; cleavage from $B^{25}$ to $B^{27}$ results in only about 30% activity, but the negative cooperativity of the molecule disappears. This region, when it becomes buried in the hydrophobic interior of the molecule, is also the hydrophobic site responsible for the dimerization of insulin (see Fig. 5.17). Therefore, dimerization, the result of a higher hormone concentration, causes a loss of negative cooperativity in the same way as if the amino acids involved were cleaved off. The dimer, however retains full biological activity (see De Meyts et al., 1978).

The oligosaccharide portion of the receptor (see below) plays a part in both the recognition of insulin and the transduction of its signal. Neuraminidase, which splits off sialic acid, eliminates the action of insulin without affecting binding of the hormone. Actually, recognition of the ligand may be even more complicated than this, since other glycoproteins, not necessarily part of the receptor, can interact in the binding process (Andreani and DePirro, 1981).

*Biochemical Effects.* The biochemical effects of insulin include:

1. Facilitation of glucose transport into cells
2. Enhancement of intracellular glucokinase activity
3. Enhancement of amino acid incorporation into proteins
4. Stimulation of DNA translation
5. Increased lipid synthesis
6. Stimulation of $Na^+$, $K^+$ and $P_i$ transport into cells

Insulin is therefore seen as a hormone promoting anabolism rather than catabolism, since it promotes the synthesis of glycogen, proteins, and lipids. It is quite clear from the preceding list of insulin actions on metabolic functions that the hormone has a much broader biochemical role than the regulation of carbohydrate metabolism and removal of free circulating glucose. In the absence of insulin, there are therefore profound changes in the entire metabolic pattern. Since carbohydrate, the principal nutrient in most diets, cannot be utilized properly, the energy requirements of the diabetic organism must be met in other ways. One of these is an increased gluconeogenesis from protein, with the consumption of body tissues; there is also an increased excretion of nitrogen and diminished protein synthesis. The glucose produced in such a tedious and harmful way is also excreted and wasted. Lipolysis occurs, resulting in lipemia (an excess of circulating lipids), and ketone bodies such as acetoacetate and acetone are produced excessively.

*Cellular and Molecular Mechanisms of Action.* In recent years, tremendous progress has been achieved in the elucidation of the cellular and molecular mechanisms of insulin action (for reviews, see Czech, 1985; Gammeltoft, 1984; Kahn et al., 1986; Simpson and Cushman, 1986).

The *insulin receptor* has been isolated and characterized as a membrane glycoprotein (MW $\sim$ 450 kD) of an $\alpha_2\beta_2$ subunit structure, linked by disulfide bonds. The $\alpha$ subunits bind insulin, whereas the $\beta$ subunits function as the tyrosine kinase. Tyrosine autophosphorylation activates receptor kinase activity—that is, the receptor acts as its own tyrosine kinase—but the significance of this property is unknown. Perhaps affinity regulation and receptor internalization are regulated in this way. Also present in the receptor complex is a noncovalently bound serine phosphorylase whose role is also unknown (van Obberghen and Gammeltoft, 1986). It is notworthy that insulin induces protein phosphorylation *directly*, not through any second messenger. Steroid hormones, growth factors, and related oncogenes (Chap. 2, Sec. 5.5) share this property.

The acute cellular action of insulin is initiated by rapid clustering of occupied receptors on the cell surface, formation of coated pits, and internalization (Fig. 2.10). Within three minutes, a redistribution of glucose transporters from the cytoplasm to the plasma membrane can be measured (Simpson and Cushman, 1986). Lipolysis is also increased, and seems to involve a novel mechanism recently characterized by Cuatrecasas and his co-workers (see Saltiel et al., 1986). These investigators found a phosphoinositol-glycan containing glucosamine, which seems to provide for the activation of protein kinase C and cAMP phosphodiesterase, in addition to the activation of phospholipase C.

The second level of control by internalized insulin–receptor complexes is exerted on receptor dephosphorylation, resulting in receptor recycling to the membrane. The number of glucose transporters can also be regulated in some unknown way.

*Hypoglycemic Agents.* Hypoglycemic agents are drugs that lower the blood glucose level and substitute for insulin that is missing for any reason (insufficient production, increased destruction, or the presence of anti-insulin antibodies). In juvenile diabetes a strict diet and insulin administration are the only possible treatments. In "maturity onset" diabetes, other drugs that can be taken orally may be used, a great convenience as compared to insulin injections.

Some sulfonamide antibacterial agents have shown unwanted side effects, causing convulsions of hypoglycemic origin. Through drug modification, the antibacterial effects of these agents were eliminated and their hypoglycemic side effects enhanced. In this way the family of sulfonylurea-type hypoglycemic agents was developed (Fig. 5.18). The prototype of these compounds is **tolbutamide** (5-61a), which, with its congeners such as **acetohexamide** (5-61b) and the long-acting **chlorpropamide** (5-61c; half-life: 33 hours), are widely used in the treatment of late-onset diabetes. Another group of hypoglycemic agents, including **glymidine** (5-62) and **glipizide** (5-63), is based on the pyrimidine nucleus. All of these compounds act by mobilizing available insulin from pancreatic cells, and can therefore be used only if the patient has a somewhat functional pancreas.

An entirely different hypoglycemic agent is **pirogliride** (5-64), a guanidine derivative. Its mode of action is controversial, but it enhances insulin secretion in the presence of glucose only, and may have an additional, nonpancreatic mechanism of action (Rasmussen et al., 1981). There are also many minor or experimental hypoglycemic agents (see Chap. 31 of Burger/Wolff, cited in the Preface).

*Managing the Complications of Diabetes.* In addition to the metabolic problems, there are numerous neurological, circulatory, and renal complications in type I (insulin-dependent) diabetes, even when the blood glucose level is properly controlled. The main reason is the unnatural administration of insulin by injection,

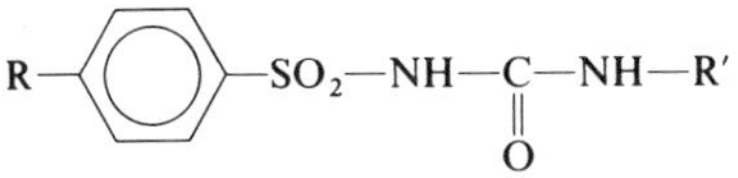

5-61

a) Tolbutamide (R = $CH_3$, R' = $n$-Bu)

b) Acetohexamide (R = $CH_3CO$-, R' = cyclohexyl)

c) Chlorpropamide (R = Cl, R' = $n$-Pr)

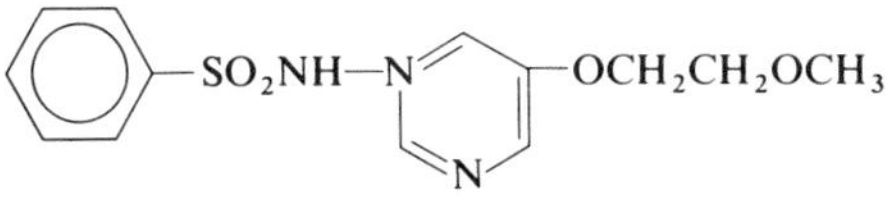

Glymidine
5-62

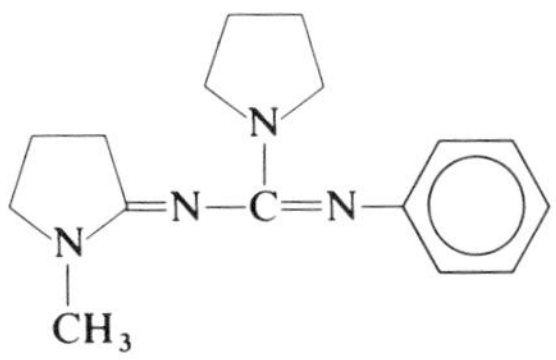

Pirogliride
5-64

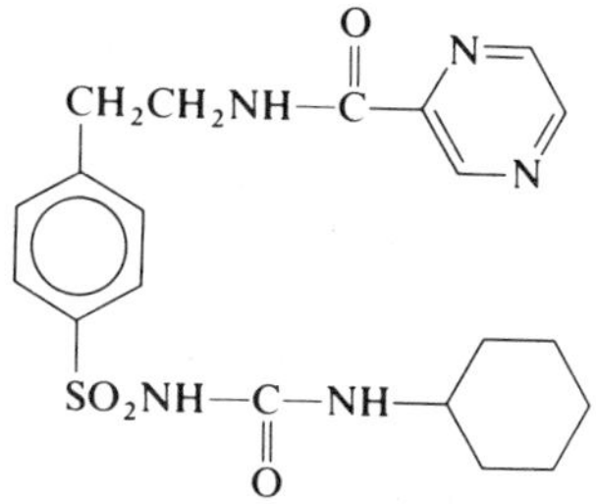

Glipizide
5-63

**Fig. 5.18.** Some synthetic hypoglycemic agents.

instead of the constant secretion by the pancreas in response to changing blood glucose levels. Continuous "minipumps" (Chap. 8, Sec. 5), which release insulin at a steady rate, are being successfully utilized.

Another complication of type I diabetes is blindness, which is due to cataract formation (proliferative retinopathy); this accounts for about 12% of all blindness. Under normal conditions, the glucose metabolism includes the pathway:

$$\text{Glucose} \xrightarrow[\text{aldose reductase}]{\text{NADPH} \quad \text{NADP}^+} \text{Sorbitol} \xrightarrow{\text{NADP}^+ \quad \text{NADPH}} \text{Fructose}$$

In hyperglycemia, however, fructose is only slowly metabolized, and sorbitol accumulates in tissues. Because aldose reductase is found in cornea, kidneys, optic nerve, and peripheral neurons (Lipinski and Hutson, 1984), cataracts and painful neuropathies develop in poorly controlled or long-standing diabetes, as a result of sugar alcohol (sorbitol) accumulation. *Aldose reductase inhibitors*, like **tolrestat** (5-65) or **sorbinil** (5-66), show great promise in controlling these additional symptoms of diabetes (Kador et al., 1985).

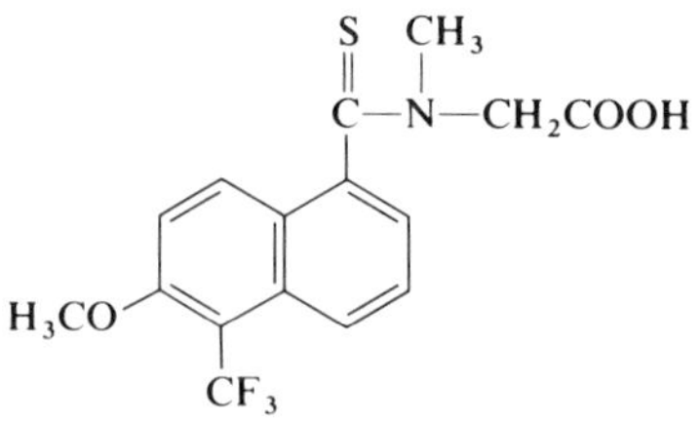

| Tolrestat | Sorbinil |
| 5-65 | 5-66 |

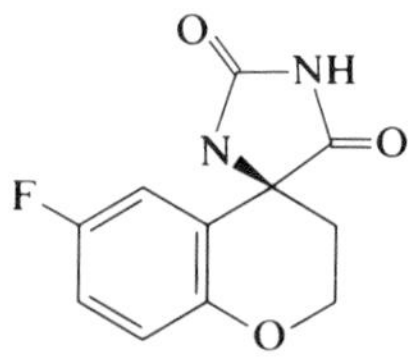

### 2.7.2. Glucagon

Glucagon, the second pancreatic hormone, was discovered as an impurity in early insulin preparations. It is a peptide containing 29 amino acids, and is perhaps also biosynthesized as proglucagon. The hormone activates receptors in liver cell membranes and acts through adenylate cyclase and cAMP. It triggers glycogenolysis and thus elevates blood glucose levels, and also activates protein phosphorylation in cell organelles.

The glucagon receptor has been characterized (Iyengar, 1986), but not nearly to the extent that the insulin receptor has. It is a 63-kD dimeric protein with a disulfide bond necessary for activity. A 20-kD region contains the hormone-binding site and the site that interacts with the $G_s$ nucleotide binding protein.

The hyperglycemic action of glucagon is believed to play a role in diabetes. The hormone is produced by the α cells of the islets of Langerhans, which are not impaired in diabetes. Therefore, the decrease in insulin production is not matched by a similar decrease in glucagon synthesis, thus aggravating the hyperglycemia in diabetes.

Glucagon derivatives and analogues are not therapeutically useful, and no antagonists of these agents have yet been discovered. The only way to decrease glucagon production is by using somatostatin (Sec. 2.3.4) analogues, which are specific for reducing glucagon synthesis. This is a new and promising development in the improved treatment of diabetes.

### Selected Readings

D. Andreani and R. DePirro (Eds.) (1981). *Current Views on Insulin Receptors.* Academic Press, New York.

W. R. Butt (1975). *Hormone Chemistry*, 2nd ed., Vol. 1. Wiley, New York, Chap. 9.

M. P. Czech (1985). The nature and regulation of the insulin receptor: structure and function. *Annu. Rev. Physiol. 47*: 357–381.

P. De Meyts, E. van Obberghen, J. Roth, A. Wollmer, and D. Brandenberg (1978). Mapping of the residues responsible for the negative cooperativity of the receptor-binding region of insulin. *Nature 273*: 504–509.

S. Gammeltoft (1984). Insulin receptors: binding kinetics and structure–function relationship of insulin. *Physiol Rev. 64*: 1321–1378.

R. Iyengar (1986). Structural characterization of the glucagon receptor. In: *Hormonal Control of Gluconeogenesis* (N. Kraus-Friedman, Ed.), Vol. 2. CRC Press, Boca Raton, FL, pp. 21–34.

P. F. Kador, J. H. Kinoshita, and N. E. Sharpless (1985). Aldose-reductase inhibitors: a potential new class of agents for the pharmacological control of certain diabetic complications. *J. Med. Chem. 28*: 841–849.

C. R. Kahn, F. Grigorescu, S. Takayama, and M. White (1986). The insulin receptor as a protein kinase. In: *The Role of Receptors in Biology and Medicine* (A. M. Gotto and B. W. O'Malley, Eds.). Raven Press, New York.

C. A. Lipinski, and N. J. Hutson (1984). Aldose-reductase inhibitors as a new approach to the treatment of diabetic complications. *Annu. Rep. Med. Chem. 19*: 169–177.

C. R. Rasmussen, B. E. Maryanoff, and G. F. Tutwiler (1981). Diabetes mellitus. *Annu. Rep. Med. Chem. 16*: 173–188.

A. R. Saltiel J. A. Fox, P. Sherline, and P. Cuatrecasas (1986). Insulin-stimulated hydrolysis of a novel glycolipid generates modulators of cAMP phosphodiesterase. *Science 233*: 967–972.

J. A. Simpson, and S. W. Cushman (1986). Mechanism of insulin stimulatory action on glucose transport in the rat adipose cell. *Biochemical Action of Hormones* (G. Litwack, Ed.), Vol. 13. Academic Press, New York, pp. 1–31.

R. H. Unger, R. E. Dobbs, and L. Orci (1978). Insulin, glucagon and somatostatin in the regulation of metabolism. *Annu. Rev. Physiol. 40*: 307–343.

E. van Obberghen and S. Gammeltoft (1986). Insulin receptors: structure and function. *Experientia 42*: 727–734.

## 2.8. The Renin–Angiotensin System: A Blood Pressure Regulator

Blood pressure is regulated by a multitude of interrelated factors (see Table 5.6) involving neural, hormonal, vascular, and volume-related effects, as first formulated by Page over 30 years ago (see Khosla et al., 1979; Connell, 1986). Two of these factors have been dealt with in preceding chapters in sections on the adrenergic neuronal system and the hypothalamic hormone vasopressin.

Another group of peptide hormones involved in blood pressure regulation, the angiotensins, was recognized many years ago through the enzyme that activates some of them. Since this enzyme is produced in the kidneys, it was named *renin*.

The renin—angiotensin system regulates blood pressure through several feed-back mechanisms. The accompanying enzymes and the drugs influencing the system are shown in Fig. 5.19.

A decrease in blood pressure due to blood loss, sodium loss, or caused experimentally by clamping of the renal artery stimulates the juxtaglomerular cells of the kidney to secrete renin, a proteolytic enzyme. This enzyme acts on a circulating protein, **angiotensinogen**, cleaving off **angiotensin I**, which is then further cleaved in the lung and kidneys by angiotensin converting enzyme to yield **angiotensin II**. (The enzyme *tonin* can convert angiotensinogen directly into angiotensin II; see Fig. 5.19 and also Gutkowska et al., 1982.) This hormone has a powerful constrict-ing action on arterioles, and elevates the blood pressure instantly. Long-range effects are also achieved, either by angiotensin II or its cleavage product, **angiotensin III**. One or both trigger aldosterone release (Sec. 1.7), causing Na retention and an increase in fluid volume (i.e., antidiuresis) which results in elevation of the blood pressure. Since the angiotensins are quickly hydrolyzed, their effect is transitory and therefore suitable for continuous homeostatic regulation of the blood pressure.

**Table 5.6.** Pharmacological reduction of blood pressure

| Mechanism | Drug action | Example(s) |
|---|---|---|
| | Reduction of arteriolar resistance | |
| *Inhibition of sympathetic nervous system* | | |
| Reduction of central sympathetic outflow | Centrally acting $\alpha_2$ agonist | clonidine |
| Blockade of postganglionic neuron | Neuron blocker | guanethidine |
| Blockade of norepinephrine action | Postsynaptic $\alpha_1$-adrenoreceptor antagonist | prazosin |
| *Inhibition of renin/angiotensin system* | | |
| Inhibition of renal renin release | $\beta_1$-Adrenoreceptor antagonist | propranolol[a], atenolol[a] |
| Prevention of conversion of angiotensin I to angiotensin II | Angiotensin converting enzyme inhibitor | captopril[a] |
| Blockade of action of angiotensin II | Angiotensin II receptor antagonist | saralasin |
| *Direct arteriolar vasodilation* | | |
| Blockade of voltage dependent $Ca^{2+}$ entry to smooth muscle, i.e. inhibit excitation/contraction coupling | Calcium entry blocker | nifedipine[a] |
| Activation of guanylate cyclase; increased cGMP leads to dephosphorylation of myosin light chain | ? Activators of guanylate cyclase | sodium nitroprusside, ? hydrallazine atriopeptins |
| *Mechanism of action known* | | |
| ? Altered transmembrane cation distribution | Thiazide diuretics | bendroflumethiazide[a] |
| | Reduction of intravascular volume/cardiac output | |
| *Initial increase in renal sodium excretion* | | |
| Inhibition of active $Na^+$ transport in loop of Henle | Loop diuretics[b] | frusemide |
| Inhibition of $Na^+$ transport in distal tubule | Thiazide diuretics[b] | bendroflumethiazide[a] atriopeptins |
| *Reduction in cardiac output* | | |
| Reduction in heart rate and contractility | $\beta_1$-Adrenoreceptor antagonist | propranolol[a], atenolol |

*Note*: The mechanism of action of some of the agents given as examples is uncertain.

[a] These drugs are most commonly used in clinical practice.

[b] Diuretics are most effective when used in conjunction with angiotensin converting enzyme inhibitors.

Reproduced by permission from Connell (1986).

302

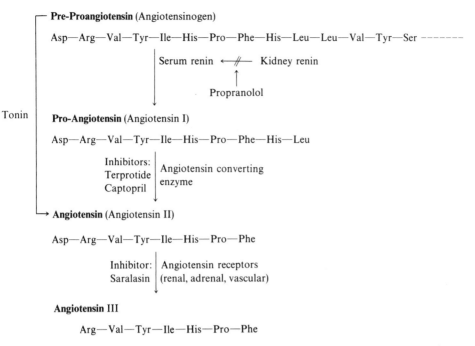

**Fig. 5.19.** The renin–angiotensin system, and drugs that interfere with it at various points.

A number of other factors regulating blood pressure are also cited in Table 5.6. These include:

1. Hormones such as vasopressin, cortisol, and estrogens, which influence Na metabolism
2. Vasopressin, adrenocorticoids, and catecholamines, including $\beta$ blockers, which affect renin production
3. Potassium, which influences aldosterone secretion
4. Prostaglandins and thromboxane (Sec. 4)

All of these factors cooperate to maintain normal blood pressure in healthy individuals, but their interplay is disrupted in pathological conditions. Since the regulatory system is very complex, many factors can produce hypertension, but the hypertension of renal origin is unquestionably one of the most important and is therefore described in detail in the following section.

### 2.8.1. Characterization of Components

**Angiotensinogen** (pre-proangiotensin) is produced by the liver and probably by the renal cortex. It is a glycoprotein of unknown structure, with a molecular weight varying from 60,000 to 100,000. Its production is under endocrine control by adrenocorticoids and estrogens (with the latter being prominent during pregnancy). The amino acid sequence of the N-terminus of angiotensinogen—the first 14 amino acids—is shown in Fig. 5.19.

**Renin** (E.C. 3.4.99.19), a highly specific endopeptidase, is a glycoprotein with a molecular weight of 37,000–43,000. It cleaves angiotensinogen between leucine residues 10 and 11, to yield the inactive decapeptide **angiotensin I** (proangiotensin). Renin has a precursor, prorenin, which seems to be activated by pepsin or trypsin. Prorenin has a molecular weight of about 60,000 but its properties are not well known.

Renin secretion is strictly regulated by (1) renal vascular baroreceptors which sense the vessel wall tension in arterioles; (2) by $Na^+$ or $Cl^-$ receptors; (3) by an angiotensin feedback mechanism; and (4) by the CNS, through catecholamines (see Reid et al., 1978).

**Angiotensin II** (angiotensin), the octapeptide cleavage product of angiotensin I, is produced by *angiotensin-converting enzyme*. This widely distributed exopeptidase, which is most abundant in the lungs and kidneys, is a glycoprotein containing a $Zn^{2+}$ ion. Angiotensin II is a strong vasoconstrictor, and its effect depends on its C-terminal Phe[8] residue. If this terminal amino acid is replaced by any aliphatic amino acid, the activity of the hormone is lost; replacement by threonine leads to antagonist action.

The *angiotensin receptors* are integral membrane proteins; they have been solubilized from blood vessel walls, but have an unknown structure. The high-affinity receptor site has a $K_D$ of $0.5-20 \times 10^{-9}$ M, depending on the organ. The receptor seems to regulate a $Ca^{2+}$ channel rather than adenylate cyclase, and the resultant changes in the Ca concentration are responsible for muscular contraction in the blood vessel wall. Angiotensin II also stimulates aldosterone release from the adrenal gland, but this effect may be due to its degradation product **angiotensin III**, following cleavage of the Asp[1] of the precursor. This peptide has quite specific receptors in the adrenal cortex, with a $K_D$ of $0.5-1.5 \times 10^{-10}$ M.

### 2.9. Inhibitors of the Renin–Angiotensin System

There are several points in the renin–angiotensin system that have proven to be amenable to inhibition. Renin, angiotensin converting enzyme, and the angiotensin receptors are the most important sites of regulation.

### 2.9.1. Renin Inhibitors

Renin inhibitors have been found among naturally occurring phospholipids and synthetic phosphatidylethanolamine derivatives such as (5-67). **Pepstatin** (5-68), isolated from *Streptomyces* strains, is a pentapeptide with an acylated N-terminus and the unusual 4-amino-3-hydroxy-6-methylheptanoic acid (**AHMH, statin**) residues. It is a general protease inhibitor that leads to a significant, short-duration lowering of the blood pressure following intravenous injection. A number of highly active analogues are reviewed by Smith and Regan (1986).

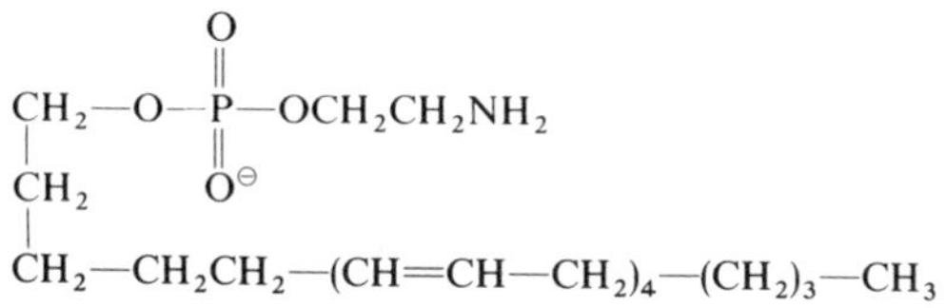

$$CH_2-O-\overset{\overset{O}{\|}}{P}-OCH_2CH_2NH_2$$

$$CH_2 \quad \overset{\ominus}{O}$$

$$CH_2-CH_2CH_2-(CH=CH-CH_2)_4-(CH_2)_3-CH_3$$

5-67

| Iva | Val | Val | Statin | Ala | Statin |

$H_2N-CH_2CO-NHCHCO-NHCHCO-NHCHCHCH_2CO-NHCHCO-NHCHCHCH_2COOH$

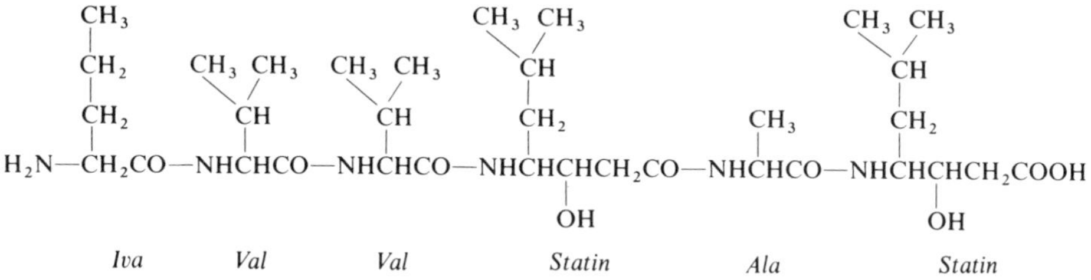

Pepstatin
5-68

## 2.9.2. Angiotensin Converting Enzyme Inhibitors

Angiotensin-converting enzyme inhibitors are the most effective antihypertensive drugs. The first such drug to be developed was **teprotide** (5-69), a nonapeptide identical in sequence to peptides isolated from the venom of a Brazilian viper. The four prolines and the pyroglutamate in this peptide make it degradation resistant, with the result that is has a long-lasting action but is not hypotensive in normal animals. Teprotide competitively inhibits the degradation of angiotensin I by the converting enzyme.

pyroGlu—Trp—Pro—Arg—Pro—Glu—Ile—Pro—Pro—OH

Teprotide
5-69

In 1977, a new drug, **captopril** (5-70)—(2S)-1-(3-mercapto-2-methylpropionyl)-L-proline—was developed by Ondetti and Cushman (1978). It shares many of the actions of teprotide but is more than 10 times as active, with a $K_D$ of $1.7 \times 10^{-9}$ M. It was designed to fit the known active site of carboxypeptidase A, an exopeptidase very similar to the converting enzyme. The mercapto group binds to the Zn ion; the amide carbonyl, to a hydrogen bonding site; and the proline carboxylate, to an electrophilic center. **Enalaprilate** (5-71), unlike captopril, is orally active because it is not hydrolyzed as easily, and its more polar nature prevents any central effects. Ease of administration and lack of side effects make enalaprilate a superior drug.

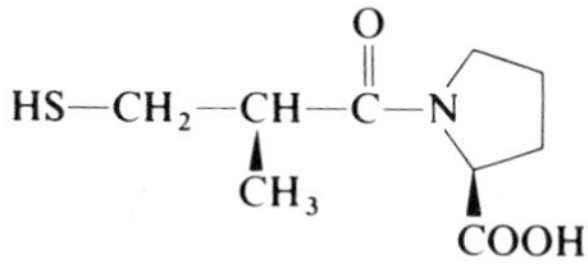

Captopril
5-70

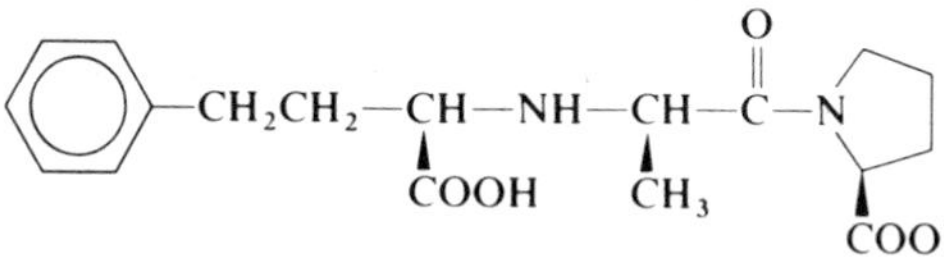

Enalaprilate
5-71

### 2.9.3. Angiotensin Receptor Antagonists

Angiotensin receptor antagonists were developed through modification of the angiotensin molecule; Devynck and Meyer (1978) give detailed SAR charts for these drugs. The most effective compound is the [Sar$^1$, Val$^5$, Ala$^8$] angiotensin II, called **saralasin** (5-72) (sar = sarcosine = $N$-methylglycine). Unfortunately, the angiotensin receptor antagonists are active only during continuous intravenous infusion, and are therefore restricted to use on hospitalized patients.

Sar—Arg—Val—Tyr—Val—His—Pro—Ala-COOH       (Sar = $CH_3$—NH—$CH_2$—CO—)

Saralasin

5-72

There have been experiments with *renin antibodies* (see Hofbauer and Wood, 1985).

### Selected Readings

J. M. Connell (1986). Essential hypertension: rational pharmacotherapy. *Trends Pharmacol. Sci. 7*: 412–418.

D. W. Cushman and M. A. Ondetti (1980). Control of blood pressure by angiotensin blockage. *Trends Pharmacol. Sci. 1*: 260–263.

D. W. Cushman and M. A. Ondetti (1980). Inhibitors of angiotensin-converting enzyme. In: *Progress in Medicinal Chemistry* (P. G. Ellis and G. B. West, Eds.), Vol. 17. Elsevier/North Holland, New York, pp. 41–104.

M. A. Devynck and P. Meyer (1978). Angiotensin receptors. In: *Advances in Pharmacology and Therapy*, Vol. 1: *Receptors* (J. Jacob, Ed.). Pergamon Press, New York, pp. 279–289.

J. Gutkowska, P. Corvol, G. Thibault, and J. Genest (1982). Tonin as an activator of renin. *Can J. Biochem. 60*: 843–846.

K. G. Hofbauer and J. M. Wood (1985). Inhibition of renin: recent immunological and pharmacological advances. *Trends Pharmacol. Sci. 6*: 173–177.

M. C. Khosla, I. H. Page, and F. M. Bumpus (1979). Interrelations between various blood pressure regulatory systems and the mosaic theory of hypertension. *Biochem. Pharmacol. 28*: 2867–2882.

M. A. Ondetti and D. W. Cushman (1978). Inhibitors of the renin–angiotensin system. *Annu. Rep. Med. Chem. 13*: 82–91.

T. A. Reid, B. J. Morris, and W. F. Ganong (1978). The renin–angiotensin system. *Annu. Rev. Physiol. 40*: 377–410.

R. D. Smith, and J. R. Regan (1986). Antihypertensive agents. *Annu. Rep. Med. Chem. 21*: 63–72.

## 3. ENKEPHALINS, ENDORPHINS, AND OPIATE ANALGESICS

Relief from pain has been an age-old aspiration of humankind. Natural substances—opium alkaloids in the latex of the poppy (*Papaver somniferum*, the "sleeping-bringing poppy")—have been used since ancient Chinese and classical Greek times to modify pain perception, but also misused for their euphoric effect. The opium alkaloids, called opiates and including morphine, are centrally acting *analgesics* (pain relievers) that have a strong narcotic action, producing sedation and even loss of consciousness. In contrast, the minor pain relievers like aspirin are nonnarcotic, act peripherally, and are better called *antalgics*. They will be discussed in Sec. 4.2.

At the time of the discovery of opiate receptors in 1973, it was assumed that only opiates of plant origin exist, as the opium alkaloids were well known and widely used. Thus, the presence, in animals, of very specific receptors for a substance of plant origin was puzzling. Furthermore, these receptors were also found in organs and brain regions not implicated in pain perception. The answer to these problems started to emerge in 1975, with the discovery of the endogenous opioid peptides (Sec. 3.5), the natural analgesics of animal organisms whose receptors, fortuitously, also bind opiate alkaloids. There are two types of such peptides (see Table 5.10 in Sec. 3.5): the large *endorphins* (*end*ogenous m*orphins*) isolated from the pituitary; and the small peptides, most importantly the pentapeptide *enkephalins* [*kephalos* (Gr.) = "head"]. It eventually became evident that the opioid peptides are neurohormones that are involved not only in pain perception, but in a number of other physiological activities as well (Table 5.7; therefore, the presence of opiate receptors in a large variety of tissues is explicable. In 1985, two laboratories reported isolation and unequivocal identification of morphine from animal tissue (i.e., bovine brain, toad skin; see James, 1986), and thus the supposition that morphine was an exclusive plant product was finally disproved. It is not known whether endogenous morphine itself has any physiological role. On the basis of their multiple neuronal effects, opioids could be considered peptide neurotransmitters (discussed in Chap. 4). That idea, however, might be an oversimplification; hence this topic has been considered in this chapter, dealing with neurohormones.

### 3.1. Enkephalinergic Neuronal Pathways

The multiple actions of enkephalins notwithstanding, the principal interest of medicinal chemists is still the analgesic effect of opiate alkaloids and their analogues. An understanding of pain and the central pain pathways is therefore essential to the study of these agents, and the distinction between pain and pain perception must be made. There is a considerable personal and psychological component involved in this phenomenon, aptly called the "puzzle of pain" by Melzack (1973). Two major pathways are involved: the first is the neospinothalamic path mediating sharp localized pain; the second is the paleospinothalamic path involved in the dull, burning pain that responds well to opiates. The dorsal horn of the spinal cord is involved in collecting the nociceptive (pain) stimuli. However, these stimuli are

**Table 5.7.** Proposed physiological effects of opioid peptides

| | |
|---|---|
| Analgesia | Respiratory depression |
| Sedation | Memory effects? |
| Euphoria/dysphoria | Decrease in locomotion |
| Hypotension | Endocrine effects: |
| Inhibition of gastric secretion | Increase: growth hormone |
| Feeding stimulation | prolactin |
| Temperature regulation | Decrease: thyrotropin |
| | ACTH |
| | lutropin |
| | follitropin |

experienced by cortical centers and interpreted emotionally by the limbic system (Besson and Caouch, 1986).

The distribution of enkephalin–opiate receptors along these pathways has been demonstrated, and the intracerebral or epidural injection of opiates in very small doses can produce long-lasting analgesia. Electrical stimulation of the central and periventricular gray matter leads to the same effect.

There are many peripheral organs that possess enkephalin–opiate receptors: the *ileum*, the most distal part of the small intestine; and the *vas deferens* are the most significant. The receptors in the ileum are responsible for the antidiarrheal activity of opiates. The peripheral effects of these drugs were reviewed concisely by Hughes (1981) and by Makhlouf (1985).

### 3.2. Physiological and Pharmacological Effects of Enkephalins and Opiates

The principal opiate effect in mammals is analgesia—the reduction of pain perception. The endogenous peptides probably modulate SP, dopaminergic, and cholinergic limbic and cortical centers, causing the opening of potassium channels, which results in direct hyperpolarization. Snyder (see Feinberg et al., 1976) has proposed that enkephalin also acts on the presynaptic receptors of an excitatory neuron, resulting in its partial depolarization, which would then result in decreased neurotransmitter release following a nerve impulse. Thus, the excitatory presynaptic modulation of an excitatory neuron results in postsynaptic inhibition. Opiates first increase then decrease enkephalin release, a possible physiological explanation for the tolerance and addiction to these drugs; however, there is also a biochemical explanation for opiate dependence: low levels of neurotransmitter (ACh or DA) seem to inhibit enkephalin release, fulfilling a feedback role and thus preventing complete *physiological analgesia*, which could be life-threatening in view of the important warning function of pain.

Most opiates have a number of undesirable side effects. First and foremost are the addictive narcotic effects: euphoria and sedation in humans, apes, and dogs; excitation, fright, or convulsions in cats, cattle, and horses. It might be argued that sedation and euphoria are useful components rather than undesirable effects in alleviating the anxiety that accompanies pain. Indeed, high doses of morphine cause a deep narcosis. Another dangerous side effect is respiratory depression, which involves a decrease in the $CO_2$ sensitivity of the respiratory center, causing $CO_2$ retention and cerebral vasodilation. This is the potential cause of death when opiates are overdosed. Some humans suffer from nausea and emesis upon morphine administration; species that are sedated show miosis (contraction of the pupil), whereas those that are excited show mydriasis (expansion of the pupil).

*Withdrawal symptoms* are experienced when exogenous opiate agonists are abruptly discontinued in persons dependent on or addicted to opiates. Dependence is a physiological (as well as psychological) adaptive state. It varies with the particular drug, its dosage, and the duration of addiction; the symptoms of abstinence may therefore also vary in severity. These symptoms become manifest as a loss of appetite and weight, mydriasis, chills and sweating, abdominal cramps, muscle spasms, tremor, and piloerection ("gooseflesh"—hence the slang "cold turkey" for opiate withdrawal symptoms). Lacrimation, an increased heart rate and blood pressure, and in animals the descriptive "wet-dog shakes" are also characteristic. Withdrawal symptoms can be immediately precipitated in addicts by the administration of narcotic antagonists.

### 3.3. Biochemical Effects of Opiates

The principal biochemical effect of opiates is the inhibition of adenylate cyclase, which decreases cAMP production. If phosphodiesterase, the enzyme that hydrolyzes cAMP, is inhibited by theophylline, hyperalgesia (increased sensitivity to pain) will result; the same effect is seen with prostaglandin $E_2$ (see Fig. 5.25), which stimulates the activity of adenylate cyclase.

Current hypotheses concerning the mechanisms of *tolerance*, *addiction*, and *withdrawal* symptoms are based on adaptive syndromes. In tolerance without dependence, receptors become uncoupled from the $G_i$ unit of the receptor complex, and higher doses are necessary to produce analgesia. When dependence develops, the chronic deficit of cAMP triggers a compensatory feedback loop to rectify the low cAMP levels, which are due to inhibition of AC (i.e., its low turnover number). Compensation can be achieved either by an increased synthesis of new AC molecules or perhaps by an increase in turnover number of the existing AC (Wüster et al., 1985). The consequently increased amount of AC will, even in its inhibited state, produce sufficient cAMP to meet all of the requirements for the cells. In this condition, the organism will function normally *only* in the presence of opiate, and physiological habituation results. When the level of opiate drops, either because of drug withdrawal, or the administration of an antagonist, the previously inhibited AC produces cAMP at a normal rate; the concentration of AC increases to higher than normal levels, causing sudden flood of cAMP; this triggers the multiple and diffuse withdrawal symptoms, which may even prove fatal.

It is proper to distinguish between habituation and addiction. The former is a biochemical–physiological process, whereas the latter has very considerable psychological and socioeconomic components as well.

### 3.4. Properties of the Opioid Receptors

Biochemical evidence for the existence of opioid receptors—or more precisely, for stereospecific opioid recognition sites—has been available since the mid-1970s; however, we know next to nothing about the effector process, except that adenylate cyclase is involved. Crude synaptosome membrane preparations containing receptors can be easily obtained from rat brain, and classical binding experiments were first performed on these by Goldstein and his co-workers. When nonspecific binding sites are protected by the inactive (+)-dextrorphan, its active (radiolabeled) enantiomer [$^3$H]levorphanol (5-80, Fig. 5.22) will bind stereospecifically to the receptor, giving a direct measure of specific binding. The Snyder group improved this method by using the displacement of [$^3$H]naloxone (5-95, Fig. 5.23), an opiate antagonist, as a measure of agonist binding.

Snyder and his group published a series of papers characterizing the receptor, and proposed that it is an allosteric lipoprotein modulated by $Na^+$ ions. In the presence of $Na^+$, the receptor preferentially binds an antagonist, whereas in the absence of $Na^+$, the agonist binding form of the receptor predominates. Lithium ions can replace sodium, but potassium is inactive. The discovery of this "Na-shift," expressed as a ratio, was of great importance because it allowed the rapid *in vitro* testing of opiate or enkephalin analogues and the determination of their agonist or antagonist nature (Table 5.8). The binding of an agonist can be increased by $Cu^{2+}$, $Mg^{2+}$, $Mn^{2+}$, and $Ni^{2+}$, whereas the binding capacity is destroyed by —SH reagents like $N$-ethylmaleimide, indicating the presence of an essential —SH group near the binding site.

After many frustrating attempts, the opioid receptor was isolated in the laboratories of Simon (1981) and of Abood (Bidlack et al., 1981) by very careful

**Table 5.8.** Effects of sodium ions on the inhibition of [$^3$H]naloxone binding on different opiate agonists and antagonists. The sodium index indicates immediately to which group a compound belongs.

| Nonradioactive opiate | IC$_{50}$ of stereospecific [$^3$H]naloxone binding (nM)[a] | | Sodium shift IC$_{50}$ ratio, + NaCl/− NaCl |
|---|---|---|---|
| | No NaCl | 100 mM NaCl | |
| Pure antagonists | | | |
| Naloxone | 1.5 | 1.5 | 1.0 |
| Naltrexone | 0.5 | 0.5 | 1.0 |
| Diprenorphine | 0.5 | 0.5 | 1.0 |
| Mixed agonists-antagonists | | | |
| Cyclazocine | 0.9 | 1.5 | 1.7 |
| Levallorphan | 1.0 | 2.0 | 2.0 |
| Nalorphine | 1.5 | 4.0 | 2.7 |
| Pentazocine | 15 | 50 | 3.3 |
| Metazocine | 10 | 60 | 6.0 |
| Pure agonists | | | |
| Met-enkephalin | 50 | 450 | 9 |
| Etorphine | 0.5 | 6.0 | 12 |
| Phenazocine | 0.6 | 8.0 | 13 |
| Meperidine | 3,000 | 50,000 | 17 |
| Leu-enkephalin | 12.0 | 225 | 19 |
| D-Ala$^2$-enkephalin | 5.3 | 140 | 26 |
| Methadone | 7.0 | 200 | 28 |
| Morphine | 3.0 | 110 | 37 |
| (±)-Propoxyphene | 200 | 12,000 | 60 |

[a] IC$_{50}$ = median inhibitory concentration.

Reproduced by permission from Creese in Yamamura et al. (Eds.), *Neurotransmitter Receptor Binding*, Raven Press, New York.

**Table 5.9.** Opioid receptor subtypes

| Subtype | Principal agonists | Endogenous ligand | Na shift | Effect mediated |
|---|---|---|---|---|
| Mu$_1$ ($\mu_1$) | morphine morphiceptine[a] DAGO[a] | $\beta$-endorphin | 20–100 | central analgesia |
| Mu$_2$ ($\mu_2$) | same, very high affinity | same | | respiratory depression |
| Delta ($\delta$) | Leu-enkephalin[a] | Leu-enkephalin | 5–20 | spinal analgesia emotional effects |
| Kappa ($\kappa$) | bremazocine | dynorphin[a] | < 5 | sedative analgesia vas deferens contraction feeding |
| Sigma ($\sigma$) | cyclazocine | | < 5 | hallucination naloxone-insensitive |
| Epsilon ($\varepsilon$) | $\beta$-endorphin | $\beta$-endorphin | | vas deferens contraction |

[a] See Table 5.11.

detergent solubilization, affinity chromatography, and electrophoresis. Subunits of 43, 35, and 23 kD, and a binding constant ($K_D$) of $3.8 \times 10^{-9}$ M have been reported, corresponding to a 2000-fold purification. Since then, further progress in purification has occurred (see Cotton and James, 1985), but it is difficult to sort out the biochemical properties of the numerous alleged isoreceptors among the sometimes conflicting claims.

### 3.4.1 Pharmacological Subtypes

The situation is somewhat clearer regarding the pharmacological subtypes of the opioid receptors. Table 5.9 shows the currently accepted isoreceptor types, a compromise between several schools of thought. The main problem is availability of sufficiently specific ligands, and poor correlation of receptor localization with neuronal circuits. Considerable changes in these concepts can be anticipated in the future, as clear understanding of receptor subtypes is imperative in designing the ideal nonaddictive analgesic, still an elusive goal.

The mu ($\mu$) receptor is the principal pain-modulating site in the CNS, and is morphine selective. Pasternak and Wood (1986) proposed the distinction between $\mu_1$ and $\mu_2$ receptor subtypes. The former is speculated to be the analgesic site; the latter is assumed to be responsible for the respiratory depression.

There is considerable interest in the kappa ($\kappa$) receptor, which mediates a sedating analgesia with decreased addiction liability and respiratory depression, and which allows for some structural flexibility. Unfortunately, the $\kappa$ receptor seems to be coupled to the sigma ($\sigma$) receptor, implicated in psychotomimetic and dysphoric side effects (Cowan and Gmerek, 1986).

### 3.5. Endogenous Opioid Peptides

In 1974, Liebeskind showed the existence of a central pain-suppressive system, and was able to produce analgesia by electrical stimulation of the periventricular gray matter. This electroanalgesia could be reversed by opiate antagonists, and showed a cross-tolerance with morphine-induced analgesia. These results indicated the existence of a neuronal system that uses an endogenous neuromodulator or neurotransmitter with opiate-like properties.

The isolation of such endogenous opiates was reported simultaneously by four laboratories: those of Goldstein in Palo Alto, Hughes in Aberdeen, Snyder in Baltimore, and Terenius in Uppsala (cf. Beaumont and Hughes, 1978). Acid acetone extracts of pig, calf, and rat brains yielded, after purification, two pentapeptides, called **enkephalins** ("in the brain"), with the structures $NH_2$-Tyr—Gly—Gly—Phe—Met-COOH (**Met-enkephalin**) and $NH_2$-Tyr—Gly—Gly—Phe—Leu-COOH (**Leu-enkephalin**), present in a 4:1 ratio in pig brain but a 1:4 ratio in cattle brain (Table 5.10). Since the genetic code for methionine (AUG) differs by only one base from one of the leucine codons (CUG), a point mutation or genetic drift may account for this difference. The ratio of the two enkephalins also varies in different brain regions.

The Goldstein group also isolated from the pituitary gland a larger peptide that proved to be identical to a fragment of **β-lipotropin**, a pituitary peptide hormone with a questionable physiological role. This peptide was called **β-endorphin** ("*en-do*genous *morphine*"), and was almost 50 times more analgetic than morphine if injected directly into the brain. Other endorphins and a smaller peptide, **dynorphin**, were also found in the pituitary (Table 5.10). The heptapeptide **dermorphin** was isolated from the skin of the frog *Phyllomedusa*.

#### 3.5.1. Biogenetic Relationships

The biogenetic relationship of the opioid peptides has also begun to be clarified. A family of peptides is produced in the hypothalamus as well as the anterior pituitary, originating in the form of a single large protein, **pro-opiomelanocortin**, consisting of

**Table 5.10.** Various endogenous opioid peptides. The Met-enkephalin sequence (italics) appears in β-endorphin.

| Peptide | Amino acid sequence |
|---|---|
| Met-enkephalin | Tyr—Gly—Gly—Phe—Met |
| Leu-enkephalin | Tyr—Gly—Gly—Phe—Leu |
| Dermorphin | Tyr—D-Ala—Phe—Gly—Tyr—Pro—Ser-NH₂ |
| Dynorphin | Tyr—Gly—Gly—Phe—Leu—Arg—Arg—Ile—Arg—Pro—Lys—Leu—Lys—Trp—Asp—Asn—Gln-OH |
| β-Casomorphin (an "exorphin") | Tyr—Pro—Phe—Pro—Gly—Pro—Ile |
| β-Endorphin | Lys—Arg—*Tyr—Gly—*Gly—*Phe—Met*—Thr—Ser—Glu—Lys—Ser—Glu—Thr—Pro—Leu—Val—Thr—Leu—Phe—Lys—Asn—Ala—Ile—Ile—Lys—Asn—Ala—Tyr—Lys—Lys—Gly—Glu |

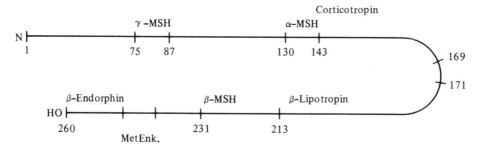

**Fig. 5.20.** Pro-opiomelanocortin, the common precursor of several neurohormones. The numbers indicate the boundaries of hormones that are produced from this precursor.

over 260 amino acids. This protein, whose sequence was determined from its cloned mRNA, is cleaved into numerous peptide neurohormones, whose succession is shown in Fig. 5.20. The complete amino acid sequence is provided in the paper of Bloom (1981). It consists of three different melanophore-stimulating hormones, ACTH, β-lipotropin (a weak lipid-degrading hormone); and β-endorphin, incorporating the met-enkephalin sequence. The enkephalins are very widespread in cells of the CNS and in the ileum, but are not derived from the cleavage of endorphin, which is found in the hypothalamus and pituitary only. Instead, the enkephalins are synthesized by the enkephalinergic neurons themselves from the precursor **proenkephalin A** and **prodynorphin** (proenkephalin B), which are encoded by two different genes. Prodynorphin contains dynorphin A, dynorphin B, and also two neoendorphins that are endorphin peptides extended on the C-terminus. Other minor peptides are being discovered continuously, but their significance is not clear. The leu-enkephalin sequence is also incorporated in the small **dermorphin** and **dynorphin**. Dermorphin is highly resistant to proteolysis, and is 30–60 times more active than morphine; if injected intracerebrally, however, its activity is about a thousand times greater than that of morphine. The D-Ala is essential for its activity and stability.

### 3.5.2. Pharmacological Activity

Although the pharmacological activity of the endorphins and enkephalins is morphine-like, the endorphins have a wider range of effects. Among others effects, they induce catatonic rigidity in animals, reversible by opiate antagonists. The hope of connecting endorphins to schizophrenia was not fulfilled; nor was the expectation of finding the ideal, nonaddicting analgesic among these peptides. Strangely, and unfortunately, repeated doses of endorphin or enkephalin give rise to addiction and withdrawal symptoms. There are indications that acupuncture analgesia operates through enkephalin mobilization, since it is reversed by opiate antagonists.

### 3.5.3. Structure–Activity Correlations

Naturally, many structure–activity investigations have been reported on molecularly modified enkephalins. Besides increasing opiate activity, the principal goal of this work is to prevent the rapid hydrolysis between Tyr and Gly, the way in which all enkaphalins become deactivated. Removing the Tyr[1] from enkephalins or interfering with its phenolic hydroxyl or amino groups abolishes the activity of these substances. When the natural L-Tyr[1] is replaced by its D-enantiomer, activity is lost in the enkephalins as well as endorphins. On the other hand, replacement of the Gly[2] residue with D-Ala renders the peptide resistant to hydrolysis, to the extent that some of the synthetic enkephalin analogues retain their activity when taken orally. D-Ala[2] analogues combined with modifications of Met[5] have produced the most potent derivatives, as shown in Table 5.11. Met[5]-amides are also resistant to hydrolysis, and some potent compounds have been discovered among them. The cyclized enkephalin analogue 5-73, which contains penicillamine (dimethylcysteine), shows the highest delta ($\delta$) affinity, whereas the cyclic analogue 5-74 is mu ($\mu$) selective.

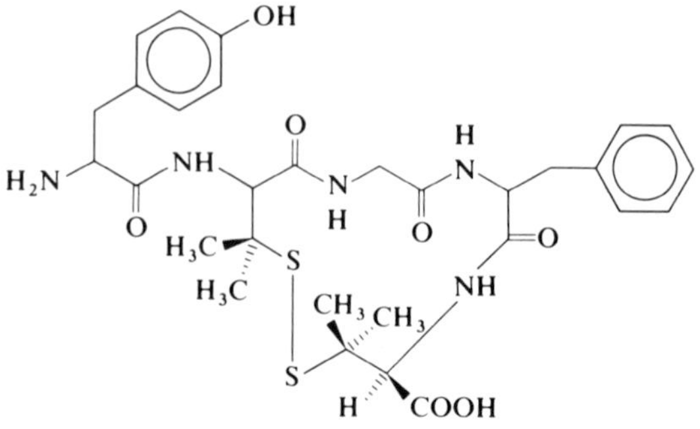

5-73

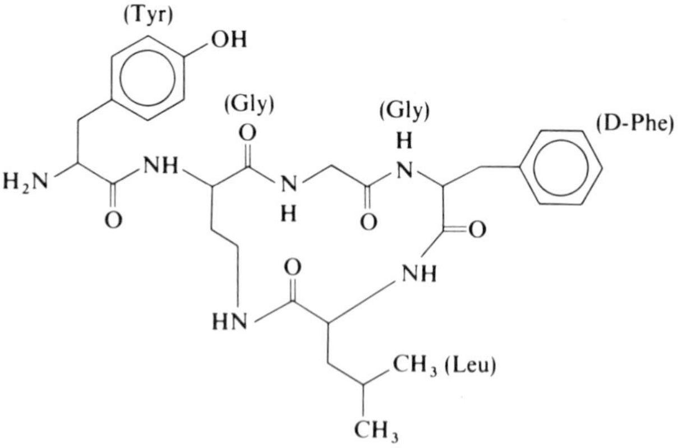

5-74

***Recent Developments.*** The constant flow of new information in this field can be followed through the reviews published each year in the *Annual Reports in Medicinal Chemistry* (see also Casey and Parfitt, 1986). Outstanding among the many exciting developments in the field of analgesic peptides and their interactions is the purification and study of the enkephalinase enzymes (Schwartz et al., 1985; Chipkin,

**Table 5.11.** Relative potency of some enkephalin analogues

| | Relative receptor activity | Tail flick, $ED_{50}{}^a$ | | Relative molar dose $1/$i.c.v. |
| --- | --- | --- | --- | --- |
| | | i.c.v. ($\mu$g/mouse) | i.v. (mg/kg) | |
| Morphine | 2 | 1 | 1.8 | 33 |
| $\beta$-Endorphin | 4 | 0.3 | NT | 1,200 |
| Met-enkephalin | 1 | 68 | 153 | 1 |
| Gly$^2$—Met$^5$—ol | NT | 43 | 52 | 2 |
| D-Ala$^2$-enkephalin | 1 | 0.7 | 51 | 100 |
| D-Ala$^2$—Met$^5$—ol | 5 | 0.04 | 12 | 1,600 |
| D-Ala$^2$—Met$^5$—S-oxide—ol | 2 | 0.01 | 12 | 10,000 |
| D-Ala$^2$—N-Me-Phe$^4$—Met$^5$—S-oxide—ol | 11 | 0.002 | 0.7 | 30,000 |
| D-Ala$^2$—D-Leu$^5$—enkephalin (DADL) | 10 | — | — | — |
| Tyr—D-Ala—Gly—N-Me-Phe—Gly—ol (DAGO) | 14 | — | — | — |
| Tyr—Pro—N-Me-Phe—D-Pro-NH$_2$ (morphiceptin) | 14 | — | — | — |

$ED_{50}$, median effective dose; i.c.v., intracerebroventricular.

1986). Several inhibitors of these Zn-containing peptidases have been synthesized; one of these **thiorphan**, $N$-($\alpha$-benzyl-$\beta$-thio-propionyl)glycine, has a $K_I$ of 4.7 nM, potentiates D-Ala$^2$—Met-enkephalin, and under some conditions has an analgesic effect of its own. The intrathecal administration of $\beta$-endorphin directly into the cerebrospinal fluid has provided rapid and long-lasting (33 hours on the average) pain relief in cancer patients with otherwise intractable pain (see Gesellchen and Zimmerman, 1981; Besson and Caouch, 1987). Morphine can be used similarly with no risk of addiction or tolerance, for over one year.

### 3.6. Opium Alkaloids

The dried latex of *Papaver somniferum* (opium), or the seed capsule of the plant itself, are the sources of almost 25 alkaloids. Some simple isoquinolines from opium, like **papaverine**, are antispasmodics. The principal alkaloid ($\sim 10\%$ of the total) is **morphine** (5-75), which is also an isoquinoline (rings C and E), but can additionally be considered a phenanthrene derivative (rings A, B, and C).

Even though pure morphine has been available since 1803, its structure was finally elucidated only in 1925, by Sir Robert Robinson. Structure (5-75) in Fig. 5.21. shows that ring numbering and clearly illustrates the isoquinoline character and T shape of the molecule. Detailed structure–activity relationships have been worked out during the many years of study of morphine and its derivatives.

The *phenolic hydroxyl on C-3* is a very important functional group on an essential ring. There are very few active opiates without the aromatic ring or the phenolic hydroxyl group. The latter probably amplifies the van der Waals binding to the receptor of the aromatic ring through hydrogen bonding. Masking this hydroxyl by acetylation or methylation, as in **heroin** (5-76) or **codeine** (5-77), changes the narcotic

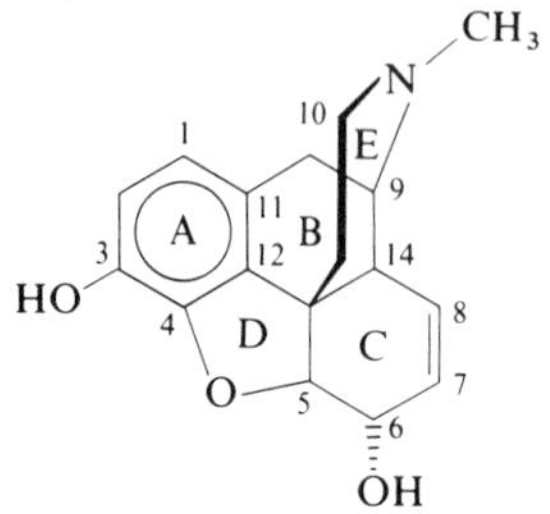


Morphine
5-75

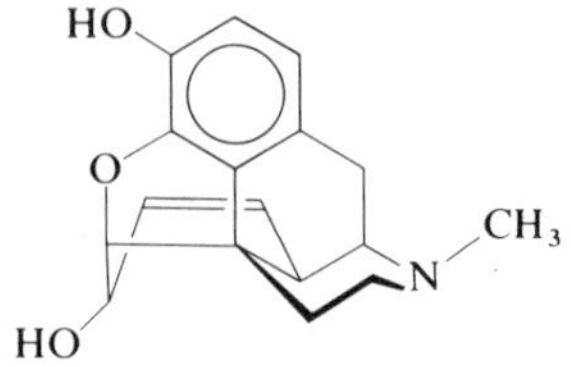


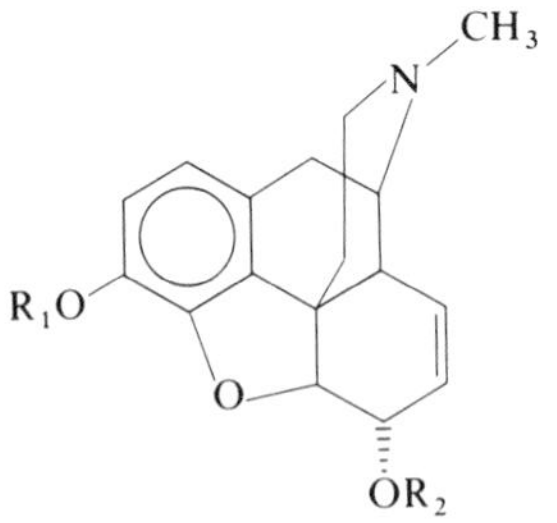


| $R_1$ | $R_2$ | | |
|---|---|---|---|
| $-CH_3$ | $-H$ | 5-76 | Codeine |
| $-C(=O)-CH_3$ | $C(=O)-CH_3$ | 5-77 | Heroin |

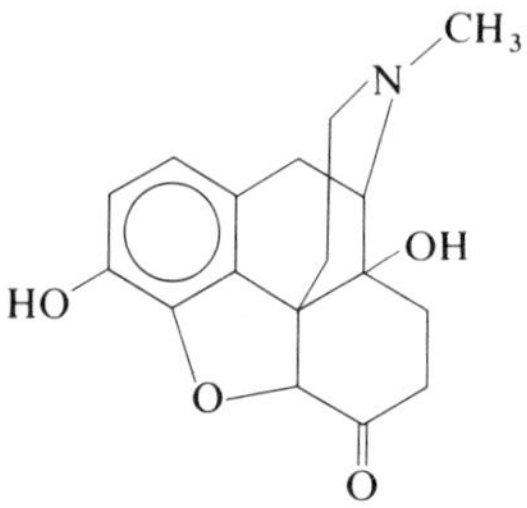


14-Oxymorphone
5-78

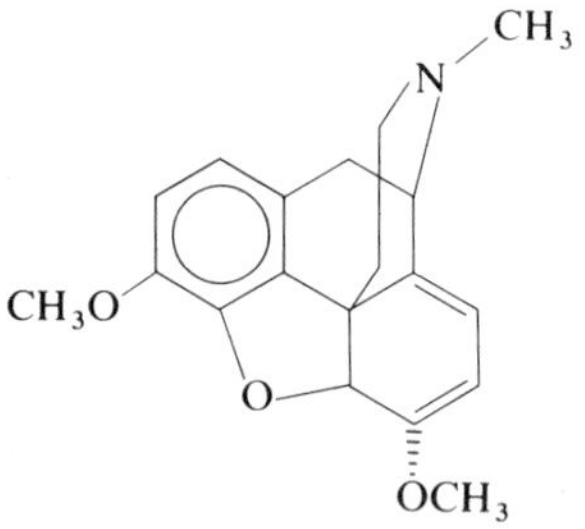


Thebaine
5-79

**Fig. 5.21.** Morphine and some of its derivatives.

analgesic effect. Codeine, which is a natural alkaloid, has only about one-tenth the effect of morphine if given parenterally, because it must be partially demethylated in the liver and thus transformed to morphine; thus given intracerebrally it is totally inactive as an analgesic. It is widely used as an antitussive (cough supressant) drug. Weakening of the electron density at the 3-OH position by the introduction of electron-attracting substituents (e.g., $-NO_2$) in position 1 also has an inactivating effect.

Heroin (diacetylmorphine) is a highly addictive drug, being twice as active as morphine. The reason may be that its increased lipophilic character results in better transport characteristics, and the increased activity due to esterification of the 6-OH group compensates for the loss of potency due to masking of the 3-OH by the acetyl group.

The *alcoholic hydroxyl at C-6* can be modified (as in heroin) or even omitted. For instance, **heterocodeine** (morphine-6-methylether) is about five times more active than morphine. The 6-keto derivative (**morphinone**) and the 6-methylene analogue are both active analgesics.

The $\Delta^{7,8}$ *double bond* is also nonessential to the activity of morphine. Dihydromorphine or dihydromorphinone are active compounds with a reduced duration of action but increased activity.

The *N-methyl substituent* on morphine is not absolutely essential to its analgesic activity; such activity is instead more a question of the proper partition coefficient. Thus *N*-normorphine, the secondary amine, has only one-eighth of the central activity of morphine, but is equiactive with the latter on the guinea pig ileum, indicating that it cannot cross the blood–brain barrier because of its polarity. Higher alkyl substituents usually render the molecule less active, although the activity rises dramatically if the side chain carries an aromatic ring. For instance, the **N-furfurylmethyl-** and **N-phenethyl-normorphine** derivatives and their analogues can be up to 50 times more active than morphine as a result of the involvement of auxiliary binding sites (see Fig. 5.24).

Nonbasic morphine derivatives, such as morphine *N*-oxide or the quaternary methiodide, are inactive parenterally. The latter cannot cross the blood brain barrier because of its ionic charge, but shows full activity if injected intracisternally.

The most important and dramatic change results when the *N*-methyl group of morphine is replaced by an *N*-alkene or *N*-cyclopropylmethyl group. The resulting compounds show antagonist properties (see Sec. 3.8).

Derivatives with a *C-14 hydroxyl group* such as **oxymorphone** (7, 8-dihydro-14-hydroxymorphine-6-one) (5-78) show increased potency (up to five times that of morphine), probably as a result of the introduction of an additional hydrogen-bonding substituent. The stereochemistry of this hydroxyl is of considerable importance in terms of activity.

### 3.7. Synthetic Morphine Analogues

Molecular modifications in the form of simplifications of the morphine ring system have been undertaken for many years, and have resulted in the development of some spectacularly successful compounds devoid of addictive properties (Fig. 5.22).

#### 3.7.1. Morphinans

Omission of the furan ring (i.e., oxygen bridge) of morphine results in compounds known as *morphinans*, such as **( − )levorphanol** (5-80), which is five to six times more active than morphine. Loss of the phenolic hydroxyl decreases the activity of this compound to 20% that of morphine. The ( + ) isomer, **dextrorphan**, is totally devoid of analgesic acitivity because it cannot bind to the receptor. However, dextrorphan has antitussive properties that are valuable, and is used in the form of its methyl ether (dextromethorphan). Shifting of the phenolic hydroxyl group to position 2 or 4 results in a total loss of activity. The **N-phenethyl** derivative of levorphanol shows an analgesic effect about twenty times greater than that of the parent compound.

### 3.7.2. Benzomorphan and Metazocine Series

Opening of the C ring as well as omission of the D ring of morphine results in the **6,7-benzomorphan** series. To indicate ring C, two methyl groups were retained on ring B. Study of the stereochemistry of these methyl groups has led to the major discovery of nonaddictive analgesics of the **metazocine** series (5-82, 5-83). Compounds in which the two alkyl substituents are *cis* (as are the corresponding carbon

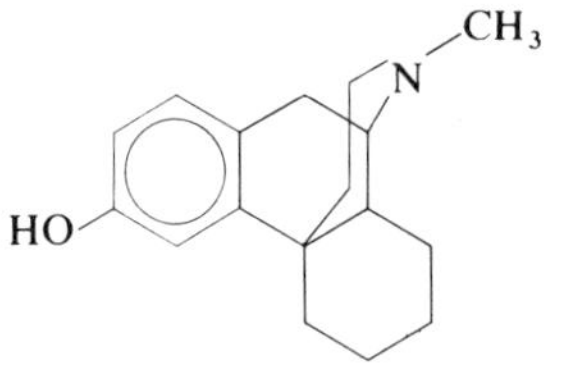

Levorphanol
5-80

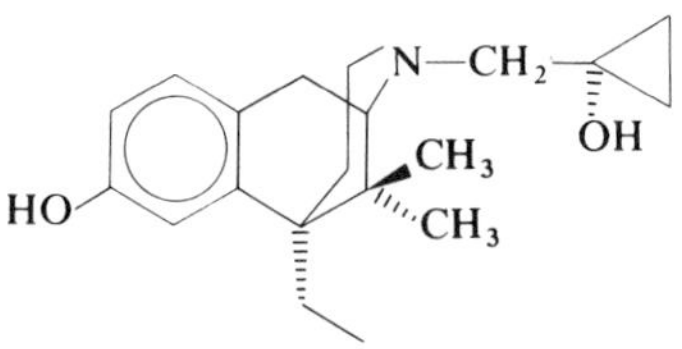

Bremazocine
5-81

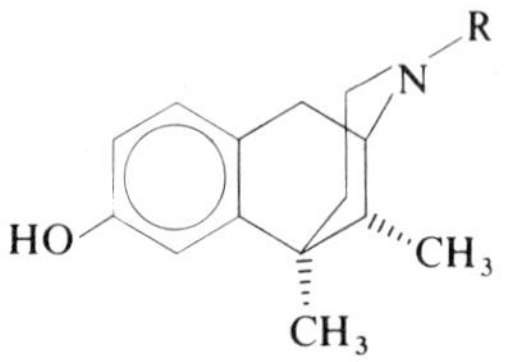

6,7-Benzomorphans

R = —CH$_3$                5-82  Metazocine

R = —CH$_2$—CH$_2$—⟨ ⟩      5-83  Phenazocine

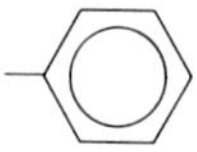

R = —CH$_2$—CH=C⟨CH$_3$/CH$_3$⟩   5-84  Pentazocine

R = —CH$_2$—◁            5-85  Cyclazocine

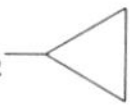

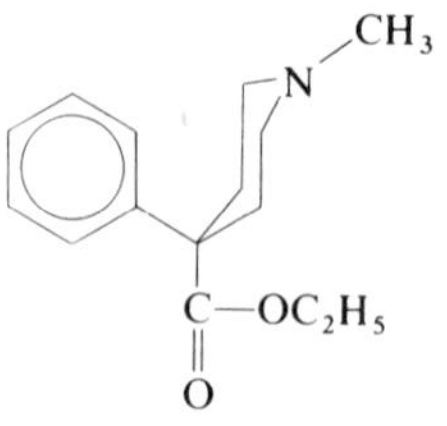

Meperidine
5-86

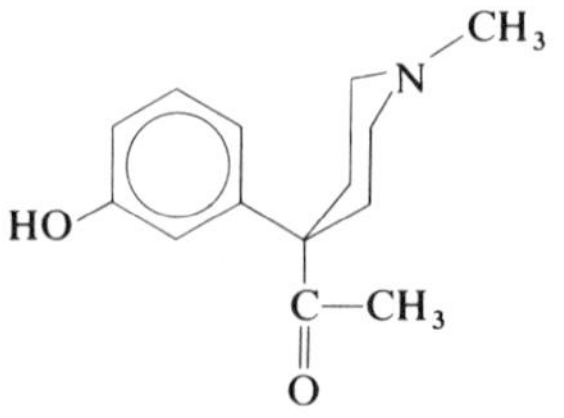

Ketobemidone
5-87

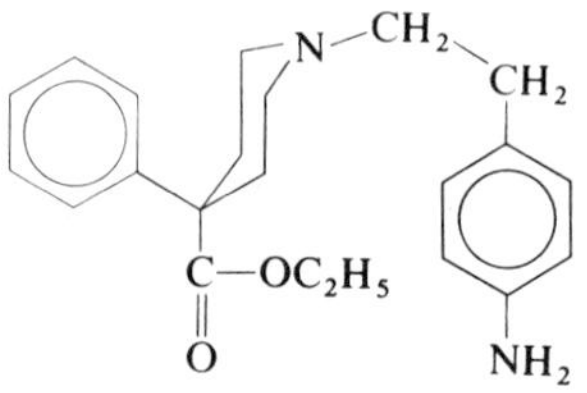

Anileridine
5-88

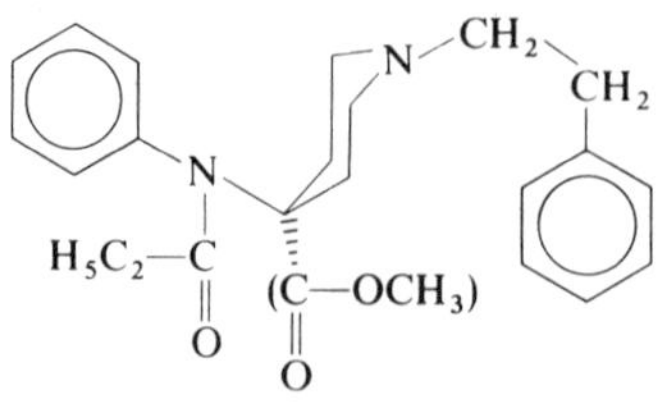

Fentanyl
5-89

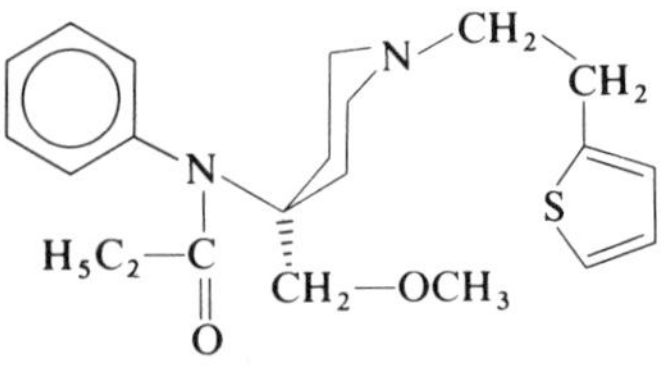

Sufentanil
5-90

**Fig. 5.22.** Morphinans, benzomorphans, and piperidine opiate analogues.

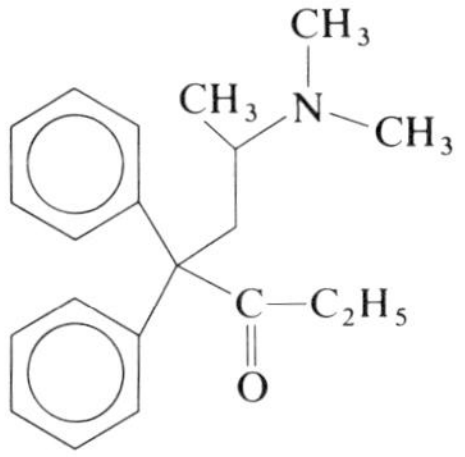

Methadone
5-91

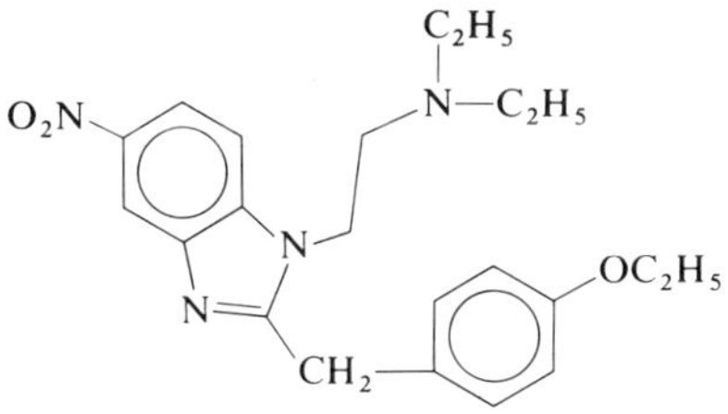

Etonitazene
5-92

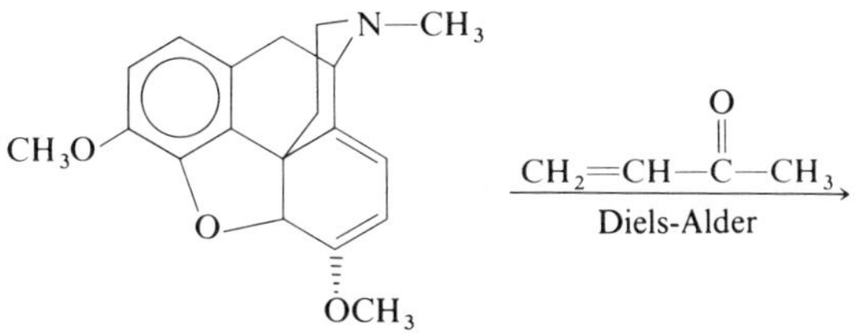

Thebaine

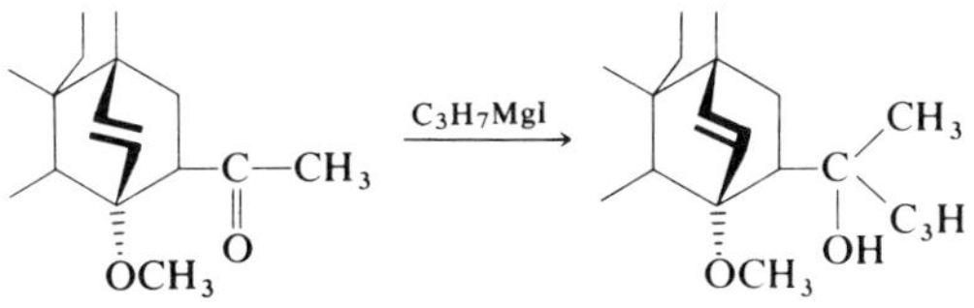

$CH_2=CH-\overset{\overset{O}{\|}}{C}-CH_3$
Diels-Alder

$C_3H_7MgI$

Etorphine
(3-OH)
(an oripavine)
5-93

**Fig. 5.22.** (*continued*)

atoms in morphine) are powerful analgesics. However, they cannot relieve withdrawal symptoms in addicted animals, meaning that these drugs are not addictive. The *trans* isomers, on the other hand, while also potent analgesics, will relieve withdrawal. The (−) isomers in this series will precipitate withdrawal symptoms in addicted animals even with an N-methyl (that is, agonist) substituent, indicating that they are mixed agonist–antagonists (or "metagonists," according to Belleau; see Sec 3.8). Although there are complicating factors and exceptions to this rule with some of the compounds, derivatives of exceptionally low addictive capacity and satisfactory analgesic potency have been prepared, such as (−) **pentazocine** (5-84), which can be used in chronic applications without the danger of addiction. Some of these derivatives, like the N-cyclopropylmethyl compound (**cyclazocine,** 5-85), cannot be used because of unpleasant dysphoric, hallucinogenic properties. A new benzomorphan, **bremazocine** (5-81), is a powerful $\kappa$ agonist (see Table 5.8) of long duration, and is devoid of addictive properties and respiratory depressant activity. On the basis of receptor binding, it is about 200 times more active than morphine and has a very low sodium shift, like all of the mixed agonist–antagonist (or metagonist) benzomorphans.

### 3.7.3. Piperidine Derivatives

Piperidine derivatives represent the ultimate simplication of the morphine skeleton. They were first developed in the 1940s, and one—**meperidine** (Demerol, 5-86)— is perhaps still the most widely used synthetic opiate in clinical practice despite its addictive properties. Drawn as in (5-86), it is obviously a morphine analogue, but only the A and E rings are retained. The addition of a 3-OH group results in the **bemidone** series, while modification of the ester group to a ketone gives the **ketobemidones** (5-87), which have more than six times the activity of meperidine. The derivative carrying the N-phenethyl side chain (**anileridine**, 5-88) has also proved to be potent. Remarkably, N-alkene or N-cyclopropylmethyl derivatives in this series do not show antagonist properties. It should be noted that the original C-13 must be quaternary in all of these compounds.

Lengthening of the N-substituent in meperidine leads to active analgesics, such as the propiophenone analogue, which is 200 times more active than meperidine. However, the butyrophenone derivative suddenly becomes a highly active neuroleptic (see Chap. 4, Sec. 4.2) without any analgesic activity. This is another interesting example of the sudden cutoff points in homologous series of drugs, discussed in Chap. 1, Sec. 3.

Perhaps the most successful modification of the 4-phenylpiperidine derivatives of morphine are the 4-anilino compounds like **fentanyl** (5-89). This drug is 50–100 times more active than morphine, owing mainly to its excellent transport across the blood–brain barrier and into the CNS, as a result of its high lipophilicity. Spectacular activities (∼5000–6000 times that of morphine) have also been achieved by introducing ether or keto substituents (**sufentanil**; 5-90), as in the meperidine or ketobemidone series. Fentanyl derivatives are very fast acting and of short duration. They are used in neuroleptanalgesia in surgery (in combination with neuroleptics or major tranquilizers like **droperidol**).

### 3.7.4. Other Analgesics

Other analgesics that are not piperidine derivatives and bear very little structural resemblance to opiates have also been prepared. **Methadone** (5-91), a diphenylpropylamine derivative, can be drawn in such a way that a charge-transfer type of interaction between the free pair of electrons on the nitrogen and the polarized carbonyl carbon can be appreciated; such an interaction—if it indeed takes place— could favor a conformation resembling a piperidine ring.

Methadone has the same activity as morphine, but has a longer duration of action, is less emetic or constipating, and leads to little euphoria. Additionally, the withdrawal symptoms are less severe than those with other addictive narcotic analgesics, such as heroin. Therefore, methadone is widely used in the maintenance and social rehabilitation of withdrawal addicts, even though it just replaces one addiction with another, somewhat less dangerous one. The methadone molecule has been extensively modified without yielding any major advantages over the parent compound.

In the way it is shown, **etonitazene** (5-92) can also be vaguely related to morphine. It is a benzimidazole derivative and is highly potent, having about 1000 times the activity of morphine. The structure is very specific, and even slight modifications result in loss of activity.

### 3.7.5. Oripavine Derivatives

Oripavine derivatives were prepared from **thebaine** (5-79, Fig. 5-21) a naturally occurring opium alkaloid which, however, is not a narcotic analgesic but a convulsant. Bentley and his colleagues in England developed a very interesting series of bridged opiates through the Diels–Alder addition of unsaturated ketones (e.g., vinylmethylketone) to the conjugated diene of thebaine, as shown in (5-93). Subsequent Grignard reaction of the ketone and hydrolysis of the 3-methoxy group to a free phenolic hydroxyl gives **etorphine** (5-93), the prototype of the 6,14-*endo*-ethenotetrahydro-oripavine-type analgesics. It is 5,000–10,000 times more potent than morphine and is active in doses as small as 0.1 mg for an adult. However, it has only about 20–30 times the affinity of morphine for the morphine receptor. Its enormous activity is due to its great lipophilicity and great ease in penetrating the blood–brain barrier. Whereas the partition coefficient ($p$) of morphine is $1 \times 10^{-4}$ (heptane/phosphate buffer, pH 7.4), that of etorphine is 1.4. Etorphine is also absorbed sublingually, a convenient mode of administration. It is used mainly for the immobilization of big game animals (like elephants), since the extremely small dose required allows a great margin of safety. A derivative with an *N*-cyclopropylmethyl substituent and *t*-butyl instead of *n*-propyl in the side chain, called **buprenorphine** (5-101, Fig. 5.23), is a mixed agonist–antagonist and the most lipophilic compound in this series ($p = 60$, versus only 1.4 for etorphine, as mentioned above). Buprenorphine is about 100 times as active as morphine as an agonist, and four times as active as nalorphine as antagonist, and is therefore nonaddicting. Because its Na index is very low, it is thus predominantly considered an antagonist.

Continuous efforts are being directed toward elucidating an analgesic "pharmacophore" that represents the essence of the molecular features recognized by the receptor. Interactive computer-graphic methods show promise in this area (Chap. 8, Sec. 3.4). One approach (Humblet and Marshall, 1980) assumes a "rigid" phenylpiperidine characterized by a polycyclic framework, where the axial phenyl and the ($-$) configuration are essential. The "flexible" pharmacophore (e.g., the meperidine type) shows an equatorial phenyl group. Quantum chemical studies (Burt et al., 1981) point toward the need for an unhindered interaction between the protonated amine with the anionic receptor site.

### 3.8. Opiate Antagonists

Opiate addiction affects the life of millions of humans in the Western hemisphere alone. Since addiction might be considered "contagious," the idea of a temporary "immunizing agent" against undertaking the misuse of opiates or against persuading others to try an initially pleasurable sensation is certainly an intriguing one.

An immunity against such initial experimentation or against relapse into narcotic use following withdrawal depends first upon the availability of a cheap, nontoxic, and preferably long-acting narcotic antagonist. The way to deliver such agents to a population considered difficult from a sociopathic point of view is a public health problem. However, considering the social cost of drug addiction, the answers to these problems will have to be found.

Although N-allyl-norcodeine was first described in 1915, the discovery of the antagonist effects produced by substituting three-carbon side chains on the morphine molecule came only in the early 1940s. **Nalorphine** (5-94) (Fig. 5.23) was the first clinically useful antagonist, having a dramatic reviving effect on patients on the verge of death from opiate-induced respiratory failure. It also precipitates withdrawal symptoms in addicts. However, although nalorphine is a mixed opiate agonist – antagonist and thus a potentially valuable nonaddicting drug, its unpleasant psychomimetic and hallucinogenic properties preclude its use as an analgesic.

The analogous compound derived from 14-hydroxymorphone, **naloxone** (5-95), was discovered in 1961 and is a pure antagonist. Its cyclopropylmethyl analogue, **naltrexone** (5-96), is even more useful, since it is longer-acting. **Levallorphan** and **cyclorphan** (5-97 and 5-98) are morphinane derivatives, whereas **oxilorphan** (5-99) is their 14-OH analogue. The cyclobutyl homologue of oxilorphan, **butorphanol** (5-100), is 4–10 times more active than morphine, has 50–70 times the activity of pentazocine as an analgesic, and 30 times the activity of pentazocine as an antagonist.

In the benzomorphan series, **cyclazocine** and **pentazocine** (5-85 and 5-84) are useful mixed agonist – antagonists. Unfortunately, the former has considerable hallucinogenic properties, although pentazocine is a very useful analgesic.

Among the oripavines, **buprenorphine** and **diprenorphine** (5-101 and 5-102) are valuable agonist – antagonists. Butorphanol and buprenorphine are examples of metagonist drugs as proposed by Belleau (1982). They are ligands that may stabilize the receptor in other than the open or closed (agonist or antagonist) conformation, allowing the specific regulation of "pain connections" of the receptor and its effector mechanism. In other words, the opiate receptor has three potential functions: agonist binding, antagonist binding, and regulation of adenylate cyclase. If a drug triggers the agonist and antagonist sites simultaneously, as for instance butorphanol does, the adenylate cyclase regulatory capacity need not be influenced and therefore addiction will not result. It seems that metagonist analgesics may be $\kappa-\sigma$ agonists coupled with a $\mu$ antagonist effect (Table 5.8). Buprenorphine, on the other hand, is a true partial agonist but also possesses $\mu$ antagonist activity.

These remarkable drugs of the "metagonist" class also offer a mode of treating narcotic addicts other than the ethically reprehensible substitution of heroin addiction with methadone addiction. Buprenorphine and butorphanol in limited dose (8 mg/day) have suppressed heroin use, and termination of maintenance with these agents does not result in withdrawal symptoms (i.e., habituation does not occur).

The recently described **chlornaltrexamine** (5-103), which acts as an irreversible alkylating affinity label on the opiate receptor, can maintain its antagonist effect for the astonishing period of 3 days. **Naloxazone**, the hydrazone of **naltrexone** (5-96), is a long-acting antagonist with a comparable duration of action.

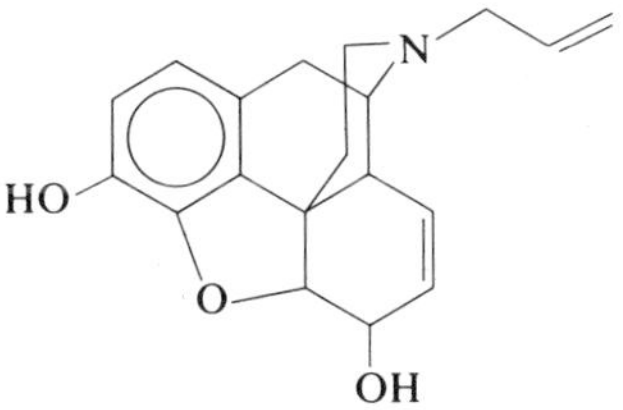

Nalorphine
5-94

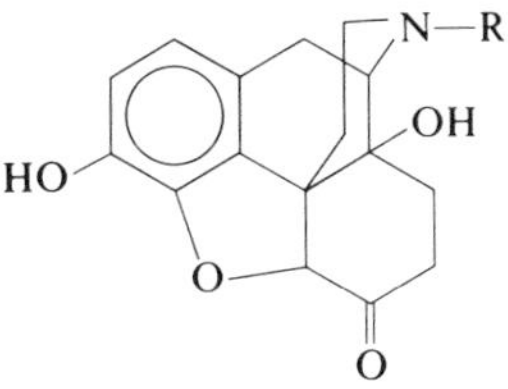

R = 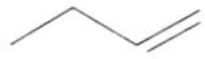

Naloxone
5-95

R = —$CH_2$— 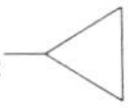

Naltrexone
5-96

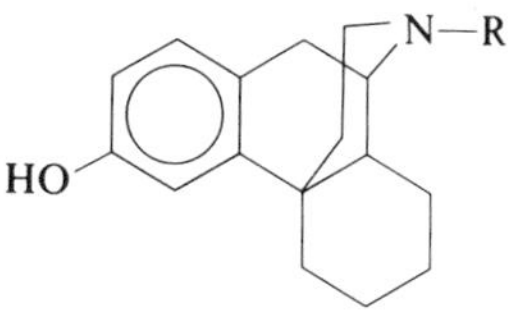

R = 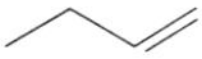

Levallorphan
5-97

R = —$CH_2$— 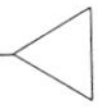

Cyclorphan
5-98

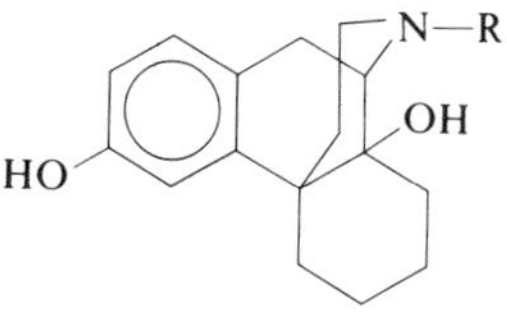

R = $CH_2$— 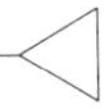

Oxilorphan
5-99

R = —$CH_2$— 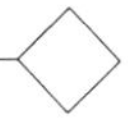

Butorphanol
5-100

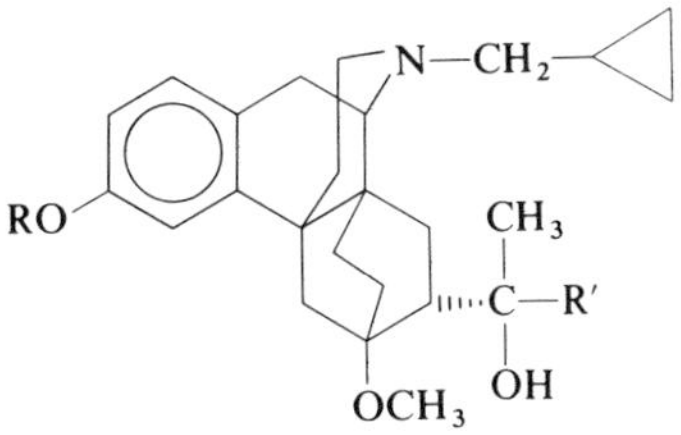

R = H, R′ = *t*-Bu

Buprenorphine
5-101

R = $CH_3$, R′ = $CH_3$

Diprenorphine
5-102

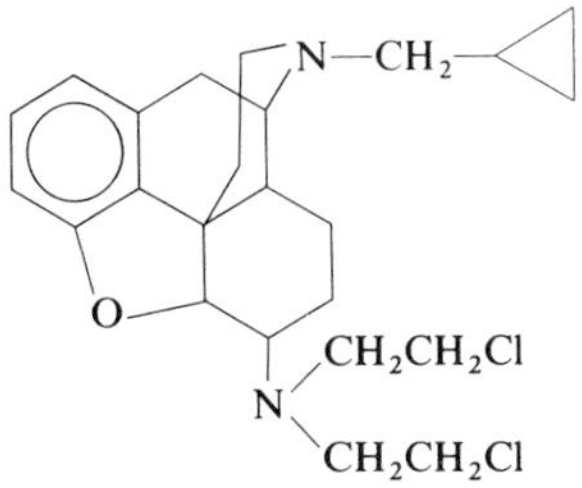

Chlornaltrexamine
5-103

**Fig. 5.23.** Opiate antagonists.

### *3.8.1. Differential Binding of Antagonists*

The minor structural differences between opiate agonists and antagonists have provoked many hypotheses concerning the receptor topography that might explain the differential binding of these compounds. The simplest explanation is the assumption of separate binding sites for agonist and antagonists. This idea has some experimental support; the *N*-methyl analogue of **chlornaltrexamine** (5-103), a long-acting agonist, is antagonized by naloxone but is not displaced from the receptor site (cf. Gesellchen and Zimmerman, 1981).

The quantum-chemical investigations of Burt et al. (1981) concluded that the interaction of the protonated amine of the opiate antagonist must be prevented from interacting with the anionic site in the receptor. Bulky substituents on C-7 or C-9 may achieve this, and a C-14 hydroxyl can cause a conformational change at the anionic site of the receptor, leading to a stable complex.

This finding supports an earlier hypothesis of Snyder and his co-workers (Feinberg et al., 1976), who proposed a single schematic representation of the receptor that might explain the startling fact that minor structural differences transform an agonist into an antagonist. Snyder retains the "flat area"—or lipophilic phenyl binding site (L) of the receptor—as well as the anionic site of previous models, but adds two additional binding areas: an agonist (AG) and antagonist (AN) site (Fig. 5.24). Agonists such as morphine would bind to the L site, with their N-methyl group very weakly interacting with the AG site. A strong agonist like **phenazocine** (5-83, Fig. 5.22) would bind to the L site as well as to the AG locus through its phenethyl side chain, as would a weak agonist like methadone, which cannot assume the critical orientation needed for strong binding. Even "anomalous" agonists such as fentanyl or etonitazene can be accommodated by this model.

In the model, the alkyl or cyclopropylmethyl side chains of the antagonist bind specifically to the AN site, but can also bind to the AG site. In the case of a mixed agonist–antagonist like nalorphine, this side chain can assume an axial or an

**Fig. 5.24.** Model of the opiate receptor as proposed by Snyder and co-workers. The top figure shows the agonist conformation, with the —SH group exposed and the accessory binding site occupied by the phenyl group of N-phenethylmorphine. The lower schematic shows the antagonist form, where the Na site and the antagonist site are occupied but the—SH group is protected and unavailable.

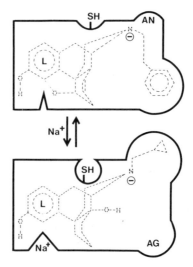

equatorial conformation, and some molecules of the substance will therefore be in the agonist form whereas others will be in the antagonist form. In the "pure" antagonists, the side chain will be forced into the equatorial, antagonist conformation by the steric hindrance of the 14-OH group, and will thus be prevented from tilting to the AG site. If one considers that availability of the AG and AN sites is also influenced by conformational changes due to the Na effect, the foregoing hypothesis, although not providing structural clues (which is impossible without knowledge of the receptor lipoprotein), becomes amenable to testing with rigid analogues and affinity labels. It should be kept in mind that the antagonist conformation is physiologically preferred because $Na^+$ is always present in all tissues. This may account for the much lower $ED_{50}$ of most antagonists as compared to that of the average agonist.

It also seems that the orientation of the N-electron pair of rigid opiates plays an important stereoelectronic role in the proton-transfer reaction between the protonated tertiary amine of these molecules and the anionic binding site of the receptor, which may be necessary for analgesic action (DiMaio et al., 1979). Some recent findings indicate that the aromatic ring may also play a role in opiate binding orientation (Salva et al., 186th ACS National Meeting, Washington, Sept. 1983).

## Selected Readings

A. Beaumont and J. Huges (1978). Biology of opioid peptides. *Annu. Rev. Pharmacol. Toxicol.19*: 245–267.

B. Belleau (1982). Stereoelectronic regulation of the opiate receptor: some conceptual problems. In: *Chemical Regulation of Biological Mechanisms* (A. M. Creighton and S. Turner, Eds.). Royal Society of Chemistry, London.

J.-M. Besson and A. Caouch (1987). Peripheral and spinal mechanisms of nociception. *Physiol. Rev. 67*: 67–243.

J. M. Bidlack, L. G. Abood, P. Osei-Gymah, and S. Archer (1981). Purification of the opiate receptor from rat brain. *Proc. Natl. Acad. Sci. USA 78*: 636–639.

F. E. Bloom (1981). Neuropeptides. *Sci. Am. 245*(4): 148–168.

M. C. Braude, L. S. Harris, E. L. May, J. P. Smith, and J. E. Villareal (Eds.) (1973). *Narcotic Antagonists*. Raven Press, New York.

S. K. Burt, G. H. Loew, and G. M. Hashimoto (1981). Quantum chemical studies of molecular features and receptor interactions that modulate opiate agonist and antagonist activity. *Ann. N.Y. Acad. Sci. 367*: 219–239.

A. F. Casey and R. T. Parfitt (1986). *Opioid Analgesics: Chemistry and Receptors*. Plenum Press, New York.

R. E. Chipkin (1986). Inhibitors of enkephalinase: the next generation of analgesics. *Drugs of the Future 11*: 593–606.

R. Cotton and R. James (1985). Analgesics, opioids and opioid receptors. *Annu. Rep. Med. Chem. 20*: 21–30.

A. Cowan and D. E. Gmerek (1986). In vivo studies on kappa opioid receptors. *Trends Pharmacol. Sci. 7*: 69–72.

B. M. Cox (1982). Endogenous opioid peptides: a guide to structures and terminology. *Life Sci. 31*: 1645–1658.

J. DiMaio, F. R. Ahmed, P. Schiller, and B. Belleau (1979). Stereo-selective control and decontrol of the opiate receptor. In: *Recent Advances in Receptor Chemistry* (F. Gualtieri, M. Gianella, and C. Melchiorre, Eds.). Elsevier, New York, pp. 221–234.

A. P. Feinberg, I. Creese, and S. H. Snyder (1976). The opiate receptor: a model explaining structure–activity relationships of opiate agonists and antagonists. *Proc. Natl. Acad. Sci. USA 73*: 4215–4219.

P. D. Gesellchen and D. M. Zimmerman (1981). Analgesics, endogenous opioids and their receptors. *Annu. Rep. Med. Chem. 16*: 41–50.

M. Gordon (1974). Abuse of CNS agents. *Annu. Rep. Med. Chem. 9*: 38–49.

A. Herz (Ed.) (1978). *Developments in Opiate Research*. Marcel Dekker, New York.

J. Hughes (1981). Peripheral opiate mechanisms. *Trends Pharmacol. Sci. 2*: 21–24.

J. Hughes, H. O. J. Collier, M. J. Rance, and M. B. Tyers (Eds.) (1984). *Opioids. Past, Present and Future*. Taylor and Francis, London.

C. Humblet and G. R. Marshall (1980) Pharmacophore identification and receptor mapping. *Annu. Rep. Med. 15*: 267–276.

A. E. Jacobsen (1978). Analgesics and their antagonists: structure–activity relationships. In: *Handbook of Psychopharmacology* (L. Iversen, S. Iversen, and S. H. Snyder, Eds.), Vol. 12. Plenum Press, New York, pp. 39–94.

R. James (1986). Analgesics, opioids and opioid receptors. *Annu. Rep. Med. Chem. 21*: 21–30.

M. J. Kuhar and G. W. Pasternak (Eds.) (1984). *Analgesics: Neurochemical, Behavioral and Clinical Perspectives*. Raven Press, New York.

G. M. Makhlouf (1985). Enteric neuropeptides: role in neuromuscular activity in the gut. *Trends Pharmacol. Sci. 6*: 214–218.

R. Melzack (1973). *The Puzzle of Pain*. Basic Books, New York.

J. S. Morley (1980) Structure–activity relationships of enkephalin-like peptides. *Annu. Rev. Pharmacol. Toxicol. 20*: 81–110.

G. W. Pasternak and P. J. Wood (1986). Multiple mu opiate receptors. *Life Sci. 38*: 1889–1898.

J.-C. Schwartz, J. Costentin, and J.-M. Lecomte (1985). Pharmacology of enkephalinase inhibitors. *Trends Pharmacol. Sci. 6*: 472–476.

E. J. Simon (1981). Opiate receptors: some recent developments *Trends Pharmacol. Sci. 2*: 155–158.

T. W. Smith and S. Wilkinson (1982). The chemistry and pharmacology of opioid peptides. In: *Chemical Regulation of Biological Mechanisms* (A. M. Creighton and S. Turner, Eds.). The Royal Society of Chemistry, London.

S. H. Snyder (1977). The opiate receptor. *Sci. Am. 236*(3): 44–56.

L. Terenius (1978). Endogenous peptides and analgesia. *Annu. Rev. Pharmacol. Toxicol. 18*: 189–204.

M. Wüster, R. Schulz, and A. Herz (1985). Opioid tolerance and dependence: re-evaluating the unitary hypothesis. *Trends Pharmacol. Sci. 6*: 64–67.

## 4. PROSTAGLANDINS AND THROMBOXANES

In 1934, von Euler in Sweden discovered a group of polyunsaturated fatty acids which had a powerful effect on smooth muscle and blood pressure. They were isolated from seminal fluid, seminal vesicles, and the prostate, and were named prostaglandins. Their structure was elucidated by the Samuelsson group (cf. Samuelsson et al., 1978) in Stockholm who, in 1975, also discovered even more potent fatty acid metabolites, the thromboxanes and prostacyclin. The effect of these compounds on blood platelet aggregation and the contraction of the vascular wall connected them to the etiology of stroke and cardiovascular disease, and their structural analysis and synthesis had an inherent challenge. Additionally, it was discovered that both steroidal and nonsteroidal antiinflammatory agents act through the prostaglandin system, adding further impetus to the research in this area. Consequently, the field became a center of such intense interest in the past decade that recent developments are almost impossible to review. Prostaglandins, thromboxane, and the leukotrienes are collectively called *eicosanoids*, since they are all derived from the C-20 fatty acid, arachidonic acid [*eicosa* (Gr.) = twenty].

## 4.1. Prostanoids

### 4.1.1. Structure and Biosynthesis

The structure and biosynthesis of prostaglandins and thromboxanes are shown in Fig. 5.25. Arachidonic acid, obtained from its phospholipid form in just about any tissue by the action of phospholipase A on phospholipids, is cyclized to **prostaglandin endoperoxide** in the form of PGG, (a side-chain peroxide), from which $PGH_2$ (a side-chain hydroxyl) is obtained. **Interleukin-1**, a polypeptide produced by leukocytes and possessing multiple immunological roles (Dinarello, 1984), mediates inflammation by increasing phospholipase activity, and thus prostaglandin synthesis. The first reaction is catalyzed by PG cyclooxygenase in the presence of $O_2$ and heme; thus the cyclooxygenase has been recognized as a cytochrome P-450 type enzyme (Ullrich and Graf, 1984; see Chap. 7, Sec. 2.1). The second reaction requires tryptophan, probably as a source of electrons. Prostaglandin cyclooxygenase has been purified to homogeneity and shows some peculiar characteristics, such as self-destruction (see Gorman, 1978).

The endoperoxide $PGH_2$ then undergoes a variety of rapid changes. It can be isomerized to the various ketol derivatives, the "*primary*" *prostaglandins* designated $PGD_2$, $PGE_2$, and $PGF_2$. The endoperoxide is also transformed into the extremely unstable and potent **thromboxane $A_2$** ($TXA_2$) in blood platelets (thrombocytes). This compound has a half-life of only about 30 seconds, and its isolation and characterization were therefore an experimental tour de force of the Samuelsson group. Thromboxane $A_2$ is rapidly inactivated to the stable but inactive $TXB_2$.

Another and equally important substance produced from the endoperoxide is **prostacyclin** ($PGI_2$), synthesized in the walls of blood vessels. It has an additional tetrahydrofuran ring, which is easily opened and deactivated. Prostacyclin has a half-life of less than 10 minutes.

Prostaglandins, prostacyclin, and thromboxane are considered local (i.e., paracrine) hormones or "autocoids," synthesized in many different organs and acting locally. They are not stored like neurotransmitters or conventional hormones, but are continuously synthesized and released immediately into the circulation, where they are usually deactivated after only one passage through the lungs. The synthesis depends on the availability of the starting material, arachidonic acid, and is modulated by cAMP. Prostaglandin A, PGB, and PGC are inactive degradation products.

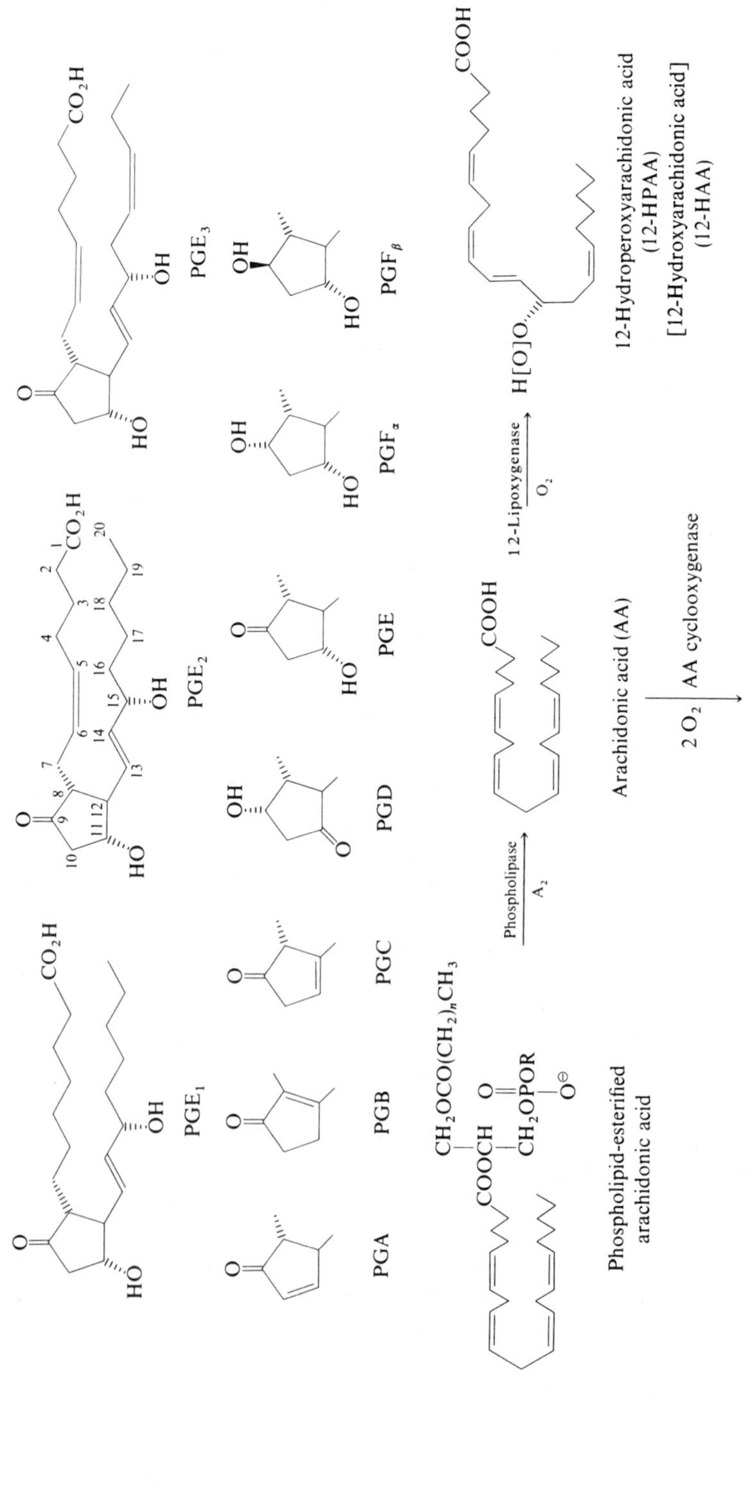

CO$_2$H

PGE$_3$

PGF$_\beta$

PGF$_\alpha$

CO$_2$H

PGE$_2$

PGE

PGD

PGC

PGB

PGA

CO$_2$H

PGE$_1$

COOH

H[O]O

12-Hydroperoxyarachidonic acid
(12-HPAA)
[12-Hydroxyarachidonic acid]
(12-HAA)

12-Lipoxygenase
O$_2$

Arachidonic acid (AA)

AA cyclooxygenase

2 O$_2$

Phospholipase
A$_2$

CH$_2$OCO(CH$_2$)$_n$CH$_3$

COOCH   O

CH$_2$OPOR

O$^\ominus$

Phospholipid-esterified
arachidonic acid

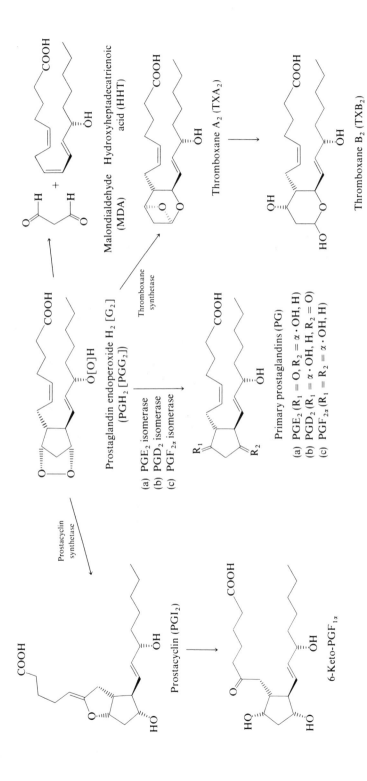

**Fig. 5.25.** The structure of various prostaglandin systems and the arachidonic acid cascade. (After Nicolaou and Smith, 1979)

Hydroxyheptadecatrienoic acid (HHT)

Malondialdehyde (MDA)

Thromboxane A₂ (TXA₂)

Thromboxane B₂ (TXB₂)

Prostaglandin endoperoxide H₂ [G₂] (PGH₂ [PGG₂])

(a) PGE₂ isomerase
(b) PGD₂ isomerase
(c) PGF₂ₐ isomerase

Thromboxane synthetase

Primary prostaglandins (PG)

(a) PGE₂ (R₁ = O, R₂ = α·OH, H)
(b) PGD₂ (R₁ = α·OH, H, R₂ = O)
(c) PGF₂ₐ (R₁ = R₂ = α·OH, H)

Prostacyclin synthetase

Prostacyclin (PGI₂)

6-Keto-PGF₁ₐ

329

### 4.1.2. Pharmacological Effects

The pharmacological effects of the prostaglandins and $TXA_2$ comprise many different activities—in fact too many. The lack of specificity of their activities implies a number of side effects which preclude the clinical application of several highly active natural prostaglandins, necessitating the development of selective synthetic compounds. The following effects of prostaglandins are known and are summarized in Table 5.12.

**Vasodilation and Constriction.** $PGE_2$ and especially $PGI_2$ (prostacyclin) are powerful, short-acting vasodilators, probably involved in blood pressure regulation. Prostaglandin $F_2$ and $TXA_2$, on the other hand, are potent vasoconstrictors.

**Blood Platelet Aggregation.** Blood platelet aggregation is an all-important mechanism in normal blood-clot (thrombus) formation, and therefore also highly significant in the pathophysiology of cardiovascular disease, stroke, coronary occlusion, and other circulatory catastrophes. Thromboxane $A_2$, formed in platelets, promotes their aggregation, whereas $PGI_2$ has the opposite effect, just as in their vaso-activity. It seems that a very efficient homeostatic control system exists: the endoperoxide $PGH_2$, a precursor of both compounds, is converted in the platelets to $TXA_2$, but is used to produce $PGI_2$ in the blood vessel wall, which does not have thromboxane synthetase. Prostacyclin dilates the vessel and increase the cAMP concentration, which in turn reverses the platelet aggregation caused by $TXA_2$ (which inhibits adenylate cyclase).

Recently, a **platelet activating factor** has been discovered (Venuti, 1985). It is a 1-alkyl-2-acetyl-phosphatidylcholine, produced by leukocytes during anaphylactic shock; it also appears to be an important mediator of inflammation and allergic responses. In concentrations as low as $10^{-11}$ M, it causes platelets to change their shape, aggregate, and release their granule content.

**Oxytocic Activity.** The oxytocic activity of prostaglandins is used clinically. Prostaglandin $E_2$ can induce labor at term in pregnant women, while $PGF_2$ and its methyl ester are used for terminating pregnancies in the second trimester when administered by the intrauterine (intraamniotic) route. The activity of $PGF_2$ and its ester is probably due to a direct effect on uterine muscle, since in late pregnancy, progesterone is already being produced by the placenta rather than by the *corpus luteum* (see Sec. 1.5). In earlier pregnancy, the $PGF_2$ causes abortion by luteolysis

**Table 5.12.** Pharmacological effects of prostaglandins and thromboxane

| Compound | Blood vessels | Platelets | Bronchi | Uterus |
|---|---|---|---|---|
| Prostaglandin $E_2$ | Dilation | | Dilation | Oxytocic |
| Prostacyclin ($PGI_2$) | Dilation | Aggregation inhibitor | | |
| Prostaglandin $F_2$ | Constriction | | Constriction | Oxytocic |
| Thromboxane ($TXA_2$) | Constriction | Aggregation | | |

and a decreased production of progesterone. Abortions initiated in this way are safe, but gastrointestinal side effects (vomiting, diarrhea) are not uncommon.

***Other Effects.*** Other effects of the prostaglandins include *bronchodilation* by the PGE series and *constriction* by $PGF_2$, as well as *antiulcer* and antisecretory effects of some synthetic analogues in the stomach. It has been suggested that alcohol facilitates arachidonic acid release and subsequent $PGE_2$ synthesis, and is indeed responsible for *hangover headaches.* The analgesic tolfenamic acid is alleged to have a preventive effect (Kaivola et al., 1983). Hope springs eternal . . .

### 4.1.3. Prostaglandin Receptors

Prostaglandin receptors have been demonstrated, even if the extent of binding does not always correlate with physiological activity. It has been proposed (Coleman et al., 1984) that each natural prostaglandin, as well as thromboxane, has its own receptor. Prostaglandin binding sites have been reported in adipocytes, the corpus luteum, blood platelets, and the uterus, skin, stomach, and liver (see Gorman, 1978; Samuelsson et al., 1978). The $K_D$ values for these vary from $10^{-8}$ to $10^{-11}$ M, indicating a high affinity; however, correlation of these values with the activation of adenylate cyclase is not very convincing. Prostaglandin receptors have also been found in plasma membranes of the *corpus luteum*, with a $K_D$ of $5-8 \times 10^{-8}$ M. The receptors have been isolated and partially characterized.

### 4.1.4. Structure–Activity Correlations

Study of the structure–activity correlations and synthetic modifications of prostaglandins and thromboxane are extremely active fields. Although a large number of analogues have been synthesized (Adaiken and Kottegoda, 1985), pharmacological information on these analogues is spotty.

*Expansion* or *contraction* of the *cyclopentane ring* or its replacement by heteroaromatic rings provide less active compounds in the $E_2$ and $F_2$ series. The replacement of the ring oxygen in prostacyclin gives more stable analogues, such as the **thia-PGI$_2$** (5-104), which shows about half of the platelet aggregation-inhibiting effect of $PGI_2$ but acts as a vasoconstrictor instead of a vasodilator. Analogues containing a nitrogen atom at the same position are very stable and good mimics of $PGI_2$. The aromatic **pyridazo analogue** (5-105) is an excellent vasodilator.

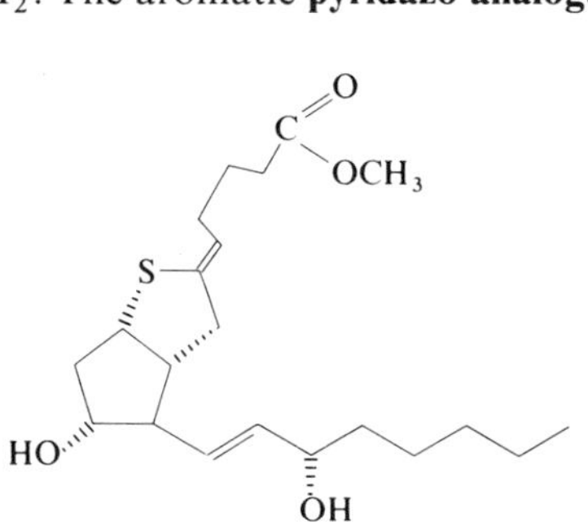

5-104 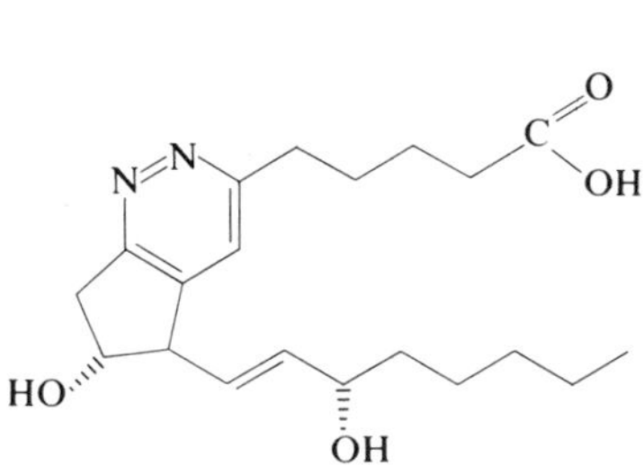                    5-105

The *carboxylic acid side chain* has also been modified extensively. More notable compounds are the **7-thia** and **7-oxa analogues**, which show antagonist activity in isolated smooth-muscle preparations. The **phenoxy derivative** (5-106) is 10 times more active than PGE as an inhibitor of platelet aggregation, but is a weak smooth-muscle spasmogen. The **sulfonamide** (5-107) is 10–30 times more active than $PGE_2$ as an antifertility agent, and has few side effects.

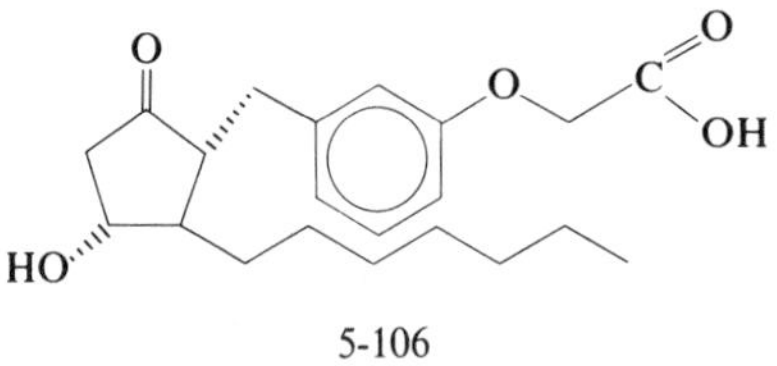

5-106

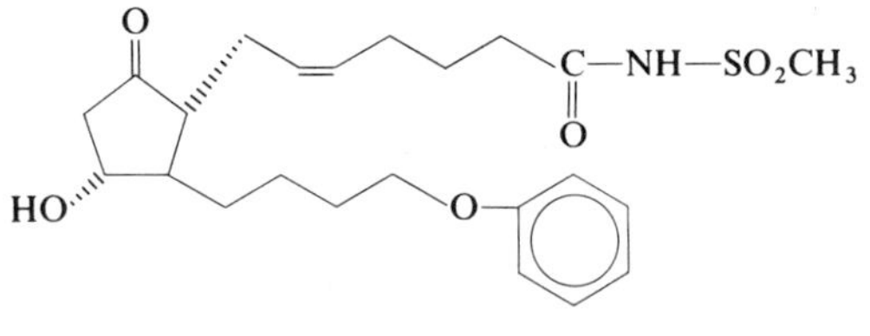

5-107

This compound has also undergone *ω-chain* (carbinol chain) modification.

Successful derivatives were mainly found among the 15- and 16-methyl $PGE_2$ analogues: **carboprost** (5-108) and compound (5-109) possess very high uterotonic activity and are used to induced abortions. They probably inhibit the dehydrase that inactivates the prostaglandins by removing the 15-OH group. **Nileprost** (5-110), a stable prostacyclin analogue, is an experimental antiulcer agent.

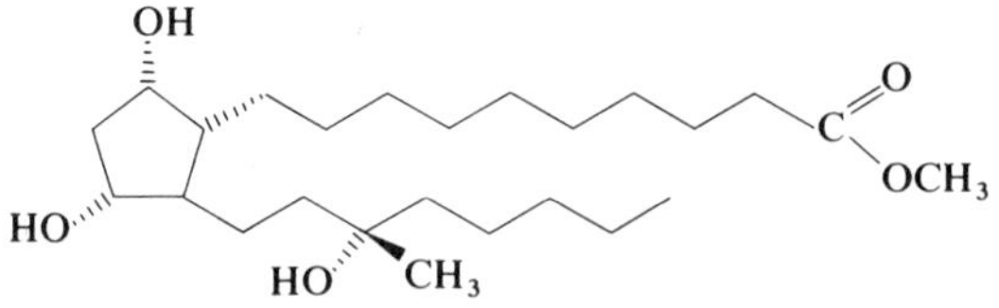

Carboprost
5-108

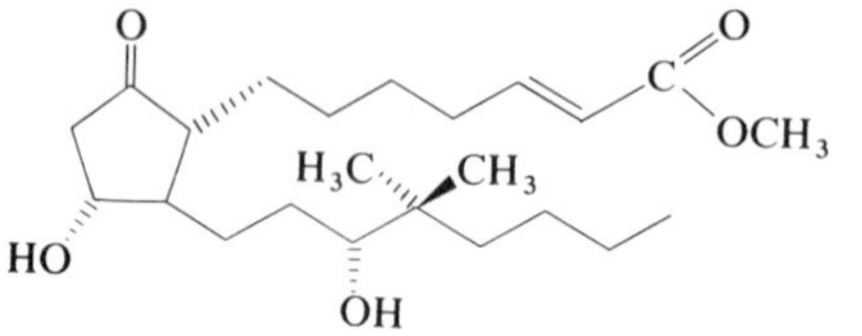

5-109

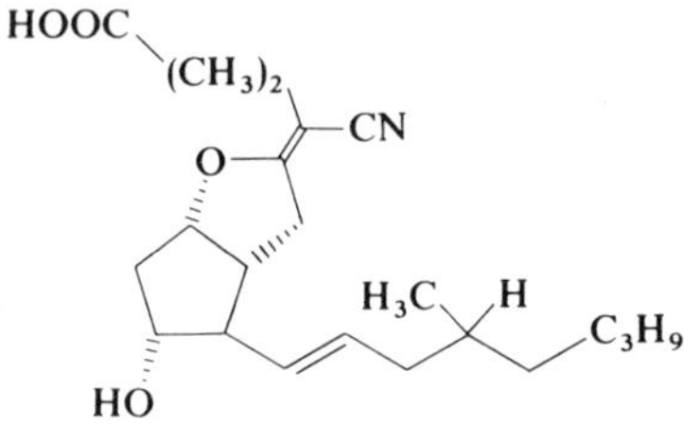

Nileprost
5-110

Among thromboxane analogues, **pinane-thromboxane** ($PTA_2$) (5-111) is a stable, selective thromboxane synthetase inhibitor, inhibiting neither cyclooxygenase nor prostacyclin synthetase. It could therefore become a very useful antithrombotic drug if further testing gives positive results. Compound (5-112) is a new experimental drug, also inhibiting thromboxane synthesis (Kucher and Rejholec, 1986).

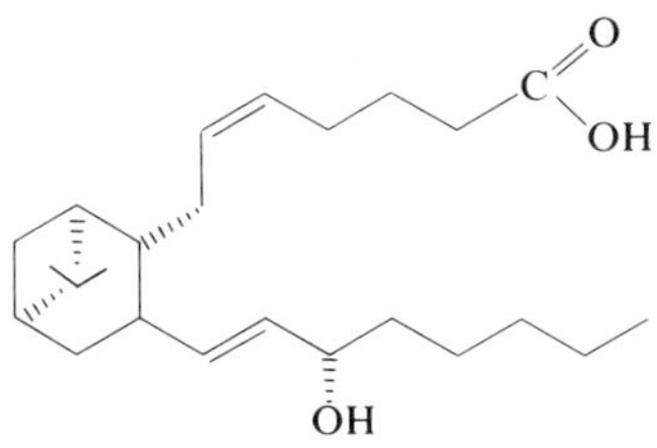

5-111 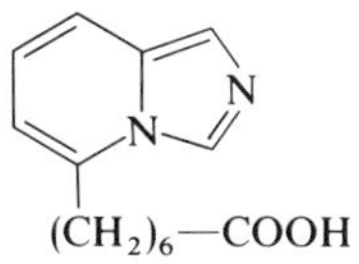                                      5-112

### 4.1.5. Leukotrienes

Leukotrienes (LT)—which were formerly called slow-reacting substance (SRS)—are compounds intimately involved in inflammation, anaphylactic reactions, allergic reactions, and asthma. Like the prostaglandins, they are lipids, also formed from arachidonic acid by lipoxygenase and glutathione. Their structure and biosynthesis were also elucidated by the Samuelsson group (Samuelsson, 1980), and are shown in Fig. 5.26. A review of these compounds was published by Kreuter and Siegel (1984), and a whole volume (Pike and Morton, 1985) is devoted to a comprehensive review of PGs and LTs. Since $LTC_4$ and $LTD_4$ are such important mediators of respiratory disorders, considerable research into LT receptor antagonists and LT biosynthesis inhibitors is under way (Gleason et al., 1986).

### 4.2. Antiinflammatory Agents and Minor Analgesics

Antiinflammatory agents are believed to act by disrupting the arachidonic acid cascade. These drugs are widely used for the treatment of minor pain and also for the management of edema and the tissue damage resulting from arthritis. Some of them are antipyretics (reduce fever) in addition to having analgesic and antiinflammatory actions.

The *adrenal steroids* (corticosteroids, Sec. 1.7) probably act by blocking phospholipase $A_2$, the enzyme that liberates arachidonic acid from phospholipids. These steroids also inhibit collagenase, an enzyme responsible for damage to the cartilaginous tissue in joints affected by arthritic diseases.

The *nonsteroidal* antiinflammatory agents block the cyclooxygenase that converts arachidonic acid to $PGG_2$ and $PGH_2$. Since the cyclic endoperoxides are the precursors of all prostaglandins, the synthesis of the latter is interrupted. Prostaglandin $E_1$ is known to be a potent pyrogen (fever-causing agent), and $PGE_2$ causes pain, edema, erythema (reddening of the skin), and fever. The prostaglandin endoperoxides ($PGG_2$ and $PGH_2$) can also produce pain, and inhibition of their synthesis can thus account for the action of the nonsteroidal antiinflammatory agents.

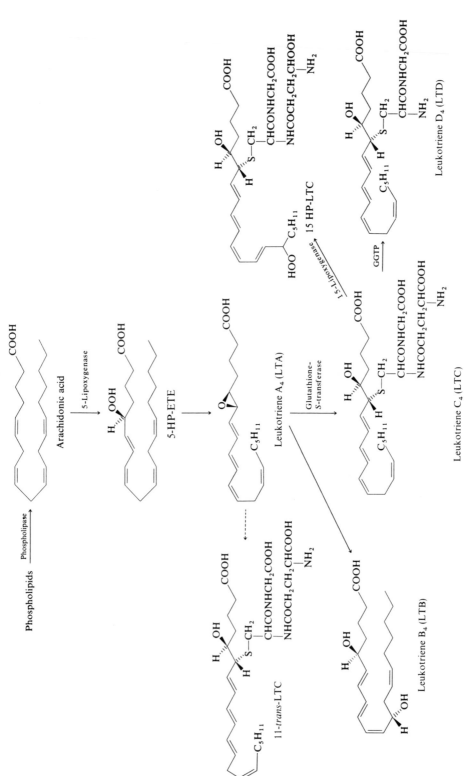

**Fig. 5.26.** Biosynthetic correlations of leukotrienes, the "slow reacting factors" in inflammation. (Reproduced by permission from Samuelsson (1980), Elsevier/North Holland Biomedical Press, Amsterdam)

It must be understood that this rationale is probably only one of several reasons for the activity of nonsteroidal antiinflammatory drugs, since some of these compounds (e.g., indomethacin) have no effect on cyclooxygenase. Facile correlations had to be abandoned as more was learned about prostaglandin activity: for example, inhibition of the vasoconstricting $TXA_2$ should really increase inflammation and erythema, not reduce it. Nevertheless, we have at least a partial explanation of the mode of activity of such extremely widely used drugs as aspirin.

An active site for prostaglandin cyclooxygenase, that could accommodate arachidonic acid as well as indomethacin and other carboxylic acids (Fig. 5.27), was derived through computer-assisted molecular modeling (Gund and Shen; see Gund et al., 1980). There is, however, no experimental evidence on the true structure of the cyclooxygenase active site, and the fit of indomethacin may be fortuitous, since it does not seem to be an inhibitor of this enzyme. Nevertheless, the attempt merits attention. For a review, see Salmon (1987).

Most of the nonsteroidal antiinflammatory drugs (Fig. 5.28) are carboxylic acids. **Aspirin** (acetylsalicylic acid, ASA) (5-113), has been used since the turn of the century to reduce pain and fever, but the parent compound, **salicylic acid**, has been known and used since antiquity, owing to its common occurrence as a glycoside in willow bark. Acetylation merely decreases its irritating effect. Among the numerous other salicylates known and used, **flufenisal** (5-114) has a longer duration of activity and fewer side effects than aspirin. **Mefenamic acid** (5-115) and **flufenamic acid** (5-116) are derivatives of anthranilic acid, while **ibuprofen** (5-117) and **naproxen** (5-118) are derivatives of phenylacetic and naphthylacetic acids, respectively.

Among indoles derivatives, **indomethacin** (5-119) is very widely used despite serious side effects. Its indene analogue **sulindac** (5-120), is a pro-drug, the active form being its —SH derivative. **Piroxicam** (5-121) is a long-lasting antirheumatoid agent but can have serious gastrointestinal side effects.

There is no clear-cut statistical evidence for the superiority of one or another of these useful drugs. Individual patients may do better with some than with others, and there are differences in side effects, primarily gastric bleeding and renal toxicity, which can be especially serious with the prolonged administration of high doses— necessary in chronic diseases like rheumatoid arthritis. Some of these compounds (e.g., **indoprofen**, 5-122) are powerful enough to be effective against the major pain caused by malignancies. The once widely used **phenylbutazone** derivatives have too many side effects and have fallen into disrepute.

The antiinflammatory and *antalgic* (analgesic) effect of these compounds is, however, symptomatic only. The remission (not cure) of rheumatoid or osteoarthritis requires corticosteroid treatment, often combined with the use of penicillamine, antimalarials, gold compounds (e.g., **auranofin**, 5-123), or immunosuppressive agents. There is an enormous literature on the subject, including detailed monographs (e.g., Hart and Huskinson, 1984; Lands, 1985; Rainsford, 1985) and the periodic reviews in the *Annual Reports in Medicinal Chemistry*, which provide details on new developments in this field.

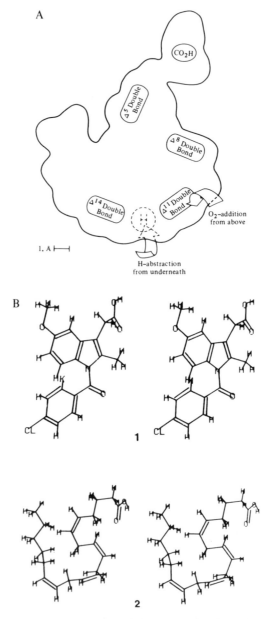

**Fig. 5.27.** (**A**) Hypothetical receptor map of the prostaglandin cyclooxygenase active site, showing points of binding. Indomethacin fits this site. (Reproduced by permission from Gund et al. (1977), *J. Med. Chem.* **20**: 1146, American Chemical Society, Washington, DC) (**B**) Stereoscopic view of indomethacin (1) and arachidonic acid (2). To obtain a three-dimensional image without a stereoscopic-viewer, cross your eyes to fuse the two images in the middle; then relax your eyes while staring at the fused image. In a few seconds you should be able to see a three-dimensional structure. (Reproduced by permission from Gund et al. (1980), AAAS, Washington, DC)

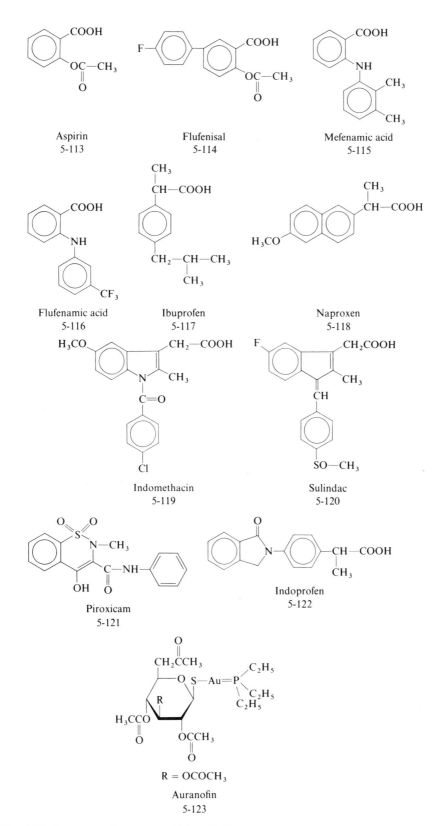

**Fig. 5.28.** Representative nonsteroidal antiinflammatory agents.

## Selected Readings

P. G. Adaikan and S. R. Kottegoda (1985). Prostacyclin analogs. *Drugs of the Future 10*: 765–774.

M. R. Bell, F. H. Batzold, and R. C. Winneker (1986). Chemical control of fertility. *Annu. Rep. Med. Chem. 21*: 169–177.

K. Brune and K. D. Rainsford (1980). New trends in the understanding and development of anti-inflammatory drugs. *Trends Pharmacol. Sci. 1*: 95–97.

D. A. Clark and A. Marfat (1982). Structure elucidation and the total synthesis of leukotrienes. *Annu. Rep. Med. Chem. 17*: 291–300.

R. A. Coleman, P. P. A. Humphrey, I. Kennedy, and P. Lumley (1984). Prostanoid receptors—the development of a working classification. *Trends Pharmacol. Sci. 5*: 303–306.

C. A. Dinarello (1984). Interleukin-1: an important mediator of inflammation. *Trends Pharmacol. Sci. 5*: 420–422.

J. G. Gleason, C. D. Perchonock, and T. J. Torphy (1986). Pulmonary and antiallergy agents. *Annu. Rep. Med. Chem. 21*: 73–83.

R. R. Gorman (1978). Prostaglandins, thromboxanes and prostacyclin. In: *Int. Rev. Biochem.* (H. V. Rickenberg, Ed.), Vol. 20. University Park Press, Baltimore, pp. 81–107.

P. Gund, J. D. Andose, J. B. Rhodes, and G. M. Smith (1980). Three-dimensional molecular modeling and drug design. *Science 208*: 1625–1731.

F. D. Hart and E. C. Huskinson (1984). Non-steroidal antiinflammatory drugs: current status and rational therapeutic use. *Drugs 27*: 232–255.

S. Kaivola, J. Paratainen, T. Österman, and H. Tiomonen (1983). *Cephalagia 3*: 31–36; as quoted in *Trends Pharmacol. Sci. 6*: 435 (1985).

W. Kreutner and M. I. Siegel (1984). Biology of leukotrienes. *Annu. Rep. Med. Chem. 19*: 241–251.

M. Kucher and V. Rejholec (1986). Antithrombotic agents. *Drugs of the Future 11*: 687–701.

W. E. M. Lands (1985). Mechanism of action of antiinflammatory drugs. *Adv. Drug Res.* (B. Testa, Ed.) Vol. 14. Academic Press, New York, pp. 147–164.

R. Nickander, F. G. McMahon, and A. S. Ridolfo (1979). Non-steroidal anti-inflammatory agents: *Annu. Rev. Pharmacol. Toxicol. 19*: 649–690.

K. C. Nicolaou and J. B. Smith (1979). Prostacyclin, thromboxane and the arachidonic acid cascade: *Annu. Rep. Med. Chem. 14*: 178–187.

J. E. Pike and D. R. Morton, Jr. (Eds.) (1985). *Advances in Prostaglandin, Thromboxane, and Leukotriene Research*, Vol. 14. Raven Press, New York.

K. D. Rainsford (1985). *Anti-inflammatory and Anti-rheumatic Drugs*, 3 vols. CRC Press, Boca Raton, FL.

J. A. Salmon (1987). Inhibition of prostaglandin, thromboxane and leukotriene biosynthesis. *Adv. Drug Res. 15*: 111–167. Academic Press, New York.

B. Samuelsson (1980). The leukotrienes: a new group of biologically active compounds including slow reacting substance. *Trends Pharmacol. Sci. 1*: 227–230.

B. Samuelsson, P. W. Ramwell, and R. Paoletti: (1976). *Advances in Prostaglandin and Thromboxane Research*, 18 vols. Raven Press, New York.

B. Samuelsson, M. Goldyne, E. Granström, M. Hamberg, S. Hammarström, and C. Malmensten (1978). Prostaglandins and thromboxanes. *Annu. Rev. Biochem. 47*: 997–1029.

V. Ullrich and H. Graf (1984). Prostacyclin and thromboxane synthase as P-450 enzymes. *Trends Pharmacol. Sci. 5*: 352–355.

M. C. Venuti (1985). Platelet-activating factor: multifaceted biochemical and physiological mediator. *Annu. Rep. Med. Chem. 20*: 193–202.

## 5. ATRIAL NATRIURETIC FACTORS

de Bold and his co-workers discovered in 1981 that extracts of heart atria (but not ventricles) cause a profound natriuresis, diuresis, and hypotension in rats. It was found that secretory granules in the atria contain a series of peptides responsible for these homeostatic regulatory effects, and that the heart is de facto an endocrine organ. These *atrial natriuretic factors* (ANF) or *atriopeptins* (AP) are derived from a prohormone consisting of 151 amino acids that is produced in response to atrial stretch, high blood volume, and high sodium concentration. The prohormone is subsequently modified and cleaved into shorter segments. Among the three principal peptides, AP III is the standard, containing 24 amino acids with the following sequence:

$$NH_2\text{-Ser—Ser—Cys—Phe—Gly—Gly—Arg—Ile—Asp—Arg}$$
$$|$$
$$S$$
$$|$$
$$S$$
$$|$$
$$HOOC\text{-Tyr—Arg—Phe—Ser—Asn—Cys—Gly—Leu—Gly—Ser—Gln—Ala—Gly—Ile}$$

AP III

All other APs differ in the composition of their termini. The extension of the amine terminal increases the natriuretic effect, but not renal vasodilation, indicating a heterogeneity of receptors in vascular walls and kidney tubules (Needleman, 1986). The hypotensive effect is, however, not the result of diuresis, but rather of hemo-dynamic effects, even though the APs are the most potent diuretics known (see also Chap. 6, Sec. 3.4). The vasodilatory effect seems to be restricted to the vascular bed in the kidneys only.

In addition to their direct effects on vascular muscle and kidneys, APs also influence endocrine systems, most notably causing a reduction of angiotensin II-induced aldosterone secretion, which is obviously involved in the hypotensive effect. ACTH production is also decreased (Lappe and Wendt, 1986).

*AP receptors* have been isolated and characterized in a preliminary fashion (Vandlen et al., 1986). The possible mechanism underlying the vasorelaxant effect has been postulated to consist of (1) production of cGMP, (2) activation of kinase, (3) inhibition of $Ca^{2+}$ translocation through agonist- or receptor-operated Ca channels, (4) enhanced $Ca^{2+}$ extrusion by Ca-ATPase in the sarcolemma, or (5) interference with $Ca^{2+}$ release from intracellular stores (Winquist, 1986).

The intense interest in these peptides may lead to the discovery of more active and stable analogues for the treatement of hypertension, and also edema in congestive heart failure or renal insufficiently.

### *Selected Readings*

A. J. de Bold, H. B. Borenstein, A. T. Veress, and H. Sonnenberg (1981). A rapid and potent natriuretic response to intravenous injection of atrial myocardial extracts in rats. *Life Sci. 28*: 89–94.

M. Cantin and J. Genest (1986). The heart as an endocrine gland. *Sci. Am. 254*(2): 76–81.

R. W. Lappe and R. L. Wendt (1986). Atrial natriuretic factor. *Annu. Rep. Med. Chem. 21*: 273–281.

P. Needleman (1986). Atriopeptin biochemical pharmacology. *Fed. Proc. 45*: 2096–2100.

R. L. Vandlen, K. E. Arcuri, L. Hupe, M. E. Keegan, and M. A. Napier (1986). Molecular characteristics of receptors for atrial natriuretic factor. *Fed. Proc. 45*: 2366–2370.

R. J. Winquist (1986). Possible mechanisms underlying the vasorelaxant response of atrial natriuretic factor. *Fed. Proc. 45*: 2371–2375.

# 6

# Nonmessenger Targets for Drug Action

## 1. EXCITABLE MEMBRANES

Excitable membranes show changes in ion permeability upon electrical stimulation, whereas most other membranes (plasma membranes, the membranes of cell organelles, etc.) change their ion conductance only upon chemical stimulation.

Our present ideas about the nature of biological membranes—which are so fundamental to all physiological processes—are based on the Singer–Nicholson mosaic model, well known from basic biochemistry. A composite picture of a cell membrane is shown in Fig. 6.1. The Singer–Nicholson model of the membrane is based on a phospholipid bilayer that is, however, asymmetrical. The outside monolayer contains *phosphatidylcholine* (lecithin) only, whereas the inner monolayer on the cytoplasmic side consists of phosphatidylethanolamine, phosphatidylserine, and phosphatidylinositol. *Cholesterol* molecules are also inserted into the bilayer, with their 3-hydroxyl group pointed toward the aqueous side, and their number approximately equals that of phospholipids. The hydrophobic fatty acid tails and the steran skeleton of cholesterol form the inner, hydrophobic part of the lipid bilayer, which behaves as a liquid. Cholesterol serves to make the bilayer more rigid and less permeable. There are also *glycolipids* in the outer monolayer, but their purpose is unknown.

A large number of protein molecules are embedded in the lipid bilayer to a greater or lesser extent. There are some anchored only superficially onto an outer or inner monolayer, but there are also those spanning the entire width of the membrane. They include (1) simple *helical proteins*, like *glycophorin* of red blood cells, and many receptors in the plasma membrane; and (2) *globulins*, like the complex multisubunit ionophores (e.g., the cholinoceptors; see Chap. 4, Sec. 2.2), and enzymes. They are anchored in the hydrophobic interior of the bilayer by stretches of apolar amino acids that form hydrophobic bonds with the lipid hydrocarbon chains, but their hydrophilic parts protrude into the outer and inner aqueous phase and serve as the major communication link between cells. Most proteins carry oligosaccharides on their outer surfaces and are therefore called glycoproteins; the oligosaccharide

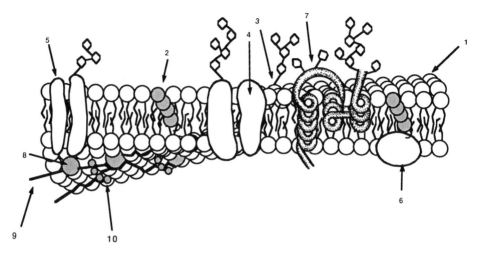

**Fig. 6.1.** A composite illustration of a plasma membrane, showing a lipid bilayer which is composed of phospholipid molecules (1), the rigid molecule, cholesterol (2), which stabilizes the tails of the phospholipids and keeps them relatively organized. Some glycolipids (3) carry oligosaccharides on the outside of the bilayer. Proteins can span the entire width of the lipid bilayer, as do the globulin (4) or band 3 protein (5) ionophores; or be situated on only one side of the membrane, like many enzymes (6); or be contained within or span the membrane, as helices, like the sugar carrier, glycophorin (7). Most of these membrane proteins are glycoproteins, carrying oligosaccharides. On the underside (inside) of the membrane, ankyrine (8) on band 3 proteins anchors filamentous spectrin (9), which imparts both rigidity and flexibility to membranes that are subjected to continuous flexing (e.g., erythrocyte membranes). Very short actin chains (10) anchor the long spectrin molecules. Not all of these components are present in all types of membranes.

"antennae" often serve as a recognition structure—for example, erythrocyte blood type factors.

There are also a number of other structures involved in cell–cell communication: for example, *gap junctions*, consisting of six protein helices, are present on each adjacent membrane of communicating cells. When twisted open, these protein helices permit unhindered passage of relatively large (i.e., 200-nm diameter) molecules.

Another transport structure, used for the passage of large molecules such as hormone–receptor complexes, is the *coated pit*, discussed previously (Chap. 2, Sec. 5.3.6). The coat is formed by a network of *clathrin* molecules in a regular pattern. After binding of the ligand—and, frequently, clustering of ligand–receptor complexes—the coated pit deepens, invaginates, and undergoes endocytosis, to become a *coated vesicle*. The clathrin coat is shed, and the vesicle fuses with a preformed endosome, which releases the enclosed molecules, often to a lysosome. The receptors, still embedded in the internalized membrane, can then be recycled to the plasma membrane via vesicles pinched off the endosome. Alternately, internalized ligand–receptor complexes can be transported intact to the other end of the cell and the ligand released by exocytosis. This process is depicted in Fig. 2.10.

Cells that have to withstand severe deformation—such as red blood cells, which must squeeze through narrow capillaries—contain a scaffold directly below the plasma membrane (see Fig. 6.1). Large glycoprotein molecules, known as *band-3* proteins (indicating their electrophoretic position), hold small *ankyrin* molecules on the inside. Long filaments of *spectrin* connect the ankyrins, and short *actin* chains secure the crossover points of spectrin molecules like braces. This subsurface mesh reinforces the delicate plasma membrane, providing the toughness and flexibility required during the long life of erythrocytes. The mesh can also anchor enzymes, like protein kinases. Membrane structure is discussed in many excellent textbooks of biochemistry and in specialized monographs, such as those by Weissman and Claiborne (1975), Harrison and Lunt (1975), de Duve (1984), Chapman (1984), and reviews such as the superbly illustrated paper of Bretscher (1985).

Some drugs, such as general and local anesthetics and barbiturate hypnotics, interfere with neuronal conduction at the membrane level and exert their pharmacological effect by blocking ion conduction through the ionophore channel. Thus, they interfere with the effect of neurotransmitters in a noncompetitive fashion, since they do not block receptors. Many other interactions are also feasible (Goldstein, 1984).

### 1.1. Sodium and Potassium Channels of Neuronal Membranes

As discussed in the outline of neurophysiology in Chap. 4, Sec. 1, $Na^+$ and $K^+$ ions are transported independently across neuronal membranes. The discovery of the highly specific effects of some channel-blocking neurotoxins has helped to clarify our concept of these ion channels. **Tetrodotoxin** (TTX, 6-1) is found in the liver and ovaries of the puffer fishes and in the eggs of some amphibians (e.g., the newt *Taricha*). The potential dangers notwithstanding, the flesh of the puffer fish (Fugu fish) is a delicacy in Japan, prepared by licensed chefs. Yet despite their skills, there are quite a few losers every year in this gastronomic game of Russian roulette. **Saxitoxin** (STX, 6-2) is the toxin produced by marine dinoflagellates of the genera *Gonyaulax* and *Gymmodinium*, important members of the phytoplankton. Under certain environmental conditions they multiply explosively, causing "red tides." These algae are consumed by shellfish, which remain unaffected by the poison, but can produce extremely toxic effects known as paralytic shellfish poisoning in humans who consume as little as 1 mg of STX in the shellfish.

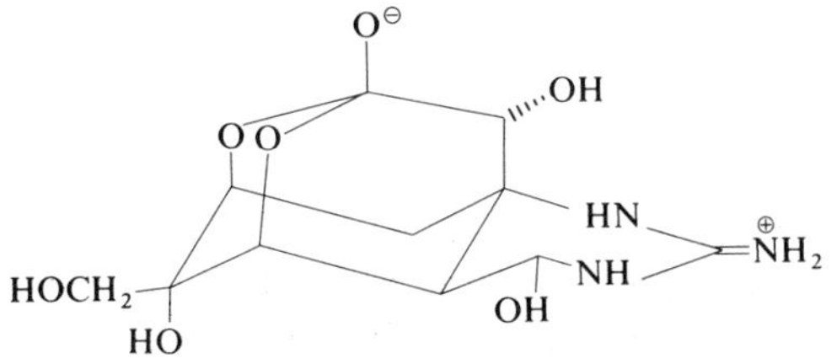

Tetrodotoxin
6-1

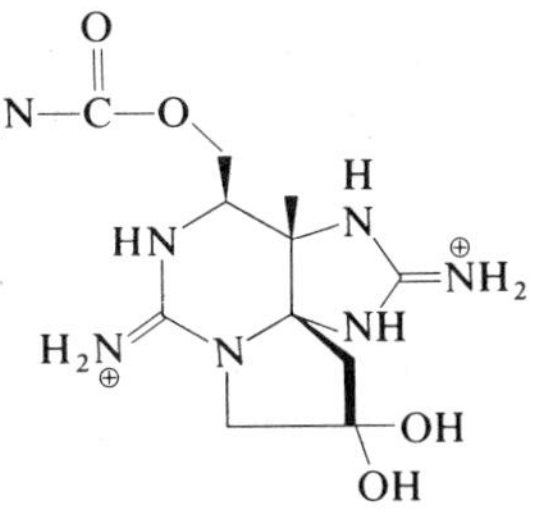

Saxitoxin
6-2

Both TTX and STX exert their effect by blocking the inward $Na^+$ current during neuronal depolarization, while not affecting the outward $K^+$ current. The toxins are effective only if applied from the outside; they are ineffective if perfused into the axon. Both toxins seem to be Na-channel-specific by virtue of their guanidinium groups, since guanidinium ions can pass through Na channels, and interact with the open ion channel only. The affinity of both toxins is very high, ranging from 2 to 8 nM.

The potassium channel, on the other hand, is blocked by both **tetraethyl-ammonium** salts (TEA) and **nonyl-triethyl ammonium** salts indicating the presence of a hydrophobic binding site that accommodates the nonyl group. Both blocking agents must be applied intraaxonally, which is understandable if one considers that the K current is always directed outward.

The ion channels are the key elements in neuronal signal propagation; their functional role, elucidated primarily by Hille (1984), is discussed in Chap. 4, Sec. 2.2.

### 1.1.1. Characterization of the $Na^+$ Channel

The voltage-sensitive $Na^+$ channel has been purified and reinserted into artificial membranes (Noda et al., 1984). It has been proposed (see Catterall, 1985) that the channel is a six-segmented protein, of which four segments span the axon membrane. A sequence of 200 amino acids between domains 2 and 3 has been shown to have four clusters of negatively charged side chains (carboxylates), acting in concert with domains of positively charged residues (ammonium and guanidinium ions of lysine and arginine), to provide the gating charges. There are two "*gates*": the M gate in the middle, and the H gate at the inner opening. Both gates must be open to allow passage of the ion. In a depolarized membrane, the channel's M-gate is closed; whereas a depolarized channel that is also refractory and inactive has the H-gate closed—a different conformation. The *ion-selective filter* is 0.3–0.5 nm in diameter, and consists of an ionized acidic group that attracts the cations but repels anions, and lowers the free energy required to dehydrate the $Na^+$ ion, a necessary criterion of passage. The oxygen lining of the selective slit substitutes for the lost water of hydration of the ion during transit. Direct evidence for the validity of this model is that a nucleophile (e.g., an amine or alcohol), in the presence of a condensing agent such as a carbodiimide, abolishes Na transport or TTX binding owing to amide or ester formation with the carboxylate anions of the acidic groups. Both TTX and STX slide into the open ion-slit with their guanidinium group and block it with the rest of the molecule, binding to the edges of the channel. The model of this process is shown in Fig. 6.2.

### 1.1.2. Characterization of the $K^+$ Channel

Several $K^+$ channels have been identified, responding to different stimuli: voltage activation, intracellular $Ca^{2+}$ ions, or neurotransmitters. Some of these ion channels can be permanently modified by cAMP-dependent protein kinases (Reichardt and Kelly, 1983).

The $K^+$-selective filter has an entrance pore of about 0.8 nm, which is the size of a

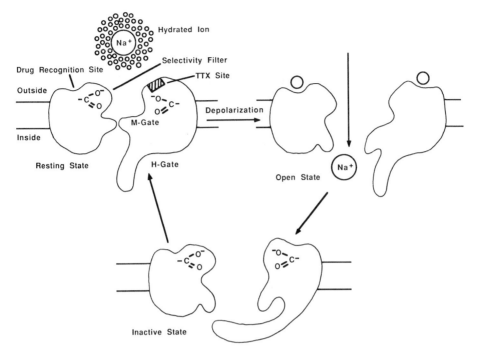

**Fig. 6.2.** Model of an ion-selective Na channel. In the resting state, the M gate is closed but the H gate (inactivation gate) is open. On activation, both the M and H gates open, the $Na^+$ ion is dehydrated (i.e., loses its water of hydration), passes the selectivity filter, and crosses the channel. On depolarization, the H gate closes and prevents passage of ions already in the ion channel. This stage then reverts to the other depolarized (resting) state.

hydrated $K^+$ ion or an ammonium group. However, the selectivity gate or channel is only 0.4 nm in diameter, forcing the $K^+$ ion to dehydrate—which the ammonium ion cannot do. Upon opening of the M-gate due to a decrease in the membrane potential (i.e., depolarization), the ions are swept in, and the K-current will flow until the nonyl-trimethylammonium ion blocks it, with the long alkyl chain of the latter ion becoming buried in the lipid layer. Shortening of the alkyl chain decreases the effect. The blockage is reversible, and the ion channel becomes free after it closes down; that is, the nonyl-trimethylammonium ion leaves and K-current flows again.

It is significant that TTX and TEA are effective only on the electrically excitable neuronal membrane; the chemically activated neuromuscular endplates are insensitive to these agents, indicating that the endplate and membrane differ basically in their ion-channel structure.

### 1.2. Calcium Channels and Their Agonists and Antagonists

The contraction of cardiac muscle is based on the interaction of the proteins actin and myosin, which converts the energy of ATP into mechanical work. ATP hydrolysis is mediated by the enzyme adenosine triphosphatase (ATPase; see

Sec. 3.3), which requires the binding of $Ca^{2+}$ ions to regulatory proteins, the tropomyosin–troponin complex. When $Ca^{2+}$ is pumped out of the cytosol, contraction ceases and the muscle relaxes. The cardiac troponin complex is a substrate for a cAMP-dependent protein kinase, leading to desensitization that counteracts catecholaminergic stimulation. The uptake of $Ca^{2+}$ from the cytosol into the sarcoplasmic reticulum of cardiac muscle is regulated by an ATP-dependent $Ca^{2+}$ pump, which allows for Ca pooling until $Ca^{2+}$ ions are needed again.

During the excitation–contraction coupling of heart muscle there is (1) a *fast influx* of $Na^+$ ions, causing a rapid action potential; and (2) a *slow influx* of $Ca^{2+}$, causing a plateau phase of the action potential. The slow Ca channel causes depolarization and a rise in $Ca^{2+}$ concentration, and triggers $Ca^{2+}$ release from the endoplasmic reticulum. In addition, there is also a 3 $Na^+$:1 $Ca^{2+}$ port–antiport system, which moves three positive charges out of the cell for each $Ca^{2+}$; the resulting negative inside charge of the resting cell therefore favors Ca efflux.

Unfortunately, little is known about the structure of the ion channel, but it is believed to be similar to the channel model shown in Fig. 6.2.

### 1.2.1. Ca Channel Antagonists

The great interest in Ca channel antagonists as therapeutic agents is due to the recognition of Ca as a major regulator of smooth muscle contraction. Thus these drugs are useful as

1. *Antiarrythmics* (see Sec. 1.4), regulating the timing of heart muscle contraction
2. *Hypotensives*, relaxing heart muscle
3. *Antianginal agents*, counteracting the painful ischemic contraction of coronary arteries

Vasodilation may also be due to $\alpha_2$-adrenoceptor mediation. Ca blockers also inhibit atheroma formation without decreasing serum cholesterol levels (Saini, 1984).

Currently, three groups of Ca blockers are distinguished (Spedding, 1985) (Fig. 6.3): (1) dihydropyridines (**nifedipine**, 6-3), (2) amines (**verapamil**, 6-4; **diltiazem**, 6-6), and (3) diphenylalkylamines (**fendiline**, 6-7; and analogues). The groupings are based on the demonstration of different binding sites. The dihydropyridines bind with high affinity ($K_D = 0.1-3$ nM) to vascular voltage-operated channels, but not to those in the myocardium. The structurally unrelated **verapamil** (6-4) reduces, whereas **diltiazem** (6-6) increases nifedipine binding on this site. Therefore, they must bind on a different, probably allosteric, site. The lipophilic diphenylalkylamines displace dihydropyridines completely, perhaps by a nonspecific mechanism.

Nifedipine is selective for vascular smooth muscle and is therefore an excellent hypotensive. However, it can cause tachycardia (i.e., an increase in heart rate), and is therefore prescribed with $\beta$ blockers. Verapamil and diltiazem have a direct effect on the heart, do not cause tachycardia, and are therefore the ideal antianginal agents. Diphenylalkylamines need a 1- to 2-week lag period until their antianginal effect is evident. **Flunarizine** (6-9) is a Ca blocker that is used in migraine

## Group I Antagonist

**Nifedipine**
**6-3**

## Group II Antagonists

**Verapamil (Gallopamil)**
**6-4**

**Diclofurime**
**6-5**

**Diltiazem**
**6-6**

## Group III Antagonists

**Fendiline**
**6-7**

**Cinnarizine**
**6-8**

**Flunarizine**
**6-9**

## Agonist

**BAY K 8644**
**6-10**

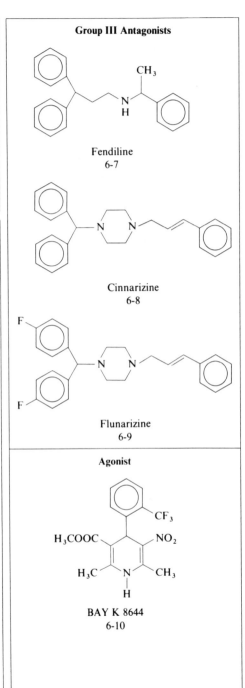

**Fig. 6.3.** Structures of some calcium antagonists and an agonist.

prophylaxis and also vertigo (dizziness). As its structure suggests, it is also an $H_1$ histamine antagonist. Structure–activity relations of all antagonist groups have been studied extensively by Meyer (1983) and summarized by Wehinger and Gross (1986).

### 1.2.2. Ca Channel Agonists

Recently, Ca channel agonists were found among dihydropyridines. **BAY K 8644** and **PN 202791** cause vasoconstriction and also positive inotropy (an increase in the force of contraction). The latter occurs because the increase in Ca influx prolongs the plateau phase of the cardiac action potential. Such compounds are potential drugs for the treatment of congestive heart failure (Chap. 6, Sec. 3.3). Both these compounds are extremely stereoselective: the $S$ enantiomers are agonists, whereas the $R$ enantiomers are antagonists of calcium channels (Triggle and Janis, 1987). As outlined in Chap. 1. Sec. 5.1, such opposite effects are unusual, because normally one enantiomer is the eutomer, the other one the inactive distomer, not an antagonist.

### 1.2.3. Calmodulin

The calcium in various tissues other than skeletal and cardiac muscle is bound to an activator protein, calmodulin. This 17-kD monomeric globulin has a high content (about 35%) of aspartate and glutamate, which participate in binding four Ca– atoms with a dissociation constant in the micromolar range. Calmodulin (CaM) is involved in the regulation of many enzymes (adenylate cyclase, phosphodiesterase, tryptophan hydroxylase, prostaglandin 15-hydroxydehydrogenase, and others), as well as many important processes (neurotransmitter release, endo- and exocytosis, DNA synthesis, α-adrenergic functions, etc.; see Mannhold, 1984). Although a very large number of drugs bind to CaM (neuroleptics, local anesthetics, Ca antagonists, antitumor agents), their specificity is restricted, and thus their usefulness as pharmacological agents acting through CaM is still questionable.

### 1.2.4. A General Classification of Channel Families

Sufficient information has emerged regarding the structure of channels that attempts are being made to classify them (Stevens, 1987). We have already seen that functional classification distinguishes two groups: (1) the ligand-gated channels of postsynaptic membranes which respond to neurotransmitters; and (2) voltage-gated channels of excitable membranes, which sense electric fields and open ion pores. The elucidation of the amino acid sequence of many channels (see Stevens, 1987, for review) now allows structural and even phylogenetic comparisons. The nicotinic AChR Na-channel (Chap. 4, Sec. 2.2), the $GABA_A$ ionophore (Schofield et al., 1987; see Chap. 4, Sec. 8.1), and the glycine receptor (Chap. 4, Sec. 8.2)—the latter two are both $Cl^-$ channels—show about 50% homology, and the subunits are closely related within a particular channel: they all have four transmembrane domains.

The other channel family is the voltage-gated ionophores. The sodium and calcium channels have four structurally similar regions consisting of six transmembrane domains, each containing an $(Arg-X-X)_7$ region, where X is a hydrophobic amino acid. This region is thought to be the voltage sensor that transforms electric fields into conformational changes. The potassium channel is much smaller, but part of it is 50% homologous with the voltage sensor of the Na and Ca channels. The K channel is believed to be the most ancient one and may have developed into the more modern structures by gene duplication.

### 1.3. Local Anesthetics

**Local anesthetics** block postsynaptic neural membranes in a noncompetitive, non-selective manner. This means that they do not inhibit acetylcholine binding and are not strictly structure-specific, or able to distinguish between neurons. Local anesthetics seem to act by two mechanisms, which may occur concurrently or separately:

1. The plugging of ion channels
2. The exertion of an allosteric effect on the ion channel

It seems that most local anesthetics act through the first mechanism. They plug both the $Na^+$ and $K^+$ channels and can act either from the outside or from the inside of the neuron. Ionized species (most local anesthetics are protonated tertiary amines) can act only from the outside, since they cannot penetrate the membrane. Their rapid onset of action also supports this model. Only ionized species can enter the open $Na^+$ channel and block it, converting it to the desensitized state; those with a lower $pK_a$ have a faster onset time (Concepcion and Covino, 1984). Nonionized compounds that can penetrate the membrane are capable of blocking from inside.

On the other hand, some local anesthetics, like **lidocaine** (6-15), have no effect on TTX binding, in contrast to what one would expect from the foregoing model. It might therefore be assumed that the gating mechanism and selectivity filter of the channel are independent entities: since TTX binds to the gate, lidocaine may act on the filter.

Quaternary anesthetics or, rather, quaternerized forms of the tertiary amines can enter the sodium as well as the potassium channel from the inside of the neuron only. They are inactive if applied externally, and since they cannot penetrate the membrane, their intraneuronal application is required in experimental situations. This is possible only when using the giant axon of the squid *Loligo*—a favorite tool of neurophysiologists—since it is over 1 mm in diameter. Normally, of course, the tertiary amines, with $pK_a$ values of around 7–9, exist in both ionized and un-ionized forms. Only the latter can cross the neuronal membrane, become ionized, and block the $Na^+$ or $K^+$ channel from the inside. Experiments with affinity labeling of the $K^+$ channel and TEA binding site (Hucho et al., 1979) (Fig. 6.4) support these findings. When the photoaffinity label of the $K^+$ channel is applied from the outside, it blocks the K current reversibly in the dark and irreversibly after irradiation. $Na^+$ conductance is not affected. If the label is applied from the inside, the initial blocking effect fades upon irradiation, indicating labeling of an axonal protein and not the channel, and resulting in subsequent removal of the blocking agent.

Other local anesthetics, like **procaine** (6-13), can act in an un-ionized form, perhaps by influencing shut channels. Blanchard et al. (1979) suggested that local

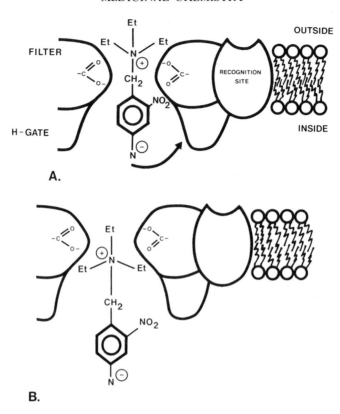

**Fig. 6.4.** Model of a potassium ion channel. In (**A**), a photoaffinity label is applied from the outside. It is capable of reacting covalently with the H gate (arrow), and blocks the K conductance. In (**B**), the same compound is applied from the inside, and the nitrene affinity label cannot react. The initial blocking effect will fade, indicating a lack of channel labeling. (After Hucho et al., 1979)

anesthetics, and especially those that are uncharged and cannot ionize, such as benzyl alcohol, interact with the lipid "annulus" or ring immediately surrounding the ionophore, and exert a direct influence on the channel proteins. The liquid crystalline state of this lipid prevents the relaxation of the sodium channel (e.g., allows it to open). The local anesthetic, by perturbing the lipid structure, triggers a fluidization of the lipid bilayer, allowing a conformational change in its protein constituents that results in "relaxation." This increases the energy barrier to ion translocation and effectively shuts the ion channel. Such a mechanism would be tantamount to an allosteric regulation, suggested as alternative (2) in the list on the preceding page.

The hypothesis outlined above is very similar to some contemporary concepts about the mode of action of *general anesthetics* and *barbiturates* (Chap. 1, Sec. 3). Polar aliphatic alcohols, including ethanol, increase the amplitude of miniature endplate potentials (MEPP)—a measure of ion-channel lifetime. The apolar octanol, ether, halothane, and chloroform have the opposite effect and shorten the channel lifetime. Since the effect of these general anesthetics is proportional to their partition coefficients, fluidization of the lipid phase and general distortion of the ion channels, resulting in their closure, is a plausible concept.

Barbiturates also reduce miniature endplate potentials. They can act as sleep-inducing hypnotics and as anticonvulsants (Chap. 4, Sec. 8.16) influencing GABAergic neurotransmission. Some are general intravenous anesthetics, such as **thiopental** (1-9) or **methohexital** (1-10, Fig. 1.3), but can also act as local anesthetics. However, unlike most local anesthetics, these agents act in their uncharged form, gaining access to the inside of neurons (an effect independent of the miniature endplate potential (MEPP). Since their activity *follows* that of applied pulses (electrical in the neuron, chemical at the endplate), it seems more likely that they block open ion channels rather than fluidize the lipid environment of the receptor–ionophore channel at the endplate or the ion channel of the neuronal membrane.

### 1.3.1. Structure–Activity Correlations

The structure–activity correlations of local anesthetics have their origin in the chemistry of the alkaloid **cocaine** (6-11), which has been known for centuries by South American Indians. Besides being a centrally acting reuptake inhibitor of norepinephrine, and therefore a psychotomimetic drug, cocaine has been used as a local anesthetic for almost a hundred years, on the recommendation of Sigmund Freud. Cocaine is an unstable drug that gives rise to side effects, and is a restricted narcotic (although it is physiologically nonaddictive). Analogues with more favorable properties have been synthesized, some of which are very widely used. Through this work, it has been realized that basic esters of aromatic carboxylic acids constitute the pharmacophoric group in cocaine and its analogues. Some of these analogues, like **benzocaine** (6-12), are useful only topically; the prototype of injectable local anesthetics, **procaine (novocaine, 6-13)**, was introduced in 1909. Its relatively short duration of action and low potency can be increased by administering it together with epinephrine, which, by virtue of its vasoconstricting effect, prevents the removal of procaine from the site of injection by the bloodstream and thus prolongs its effect. Among the many analogues of procaine, **tetracaine** (6-14) represents an advance, due to its higher lipid solubility.

The synthesis of **lidocaine** (6-15) in 1946 was a major breakthrough in the field of anesthetic research. Since this drug is an amide instead of an ester, it is stable in

Cocaine
6-11

Benzocaine
6-12

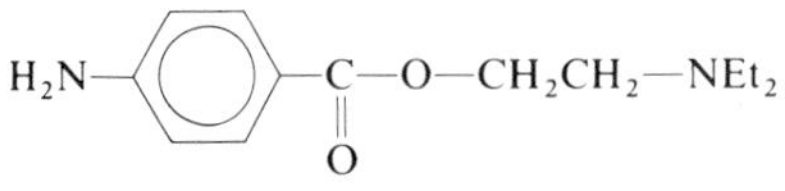

Procaine
6-13

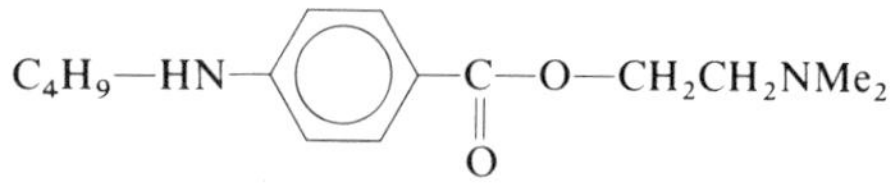

Tetracaine
6-14

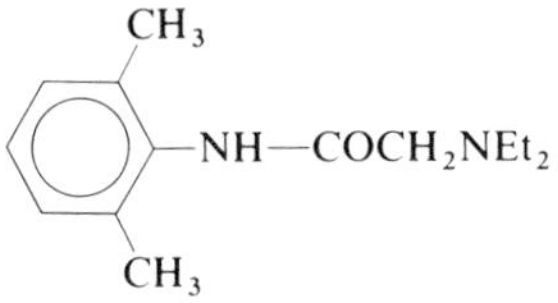

Lidocaine
6-15

aqueous solution, permitting its heat sterilization. It is more potent than the esters, less toxic, and generally a more versatile drug. There are numerous analogues and congeners of lidocaine in use, as well as some basic ethers and ketones, none of which, however, offers major advantages over the parent drug.

## 1.4. Antiarrhythmic Drugs

The electrophysiology of cardiac rhythm and its disorders is a complex topic beyond the confines of this book, and the reader is referred to pharmacology texts (e.g., Lucchesi and Lynch, in Craig and Stitzel, 1986, chap. 27). In a brief and superficial way, normal cardiac rhythm is maintained by:

1. Specialized cells in the *atrium* of the heart, the sinoatrial "pacemaker cells," which maintain *automaticity*—that is, the ability to alter resting membrane potential without an external stimulus
2. Fast and uniform conduction of electric impulses along a predetermined pathway
3. Uniform and long action potentials and refractory periods

If any of these prerequisites are not met, life-threatening arrhythmia leads to ventricular fibrillation (useless rapid twitching of the ventricles that do not pump blood), and sudden cardiac death ensues. The very large number of people dying this way suffer from chronic ischemic heart disease, and their myocardium is electrically unstable.

Antiarrhythmic drugs are classified according to their prevalent physiological mode of action, but many of them have mixed effects. Table 6.1 presents the classification, typical properties, and representative drugs whose structures are also given here or elsewhere in this book. Reiser and Sullivan (1986) and Steinberg et al. (1986) provide brief summaries.

### *Selected Readings*

S. G. Blanchard, J. Elliott, and M. A. Raftery (1979). Interactions of local anesthetics with *Torpedo californica* membrane-bound acetylcholine receptor. *Biochemistry* *18*: 5580–5585.

**Table 6.1.** Classification of antiarrhythmic agents

| Class | Dominant cellular properties | Prototypic drugs |
|---|---|---|
| I.A. | Decrease in rate of depolarization; prolonged duration of action potential | **Quinidine** (6-16) **Procainamide** (6-17) |
| I.B. | Little effect on depolarization; shorter duration of action potential | Lidocaine (6-15) Phenytoin (4-224) |
| I.C. | Decrease in action potential; no effect on action potential duration; great reduction of conduction velocity | **Flecainide** (6-18) |
| II. | $\beta$-Adrenergic blockers, with membrane stabilizing effect | Propranolol (4-91) Sotalol (4-85) |
| III. | Prolong duration of action potential; no effect on resting potential | Bretylium (4-53) **Amiodarone** (6-19) |
| IV. | Ca-channel blockers, which inhibit automaticity and slow-response action potential | Verapamil (6-4) Diltiazem (6-6) |

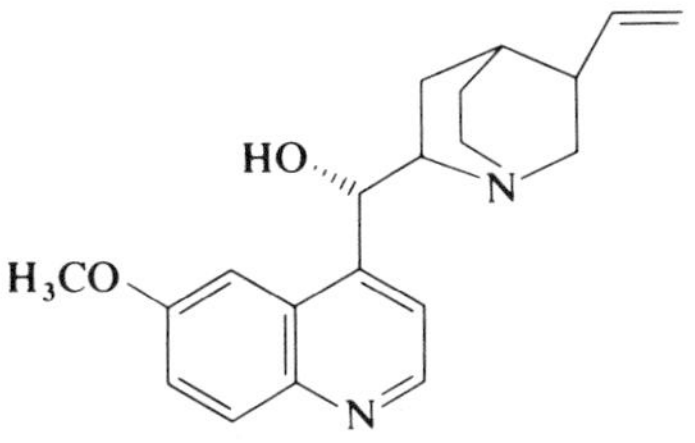

Quinidine
6-16

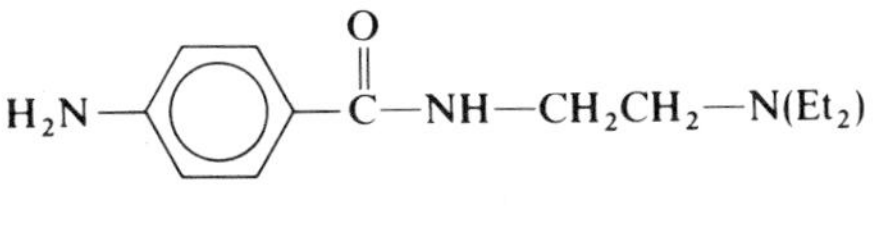

Procainamide
6-17

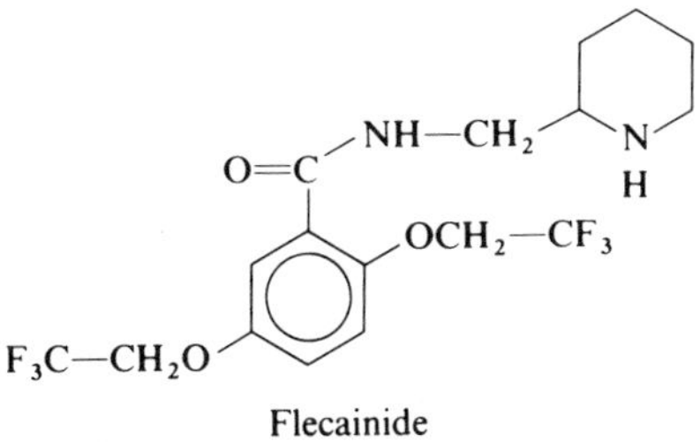

Flecainide
6-18

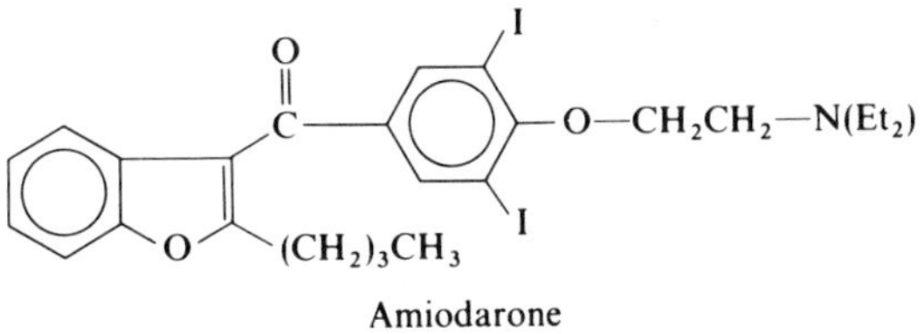

Amiodarone
6-19

M. S. Bretscher (1985). The molecules of the cell membrane. *Sci. Am. 253*(4): 100–108.

W. A. Catterall (1985). The electroplax sodium channel revealed. *Trends Neurosci. 8*: 39–41.

D. Chapman, (Ed.) (1984). *Biomembrane Structure and Function, Topics in Molecular and Structural Biology*, Vol. 4. Verlag Chemie, Weinheim.

D. Colquhoun (1979). The link between drug binding and response: theories and observations. In: *The Receptors* (R. D. O'Brien, Ed.), Vol. 1. Plenum Press, New York, pp. 93–142.

M. Concepcion and B. G. Covino (1984). Rational use of local anesthetics. *Drugs 27*: 256–270.

R. A. Cone and J. E. Dowling (Eds.) (1979). *Membrane Transduction Mechanisms*. Raven Press, New York.

C. R. Craig and R. E. Stitzel (Eds.) (1986). *Modern Pharmacology*, 2nd ed. Little, Brown, Boston.

C. de Duve (1984). *A Guided Tour of the Living Cell*. Scientific American Books, San Francisco.

H. Glossman, D. R. Ferry, A. Goll, J. Striessnig, and G. Zernig (1985). Calcium channels and calcium channel drugs: recent biochemical and biophysical findings. *Arzneimittelforschung 35*: 1917–1935.

To Godfraind, R. Miller, and M. Wibo (1986). Calcium antagonism and calcium entry blockade. *Pharmacol. Rev. 38*: 321–416.

D. B. Goldstein (1984). The effect of drugs on membrane fluidity. *Annu. Rev. Pharmacol. Toxicol. 24*: 43–64.

R. Harrison and G. G. Lunt (1975). *Biological Membranes*. Blackie, London.

B. Hille (1984). *Ionic Channels of Excitable Membranes*. Sinauer, Sunderland, MA.

F. Hucho, S. Stengelin, and G. Bandini (1979). Effector binding sites and ion channels in excitable membranes. In: *Recent Advances in Receptor Chemistry* (F. Gualtieri, M. Gianella, and C. Melchiorre, Eds.). Elsevier/North Holland, New York, pp. 37–58.

R. D. Keynes (1979). Ion channels in the nerve-cell membrane. *Sci. Am. 240* (3): 126–135.

R. Mannhold (1984). Calmodulin—structure, function and drug action. *Drugs of the Future 9*: 677–691.

H. Meyer (1983). Structure–activity relationships in calcium antagonists. In: *Calcium Antagonists and Cardiovascular Disease* (L. H. Opie, Ed.). Raven Press, New York.

M. Noda, S. Shimizu, T. Tanabe, T. Takai, T. Kayano, T. Ikeda, H. Takahashi, H. Nakayama, Y. Kanaoka, and N. Minamino (1984). Primary structure of *Electrophorus electricus* sodium channel deduced from cDNA sequence. *Nature 312*: 121–127.

E. Racker (1987). Structure, function and assembly of membrane proteins. *Science 235*: 959–961.

L. F. Reichardt and R. B. Kelly (1983). A molecular description of nerve terminal function. *Annu. Rev. Biochem. 52*: 871–926.

H. J. Reiser and M. E. Sullivan (1986). Antiarrhythmic drug therapy: new drugs and changing concepts. *Fed. Proc. 45*: 2206–2208.

J. M. Ritchie (1979). A pharmacological approach to the structure of sodium channels in myelinated axons. *Annu. Rev. Neurosci. 2*: 341–362.

S. H. Roth (1979). Physical mechanisms of anesthesia. *Annu. Rev. Pharmacol. Toxicol. 19*: 159–178.

R. K. Saini (1984). Calcium antagonists. In: *Cardiovascular Pharmacology* (N. Antonaccio, Ed.). Raven Press, New York.

R. R. Schofield, M. G. Darlison, N. Fujita, D. R. Burt, F. H. Stephenson, H. Rodrigues, L. M. Rhee, J. Ramachandran, V. Reale, T. A. Glencorse, P. H. Seeburg, and E. A. Barnard (1987). Sequence and functional expression of the GABA$_A$ receptor shows a ligand-gated receptor super-family. *Nature 328*: 221–227.

M. Spedding (1985). Calcium antagonist subgroups. *Trends Pharmacol. Sci. 6*: 109–114.

M. I. Steinberg, W. B. Lancefield, and D. W. Robinson (1986). Class I and III antiarrhythmic drugs. *Annu. Rep. Med. Chem. 21*: 95–108.

C. F. Stevens (1987). Channel families in the brain. *Nature 328*: 198–199.

D. J. Triggle and R. A. Janis (1987). Calcium channel ligands. *Annu. Rev. Pharmacol. Toxicol. 27*: 347–369.

E. Wehinger and R. Gross (1986). Calcium modulators. *Annu. Rep. Med. Chem. 21*: 85–94.

G. Weissman and R. Claiborne (Eds.) (1975). *Cell Membranes. Biochemistry, Cell Biology and Pathology*. H. P. Publishing, New York.

## 2. CELL WALL SYNTHESIS INHIBITORS

The successful chemotherapeutic management of any host–parasite interaction—whether viral, bacterial, or protozoan—or of invasive malignant cells depends upon the exploitation of biochemical differences between the host and parasite or tumor cells. The greater these differences are, the better the likelihood of finding or designing drugs that exploit them and inhibit some crucial function of the parasite in order to kill it without harming the host cell. This almost utopian goal (Paul Ehrlich's "magic bullet") has been approximated very closely in the case of cell wall synthesis inhibitors, such as antibacterial agents, for the simple reason that a very fundamental difference exists between bacteria and mammalian cells: the former have cell walls and the latter do not. The rigid cell wall of bacteria encloses and strengthens the vulnerable cell membrane, which is subjected to considerable internal osmotic pressure. If the integrity of the cell wall is impaired, the bacterial cell will undergo lysis and perish. The antibiotics that inhibit cell wall synthesis cannot find an analogous target in animal cells and are in most cases extremely nontoxic.

Cell walls are complex and variable structures but have a number of common characteristics, discussed in most biochemistry textbooks (e.g., Lehninger, 1982) and numerous monographs. The basic structural unit of the wall is the *muropeptide* [*murus* (Latin) = wall], a repeating disaccharide linked through a lactyl ether to a tetrapeptide. The peptides are, in turn, cross-linked (in *Staphylococcus aureus*) by a pentaglycine chain, as shown in Fig. 6.5. The resulting polymer, called *murein*, forms a closed sack around the bacterium and can be dissolved by the enzyme lysozyme. Other glycopeptides, such as teichoic acid and polypeptides, contribute to the antigenic properties of bacteria.

The classical division of bacteria into Gram-positive and Gram-negative groups on the basis of specific staining procedures also depends on cell wall components. The Gram-positive organisms have a rigid cell wall which is covered with an outer membrane (discussed below) containing teichoic acids, whereas the wall of Gram-negative bacteria is covered with a smooth, soft lipopolysaccharide. Most penicillins are much more effective against Gram-positive bacteria.

During the biosynthesis of the cell wall, the muropeptide is formed from UDP-acetylmuramyl-pentapeptide, which terminates in a D-alanyl-D-alanine (i.e., one alanine more than shown in the finished murein, on the left in Fig. 6.5. The synthesis of this precursor is inhibited by the antibiotic **cycloserine** (6-20), a compound produced by many *Streptomyces* fungi but which is not used clinically. During the cross-linking of the pentapeptide precursor, the terminal fifth alanine must be split off by a transpeptidase enzyme (Kelly et al., 1982). This last reaction in cell wall

Cycloserine
6-20

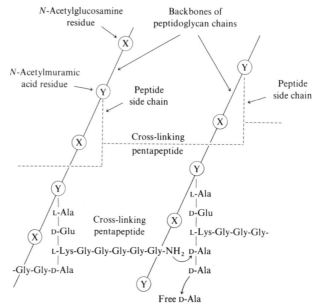

UDP-*N-acetylmuramylpentapeptide*. Note *the γ peptide linkage in the D-glutamic acid residue.*

Completion of a cross-link between two adjacent peptidoglycan chains in the bacterial cell wall. This reaction is blocked by penicillin.

**Fig. 6.5.** Bacterial cell-wall building blocks and their assembly into muropeptide, the cell wall polymer.

synthesis is inhibited by the β-lactam antibiotics, such as the **penicillins** and **cephalosporins**, after the bacterium has expended considerable biosynthetic energy. In contrast to this inhibition of the last step of a reaction sequence, the feedback inhibition of enzymatic reactions normally occurs at the first step in a sequence, avoiding any wastage of precursor substances; if it occurs late, biochemical efficiency is seriously jeopardized.

Recent investigations (see Gootz, 1985) have elucidated many details of this process. First, the antibiotic has to penetrate the outer membrane of the Gram-negative bacteria, which are less susceptible to antibiotics. This membrane consists of lipopolysaccharides, phospholipids, lipoproteins, and proteins. The $\beta$-lactam antibiotics (penicillins, cephalosporins) cross this diffusion-resistant membrane through *porin channels*, trimeric proteins that traverse the membrane. There are about $10^5$ channels per bacterial cell, and their diameter is 1.2 nm. Some bacterial genera (e.g., *Pseudomonas*) are insensitive to most $\beta$-lactam antibiotics because the majority of their porin channels are not functional. The next hurdle the antibiotic has to surmount involves the *$\beta$-lactamase* enzymes (see below) in the periplasmic space, between the outer and inner membranes (Fig. 6.6); these can deactivate the antibiotic (Gram-positive bacteria excrete the lactamase into the medium). Beyond that is the peptidoglycan *cell wall* with the associated *penicillin-binding proteins*, which are the essential transpeptidases, transglycosylases, and D-alanine carboxykinases involved in cell wall synthesis (Waxman and Strominger, 1983).

**Fig. 6.6.** Structure of the cell membrane of Gram-negative bacteria. The outer membrane consist of hydrophilic lipopolysaccharides (1) and the protein porin channels (2): the inner half is composed of phospholipids. The periplasmic space (3) contains the lactamase molecules (4) and the peptidoglycan cell wall (5). The inner membrane (6) is a normal membrane, consisting of phospholipids and protein, but on the outer bilayer surface, penicillin-binding protein molecules (7) are bound.

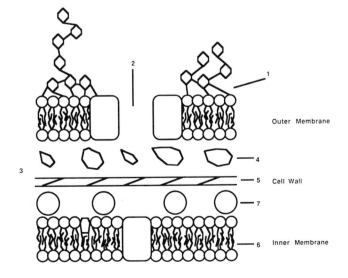

**Penicillin** inactivates these by acylation of the active sites, as a "suicide substrate" (see Sec. 3.1). The indicated hydrogen bonds are hypothetical. Different trans-peptidases have different roles in the cell, and their selective inactivation can lead to cell lysis, or production of deformed (spherical or threadlike) cells besides cell wall synthesis inhibition.

## 2.1. Penicillins

The penicillins (or penams) were discovered in 1929 by Sir Alexander Fleming, and developed by Florey, Chain, and Abraham at Oxford University. The history of penicillin became a story of legendary proportions, illustrating the case of a serendipitous discovery combined with brilliant development; it also marks the beginning of the modern chemotherapy of infectious diseases (Hare, 1970).

The penicillins produced by the molds *Penicillium notatum* and *P. chrysogenum* are shown in Table 6.2. Through the use of different culture media or biosynthetic precursors (e.g., phenylacetic acid), a number of biosynthetic ("natural") penicillins have been isolated, distinguished by Roman numerals in Britain and by letters in the United States. The most important and still widely used among these is **benzylpeni-cillin** or **penicillin G**, a singularly nontoxic compound highly active against Gram-positive infections such as staphylococcal sepsis, meningitis, and gonorrhea.

Structurally, the $\beta$-lactam ring fused with the thiazolidine ring is most unusual, since $\beta$-lactam rings were unknown before the discovery of penicillin. Consequently, the structure elucidation of penicillin during World War II, a top-secret joint Anglo-American project, was a difficult undertaking, ultimately settled by x-ray crystal-lography. The penam ring can be considered as a dipeptide composed of a cysteine

**Table 6.2.** Biosynthetic penicillins

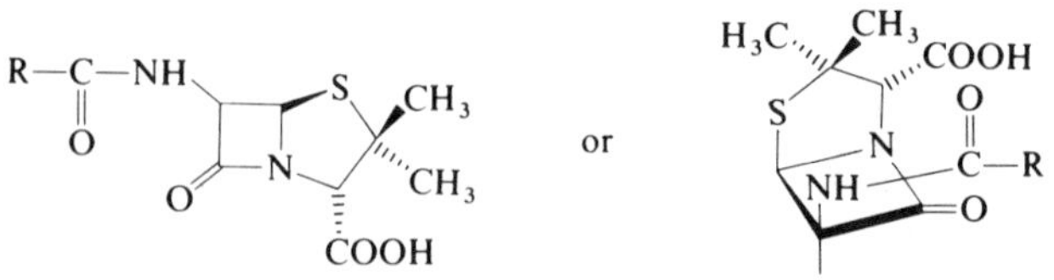

| Penicillin | R |
|---|---|
| I or F | $CH_3CH_2-CH=CH-CH_2-$ |
| II or G | 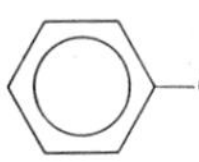 ⬡$-CH_2-$ |
| III or X | $HO-$⬡$-CH_2-$  |
| IV or K | $CH_3(CH_2)_6-$ |
| V | 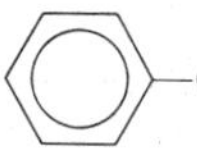 ⬡$-O-CH_2-$ |

and a valine residue:

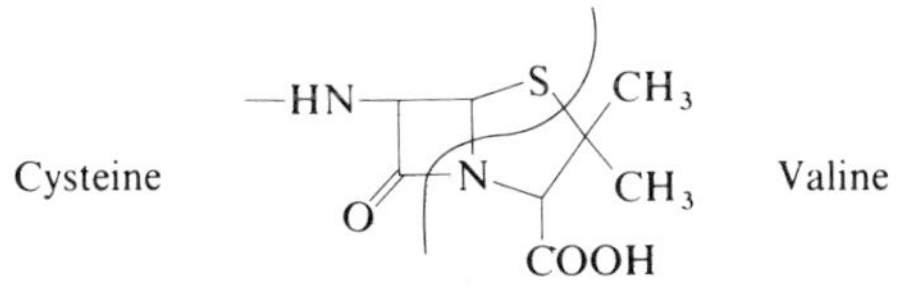

Owing to the strain of the four-membered $\beta$-lactam, the ring is easily cleaved by acid hydrolysis and alcoholysis, and by heavy metals such as $Zn^{2+}$, $Cu^{2+}$, and $Pb^{2+}$. The resulting penicilloic acid (6-21) is inactive, and undergoes a complex series of rearrangements (see Stenlake, 1979, Vol. 1, p. 560). The acid sensitivity of penicillins varies with their structure. For example, phenoxymethyl penicillin is more resistant to acid cleavage than benzylpenicillin, and is therefore more suitable for oral use. Even so, considerably higher peroral doses are required than parenteral ones. Only among the semisynthetic penicillins does one find good acid resistance. The high reactivity of the $\beta$-lactam ring is the key to the biological activity of the $\beta$-lactam antibiotics. It acts as an irreversible inhibitor of the bacterial transpeptidase because it acylates the enzyme protein near the active site through opening of the lactam ring.

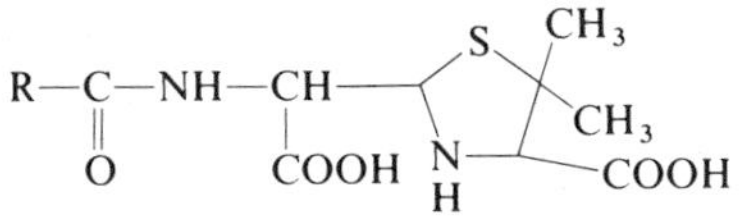

6-21

### 2.1.1. Mechanisms of Bacterial Resistance

The most serious threat to antibiotic therapy, and to the use of $\beta$-lactam antibiotics in particular, is the emergence of resistant bacterial strains. The primary reason for this resistance is the production of an enzyme, $\beta$-lactamase (penicillinase), which in Gram-positive bacteria is excreted into the growth medium, but in Gram-negative bacteria remains contained in the cell. Thus, Gram-positive organisms quickly destroy the antibiotic in the surrounding solution by hydrolysis, converting it to the inactive penicilloic acid. Since production of penicillinase enzymes is under R-plasmid control, resistant bacteria can transfer their resistance through mating. Hence, bacterial species that were in the past easily controlled with penicillin have increasingly become a serious medical problem.

Gram-negative bacteria can also produce acylase enzymes capable of cleaving the side-chain amide bond of penicillin.

*Inhibitors of $\beta$-lactamase* are known. The synthetic **sulfone** (6-22) and **clavulanic**

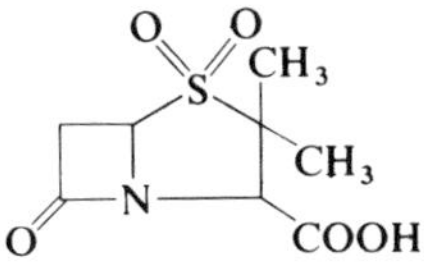

6-22

**acid** (6-25) both have weak antibacterial activity besides $\beta$-lactamase inhibitory activity, and can be used in combination with vulnerable antibiotics.

## 2.2. Semisynthetic Penicillins and Cephalosporins

However useful they may be, natural penicillins have several drawbacks. They have a relatively narrow activity "spectrum," primarily inhibiting Gram-positive bacteria only. They are acid- and lactamase-sensitive, and in a small percentage of patients cause allergic side effects. All of these limitations could potentially be overcome by molecular modifications during the biosynthesis of these drugs. Unfortunately, however, the fermentation process used in penicillin production is not very flexible and does not permit the incorporation of too many amide side chains into the molecule.

### 2.2.1. Semisynthetic Penicillins

Although the total synthesis of penicillins was accomplished by Sheehan and his co-workers in 1953, and although some other approaches have also been successful (Christensen and Radcliffe, 1976), the syntheses are of limited practical value; nevertheless, they do allow modification of the ring system.

On the other hand, the discovery of the parent amine **6-aminopenicillanic acid (6-APA**; 6-23) in fermentation products constituted a major breakthrough in penicillin synthesis. It is formed by acylases that cleave off the side chain of the penicillins, and can also be obtained by the selective chemical cleavage of the amide, leaving the lactam intact. After this, 6-APA can be easily acylated by any carboxylic acid, and this has yielded literally thousands of semisynthetic penicillins in the past 20 years, many showing improved stability and activity. Table 6.3 shows a number of these antibiotics now in clinical use. Some of them are lactamase resistant (**methicillin, oxacillin** and its halogenated derivatives), whereas others are broad-spectrum antibiotics, like the orally active **ampicillin**, which also inhibits Gram-negative bacteria but is sensitive to lactamase. **Carbenicillin** is particularly active against *Pseudomonas* and *Proteus* infections, which are unaffected by "natural" penicillins. **Piperacillin**, a broad-spectrum compound, is spectacularly active against *Pseudomonas*.

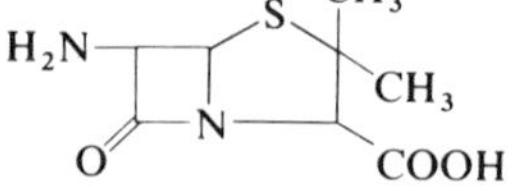

6-APA
6-23

### 2.2.2. Cephalosporins

The cephalosporins, discovered in the 1950s, are produced by various species of the mold *Cephalosporium*. **Cephalosporin C** (Table 6.4) is the prototype of these

**Table 6.3.** Some clinically used semisynthetic penicillins

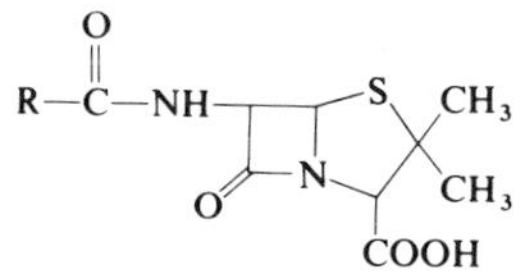

| R | Name | R | Name |
|---|---|---|---|
| 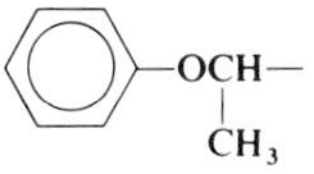 | Phenethicillin | 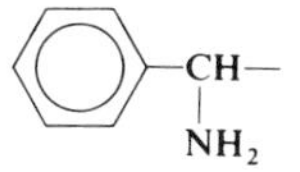 | Ampicillin |
| 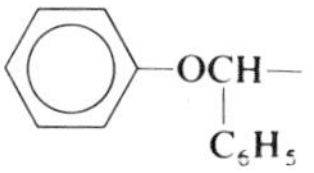 | Phenbenicillin | 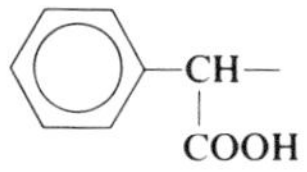 | Carbenicillin |
| 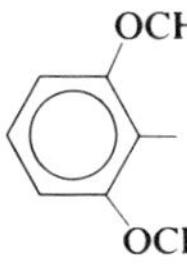 | Methicillin | 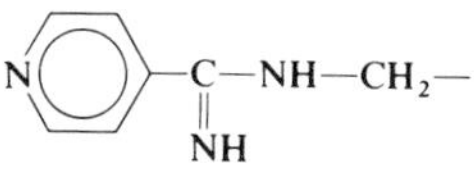 | Pirbenicillin |
| 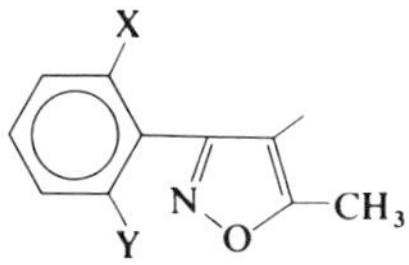  X=Y=H   Oxacillin | | | 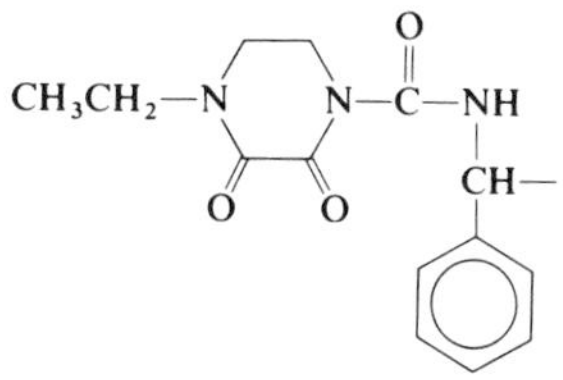 Piperacillin |
| X=H   Y=Cl   Cloxacillin | | | |
| X=Y=Cl   Dicloxacillin | | | |
| 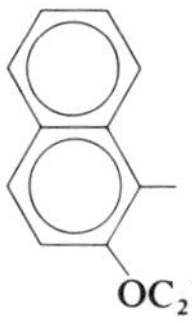 | Nafcillin | | |

antibiotics, and its structure shows a close similarity to the penam structure. The 5-thia-1-azabicyclo[4.2.0]octane ring system is therefore called the cepham ring. The parent compound carries the aminoadipate side chain, which can be cleaved to supply the 7-amino-cephalosporanic acid. This amine can easily be acylated and thus forms the basis of many useful derivatives. The 3-acetoxymethyl substituent is also amenable to modifications.

Since cephalosporin C is only one-thousandth as active as benzylpenicillin, its use is very limited. However, it is remarkably resistant to enzymatic hydrolysis, and becomes highly concentrated in the urine, which makes it useful in urinary tract infections caused by Gram-negative organisms.

Among the semisynthetic derivatives shown in Table 6.4, **cephalothin** is the most widely used, since it is a broad-spectrum antibiotic resistant to lactamase. Its main

**Table 6.4.** Some natural and semisynthetic cephalosporins

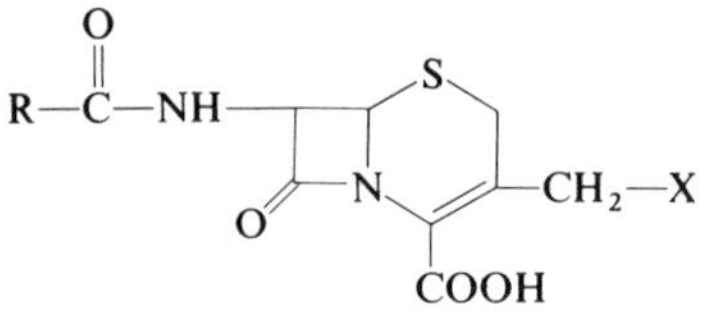

$$\text{R}-\overset{\displaystyle O}{\overset{\|}{\text{C}}}-\text{NH}-\underset{\underset{\text{COOH}}{}}{\boxed{\beta\text{-lactam / cephem nucleus}}}-\text{CH}_2-\text{X}$$

| R | X | Name |
|---|---|------|
| $\text{HOOC}-\underset{\text{NH}_2}{\text{CH}}-(\text{CH}_2)_3-$ | $-\overset{\displaystyle O}{\overset{\|}{\text{OCCH}_3}}$ | Cephalosporin C |
| thiophene$-\text{CH}_2-$ | $-\overset{\displaystyle O}{\overset{\|}{\text{OCCH}_3}}$ | Cephalothin |
| 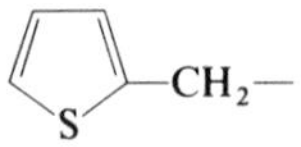 thiophene$-\text{CH}_2-$ | 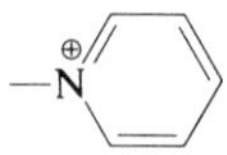 $-\overset{\oplus}{\text{N}}$ (pyridinium) | Cephaloridine |
| 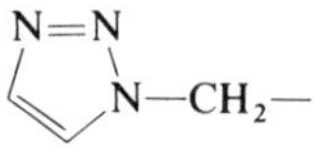 $\underset{\text{(tetrazole)}}{\overset{\text{N}=\text{N}}{}}\text{N}-\text{CH}_2-$ | 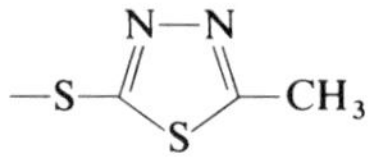 $-\text{S}-\underset{\text{(thiadiazole)}}{\overset{\text{N}-\text{N}}{}}-\text{CH}_3$ | Cefazolin |
| 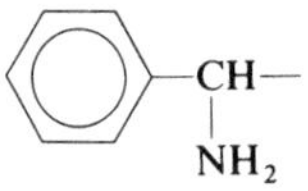 $\underset{\text{NH}_2}{\text{C}_6\text{H}_5-\text{CH}}-$ | $-\text{H}$ | Cephalexin |
| $\underset{\text{NH}_2}{\overset{\overset{\displaystyle \text{N}-\text{OCH}_3}{\|}}{\underset{\text{(aminothiazole)}}{\text{C}}}}-$ | $-\overset{\displaystyle O}{\overset{\|}{\text{OCCH}_3}}$ | Cefotaxime (Claforan) |
| $\underset{\text{HO}}{\overset{\overset{\displaystyle \text{COO}^{\ominus}}{\|}}{\text{CH}}}-$ | $-\text{S}-\underset{\underset{\text{CH}_3}{\|}}{\overset{\text{N}-\text{N}}{\text{N}}}$ (tetrazole) | Moxalactam (ring S replaced by O) |

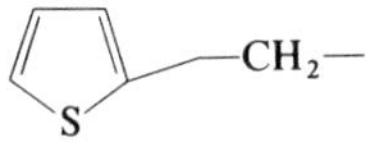

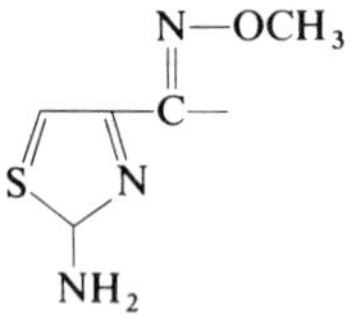

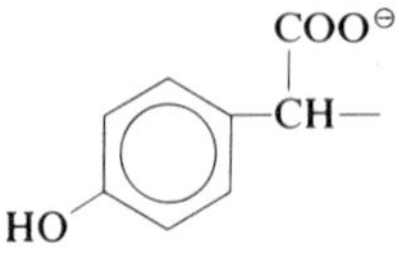

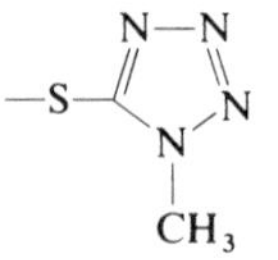

drawback is that it must be injected. **Cefazolin** and **cephaloridine** are metabolized to a lesser extent; **cephalexin** (analogous to ampicillin) is orally active and has a much higher acid stability than the penicillins. **Cefotaxime** and **moxalactam** are highly active against meningitis.

### 2.2.3. Other β-Lactam Antibiotics

Other β-lactam antibiotics found in recent years have revolutionized our understanding of the structure–activity relationships in this large group of antibiotics.

**Thienamycin** (6-24), discovered in 1976 but not yet described in detail, seems to be a very broad-spectrum antibiotic of high activity. It is lactamase resistant because of its hydroxyethyl side chain, but is not absorbed orally as it is highly polar. Unfortunately, it is very unstable and therefore unlikely to be of use in its native form. The N-formimidyl derivative overcomes this problem.

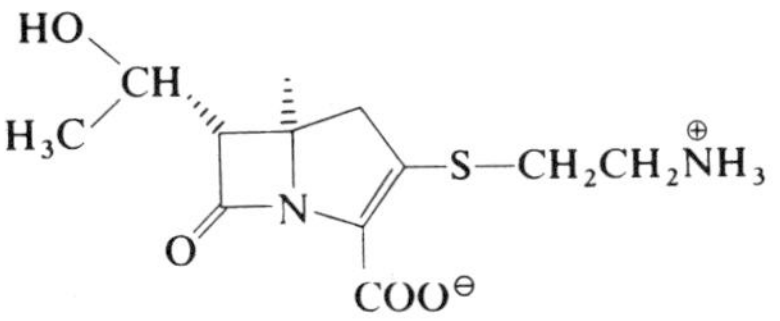

Thienamycin
6-24

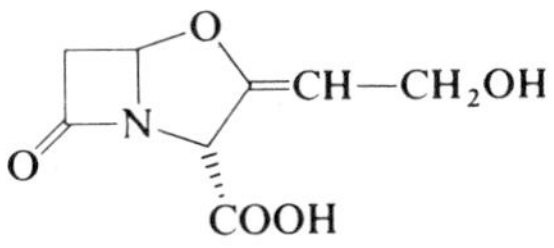

Clavulanic acid
6-25

**Clavulanic acid** (6-25), which is produced by a *Streptomyces* species, has only weak antibiotic acitivity but is a potent β-lactamase inhibitor. It can therefore protect lactamase-sensitive but otherwise potent antibiotics (e.g., ampicillin) from deactivation (see Cama and Christensen, 1978; Brown, 1985).

The monocyclic **nocardicins** (6-26) represent the ultimate "simplification" of the β-lactam structure, containing the azetidinone ring by itself, with a side chain resembling that of cephalosporin C. **Nocardicin A**, the (Z)-oxime, has limited activity against some Gram-negative bacteria. The similar **azthreonam** (6-27) (Debono and Gordee, 1982) is active against Gram-negative bacteria and *Pseudomonas*, and is lactamase resistant.

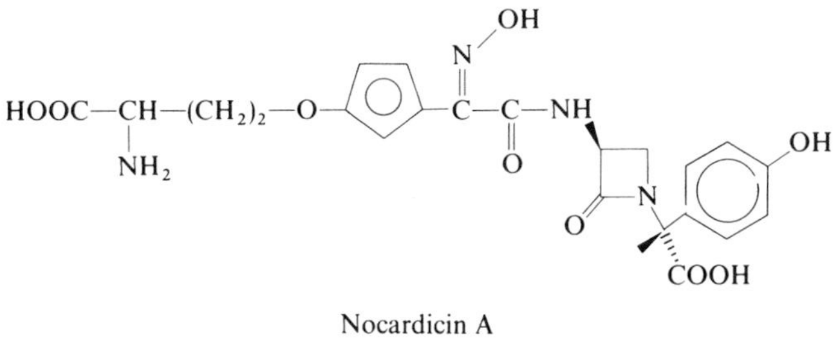

Nocardicin A
6-26

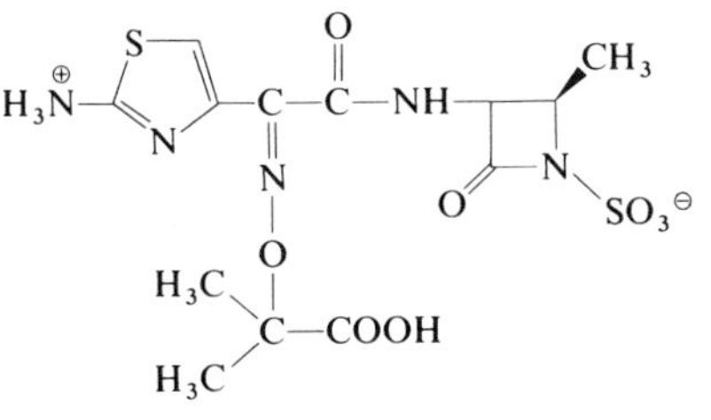

Azthreonam
6-27

## 2.2.4. Structure–Activity Correlations

Structure–activity correlations in the β-lactam antibiotic field have required drastic reevaluation in view of the novel structures described above. Apparently, only the intact β-lactam ring is an absolute requirement for activity. The sulfur atom can be replaced (moxalactam) or omitted (thienamycin), and the entire ring itself is, in fact, unnecessary (nocardicin). The carboxyl group, previously deemed essential, can be replaced by a tetrazolyl ring, which results in increased activity and lactamase resistance. The amide side chain, so widely varied in the past, is also unnecessary, as shown in the example of thienamycin. Nevertheless, there is a considerable literature analyzing the classical structure–activity relationships of the penicillin and cephalosporin groups, which has been reviewed by Christensen and Radcliffe (1976), by Cama and Christensen (1978), and also by Hoover and Dunn in Part II of Burger/Wolff (1980).

### Selected Readings

P. Actor, R. D. Sitrin, and J. V. Uri (1979). Antibiotics. *Annu. Rep. Med. Chem. 14*: 103–113.

A. G. Brown (1985). Clavulanic acid and related compounds: inhibitors of β-lactamase enzymes. In: *Medicinal Chemistry: The Role of Organic Chemistry in Drug Research* (S. M. Roberts and B. J. Price, Eds.). Academic Press, New York.

L. D. Cama and B. G. Christensen (1978). Structure–activity relationships of "non-classical" β-lactam antibiotics. *Annu. Rep. Med. Chem. 13*: 149–158.

B. G. Christensen and R. W. Radcliffe (1976). Total synthesis of β-lactam antibiotics. *Annu. Rep. Med. Chem. 11*: 271–280.

M. Debono and R. S. Gordee (1982) Antibacterial agents. *Annu. Rep. Med. Chem. 17*: 107–117.

E. F. Gale, E. Cundliffe, P. E. Reynolds, M. H. Richmond, and M. J. Waring (1981). *The Molecular Basis of Antibiotic Action,* 2nd ed. Wiley, New York.

T. D. Gootz (1985). Determinants of bacterial resistance to beta-lactam antibiotics. *Annu. Rep. Med. Chem. 20*: 137–144.

D. Gottlieb and P. D. Shaw (Eds.) (1967). *Antibiotics*, Vol. 1: *Mechanism of Action*; Vol. 2: *Biosynthesis.* Springer, New York.

E. S. Hamanaka and M. S. Kelly (1983). Antibacterial agents. *Annu. Rep. Med. Chem. 18*: 109–118.

R. Hare (1970). *The Birth of Penicillin.* Allen and Unwin, London.

J. R. E. Hoover and G. L. Dunn (1979). The β-lactam antibiotics. In: *Burger's Medicinal Chemistry*, 4th ed. (M. E. Wolff, Ed.), Part 2. Wiley-Interscience, New York, pp. 83–172.

J. A. Kelly, P. C. Moews, J. R. Knox, J. M. Frère, and J. M. Glunysen (1982). Penicillin target enzyme and the antibiotic binding site. *Science 218*: 479–481.

A. L. Lehninger (1982). *Principles of Biochemistry.* Worth, New York.

R. B. Morin and M. Gorman (Eds.) (1982). *Chemistry and Biology of β-Lactam Antibiotics.* Academic Press, New York.

A. D. Russel (1983). Design of antimicrobial chemotherapeutic agents. In: *Introduction to Principles of Drug Design* (J. Smith and H. Williams, Eds.). Wright, Bristol, England.

J. B. Stenlake (1979). *Foundations of Molecular Pharmacology.* Athlone Press, London.

D. J. Waxman and J. L. Strominger (1983). Penicillin-binding proteins and the mechanism of action of β-lactam antibiotics. *Annu. Rev. Biochem. 52*: 825–869.

## 3. ENZYMES AS DRUG TARGETS

Enzymology occupies a central role in biochemistry and in all disciplines involving biochemical principles. It is therefore very understandable that medicinal chemistry has also utilized and assimilated aspects of enzymology, especially those that explain the mode of action of drugs and help in their rational design. Additionally, the principles and concepts of enzymology, as a more mature science, have helped to shape our contemporary ideas on drug receptors and the molecular mode of their function. In previous chapters, we have come across drugs that are basically associated with some effect on an enzyme and should therefore be discussed in this chapter. A recent example is penicillin, covered in the preceding section because it interferes with bacterial cell wall synthesis. What really happens is that penicillin acts on an enzyme active in this synthetic process. Similarly, the renin–angiotensin system (Chap. 5, Sec. 2.8) could also be discussed here, since most drugs connected with that blood pressure-regulating system act on some enzyme. However, nothing would be gained by carrying the basic rationale of this book *ad absurdum*.

Nevertheless, a considerable number of enzymes do occupy a central and crucial role in the activity of drugs. Dihydrofolate reductase (Sec. 3.5), an enzyme involved in purine and amino acid biosynthesis, is the target of antibacterial sulfanilamides, which act both as bacteriostatics and antimalarials. These drugs act on the enzyme in different ways, some being so-called antimetabolites (i.e., reversible enzyme inhibitors). Some diuretics act on carbonic anhydrase which regulates proton equilibria in the kidney, whereas a group of hypotensive drugs influence the enzymatic destruction of catecholamines. Another ubiquitous enzyme to be discussed is adenylate cyclase, since so many drug receptors are coupled to it directly, and exert their activity by regulating cAMP production.

Enzymology is a large and complex subdivision of biochemistry, and no attempt will be made to cover it in this book. The reader who requires a review of enzyme kinetics and mechanisms is referred to the many currently available and excellent textbooks of biochemistry, and to the more specialized monograph of Fersht (1985).

### 3.1. Special Inhibitory Mechanisms

There are, however, two specific enzyme-inhibitor mechanisms that deserve special discussion because they are the basis of action for several important drugs. They are the transition-state analogues and the "suicide" substrates.

### 3.1.1. Transition-State Analogues

Transition-state analogues are inhibitors that mimic the transition-state structure of the substrate of an enzyme which, by definition, has the highest energy—that is, the least stable configuration or conformation. Since the transition state of an enzyme substrate is the form that is most tightly bound, its analogue should have a higher affinity and specificity for the enzyme than any substrate in the "ground state." Hence, binding constants of $10^{-15}$ M for the transition state can be expected for substrates with $K_D$'s of only $10^{-3}$–$10^{-5}$ M (see Lindquist, 1975).

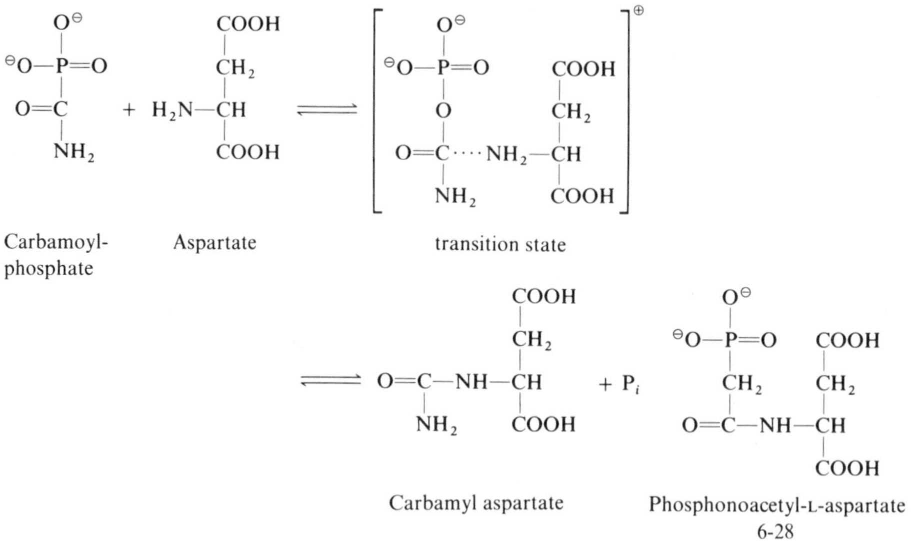

| Carbamoyl-phosphate | Aspartate | transition state |

Carbamyl aspartate          Phosphonoacetyl-L-aspartate
6-28

**Fig. 6.7.** Transition-state inhibition of aspartate transcarbamylase. Phosphonoacetyl-L-aspartate (PALA) mimics the transition state of the substrate and is an enzyme inhibitor and potential antitumor agent.

There are two problems with this potentially very powerful concept. First, the mechanism of the enzymatic reaction must be known in order to mimic the transition state of the substrate, since the structural specificity of the reaction is quite high. Second, a stable analogue of the labile transition state is, by implication, very difficult to prepare. Often one must be content with a metastable intermediate analogue of the substrate. Nevertheless, there are several successful applications of this interesting principle. The "multisubstrate" transition state in the initial step of pyrimidine synthesis is catalyzed by aspartate transcarbamylase, as shown in Fig. 6.7. The transition state has been mimicked by the analogue (6-28), in which the phosphate oxygen is replaced by a stable —CH$_2$— group. This **phosphono-acetyl-L-aspartate** (PALA) binds much more strongly to the enzyme than the two natural substrates combined. It is an antitumor agent because it blocks pyrimidine synthesis preferentially in the rapidly dividing malignant cells. It is currently undergoing clinical trials.

Penicillin, as outlined in Sec. 2, is a transition-state analogue of a distorted Gly-D-Ala-D-Ala peptide involved in the cross-linking of glycol-peptides constituting the cell walls of bacteria. Many other examples of such analogues are presented in the review by Lindquist (1975).

### 3.1.2. "Suicide" Substrates

$K_{cat}$ or "suicide" substrates, which are a variation of affinity-labeling agents, are irreversible enzyme inhibitors that bind covalently. The reactive anchoring group is, however, not intrinsically active, as in "classical" affinity labels (see Chap. 3,

Sec. 3.2), but is instead catalytically activated by the enzyme itself, through the enzyme–inhibitor complex. The enzyme thus produces its own inhibitor from an originally inactive compound, and is perceived to "commit suicide."

To design a $K_{cat}$ substrate, the catalytic mechanism of the enzyme as well as the nature of the functional groups at the enzyme active site must be known. Conversely, successful $K_{cat}$ inhibition provides valuable information about the structure and mechanism of an enzyme. Compounds that form carbanions are especially useful in this regard. Pyridoxal phosphate-dependent enzymes form such carbanions readily because the coenzyme can stabilize the anion by resonance delocalization on the heterocyclic ring (Fig. 6.8). Additionally, flavine-dependent oxidases such as monoamine oxidase (Sec. 3.7) can be attacked by acetylenes as suicide substrates. Many examples of such reactions are reviewed by Walsh (1984) and Penning (1983).

**Fig. 6.8.** Transformation of trifluoroalanine into a covalently bound "suicide substrate" by a pyridoxal coenzyme.

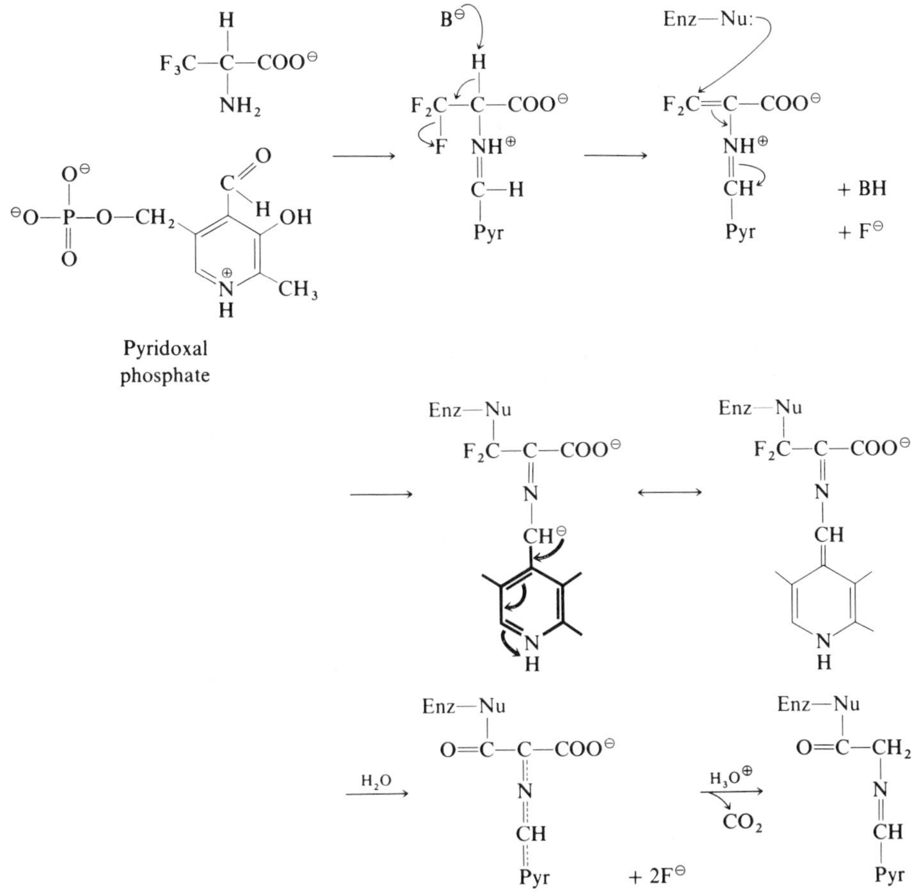

### 3.2. Acetylcholinesterase

Acetylcholinesterase (AChE) is the enzyme that hydrolyzes and thereby deactivates acetycholine (ACh) after it binds to the receptor. The enzyme is present in all peripheral and central junctional sites, in erythrocytes, the placenta, in the basement membrane of *Torpedo*, and in the *Electrophorus* electroplax (see Chap. 4, Sec. 2.2).

### 3.2.1. Physicochemical Properties

The purification of AChE and the elucidation of its physicochemical properties were achieved by using the enzyme isolated from the *Electrophorus* electroplax, the richest source of AChE, as well as from brain and erythrocytes. With high ionic strength solutions (1 M NaCl or 2 M $MgCl_2$), the extraction is selective and can be facilitated by treatment of the electroplax with collagenase. The basic unit of the enzyme is a tetramer with a molecular weight of 320,000; each of the protomers contains an active site. Normally, three such tetrameric units are linked through disulfide bonds to a $50 \times 2$-nm stem. This stem is a collagen triple helix which seems to have a structural role only (Fig. 6.9; see Taylor et al., 1987). In neural and electroplax tissue, the enzyme "trees" (stems with tetramers) are anchored in the basement membrane or neurolemma—a porous, collagen-rich structure. However, because many tissues (erythrocytes, peripheral ganglia) do not have a basement membrane, attachment of the AChE must occur in another, unknown fashion.

**Fig. 6.9.** A possible model for the association of acetylcholinesterase with the postsynaptic membrane. (Reproduced by permission from Taylor et al. (1979), in *Cholinergic Mechanisms and Psychopharmacology* (D. J. Jenden, Ed.), Plenum Publishing, New York)

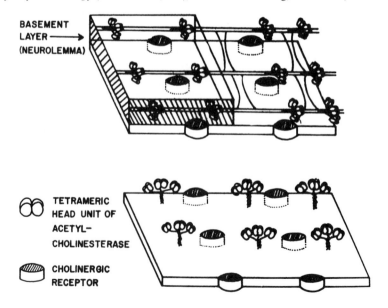

BASEMENT
LAYER ⟶
(NEUROLEMMA)

TETRAMERIC
HEAD UNIT OF
ACETYL-
CHOLINESTERASE

CHOLINERGIC
RECEPTOR

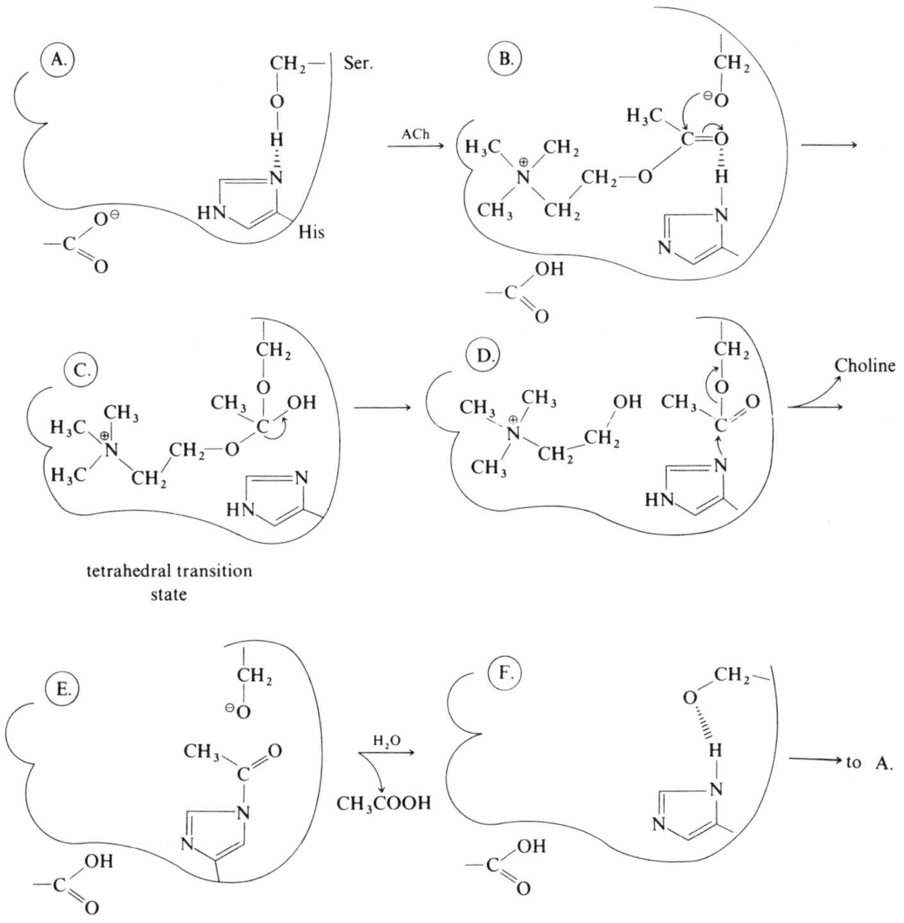

**Fig. 6.10.** Mechanism of acetylcholine hydrolysis by acetylcholinesterase.

Analysis of the amino acid composition of the enzyme shows that it bears a close similarity to the AChR in its high proportion of acidic amino acids.

In contrast to the AChR, AChE does not bind bungarotoxin or sulfhydryl reagents. It is inhibited by excess substrate ($3 \times 10^{-3}$ M), and the $K_m$ of electric eel AChE is about $10^{-4}$ M. The specific activity of the enzyme is one of the highest known: 750 nmol/mg-hr, with a turnover time of 30–60 msec and a turnover number of $2–3 \times 10^6$. It is therefore one of the most efficient and fastest enzymes known.

The *active site* of AChE has the composition of most serine esterases. It includes a charge-relay system of histidine and serine, and an acidic center, probably glutamate, which binds the choline cation. The serine hydroxyl is rendered more nucleophilic through the proton-acceptor role of histidine (Fig. 6.10A), and is capable of executing a nucleophilic attack on the carbonyl carbon of ACh (Fig. 6.10B). A tetrahedral transition state is reached (Fig. 6.10C), resulting in serine acetylation (Fig. 6.10D) and the desorption of free choline. The acetyl group is taken over by histidine as an *N*-acetate (Fig. 6.10E), which is then easily hydrolyzed, regenerating the enzyme active site. The choline is taken up into the nerve ending by an active transport system (Chap. 4, Sec. 2.1) and reused for ACh synthesis. Finally, the acetate goes into the ubiquitous acetate pool of intermediary metabolism.

On the basis of studies of the hydrolysis rate of succinyl-methylcholine isomers, Stenlake (1979) proposed the hypothesis that the AChR and AChE bind two different faces of ACh which implies that they have opposite sterochemistries at their respective anionic subsites. Simultaneous attachment to the receptor and the enzyme is therefore impossible, although a fast sequential attack of the enzyme on the receptor-bound neurotransmitter is still feasible since no ACh reorientation is necessary for this. However, there is no evidence that such a sequence of events occurs *in vivo*.

Another cholinesterase, called *serum cholinesterase* or butyrylcholinesterase, is found in serum and the liver. It is less active than AChE but also less selective, and plays an important role in drug metabolism.

### 3.2.2. Anticholinesterase Inhibitors

Anticholinesterase drugs are compounds that block AChE and inhibit the destruction of released ACh. The resultant higher neurotransmitter levels then increase the biological response. Anticholinesterases can therefore be considered as indirect cholinergic agents. Acetycholinesterase inhibitors can act by either of two mechanisms:

1. As classical competitive enzyme inhibitors, they have a high affinity for the active site, but are not substrates. The enzyme is occupied by the inhibitor for relatively long periods, and therefore cannot handle ACh efficiently, as a result of the saturation phenomenon.
2. The inhibitor acylates the serine hydroxyl of AChE, forming an ester more stable than acetate, such as a carbamate or phosphate. The hydrolysis of these esters takes a long time even if they are not irreversible, as formerly thought. Acetylcholine cannot then be hydrolyzed, since the active site is covalently occupied.

A representative of the first group of inhibitors is **edrophonium** (6-29), a short-acting drug that binds to the anionic site of the enzyme and also forms a hydrogen bond with the imidazole nitrogen of the active site. **Ambenonium** (6-30) also does not react covalently with the enzyme but has a much longer duration of action than edrophonium.

The second group of AChE inhibitors is represented by the alkaloid **physostigmine (eserine**, 6-31), isolated from the seeds of *Physostigma venenosum*, the Calabar bean, which has been used in West Africa in witch ordeals. Synthesis analogues, like **neostigmine** (6-32) and its congeners, are also active. They have a very high affinity for the enzyme and will carbamoylate the serine hydroxyl of the active site. Because the half-life of the dimethyl-carbamoyl ester is 20–30 minutes, as compared to only

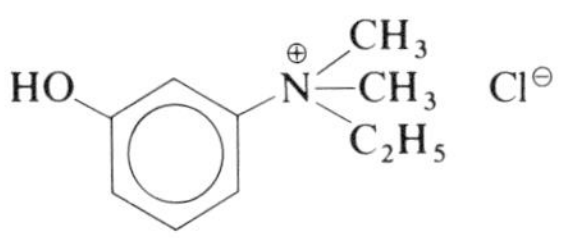

Edrophonium
6-29

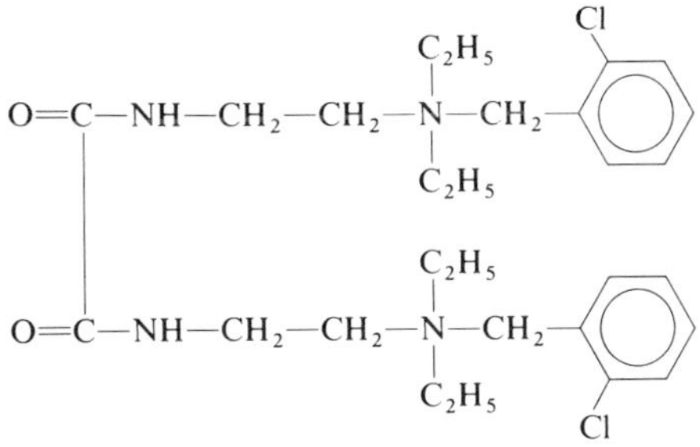

Ambenonium
6-30

microseconds for the acetylated enzyme, AChE is inhibited for hours after a single dose of such drugs.

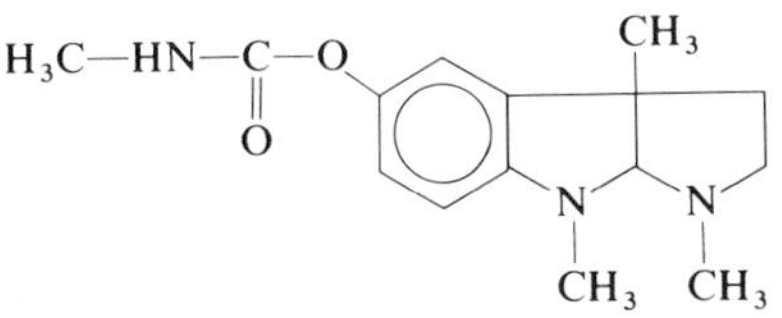

Physostigmine
6-31

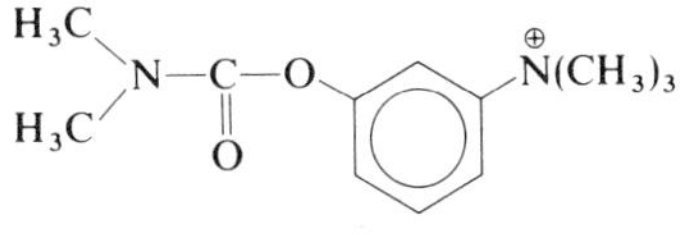

Neostigmine
6-32

*Insecticides.* Uncharged carbamates, like **carbaryl** (**sevin**; 6-33), can penetrate the CNS of insects (which do not use AChE in their neuromuscular junction), and act quite selectively as insecticides with a low toxicity to mammals [median lethal dose $(LD_{50})$ in the rat = 540 mg/kg, p.o.]

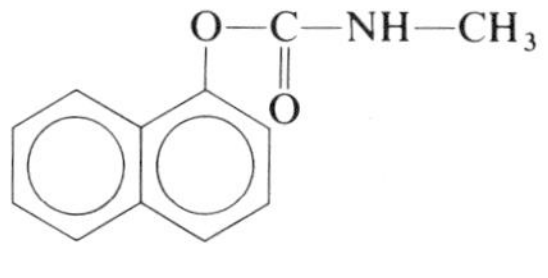

Sevin
6-33

Even longer-acting covalent AChE inhibitors are the organophosphate esters, such as **diisopropyl-phosphorfluoridate** (6-34) and **echothiophate** (6-35), developed from "nerve gases" discovered but never used in World War II. These compounds are known for their extreme toxicity and rapid dermal absorption. Many useful insecticides can also be found in this group. **Malathion** (6-36) is a pro-drug, since the thiophosphate must be bioactivated to the phosphate form—a transformation carried out by insects but not mammals. Additionally, the ester groups of malathion are rapidly hydrolyzed in higher organisms to water-soluble and inactive compounds; however, insects do not have the hydrolases that would deactivate the insecticide. Because of this double safety feature, the selectivity of malathion is high and its mammalian toxicity low: the $LD_{50}$ for rats is 1500 mg/kg, p.o. The volatile

**dichlorvos** (6-37) is used in fly-killer strips. Whereas the vapors of this compound are rapidly hydrolyzed by mammals, they are cumulative and highly toxic for insects.

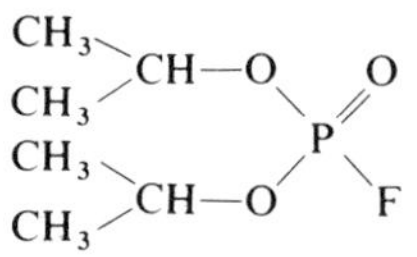

Diisopropyl-phospho-
fluoridate (DFP)
6-34

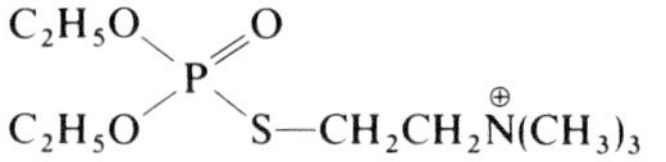

Echothiophate
6-35

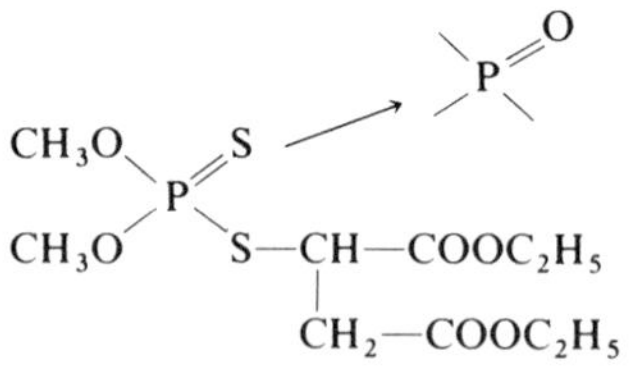

Malathion
6-36

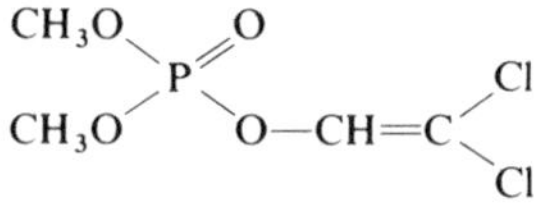

Dichlorvos
6-37

*Clinical Applications.* Both groups of AChE inhibitors are used therapeutically. One use is in *glaucoma*, in which high intraocular pressure can lead to permanent damage to the optic disk, resulting in blindness. The local instillation of **physostigmine** or **echothiophate** solution in the eye results in a long-lasting decrease in the intraocular pressure as well as myosis (contraction of the pupil). Recently, cannabinoids have been found to have similar activity (Chap. 4, Sec. 5.3.4).

The other use of anticholinesterase drugs is in *myasthenia gravis*. This is an autoimmune disease caused by the development of antibodies against the patient's own ACh receptors, accompanied by disturbed neuromuscular transmission. The disturbance is caused by a reduction in the number of nerve terminals and an increase in the width of the synaptic cleft. Normally, nicotinic ACh receptors are destroyed by endocytosis via coated pits and proteolysis in lysosomes. In myasthenia gravis, the receptors are cross-linked by antireceptor antibodies which facilitate the rate-limiting endocytosis step; receptor destruction occurs in less than half the normal time, resulting in net receptor loss (Lindstrom, 1984). The chronic disease is characterized clinically by such muscular weakness and abnormal fatigue that patients cannot even keep their eyes open. Acetylcholinesterase inhibitors increase the ACh concentration and excitation of the neuromuscular junction, resulting in increased strength and endurance. As expected, AChE inhibitors are also potent curare antidotes, because the increased ACh levels displace the blocker more readily.

The covalently binding AChE inhibitors, and especially the organophosphates, can be highly toxic, and poisoning from their improper use in agriculture does occur. Based on the knowledge of their mode of action, an antidote, **pralidoxime** (6-38; 2-*N*-methylpyridinium-2-aldoxim iodide), was designed. The quaternary pyridinium ion of this compound binds to the anionic site of the enzyme, removes the phosphate of the inhibitor from the serine active site in the form of an oxime-phosphonate (Fig. 6.11), and regenerates AChE. Atropine is also administered to

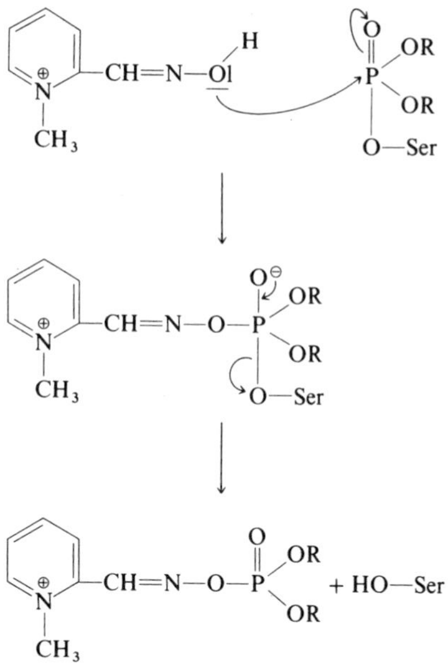

**Fig. 6.11.** Mode of action of pralidoxime, an acetylcholinesterase-inhibitor antidote.

victims of organophosphate poisoning, to relieve peripheral and central muscarinic symptoms due to excessive ACh.

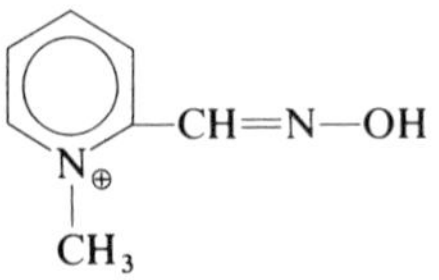

Pralidoxime
6-38

## Selected Readings

A. Fersht (1985). *Enzyme Structure and Mechanism,* 2nd ed. Freeman, San Francisco.

M. J. Jung (1978). Selective enzyme inhibitors in medicinal chemistry. *Annu. Rep. Med. Chem.* *13:* 249–260.

R. N. Lindquist (1975). The design of enzyme inhibitors: Transition state analogs. *Drug Design* (E. J. Ariëns, Ed.), Vol. 5. Academic Press, New York, pp. 23–80.

J. Lindstrom (1986). Acetylcholine receptors: structure, function, synthesis, destruction and antigenicity. In: *Myology* (A. G. Engel and B. Q. Banker, Eds.). McGraw-Hill, New York, pp. 769–790.

J. L. Neumeyer (1981). Pesticides In: *Principles of Medicinal Chemistry,* 2nd ed. (W. O. Foye, Ed.). Lea and Febiger, Philadelphia, pp. 817–836.

T. M. Penning (1983). Design of suicide substrates. *Trends Pharmacol. Sci.* *4:* 212–217.

R. R. Rando (1974). Mechanism based irreversible enzyme inhibitors. *Annu. Rep. Med. Chem.* *9:* 234–243.

M. Sandler (Ed.) (1980). *Enzyme Inhibitors as Drugs.* Macmillan, London.

J. B. Stenlake (1979). Molecular interactions at the cholinergic receptor in neuromuscular blockade. In: *Progress in Medicinal Chemistry* (G. P. Ellis and G. B. West Eds.), Vol. 16. Elsevier/North Holland, Amsterdam, pp. 257–286.

P. Taylor, M. Schumacher, K. MacPhee-Quigley, T. Friedman, and S. Taylor (1987). The structure of acetylcholinesterase: relationships to its function and cellular disposition. *Trends Neurosci. 10*: 92–96.

C. T. Walsh (1984). Suicide substrates, mechanism-based enzyme inactivators: recent developments. *Annu. Rev. Biochem. 53*: 493–535.

R. Wolfenden (1976). Transition state analog inhibitors and enzyme catalysis. *Annu. Rev. Biophys. Bioeng. 5*: 271–306.

### 3.3. Adenosine Triphosphatase

Adenosine triphosphatase (ATPase, E.C. 3.6.1.3) comprises a group of extremely widespread, membrane-bound enzymes. All cell membranes contain the $Na^+–K^+$-dependent enzyme, which is extrememly important in the establishment of the mitochondrial proton gradient and in bacterial permease systems that transport amino acids, among its other roles. As a result of the proton gradient in mitochondria, or a very steep $Na^+–K^+$ gradient, ATP synthesis rather than ATP hydrolysis can take place, according to the Mitchell chemiosmotic hypothesis. This model suggests that $Na^+–K^+$-activated ATPase located in neural membranes is responsible for maintaining transmembranal ionic asymmetry, or ion gradients, across these membranes. Adenosine triphosphatase is phosphorylated by ATP and then dephosphorylated by $K^+$. The free energy of ATP hydrolysis is used to transport sodium and potassium against their respective gradients in a 3 $Na^+$:2 $K^+$:1 ATP ratio, but $Na^+$ must be inside and $K^+$ outside the cell in order to activate the enzyme. A rich source for ATPase isolation is the electroplax of *Torpedo* or *Electrophorus* (see Chap. 4, Sec. 2.2).

### 3.3.1. Physicochemical Properties

The $Na^+–K^+$-ATPase has been purified and sequenced (see Cantley, 1986). It is a dimeric protein with an $\alpha$ subunit of 100 kD and a $\beta$ subunit of 55 kD. The $\alpha$ subunit contains the ATP hydrolysis subsite, in which an aspartate accepts the $\gamma$ phosphate of ATP. The function of the glycoprotein $\beta$ subunit is not clear. ATPases from different sources (electroplax, sheep kidney, rat, bacterial $K^+$ pump) have been shown, through cDNA cloning, to have a high degree of homology, indicating their ancient common evolutionary history. Determination of the primary sequence, proteolytic digestion, and labeling experiments have enabled investigators to elucidate the folding of the protein within the cell membrane. There are eight transmembrane domains and a large cytosolic portion. ATP binds to Lys-501, and Asp-369 is phosphorylated. The cardiac glycoside binding site, which is partially located on the outside of the $\alpha$ subunit, inhibits the ATP-driven ion transport and the $Na^+$-dependent conformational change. Investigators also discovered that the mammalian brain and heart contain distinct isozymes with different $\alpha$ subunits showing different steroid binding capabilities (Erdman et al., 1985). But all these investigations, although important, have not elucidated the nature of the coupling of ATP hydrolysis to cation movement.

The sarcoplasmic reticulum of muscle cells contains a $Ca^{2+}$-dependent ATPase, responsible for $Ca^{2+}$ concentration changes in the muscle, which are intimately connected with the contraction process.

### 3.3.2. Cardiac Steroid Glycosides

Myocardial cell membrane ATPase, the enzyme present in heart muscle, is the site of action of the cardiac steroid glycosides, which have a specific action on the heart muscle. These drugs increase the force of contraction of the muscle (positive inotropic effect) as well as its conductivity and automaticity. They are also valuable in treating congestive heart failure, in which the circulatory needs of organs are no longer satisfied, and in heart arrythmias, in which the rhythm of the cardiac contractions is upset. Due to the drug effect, the force of contraction increases and the heart rate is slowed (chronotropic effect). Consequently, the cardiac output is elevated while the size of the heart decreases.

The exterior of the myocardial enzyme, situated in the plasmalemma, is considered to be the specific binding site of cardiac steroid glycosides. It is believed (see Campbell and Danilewicz, 1978) that the positive inotropic effect is due to the inhibition of enzyme dephosphorylation and $Na^+-K^+$ exchange, which occurs before each heart contraction. The inhibition of $Na^+$ extrusion increases the $Na^+$ concentration, which in turn triggers greater $Ca^{2+}$ mobilization and controls the contraction of the heart. If, however, the $Na^+$ pump is further inhibited, toxic effects appear; indeed, cardiac steroids show considerable toxicity which, unfortunately, is due to the same mechanism as their beneficial effect. Nevertheless, it has been shown that the inotropic effect of these drugs is possible without pump inhibition. The mechanisms involved are complex, and there are several possible interpretations. The major difficulties in firmly correlating steroid glycoside action with $Na^+-K^+$-ATPase activity are the lack of a specific antagonist and the fact that many other compounds inhibit the enzyme without showing cardiac activity. In order to form a complex with a drug, the ATPase must be in the proper conformation. This conformation exists only immediately after the enzyme has transported $Na^+$ and is ready to bind $K^+$.

Cardiac steroids, or *cardenolides*, are steroid glycosides for which representative structures are shown in Fig. 6.12. Their effect has been known since the time of the ancient Egyptians. In more recent times, the foxglove (*Digitalis purpurea*) and its effect were described in 1785 by William Withering, who knew of its use in folk medicine.

*Types.* Three groups of plants produce cardenolides: the *Digitalis* species, growing in temperate climates; the *Strophanthus* species, of tropical provenance; and *Scilla* (sea onion or squill), a Mediterranean plant.

The *Digitalis* glycosides are the most widely used cardenolids. The aglycones (the steroid parts) of the molecule differ from the usual steroid structure in several points. The anellation of the A–B and C–D rings is *cis* (Z), the 3-OH is axial (β),

| Glycoside | Aglycon | Sugar |
|---|---|---|

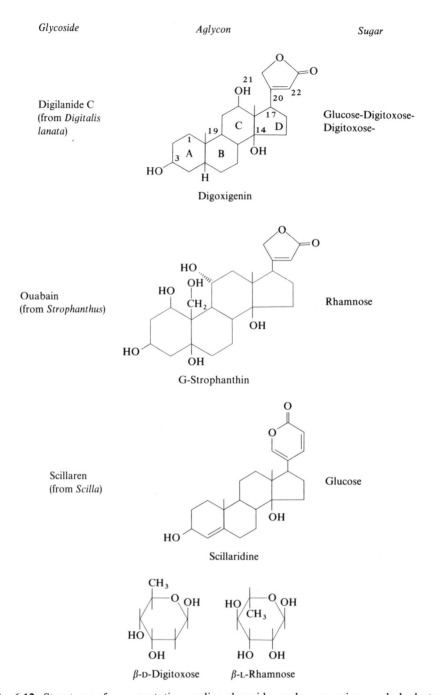

Digilanide C
(from *Digitalis lanata*)

Glucose-Digitoxose-Digitoxose-

Digoxigenin

Ouabain
(from *Strophanthus*)

Rhamnose

G-Strophanthin

Scillaren
(from *Scilla*)

Glucose

Scillaridine

β-D-Digitoxose          β-L-Rhamnose

**Fig. 6.12.** Structures of representative cardiac glycosides and some unique carbohydrates involved in their structures.

and all of these steroids carry a 14$\beta$-OH group. The C-17 side chain is an unsaturated lactone ring. The sugar part, binding to the 3-OH, is a tri- or tetrasaccharide consisting mainly of digitoxose (2,6-dideoxy-$\beta$-D-allose) and glucose.

The *strophanthin* aglycones have a 5$\beta$-OH group in addition to other hydroxyls, up to a maximum of six in ouabain. The 19-methyl is replaced by an aldehyde or primary alcohol and the sugars are the unusual rhamnose or cymarose.

The *squill* aglycones carry a six-membered lactone ring with two double bonds, and are closely related to some toad venoms (bufotalin). None is used therapeutically since all are highly toxic.

***Structure–Activity Correlations.*** The structure–activity relationships of cardenolids have been thoroughly investigated, and have undergone considerable revision in the last few years on the basis of crystallographic work and potential energy calculations by Rohrer and his group (Rohrer et al., 1979; Repke, 1985). The correlations are summarized as follows:

1. The A–B *cis* fused rings, the axial methyl group, and ring C form the rigid essential backbone of the structure.
2. Ring D has conformational flexibility, influenced by the nature of the 17$\beta$ side group. The C–D *cis* junction and 14$\beta$ conformation are essential.
3. The 14-OH group, previously considered essential, can be omitted in some structures or can be replaced by a 14$\beta$-NH$_2$ group.
4. The $\Delta^{20-22}$ double bond serves to properly orient the carbonyl oxygen. The ring itself is not as essential as previously thought, and some derivatives with side chains instead of a ring have even higher activity (see also Thomas et al., 1974). However, the side chains must be coplanar, and are much less flexible than one would expect.
5. The activity of a compound depends to a very great extent on the position of the 23-carbonyl oxygen, which is held quite rigidly by ring D and the double bond. The standard is the position of the carbonyl relative to the rigid backbone of digitoxigenin. In synthetic analogues, every 0.22-nm deviation from this position causes a loss of activity by one order of magnitude, from a maximum IC$_{50}$ of 49 nM. This correlation is highly significant, as the regression coefficient of $r^2 = 0.993$ indicates.
6. Removal of the sugar portion allows epimerization of the 3$\beta$-OH group, with a decrease in activity and an increase in toxicity due to changes in polarity.

Recently, a polypeptide of 49 amino acids was isolated from a sea anemone. This compound, **anthopleurine A**, is 30 times more potent than the *Digitalis* glycosides, and is less toxic. The discovery stimulated speculation over the possibility of a native inotropic peptide receptor, in a situation similar to the endorphin–opiate relationship, in which a plant alkaloid fortuitously fits a peptide receptor (Campbell and Danilowicz, 1978). Recently an endogenous digitalis-like activity was discovered in rat, guinea pig, and bovine heart homogenates, that inhibited Na$^+$–K$^+$-ATPase, and had an affinity for the steroid receptor one to two orders of magnitude higher than digoxin. This *cardiodigin* may be the long-sought endogenous cardioactive factor (Godfraind et al., 1982).

### 3.3.3. Miscellaneous Drugs

Among nonsteroids, **amrinone** (6-39), a bipyridine derivative, has been extensively investigated and is licensed in Europe. It enhances the inotropic effect (contractile force) by increasing Ca entry, has no effect on ATPase or adenylate cyclase, and does not produce arrhythmia (Ward et al., 1983). Its analogue **milrinone** (6-40) is about 30 times more active.

6-39   Amrinone   ($R_1$ = H; $R_2$ = $NH_2$)
6-40   Milrinone   ($R_1$ = $CH_3$; $R_2$ = CN)

The $H^+-K^+$-$ATPase$ acts as the proton pump in the parietal cells of the stomach mucosa. It transports protons and $Cl^-$ ions into the stomach via a $K^+$ antiport. Substituted benzimidazoles like **omeprazole** (6-41) inhibit this enzyme and are 2–12 times as active as **cimetidine** in inhibiting gastric stimulation (Chap. 4, Sec. 6.3). The mechanism of enzyme inhibition has been elaborated by Lindberg et al. (1986):

Omeprazole
6-41

### Selected Readings

J. A. Bristol and D. B. Evans (1981). Cardiotonic agents for the treatment of heart failure. *Annu Rep. Med. Chem. 16*: 91–102.

S. F. Campbell and J. C. Danilewicz (1978). Agents for the treatment of heart failure. *Annu. Rep. Med. Chem. 13*: 92–102.

L. Cantley (1986). Ion transport systems sequenced. *Trends Neurosci. 9*: 1–3.

E. Erdman, K. Werdan, and L. Brown (1985). Multiplicity of cardiac glycoside receptors in the heart. *Trends Pharmacol. Sci. 6*: 293–295.

T. Godfraind, A. de Pover, G. Castaneda Hernandez, and M. Fagoo (1982). Cardiodigin: endogenous digitalis-like material from mammalian heart. *Arch. Int. Pharmacodyn. 258*: 165–167.

P. Lindberg, P. Norberg, T. Alminger, A. Brändström, and B. Wallmark (1986). The mechanism of action of the gastric acid secretion inhibitor omeprazole. *J. Med. Chem. 29*: 1327–1329.

H. Lullmann, T. Peters, and A. Ziegler (1979). Kinetic events determining the effect of cardiac glycosides. *Trends Pharmacol. Sci. 1*: 102–106.

K. Repke (1985). New developments in cardiac glycoside structure–activity relationships. *Trends Pharmacol. Sci. 6*: 275–278.

D. C. Rohrer, D. S. Fullerton, K. Yoshioka, A. H. L. Frome, and K. Ahmed (1979). Functional receptor mapping for modified cardenolides: use of the Prophet system. In: *Computer Assisted Drug Design* (E. C. Olson and R. E. Christoffersen, Eds.). American Chemical Society, Washington, DC, pp. 259–279.

R. Thomas, J. Boutagy and A. Gelbart (1974). Synthesis and biological activity of semi-synthetic digitalis analogs. *J. Pharmacol. Sci. 63*: 1649–1683.

A. Ward, R. N. Brogder, R. C. Heel, T. M. Speright, and G. S. Avery (1983). Amrinone: a preliminary review of its pharmacological properties and therapeutic use.

B. Wetzel and N. Hauel (1984). Cardiotonic agents. *Annu. Rep. Med. Chem. 19*: 71–80.

### 3.4. Carbonic Anhydrase

Carbonic anhydrase (E.C. 4.2.1.1) is an enzyme located in the renal tubular epithelium and in red blood cells. It catalyzes the seemingly simple reaction

$$2 H_2O + CO_2 \rightleftharpoons H_2CO_3 \rightleftharpoons HCO_3^- + H_3O^+$$

which would be shifted far to the left without an enzyme.

In the red blood cell this reaction plays an important role in $CO_2$ transport from the tissues to the lungs. In the kidney, the proton of the $H_3O^+$ is exchanged for $Na^+$ ions, which are reabsorbed while $HCO_3$ is decomposed through a shift of the equilibrium to the left. Carbonic anhydrase therefore plays a crucial role in maintaining the ion and water balance between the tissues and urine.

When carbonic anhydrase inhibitors block the enzyme in the kidney, $H_2CO_3$ formation—and consequently the availability of $H_3O^+$ (i.e., protons)—decreases. Since the $Na^+$ ions in the filtrate cannot be exchanged, sodium is excreted together with large amounts of water, as a result of ion hydration and osmotic effects. The result is *diuresis* accompanied by a dramatic increase in urine volume. There is also failure to remove $HCO_3^-$ ions because there is no $H_3O^+$ to form $H_2CO_3$, which would decompose to $CO_2 + H_2O$. Therefore, the normally slightly acidic urine becomes alkaline. The strong carbonic anhydrase inhibitors also increase $K^+$ excretion, an undesirable effect.

Carbonic anhydrase has also been discovered in the eye and the CNS. Consequently, its inhibitors have found use in the treatment of *glaucoma*, in reducing the high intraocular pressure which can lead to blindness in this disease. They are also effective, in combination with anticonvulsant drugs, in controlling certain forms of *epilepsy* (Chap. 4, Sec. 8.1.6).

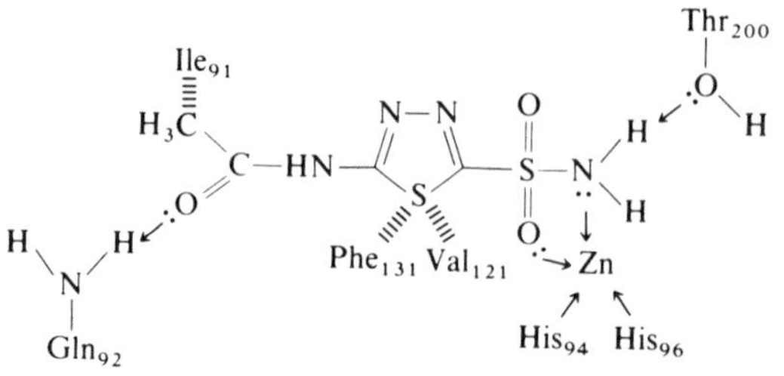

**Fig. 6.13.** Binding site map, showing amino acids of carbonic anhydrase that are critical to the binding of acetazolamide to this enzyme. Hydrogen bonds or electrostatic interactions are shown by arrows. Narrow lines of equal length ($Ile_{91}$, $Phe_{131}$, $Val_{121}$) depict hydrophobic interactions.

### 3.4.1. Enzyme Structure

The structure of carbonic anhydrase has been completely elucidated. Both the amino acid sequence and the three-dimensional structure of the crystalline enzyme are known (see Kannan et al., 1977). Actually, there are two isozymes, a low- and a high-activity form having been isolated from human erythrocytes, with the latter designated the C form (HCA-C).

The large molecule consists of a single peptide chain; 35% $\beta$-sheet and 20% helical structure are found in the folded structure. The active site is a 1.2-nm-deep conical cavity in the central pleated sheet, with a $Zn^{2+}$ ion located at its bottom. Three histidine residues hold the $Zn^{2+}$, which also binds a $H_2O$ molecule. The active-site cavity is divided into hydrophilic and hydrophobic halves. The inhibitors of the enzyme replace the water on the $Zn^{2+}$ ion and also block the fifth coordination site, where $CO_2$ should bind. The active-site region is reproduced in Fig. 6.13.

### 3.4.2. Sulfonamide Carbonic Anhydrase Inhibitors

The development of sulfonamide carbonic anhydrase inhibitors was based on the observation that antibacterial sulfanilamides (see Sec. 3.5) produce alkaline urine. This discovery led to the development of **acetazolamide** (6-42), a thiadiazole derivative. It is not an ideal drug because it promotes $K^+$ excretion and causes a very high urine pH. Since chloride ions are not excreted simultaneously, systemic acidosis also results. Much more useful are the **chlorothiazide** (6-43) derivatives, which are widely used oral diuretic drugs. These compounds differ from one another mainly in the nature of the substituent on C-3, are much weaker carbonic anhydrase inhibitors than acetazolamide, and may have another mode of action in addition to carbonic anhydrase inhibition. They are widely used in edema, hypertension, and cardiac insufficiency, in which a decrease in the amount of tissue-bound or circulating water is imperative. Some newer derivatives, like **polythiazide** (6-44), are three orders of magnitude more active, and have a duration of action of up to 24 hours. **Furosemide** (6-45) is formally a sulfanilamide but not a carbonic anhydrase inhibitor. It not only inhibits $Na^+$ reabsorption in the loop of Henle—a part of the nephron—but probably also functions as an inhibitor of $Na^+-K^+$-ATPase, which has a role in renal sodium transport as it does in other organs.

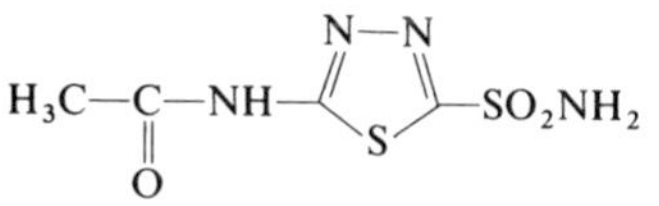

Acetazolamide
6-42

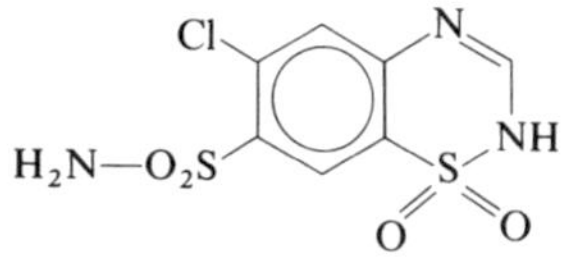

Chlorothiazide
6-43

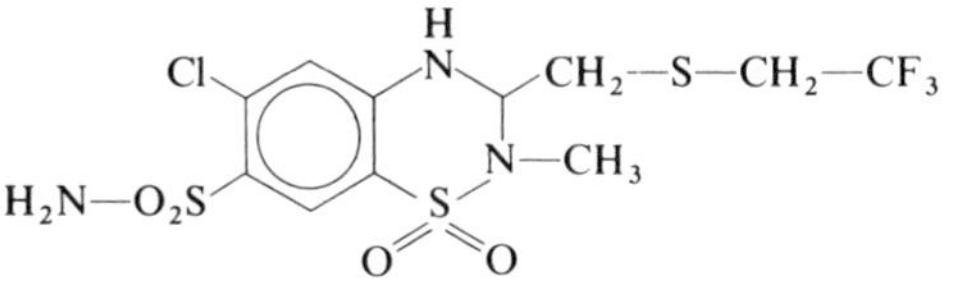

Polythiazide
6-44

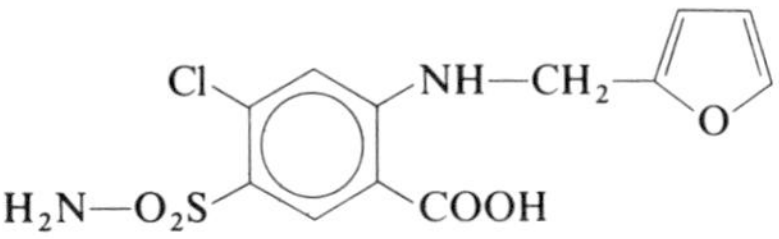

Furosemide
6-45

### 3.4.3. Miscellaneous Drugs

*Other diuretics* do not act through carbonic anhydrase inhibition and are not sulfanylamides. **Ethacrynic acid** (6-46) is based on some older Hg-containing diuretics which block enzymatic Na transport by binding to enzyme —SH groups. The unsaturated ketone of ethacrynic acid can react in a similar way, binding —SH, whereas the carboxyl group ensures the concentration of the compound in the kidneys. The halogens increase the electrophilic nature of the unsaturated ketone. Chloride ion elimination is increased together with $Na^+$ excretion, but $K^+$ and $HCO_3^-$ elimination are low, and the urine pH stays at 6. Furosemide and ethracrynic acid are powerful drugs used in patients resistant to other diuretics.

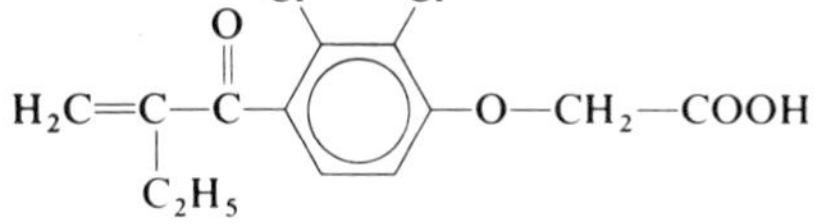

Ethacrynic acid
6-46

*Organomercury compounds,* such as **mersalyl** (6-47) are effective but obsolete drugs. However, mercury poisoning is not a concern with their use, since the drugs accumulate in the kidneys only.

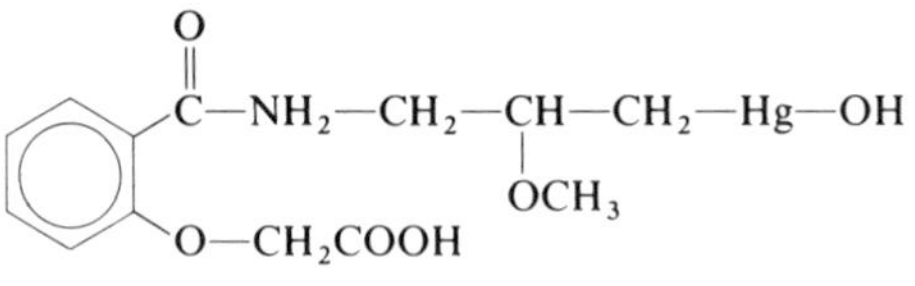

Mersalyl
6-47

### Selected Readings

E. J. Cragoe, Jr. (Ed.) (1983). *Diuretics: Chemistry, Pharmacology and Medicine.* Wiley, New York.

H. R. Jacobsen and J. K. Kokko (1976). Diuretics: sites and mechanism of action. *Annu. Rev. Pharmacol. Toxicol. 16*: 201–226.

K. K. Kannan, I. Vaara, B. Notstrand, S. Lovgren, A. Borell, K. Fridborg, and M. Petef (1977). Structure and function of carbonic anhydrase: comparative studies of sulfonamide binding to human erythrocyte carbonic anhydrases B and C. In: *Drug Action at the Molecular Level* (G. C. K. Roberts, Ed.). University Park Press, Baltimore, pp. 73–91.

R. L. Smith, O. W. Woltersdorf, Jr., and E. J. Cragoe, Jr. (1976, 1978). Diuretics. *Annu. Rep. Med. Chem. 11*: 71–79; *13*: 61–70.

### 3.5. Dihydrofolate Reductase

Dihydrofolate reductase (E.C. 1.5.1.3) is a relatively small monomeric protein (MW = 18,000–36,000) containing no disulfide bonds, and is present in all cells in different isozyme forms, depending on the organism. The best-studied form contains 159 amino acids, with 30% of them in an eight-stranded pleated sheet. The binding site is a 1.5-nm-deep cleft across the enzyme, and about a dozen amino acids are involved in binding substrate or inhibitor. Figure 6.14 shows a stereoscopic drawing of the active site.

**Fig. 6.14.** Stereoscopic drawing of the active site of dihydrofolate reductase. A bound molecule of methotrexate is also shown. To get a three-dimensional view, cross your eyes to fuse the two figures in the middle; staring at the fused image, relax your eyes. In a few seconds, the stereoscopic image will come into focus. (Reproduced by permission from Matthews et al., 1978)

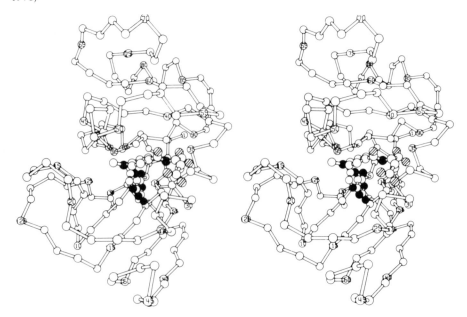

The biochemical role of dihydrofolate reductase (DHFR) is enormously important in a number of biosynthetic pathways. It reduces dihydrofolate to tetrahydrofolate, a cofactor that accepts a one-carbon fragment in several forms (Fig. 6.15) and transfers it to substrates during the synthesis of some amino acids, purines, and pyrimidines. Tetrahydrofolate—or its precursor folate—is a vitamin for humans, and must be acquired from food. It is present abundantly in the green leaves of vegetables [ *folium* (Latin) = leaf]. Bacteria, on the other hand, cannot use external folic acid, relying instead on an obligatory folate synthetic mechanism. In addition, variations in DHFR isozymes in different organisms offer a number of pharmacological targets in accord with the principle of antimetabolite therapy—that is, the use of competitive enzyme inhibitors. DHFR inhibitors have found use as

1. *Sulfanilamide antibacterial agents*, which act on the basis of folate synthesis inhibition. As mentioned above, bacteria must synthesize their own folate. Interference with this process will inhibit nucleic acid biosynthesis and result in bacteriostatic action because cell division ceases.
2. *Antimalarial agents*, of which certain categories have a very high affinity for the DHFR of the malaria parasite—the protozoan *Plasmodium*—and a low affinity for the human isozyme. They are therefore quite specific and useful drugs of moderate toxicity.
3. *Antitumor agents*, which are quite successful in certain malignancies. They primarily inhibit thymidine synthesis, and thus interfere with RNA production.

### 3.5.1. Sulfanilamide Antibacterial Agents

Discovered in 1935 by Domagk in Germany, these agents opened the modern era of bacterial chemotherapy. A tremendously rapid development of new and improved derivatives increased the antimicrobial spectrum and therapeutic ratio of these drugs, and eliminated many side effects. Although the introduction of penicillin and other antibiotics led to some loss of interest in sulfanilamides, a new emphasis began when their utility in coexistence with other antibiotics was recognized. Research in this area also triggered the formulation of many contemporary concepts of drug action, metabolism, and molecular mechanisms, specifically on competitive enzyme inhibitors.

***Mode of Action.*** The mode of action of sulfanilamides became known around 1947, when the structure and biosynthesis of folic acid were elucidated. This compound is built by bacteria from the heterocyclic pteroyl moiety, *p*-aminobenzoate, and glutamate. ***p*-Aminobenzene-sulfonamide (sulfanilamide, 6-48)** is a competitive inhibitor of the synthase enzyme, acting as an "antimetabolite" of *p*-aminobenzoate. Occasionally, the sulfanilamide can even be incorporated into the modified folate, resulting in an inactive compound and thus an inactive enzyme. This theory, proposed by Woods and Fildes in 1940, became the first molecular explanation of drug action.

***Structure.*** The structure of sulfonamides is shown for representative examples in Table 6.5. Among the several thousand compounds in existence, about 25–30 have

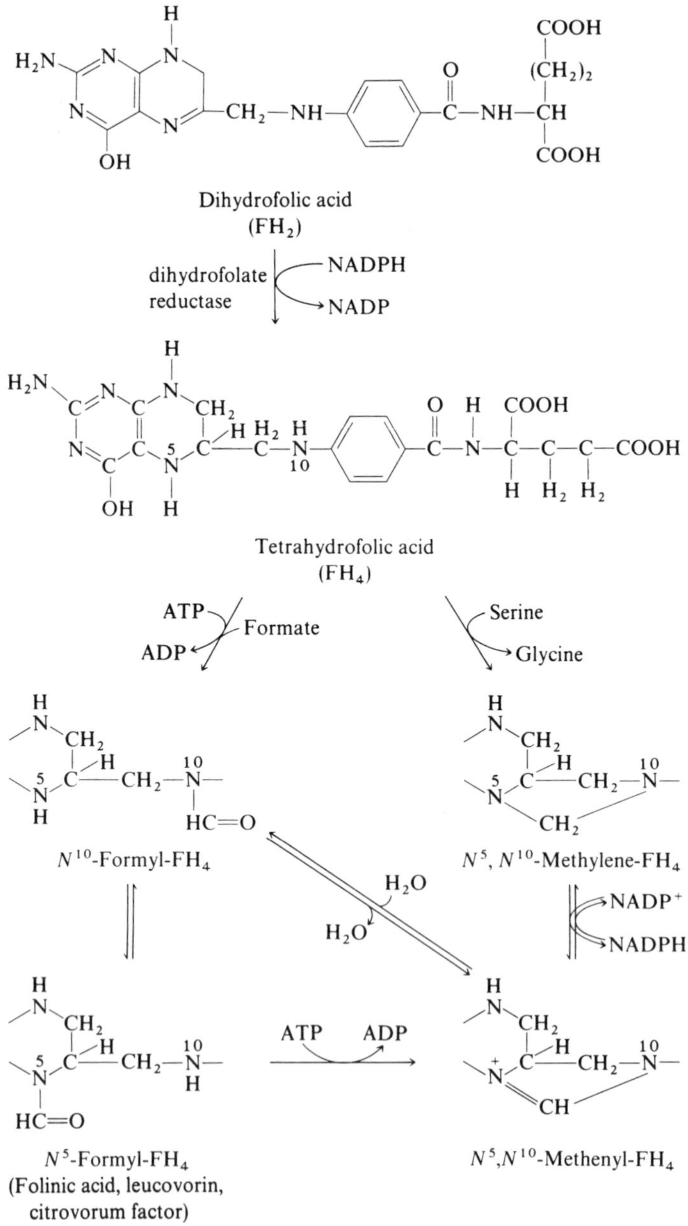

**Fig. 6.15.** Methylene-tetrahydrofolate interconversions, showing various forms of the single carbon atom tetrahydrofolate carries, to be used in purine and pyrimidine biosynthesis. (Reproduced by permission from Pratt and Ruddon (1979), Oxford University Press, New York)

**Table 6.5.** Representative sulfonamides and sulfones

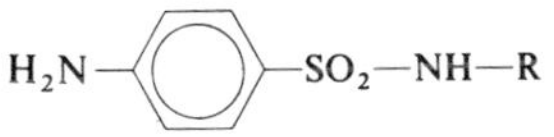

$$H_2N-\text{⟨C₆H₄⟩}-SO_2-NH-R$$

| No. | Generic name | R = | Minimum inhibitory conc. on E. coli ($\mu$M/L) | Half-life (hours) |
|-----|--------------|-----|-----------------------|-------------------|
| 6-48 | Sulfanilamide | —H | 128 | 9 |
| 6-49 | Sulfathiazole | 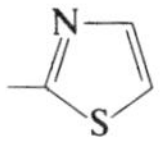 | 1.6 | 4 |
| 6-50 | Sulfisoxazole | 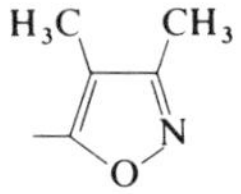 | 2.1 | 6 |
| 6-51 | Sulfamerazine | 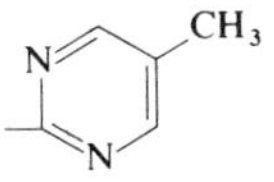 | 0.9 | 24 |
| 6-52 | Sulfamethoxine | 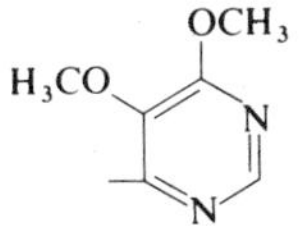 | 0.8 | 150 |
| 6-53 | Acedapsone | 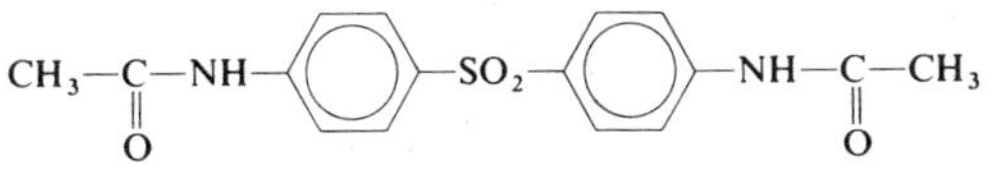 $CH_3-C(=O)-NH-\text{⟨C₆H₄⟩}-SO_2-\text{⟨C₆H₄⟩}-NH-C(=O)-CH_3$ | | 43 days |

found widespread use. Sulfanilamide itself is, by present-day standards, very inactive. It was the development of heterocyclic derivatives that produced the highly potent **sulfathiazole** (6-49). When a succinyl or phthalyl group is attached to the aniline nitrogen (see 8-39), the inactive acylanilide derivatives will not be absorbed from the intestinal tract. Slow deacylation by intestinal enzymes releases the active free sulfanilamide. Therefore, succinyl-sulfathiazole was widely used for intestinal infections. The pyrimidine derivatives, such as **sulfamerazine** (6-51), have a much longer duration of action and a broad antibacterial spectrum, including both Gram-positive and Gram-negative organisms. **Sulfisoxazole** (6-50), although less active, is better tolerated. Among the newer compounds, **sulfamethoxine** (6-52) has the spectacular half-life of about 150 hours, and requires administration only once a week.

Table 6.5 also includes a sulfone, **acedapsone** (6-53), which has found successful use in the treatment and even the clinical cure of the once dreaded leprosy (Hansen's disease), caused by *Mycobacterium leprae* (see also rifampicin, 6-98).

Some DHFR inhibitors are active against the bacterial enzyme even though originally used against the malaria parasite. For instance, **trimethoprim** (6-54; see

Fig. 6.18) is used in combination with sulfa drugs (**co-trimoxazole**), attacking both the synthesis and the proper functioning of DHFR. The therapeutic index, tolerance, delay in the development of resistance, and antimicrobial spectrum of this synergistic combination of two drugs are spectacularly greater than corresponding effects of the individual drugs, to the extent that microorganisms unaffected by either drug alone are successfully eliminated by the combination.

***Structure–Activity Correlations.*** Structure–activity correlations for the sulfa-DHFR inhibitors, derived on the basis of some 5000 compounds, have led to the

**Fig. 6.16.** Life cycle of the malaria parasite and classification of antimalarial drugs. (Reproduced by permission from Sweeney and Strube (1979), in Wolff (Ed.), *Burger's Medicinal Chemistry*, 4th ed., Part II, Wiley, New York)

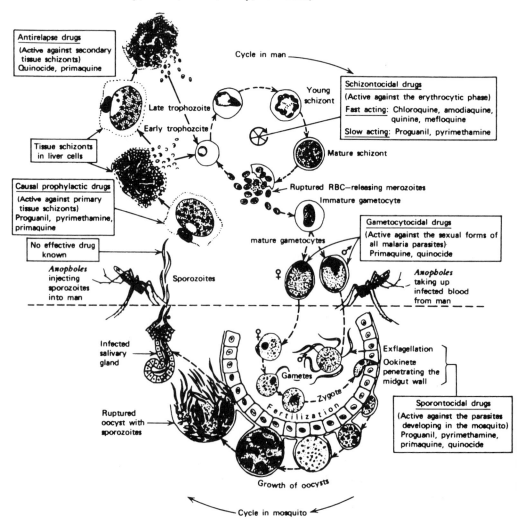

deduction of the following regularities:

1. The amino groups must be in the para position to the sulfonamide group, and this amine must be unsubstituted or become unsubstituted *in vivo*.
2. The benzene ring can be substituted in positions 1 and 4 only.
3. Sulfones, carboxamides, or ketones replacing the sulfonamide may retain activity, but often at a lower level.
4. The sulfonamide nitrogen can be monosubstituted only, and heteroaromatic substituents increase the drug activity.

It is noteworthy that sulfanilamide structural modifications have led to other valuable classes of drugs already discussed, including the hypoglycemic sulfonyl-ureas (Chap. 5, Sec. 2.7) and the diuretic carbonic anhydrase inhibitors (Sec. 3.4).

Correlations between physicochemical parameters and the antibacterial activity of sulfa drugs have been explored for a long time. Bell and Roblin, in an extensive series of drugs, found that the value of the acid dissociation constant of the sulfonamide group ($pK_a$) is essential, the optimum being between 6.0 and 7.4. This $pK_a$ depends on the electronegativity of the $-SO_2-$ group, but the degree of dissociation is also greatly influenced by the intracellular pH. For the optimal balance between intrinsic activity and penetration of the bacterial cell membrane, a 50% ionization is desirable. The hydrophobicity of the drug molecule also plays a role in its activity, as Hansch correlations have shown. Anand (1979) presents an excellent summary of these investigations.

### 3.5.2. Antimalarial Drugs

The very complex life cycle of the malaria parasites, the various *Plasmodium* species (Fig. 6.16) in mosquitoes and humans, offers several potential sites for attacking these organisms. As mentioned previously, there is a considerable difference between the drug sensitivity and affinity of DHFR in humans and *Plasmodium*. Therefore the parasite can be eliminated successfully without excessive toxic effects to the human host.

**Mechanism of Action.** The DHFR inhibitors block the reaction that transforms deoxyuridine monophosphate (dUMP) to deoxythymidine monophosphate (dTMP) at the end of the pyrimidine-synthetic pathway (Fig. 6.17). This reaction, a methylation, requires $N^5$, $N^{10}$-methylene-tetrahydrofolate (see Fig. 6.15) as a carbon carrier, which is oxidized to dihydrofolate. If the dihydrofolate cannot then be reduced back to tetrahydrofolate (THF), this essential step in DNA synthesis will come to a standstill. Although purine synthesis is also impaired by such obstruction, this factor seems to play a role only in the mode of action of some antitumor drugs (see next section).

In principle, sulfanilamides should be effective against malaria parasites by inhibiting folate synthesis. However, although some of these drugs, especially sulfa-pyrazine, are effective to a certain degree, they are much more successful as anti-malarials if used in combination with some of the drugs described below, just as their antibacterial activity is significantly improved by the same combination. The

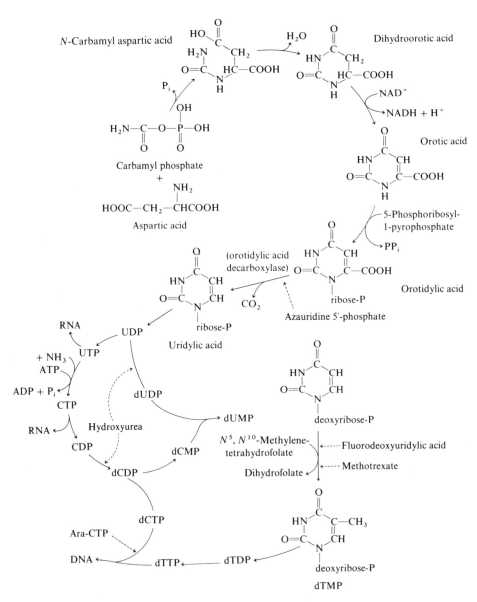

**Fig. 6.17.** The pyrimidine synthesis pathway. The dashed arrows indicate sites of drug inhibition. (Reproduced by permission from Pratt and Ruddon (1979), Oxford University Press, New York)

malaria parasites seem to be able to utilize purines from the host erythrocyte, and need not necessarily synthesize them. Protein synthesis, although also dependent on THF as a carbon carrier, is not impaired in the parasite exposed to DHFR inhibitors.

*Structure–Activity Correlations.* The DHFR-inhibitor antimalarial drugs are competitive inhibitors and therefore structural analogues of folate. As shown in Fig. 6.18, they are either diaminobenzyl- or diaminophenyl-pyrimidines (**trimethoprim**, 6-54; **pyrimethamine** 6-55) or triazines such as **cycloguanil** (6-56), which is the active form of **chlorguanide** (6-57), a pro-drug that undergoes an oxidative ring closure *in vivo.* The replacement of the folate 6-OH (or rather the lactam C=O) by an amino group in these analogues increases their affinity for the enzyme by a factor of $1 \times 10^4$ to $5 \times 10^4$ as compared to folate. Some very insoluble derivatives of cycloguanil show a protective effect for several months after a single intramuscular injection, and may play an important role in the eradication of malaria. Because of the emergence of many resistant strains of *Plasmodium*, combination chemotherapy is imperative. Eradication of the insect vector, the *Anopheles* mosquito, is also an important but elusive goal if endemic malarial areas in Africa and Southeast Asia are to be cleared. Malaria is still one of the major health problems in tropical countries, where tens of millions of people suffer from this debilitating disease.

Antimalarial drugs acting on the nucleic acids of the parasite are discussed in Sec. 5.1.1.

**Fig. 6.18.** Representative dihydrofolate reductase-inhibitor-type antimalarial drugs.

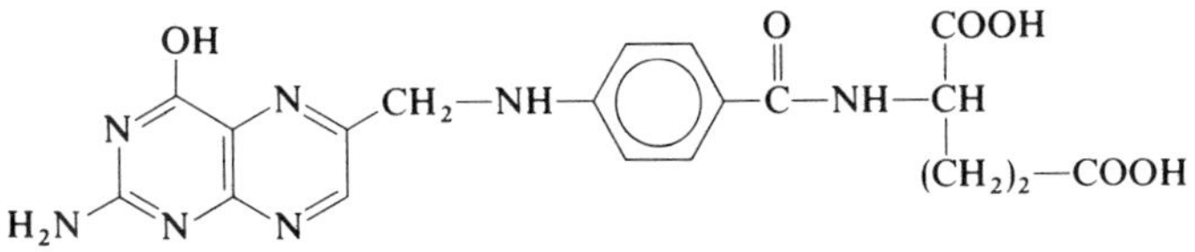

Folate

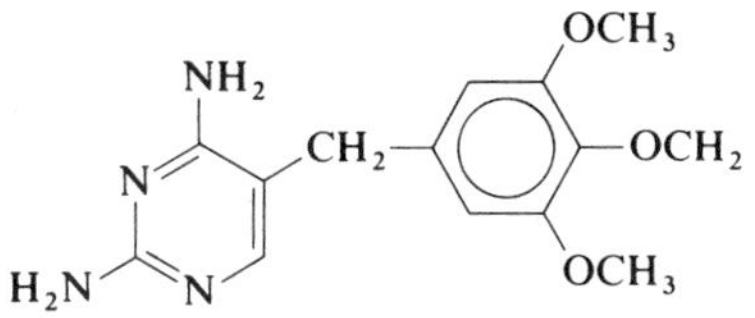

Trimethoprim
6-54

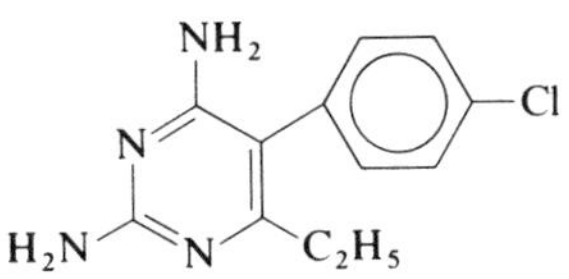

Pyrimethamine
6-55

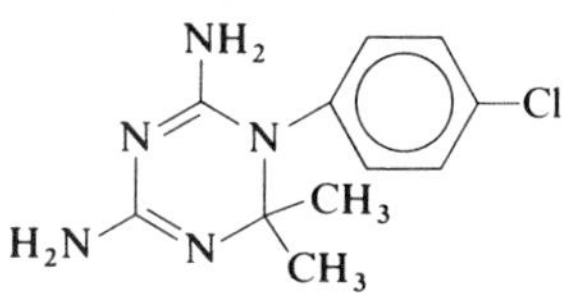

Cycloguanil
6-56

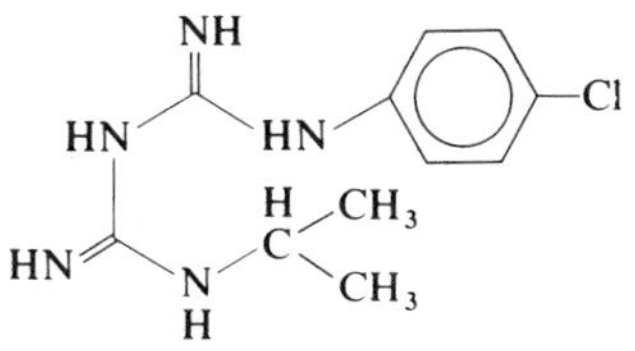

Chlorguanide
6-57

### 3.5.3. Antitumor Agents Acting on DHFR

These agents are less selective than antimalarials because they inhibit the enzyme within different cells of the same species, namely man. They primarily block thymidylate synthesis (Fig. 6.17) as competitive inhibitors, and are structurally very close to folate. **Aminopterin** (6-58) and the more widely used **methotrexate** (6-59, **amethopterin**) have a 6-$NH_2$ substituent like the antimalarials, which provides almost irreversible binding to the enzyme ($K_D = 10^{-9}$ M as compared to $10^{-6}$ M for dihydrofolate). Methotrexate is used in acute childhood leukemia and in choriocarcinoma, which show 30–50% and better than 70% remission rates, respectively. It is therefore one of the most successful antitumor agents. Because of its mode of action, the activity of methotrexate is very dependent on the cell cycle, exhibiting the highest effect during $S$ phase, when DNA synthesis occurs. Since it does not distinguish between normal and malignant cells, it has severe side effects, like most cancer chemotherapeutic agents. In a recent treatment strategy, it was found that these side effects can be partly eliminated; therefore, very high (and very effective) doses are tolerated if the patient is "rescued" after 48 hours by injecting 5-formyl-tetrahydrofolic acid. This compound is the "citrovorum factor" in Fig. 6.15, so named after the microorganism for which it is an essential cofactor. This "rescue" strategy can selectively save normal tissues, which can transform 5-formyl-THF into needed methylating agents and escape the toxic effects of methotrexate. The treatment of osteogenic sarcomas (malignancies connected with bone tissue) is therefore feasible with methotrexate in doses up to 100 times higher than previously thought possible. Recently, the 3′,5′-dichloromethotrexate was found to be equally effective but less toxic.

6-58   Aminopterin (R = H)
6-59   Methotrexate (R = $CH_3$)

### Selected Readings

N. Anand (1979). Sulfonamides and sulfones. In: *Burger's Medicinal Chemistry*, 4th ed. (M. E. Wolff, Ed.), Part 2, Wiley-Interscience, New York, pp. 1–40.

J. R. Bertino (Ed.) (1971). Folate antagonists as chemotherapeutic agents. *Ann. N.Y. Acad. Sci. 186.*

G. H. Hitchings and B. Roth (1980). Dihydrofolate reductases as targets for selective inhibitors. In: *Enzyme Inhibitors as Drugs* (M. Sandler, Ed.). Macmillan, London.

D. A. Matthews, R. A. Alden, J. T. Bolin, D. J. Filman, S. T. Freer, R. Hamlin, W. G. J. Hol, R. L. Kisliuk, E. J. Pastore, L. T. Plante, N. Xuong, and J. Kraut (1978). Dihydrofolate reductase from *Lactobacillus casei*. *J. Biol. Chem. 253*: 6946–6954.

W. B. Pratt and R. W. Ruddon (1979). *The Anticancer Drugs*. Oxford University Press, New York.

R. R. Rando (1980). New modes of enzyme inactivator design. *Trends Pharmacol. Sci. 1*: 168–171.

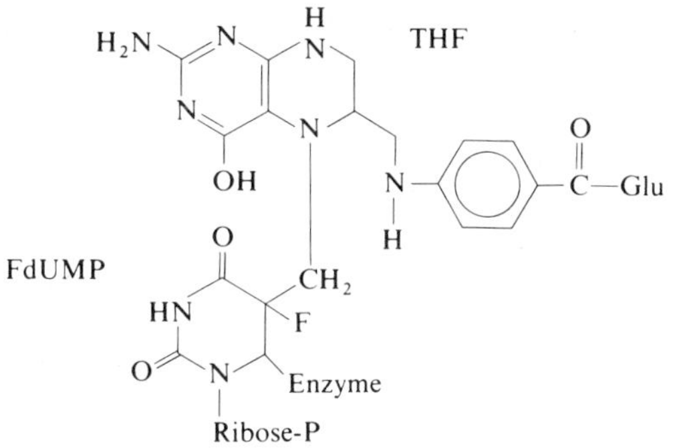

**Fig. 6.19.** Transition-state analogue complex of fluorouridine with tetrahydrofolate and thymidylate synthase. (Reproduced by permission from Langenbaum et al. (1972), *Biochem. Biophys. Res Comm. 48*: 1565)

G. C. K. Roberts (1977). Substrate and inhibitor binding to dihydrofolate reductase. In: *Drug Action at the Molecular Level* (G. C. K. Roberts, Ed.). University Park Press, Baltimore, pp. 127–150.

L. M. Werbel and D. I. Worth (1980). Antiparasitic agents. *Annu. Rep. Med. Chem. 15*: 120–129.

### 3.6. Thymidylate Synthase

Thymidylate synthase (E.C. 2.1.1.45) is the enzyme that methylates UMP to thymidine, using methylene tetrahydrofolate as the carbon carrier (Fig. 6.17). The enzyme can be inhibited directly by analogues of uracil, such as **5-fluorouracil (5-FU)**. The antimetabolite must be in the 5-fluorodeoxyuridine monophosphate (FdUMP) form to become active, and the capability of cells to achieve this transformation is a major determinant of their sensitivity to such drugs.

It was previously thought that 5-FU inhibits the enzyme by classical competitive inhibition. However, it was found (cf. Pratt and Ruddon, 1979) that 5-FU is a transition-state substrate (see Sec. 3.1), and forms a covalent complex with tetrahydrofolate and the enzyme in the same way that the natural substrate does (Fig. 6.19). The reaction, however, will not go to completion, since the fluorouridine derived from the antimetabolite remains attached to the enzyme, and the latter becomes irreversible deactivated. Recovery can occur only through the synthesis of new enzyme. Fluorouracil is used in the treatment of breast cancer and has found limited use in some intestinal carcinomas. Unfortunately, this drug has the side effects usually associated with antimetabolites. Its prodrug, **fluorocytosine** (which is also an antifungal agent) is better tolerated.

### 3.7. Monoamine Oxidase

Monoamine oxidase (MAO) (E.C. 1.4.3.4) is an enzyme found in all tissues and almost all cells, bound to the outer mitochondrial membrane. Its active site contains flavine adenine dinucleotide (FAD), which is bound to the cysteine of a —Ser—Gly—Gly—Cys—Tyr sequence. Ser and Tyr in this sequence suggest a nucleophilic environment, and histidine is necessary for the activity of the enzyme. Thiol reagents inhibit MAO. There are at least two classes of MAO binding sites, either on the same molecule or on different isozymes. They are designated as MAO-A, which is specific for 5-HT (serotonin) as a substrate, and MAO-B, which prefers phenylethylamine. Similarly, MAO inhibitors show a preference for one or the other active site, as discussed below.

Monoamine oxidase catalyzes the deamination of primary amines and some secondary amines, with some notable exceptions. Aromatic amines with unsubstituted $\alpha$-carbon atoms are preferred, but aromatic substituents influence the binding of these substrates. For example, *m*-iodobenzylamine is a good substrate, whereas the *o*-iodo analogue is an inhibitor. The mechanism of deamination is as follows:

$$\overset{\overset{\displaystyle H}{\underset{\displaystyle |}{}}\,\,\,\overset{flavin}{\nearrow}}{R-CH_2-NH_2} \rightarrow R-\overset{\oplus}{C}H-\overset{-}{N}H_2 \xrightarrow{-H^+} R-CH{=}NH \xrightarrow{H_2O} R-CH{=}O + NH_3$$

Hydrolysis of the Schiff base that results from loss of a hydride ion on an $\alpha$ proton yields an aldehyde, which is then normally oxidized to the carboxylic acid. Aromatic substrates are probably preferred because they can form a charge-transfer complex with the FAD at the active site, properly orienting the amino group of the substrate and decreasing the energy of the transition state.

Monoamine oxidase has some important physiological roles:

1. It inactivates many of the neurotransmitters in the synaptic gap or in the synapse, if the latter are not protected by synaptic vesicles. The metabolism of NE, DA, 5-HT, tyramine, and histamine is thus taken care of by MAO as well as by some other enzymes (see, for instance, Chap. 4, Sec. 3.2).
2. It detoxifies exogenous amines and may even help to maintain the blood–brain barrier, since it is also localized in the walls of the blood vessels.

### 3.7.1. Monoamine Oxidase Inhibitors

Monoamine oxidase inhibitors (MAOI) are useful as thymoleptic (antidepressant) drugs, especially since the action of some of these agents is very rapid, as compared to the lag period of days or even weeks shown by tricyclic antidepressants. All MAOIs act by increasing the available concentration of the neurotransmitters NE and 5-HT which, because they are not metabolized, accumulate in the synaptic gap and exert an increased postsynaptic effect. The drugs show hypotensive activity as a side effect, and some MAOIs are used as hypotensive drugs.

There are four structural types of MAOI, as shown in Fig. 6.20. These are hydrazines, cyclopropylamines, propargylamines, and carbolines.

*Hydrazines.* The *hydrazines* have only historic significance. The entire group of MAOIs was discovered through the euphoric side effect of **isoniazid** (isonicotinylhydrazide), a successful antituberculotic drug introduced in 1952. **Iproniazid** (6-62) is the corresponding isopropyl derivative. All of the hydrazides are highly hepatotoxic, and are no longer available.

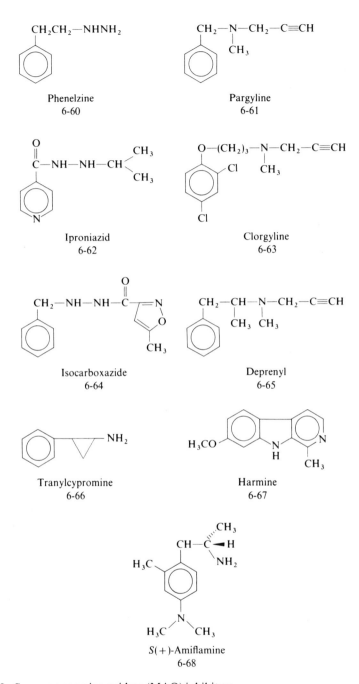

**Fig. 6.20.** Some monoamine oxidase (MAO) inhibitors.

393

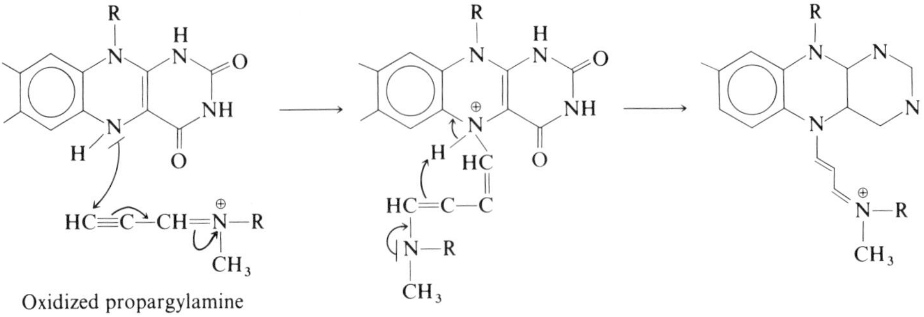

**Fig. 6.21.** Addition of a propargylamine monoamine oxidase inhibitor to the flavine coenzyme, forming a covalently bound suicide substrate.

*Cyclopropylamines.* **Tranylcypromine** (6-66) (*trans*-phenyl-cyclopropylamine) can be regarded as a ring-closed derivative of amphetamine, and therefore provides rapid stimulation as well as protracted effect. Like all MAOI drugs, it can produce severe or even fatal hypertensive crises if taken together with foods containing tyramine, such as cheese. Preventing the destruction of such a pressor amine produces a sudden increase in blood pressure, which is especially dangerous in individuals exhibiting high blood pressure. Therefore, depressive patients taking tranylcypromine must practice diet control and avoid tyramine-containing foods. Reversible MAOIs do not show this "cheese effect."

*Propargylamines.* The propargylamines are $K_{cat}$ inhibitors or suicide substrates of MAO, forming covalent derivatives with the flavine group of the enzyme, as shown in Fig. 6.21. **Pargyline** (6-61), while a MAOI, is used as a hypotensive agent, even though it is incompatible with certain foods and sympathomimetic drugs (i.e., adrenergic agonists). It may act by negative feedback on norepinephrine synthesis, and has a long duration of action.

   **Clorgyline** (6-63) is a selective MAO-A inhibitor, and does not seem to cause hypertensive crises. **Deprenyl** (6-65) is a selective MAO-B inhibitor, and produces an increase in DA levels, but does not influence NE or 5-HT concentrations. It has been proposed as an antidepressant in aging males (Knoll, 1982).

   Among the newer compounds **S(+)-amiflamine** (6-68) is noteworthy, as a nonhepatotoxic, competitive, reversible MAO-A inhibitor.

*Carbolines.* **Harmine** (6-67) and related carboline alkaloids are reversible MAO-A inhibitors, but are not used therapeutically.

### Selected Readings

J. Knoll (1982). Selective inhibition of B-type monoamine oxidase in brain: a drug strategy to improve quality of life in senescence. In: *Strategy in Drug Research* (J. A. Keverling Buisman, Ed.), Elsevier, Amsterdam.

R. A. Maxwell and H. L. White (1978). Tricyclic and monoamine oxidase inhitor antidepressants: structure–activity relations. In: *Handbook of Psychopharmacology.*

(L. L. Iversen, S. D. Iversen, and S. H. Snyder, Eds.), Vol. 14. Plenum Press, New York, pp. 83–155.

W. B. Pratt and R. W. Ruddon (1979). *The Anticancer Drugs*. Oxford University Press, New York.

M. Sandler (Ed.) (1980). *Enzyme Inhibitors as Drugs*. Macmillan, London.

M. Strolin Benedetti and P. Dostert (1985). Stereochemical aspects of MAO interactions: reversible and selective inhibitors of monoamine oxidase. *Trends Pharmacol. Sci. 6:* 246–251.

J. Van Dijk, J. Hartog, and F. C. Hillen (1978). Non-tricyclic antidepressants. *Progress in Medicinal Chemistry* (G. P. Ellis and G. B. West, Eds.), Vol. 15. Elsevier/North Holland, Amsterdam, pp. 261–320.

## 4. VITAMINS

Vitamins are low-molecular-weight organic compounds that in almost all cases, function as coenzymes. As such, they are indispensable in a multitude of biochemical reactions; however, higher organisms cannot biosynthesize many of them. Those that must be supplied in the human diet are classified in the vitamin category because their absence leads to deficiency diseases. Therefore, although strictly speaking, vitamins should be and are discussed in biochemistry textbooks, they also have therapeutic importance. Hence they are of interest to the medicinal chemist, even though they belong to a mature and slow-moving field.

The knowledge of some vitamins reaches back into folk medicine: Amerindians treated scurvy with cedar leaf tea (*Thuja*), and, from the seventeenth century on, the British Navy issued lime juice to its sailors to prevent scurvy. This is the origin of the slang-word "limey" for the British. The use of fish-liver oils for the treatment of rickets, a bone-growth disorder, has been known since the nineteenth century. These uses all predate the recognition of the essential function of these substances and the term "vitamine," coined by C. Funk in 1912, and meaning a basic substance essential for life. Since then, many nonbasic vitamins that are not amines have been found, and the letter "e" has therefore been dropped from the end of the word.

On the basis of their solubility and polarity, vitamins are divided into two categories: the water-soluble and the fat-soluble vitamins. The former cannot be stored in any biocompartment, and must be ingested regularly. In that context, it is important to caution about the many and periodically recurring fads and myths embraced by the lay public and press. Since vitamins are widely known if not understood, there is a need for educating the public about their misuse, as well as in other areas of pseudoscientific misunderstanding (Ovesen, 1984).

### 4.1. Water-Soluble Vitamins

The structures and biochemical properties of the water-soluble vitamins are shown in Fig. 6.22 and Table 6.6.

#### 4.1.1. Vitamin $B_1$ (Thiamine)

Vitamin $B_1$ (thiamine) is a pyrimidinyl methyl thiazolium derivative that occurs in bran (mostly rice hulk), beans, nuts, egg yolk, yeast, and vegetables. In its pyrophosphate form, it is a coenzyme of pyruvate dehydrogenase, which oxidatively

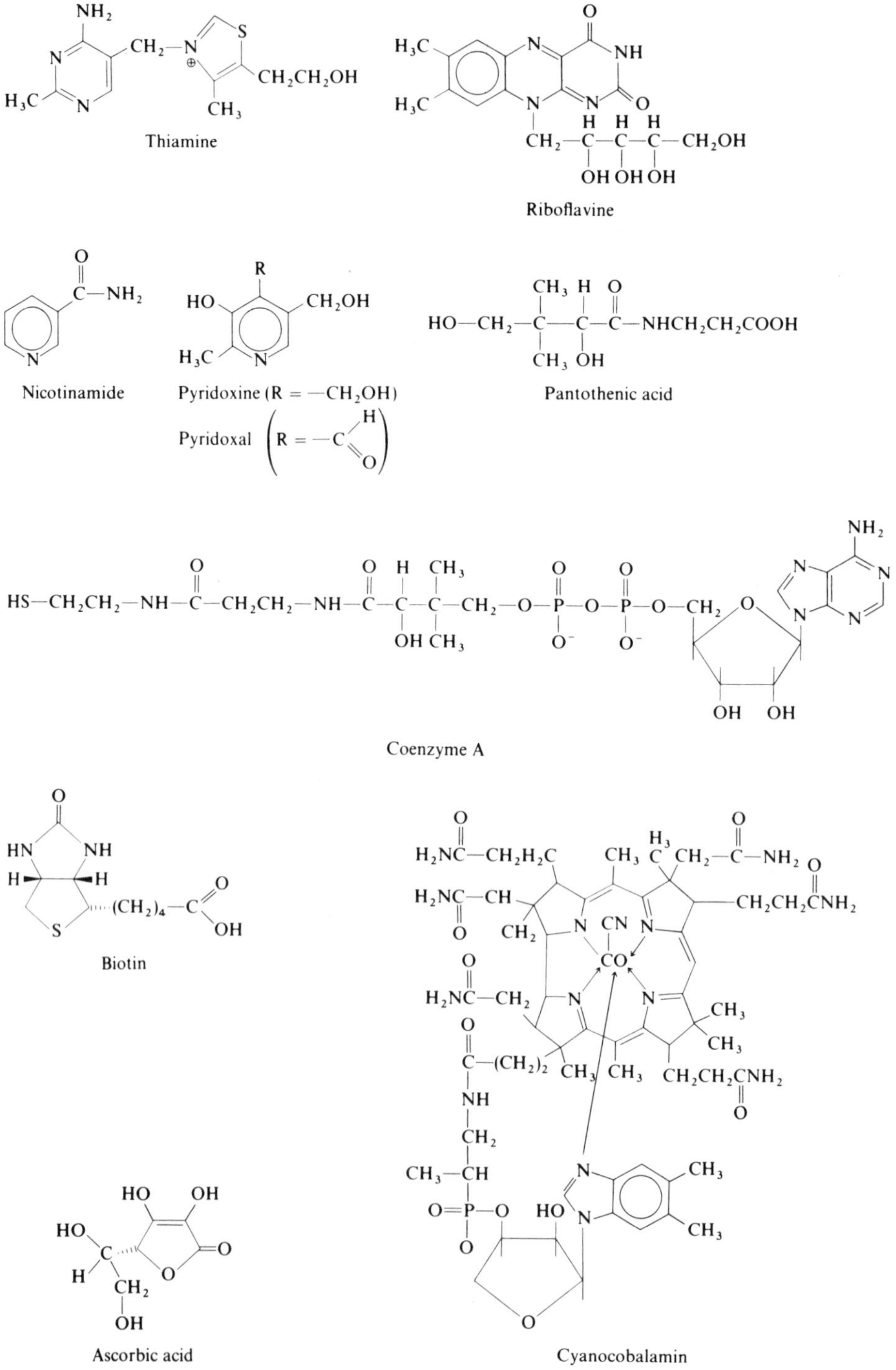

**Fig. 6.22.** Water-soluble vitamins.

396

**Table 6.6.** Water-soluble vitamins

| Vitamin | Coenzyme of | Reaction catalyzed | Group carried | Deficiency disease |
|---------|-------------|--------------------|---------------|--------------------|
| B$_1$, thiamine | Pyruvate dehydrogenase; transketolase | Pyruvate to acetyl-CoA; pentose phosphate pathway | Two-carbon groups | Beriberi |
| B$_2$, riboflavine | Flavine-adenine dinucleotide (FAD); monoamine oxidase | Electron transport; amine degradation | Electrons | Vague, dermatitis |
| Nicotinamide Niacin | Nicotinamide-adenine dinucleotide (NAD, NADP) | Electron transport | Electrons | Pellagra |
| B$_6$, pyridoxine | Transaminase; decarboxylase; racemase; aldolase | Schiff base formation; electron sink | | Similar to riboflavin and niacin deficiency |
| Pantothenic acid | Coenzyme A; fatty acid carrier protein (ACP) | Krebs cycle; lipid synthesis | Acetate; C$_2$–C$_{16}$ fatty acids | Nonspecific |
| Biotin | Carboxylases | Gluconeogenesis; fatty-acid synthesis | CO$_2$ | Unknown |
| Folate (see Sec. 3.5) | Synthases | Serine, purine, and pyrimidine synthesis | One-carbon groups | Macrocytic anemia |
| B$_{12}$, cobalamin | Cobamide enzymes | Rearrangements; methylations | | Pernicious anemia (indirectly) |
| C, ascorbate | Hydroxylases | Collagen synthesis; adrenocorticoid hydroxylation | | Scurvy |

decarboxylates pyruvate to form acetyl-coenzyme A. Consequently, it is a very important coenzyme, catalyzing the connecting step between glycolysis and the Krebs cycle. The carbon between the sulfur and nitrogen of the thiazole ring of the vitamin is very acidic and adds to the carbonyl of pyruvate, leading to decarboxylation.

Thiamine is also a coenzyme of transketolase, which transfers two-carbon fragments between carbohydrates in the pentose phosphate pathway (pentose shunt), a multipurpose metabolic reaction sequence.

Vitamin B$_1$ deficiency, known as beriberi, is seen primarily in people of Southeast Asia, whose unbalanced diet consists mainly of polished rice. The symptoms of beriberi are neurological disorders (weakness, paralysis, painful neuritis), diarrhea, loss of appetite, dermatitis, and anemia. These symptoms are due mainly to the accumulation of pyruvate and lactate.

Some analogues of thiamine are biologically active, but have not attained wide use because thiamine itself is cheaply available.

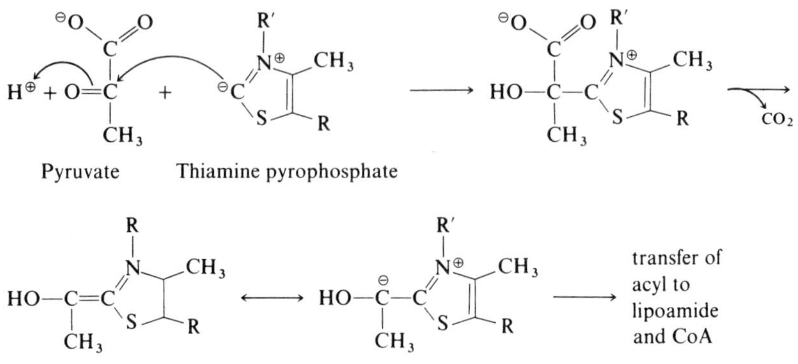

Pyruvate          Thiamine pyrophosphate

transfer of
acyl to
lipoamide
and CoA

### 4.1.2. Vitamin B₂ (Riboflavine)

Vitamin $B_2$ (riboflavine) is a benzopteridine derivative carrying a ribityl (reduced ribose) side chain. It occurs in almost all foods, the largest amounts being found in eggs, meat, spinach, liver, yeast, and milk. Riboflavin is one of the major electron carriers, as a component of flavine-adenine dinucleotide (FAD), which is involved in carbohydrate and fatty acid metabolism. A hydride ion and a proton are added to the pyrazine ring of the vitamin in the following way. In this way, FAD carries a pair of electrons and two hydrogens.

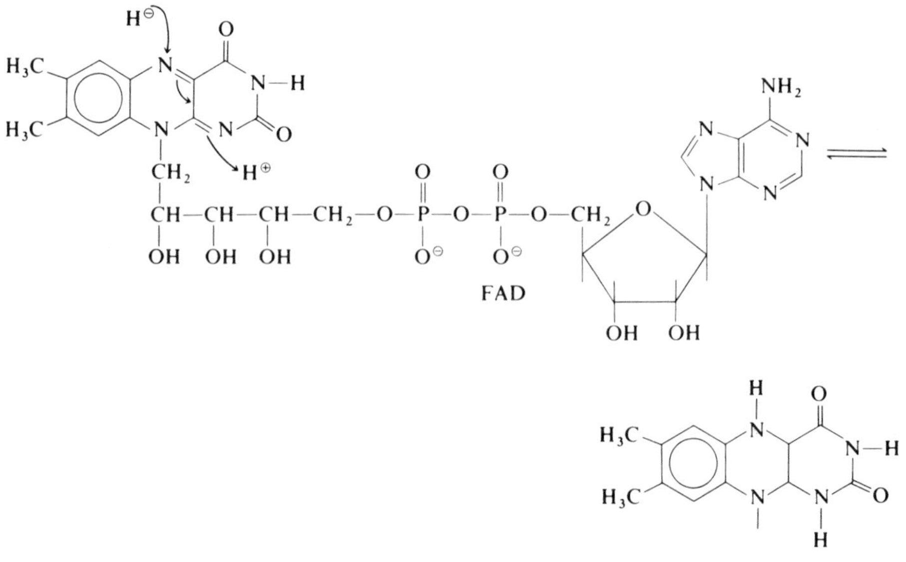

FAD

FADH₂

The monoamine oxidases (see Sec. 3.7) are also flavoenzymes. Vitamin $B_2$ deficiency is seldom seen, and its symptoms of leg ulceration (cheilosis), skin symptoms, a purplish tongue, eye disturbances, and photophobia are vague.

### 4.1.3. Nicotinamide and Nicotinic Acid (Niacin)

Nicotinamide and nicotinic acid (niacin)—which have also been referred to as vitamin $B_3$ or $B_5$—are simple pyridine-3-carboxylic acid derivatives occurring in liver, yeast, and meat. In the form of nicotinamide-adenine dinucleotide $(NAD^+)$ or its phosphorylated form $(NADP^+)$, nicotinamide is the most important electron carrier in intermediary metabolism. Unlike FAD, it adds a hydride ion (i.e., one pair of electrons and one hydrogen) only. Nicotinamide deficiency leads to pellagra [*pelle agra* (Italian) = rough skin], which manifests itself in symmetrical dermatitis, pigmentation, a red and inflamed tongue, diarrhea, and weakness. People who consume large amounts of corn in an unbalanced diet are prone to develop the disease.

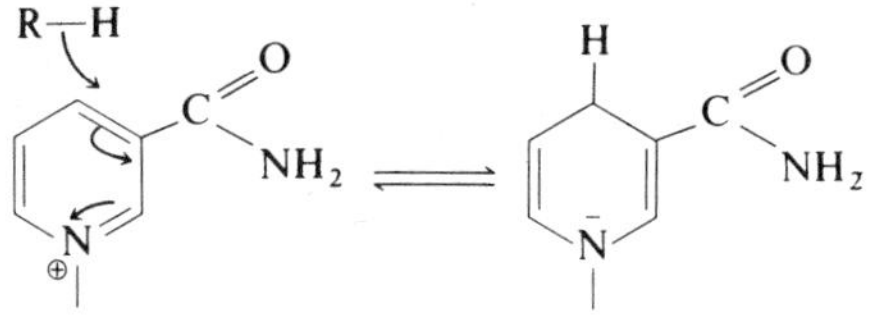

### 4.1.4. Vitamin $B_6$ (Pyridoxine)

Vitamin $B_6$ (pyridoxine) is a pyridine-alcohol, but its biologically active forms are pyridoxal 5-phosphate and the corresponding pyridoxamine. Like all the members of the vitamin B complex, it occurs in yeast, bran, wheat germ, and liver. It is a coenzyme of a remarkable number of enzymes, most of which are involved in amino acid metabolism. By forming a Schiff base, pyridoxal phosphate acts as an electron sink and weakens all three bonds about the α-carbon of amino acids, as shown in Fig. 6.23. It therefore functions in transamination, decarboxylation, deamination, racemization, and aldol cleavage, and can also influence the β and γ carbons of an amino acid.

Glutamic acid decarboxylase (Chap. 4, Sec. 8.1.1) is also a pyridoxal enzyme, and is involved in GABA synthesis. If it is inhibited by hydrazine-type compounds

**Fig. 6.23.** Pyridoxal phosphate coenzyme forms a Schiff base and labilizes three bonds of the substrate, facilitating (a) transamination, (b) decarboxylation, and (c) aldolase reaction. (After Stryer, 1981)

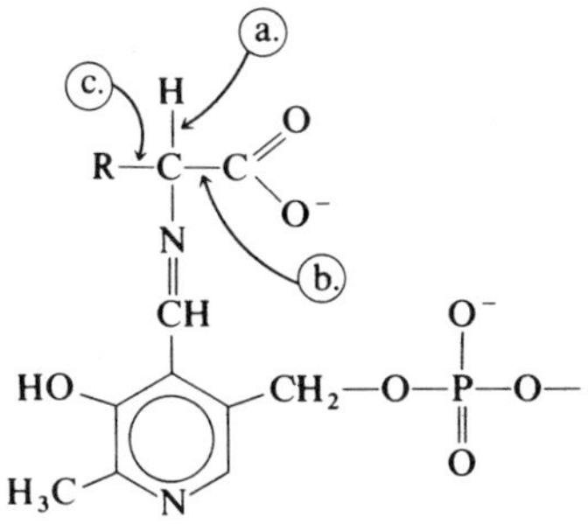

through Schiff-base formation, the resulting lack of GABA will lead to seizures. The typical $B_6$ avitaminosis is very similar to riboflavine and niacin deficiency and is manifested by eye, mouth, and nose lesions as well as neurological symptoms.

Pyridoxine is occasionally used in treating the nausea of "morning sickness" in pregnant women, as well as in parkinsonism and idiopathic epilepsy.

### 4.1.5. Pantothenic Acid

Pantothenic acid, a hydroxyamide, occurs mainly in liver, yeast, vegetables, and milk, but also in just about every other food source, which is responsible for its name [pantos (Greek) = everywhere]. It is part of coenzyme A, the acyl-transporting enzyme of the Krebs cycle and lipid syntheses, as well as a constituent of the acyl carrier protein in the fatty-acid synthase enzyme complex.

There is no record of pantothenic acid deficiency in humans, since all food contains sufficient quantities of this vitamin. Experimentally, however, neurological, gastrointestinal, and cardiovascular symptoms result from a diet lacking in pantothenic acid.

### 4.1.6. Biotin

Biotin, a thiophene-lactam, occurs in yeast, liver, kidney, eggs, vegetables, and nuts. It functions as a cocarboxylase in a number of biochemical reactions. It binds $CO_2$ in the form of an unstable carbamic acid on one of the lactam nitrogens. The carbamate carboxyl is then donated easily.

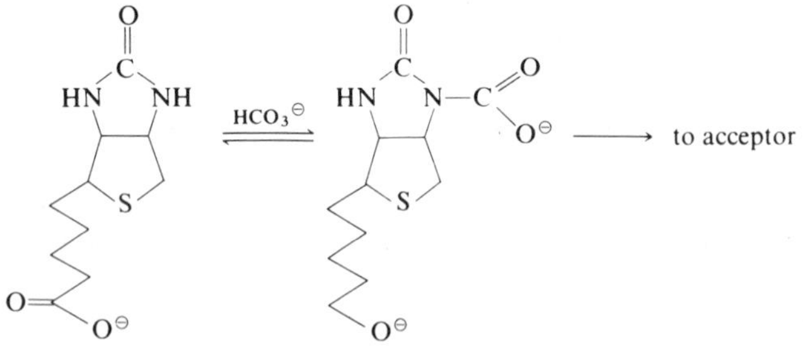

Carrier

Biotin deficiency can be triggered only experimentally, using diets rich in raw egg white. Raw egg white contains avidin, a 70-kD protein that binds biotin in an inactive form. The deficiency leads to dermatitis and hair loss in rats.

### 4.1.7. Vitamin $B_{12}$ (Cobalamin)

Vitamin $B_{12}$ (cobalamin) is an extremely complex molecule, consisting of a corrin ring system similar to heme. The central metal atom is cobalt, coordinated with a ribofuranosyl-dimethylbenzimidazole. In cyanocobalamin, the sixth coordination of cobalt can be with cyanide. However, this is an isolation artifact, and cyanide (in this nontoxic covalent form) can be replaced by other anions. Vitamin $B_{12}$ occurs in liver, but is also produced by many bacterial and is therefore obtained commercially by fermentation.

The vitamin is a catalyst for the rearrangement of methylmalonyl-CoA to the succinyl derivative in the degradation of some amino acids and the oxidation of fatty acids with an odd number of carbon atoms. It is also necessary for the methylation of homocysteine to methionine.

Lack of $B_{12}$ leads to pernicious anemia, a rare but invariably fatal disease if untreated. In 1926, Minot and Murphy discovered that about 200 g of raw liver per day could keep patients with the disease alive. The vitamin itself was isolated only in 1948. As little as $3-6 \times 10^{-6}$ g is curative, and the large amounts of the vitamin taken by some fadists are therefore totally unnecessary and without benefit in any disease other than pernicious anemia and total gastrectomy. In these patients, the uptake of the vitamin is impaired because they lack the so-called "intrinsic factor"— a gastric glycoprotein necessary for the absorption of the vitamin. The administration of very large amounts (milligrams) of vitamin $B_{12}$ forces the uptake of sufficient cobalamin to cover the need, but is only a treatment and not a cure for pernicious anemia. $B_{12}$ avitaminosis due to lack of the vitamin is unknown, since food and the intestinal flora easily cover the daily requirement of $3-4$ μg.

### 4.1.8. Vitamin C (Ascorbic Acid)

Vitamin (ascorbic acid) is a carbohydrate derivative. It occurs in citrus fruits, green pepper, and fresh vegetables, including onions, but is rapidly hydrolyzed by cooking. Only primates and the guinea pig are unable to synthesize ascorbic acid. Vitamin C is necessary for the hydroxylation of proline to hydroxyproline, reducing the Fe atom in collagen hydroxylase. Hydroxyproline is a major amino acid in the collagen that is present in all fibrous tissues, the intracellular matrix, and capillary walls. Vitamin C is also used in steroid hydroxylation in the adrenal gland, and finds a role in tyrosine metabolism.

In vitamin C avitaminosis, or scurvy, the joints become painful and the gums bleed and deteriorate, resulting in tooth loss. Gangrene and infections may also set in, and wounds do not heal. These symptoms all result from impaired collagen synthesis, and require $3-4$ months to develop.

Large doses of vitamin C are given after surgery, burns, and stress to promote healing by increasing collagen synthesis. There is controversy about its purported activity of alleviating the common cold and other viral diseases in extremely large doses. Experiments with radioisotope-labeled ascorbic acid have shown that the normal body pool is about 20 mg/kg, which can be maintained by an intake of approximately 100 mg/day (140 mg for smokers) (Hornig, 1982). The consumption of several grams per day—as advocated by believers in "megavitamin" hypotheses—seems to result in acidosis (a decreased serum pH), gastrointestinal irritation, and the excretion of excess vitamin C.

### 4.2. Fat-Soluble Vitamins

Properties of the *lipid-soluble vitamins* are summarized in Table 6.7 and Fig. 6.24, respectively.

**Table 6.7.** Lipid-soluble vitamins

| Vitamin | Physiological role | Metabolic action | Deficiency disease |
|---|---|---|---|
| A, retinol | General metabolism Vision | Rhodopsin, glycogen, and mucoprotein formation | Night blindness Xerophthalmia |
| D, calciferol | Bone formation | $Ca^{2+}$ and $PO_4^{3-}$ metabolism | Rachitis (rickets) |
| E, tocopherol | Fertility in rats | Unknown Antioxidant (?) | Unknown |
| K | Blood clotting | Glutamate carboxylation in prothrombin | Bleeding |

**Fig. 6.24.** Lipid-soluble vitamins and some vitamin K antagonists.

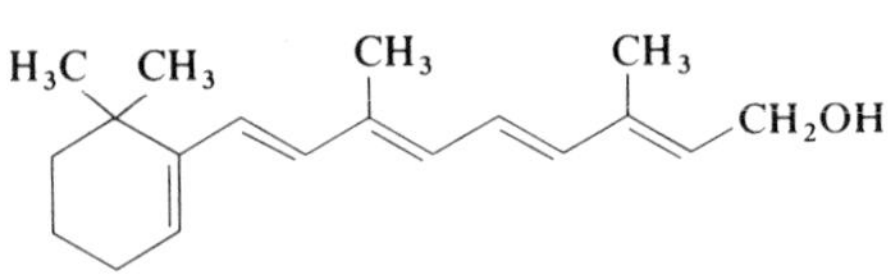

Retinol
(all *trans*)

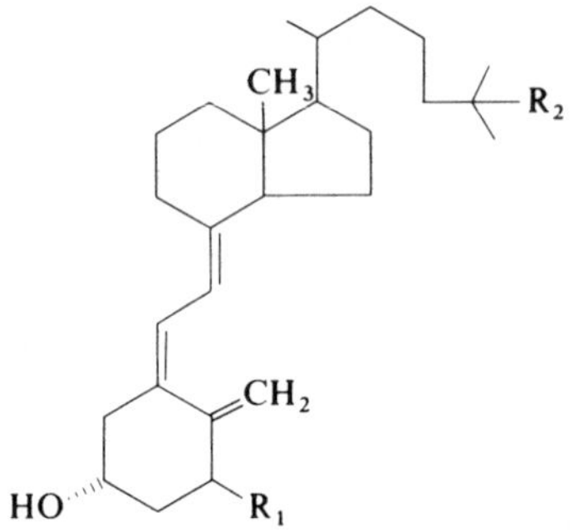

Cholecalciferol ($R_1$, $R_2$ = H)
$1\alpha$, 25-$(OH)_2D_3$ ($R_1$, $R_2$ = OH)

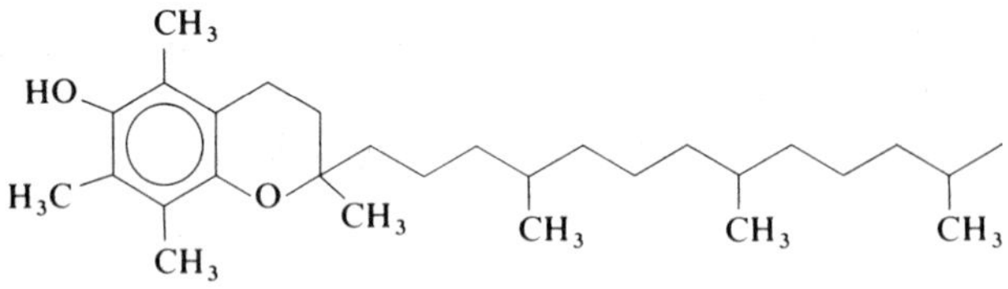

α-Tocopherol

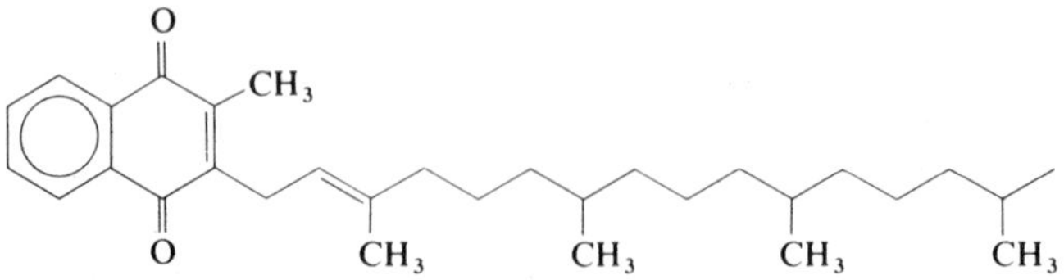

Vitamin K$_1$

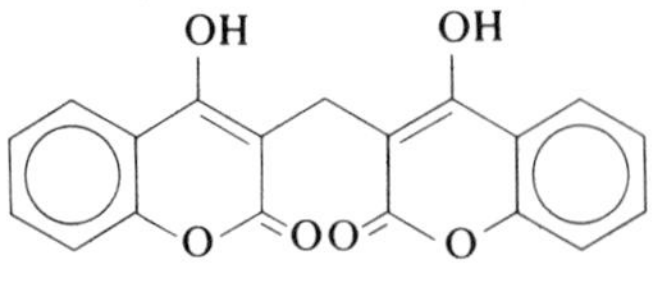

Dicumarol

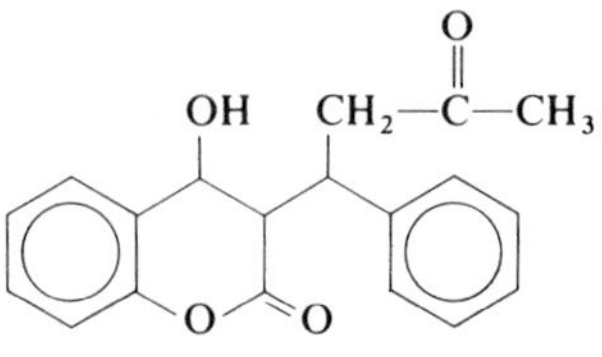

Warfarin

402

### 4.2.1. Vitamin A (Retinol)

Vitamin A (retinol) occurs in plants as the provitamin **carotene**; it is a highly unsaturated terpenoid, with all of its double bonds *trans* (*E*) to one another. Fish liver oils, milk, and eggs contain vitamin A itself, a cleavage product of $\beta$-carotene. It is necessary for the growth of young animals since it regulates bone-cell formation and the shape of bones. It also plays a vital role in vision, where *cis*-**retinal**, an isomer of retinol, is formed upon irradiation with light and becomes attached to the protein rhodopsin as a Schiff base. Light energy is thus transformed into atomic motion, resulting in hyperpolarization of the plasma membrane of the rods in the retina. A single photon can block the $Na^+$ permeability of rod membranes amplifying the response a million fold. The effect probably occurs through the intervention of cGMP, resulting in a nerve impulse. While many details of the vision process are not well understood, Stryer (1981) gives an excellent summary of it. Avitaminosis A results in the loss of night vision (nyctalopia).

In general metabolism, vitamin A influences transport processes and glycogen synthesis, and in its absence the cornea of the eye dries out (xerophthalmia). Orfanos et al. (1981) have reported considerable progress in the synthesis and use of other retinoids. The aromatic retinoid **etretoin** and its ester **etretinate** (6-71) are effective in the treatment of psoriasis, a disorder of skin cell keratinization. **13-*cis*-Retinoic acid (isotretinoin, 6-70)** produces sebaceous gland atrophy and could prove useful in the treatment of severe *acne vulgaris*. Although these compounds have toxic side effects and are not in regular use, they have opened up new therapeutic possibilities. **Retinoic acid (tretinoin, 6-69)** is currently employed in the treatment of acne.

Excessive intake of vitamin A can result in severe and even fatal toxicity.

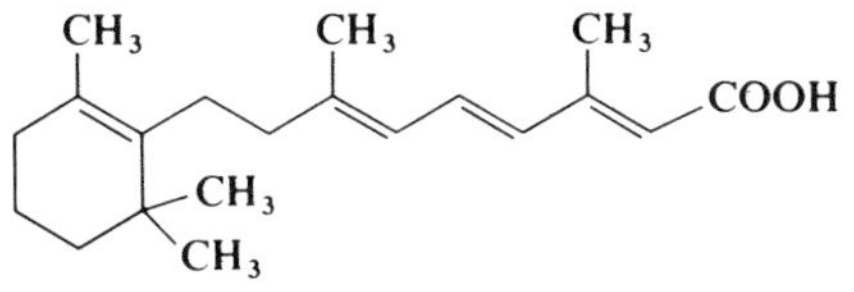

Tretinoin
6-69

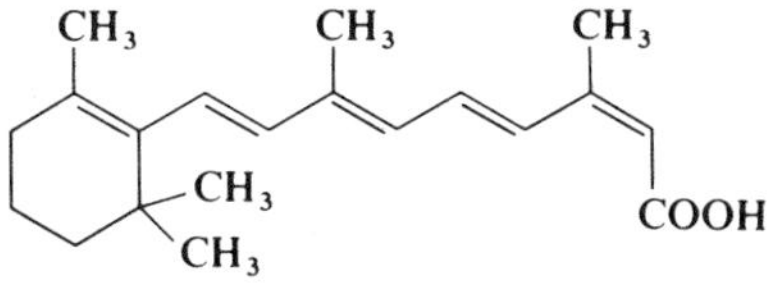

Isotretinion
6-70

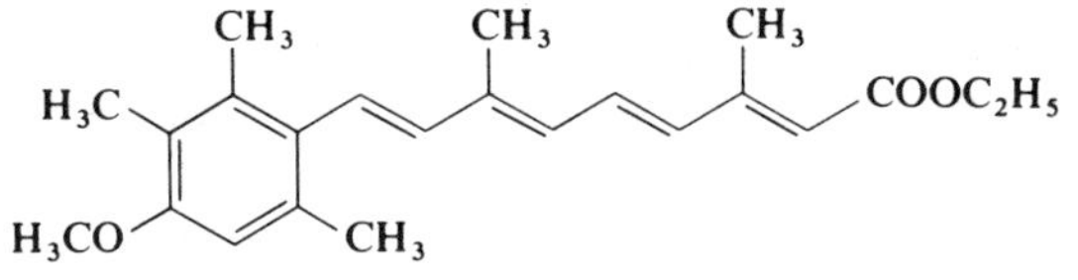

Etretinate
6-71

### 4.2.2. Vitamin D (Calciferol)

Vitamin D (calciferol) is found in fish-liver oils and milk. It is also produced in human skin from a steroid previtamin by sunlight. Ergosterol is another pre-vitamin D compound, and undergoes photo-rearrangement to vitamin $D_2$, whereas cholesterol and cholesterol derivatives produce vitamin $D_3$ (Fig. 6.25). The two

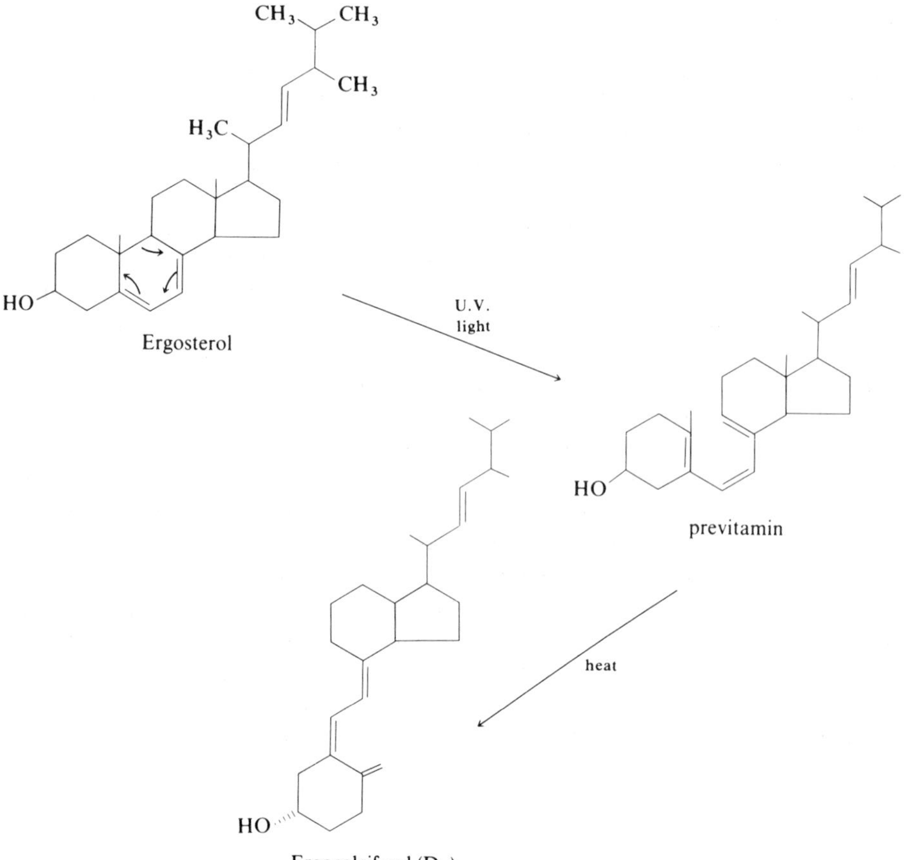

**Fig. 6.25.** Transformation of ergosterol to vitamin $D_2$. Ultraviolet light cleaves the 9,10 bond, resulting in a double-bond shift. Heat causes a *cis–trans* isomerization in the vitamin.

vitamins have the same activity in humans. These photo-rearrangements, resulting in the cleavage of the steroid B ring, are used industrially to manufacture these vitamins.

The active forms of the D vitamins are **1α, 25-dihydroxy-vitamin $D_3$** and 25-hydroxy-vitamin $D_3$. They are formed by enzymatic hydroxylation in the liver microsomes and then in the kidney mitochondria by a ferredoxin flavoprotein and cytochrome P-450 (see Chap. 7, Sec. 2.1). The 1,25-dihydroxy vitamin is then transported to the bone, intestine, and other target organs (kidneys, parathyroid gland). Consequently, it can be considered a hormone, since it is produced in one organ but used elsewhere. It mobilizes calcium and phosphate and also influences the absorption of these ions in the intestine, thus promoting bone mineralization. The hormone is also active in relieving hypoparathyroidism and postmenopausal osteoporosis, which, for example, results in the brittle bones of elderly women.

**Calcitriol** ($1\alpha$,25,25-trihydroxycholecalciferol) is a new drug used in these conditions.

The vitamin D receptor has been located in the intestinal mucosa and in skin. Preliminary purification data indicate that it is a 72-kD protein with a $K_D$ of $5-7 \times 10^{-11}$ M. De Luca and Schnoes (1984) provide a good summary of the latest results in this field as well as an overview of active analogues of the vitamin, such as $24,24\text{-F}_2\text{-}25\text{-OH-D}_3$. The O'Malley group succeeded in cloning the cDNA of the receptor protein and described its amino acid sequence, which shows homology with other steroid receptors and some oncogene products. Thus vitamin D is recognized as a true steroid by the receptor even though it is a *seco*-steroid (split steroid) (McDonnell et al., 1987). The area of vitamin D research seems to be exceptionally active in the generally quiescent vitamin field.

Overall calcium metabolism is regulated by the parathyroid gland (parathormone) and the thyroid hormone calcitonin (Chap. 5, Sec. 2.61). Parathormone regulates the synthesis of $1,25\text{-}(OH)_2D_3$, which manages bone and kidney calcium metabolism as well as promoting $Ca^{2+}$ intake in the gut. The role of calcitonin is less clear, although it works against parathormone. Serum calcium represses parathormone synthesis, completing a feedback loop (Fig. 6.26).

Vitamin D deficiency, known as rickets, was described in children as early as 1645 by Daniel Webster, in gloomy, sunless England. Arab women wearing black veils also can show osteoporosis, since they are not exposed to sunshine at all. In many Western countries, milk is fortified with about 400 units of vitamin $D_3$ (10 $\mu g$/liter), the minimum daily requirement for children or adults. Therefore, with a proper diet and sunshine, rickets should not occur in children. Pregnant women have a higher requirement, but overdosage of the vitamin leads to toxic symptoms.

### 4.2.3. Vitamin E (Tocopherol)

Vitamin E (tocopherol), a chromane, occurs in just about all vegetables as well as in oils, grain, milk, meat, and yeast. Its principal known action is the maintenance of normal pregnancy in rats kept on a special diet. Vitamin E-deficient rats reabsorb their fetuses even though developing normally in every other respect. There is no evidence for this in humans. The vitamin has antioxidant properties, probably stabilizing vitamin A and unsaturated fatty acids and preventing free-radical reactions. Free radicals, according to some hypotheses, may be involved in aging. Vitamin E deficiency is unknown in humans, but there are periodic fads promoting the intake of vitamin E. There is no known rationale for this, since vitamin E seems to be abundant in the normal diet. (Ovesen, 1984).

### 4.2.4. Vitamin K

Vitamin K is a phytyl-naphthoquinone occurring in the green leaves of most plants. The several related active compounds differ in the length of the phytyl sidechain. In the absence of the vitamin, the blood-clotting time increases, since the posttranslational carboxylation of several glutamate residues in prothrombin and other factors involved in blood clotting is impaired. In humans, this disorder is unknown

## VITAMIN D ENDOCRINE SYSTEM

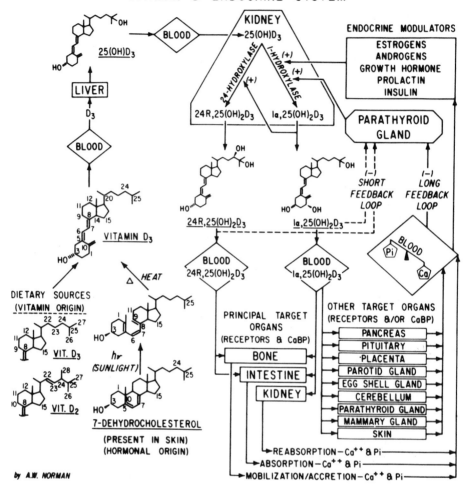

**Fig. 6.26.** Synopsis of the vitamin D endocrine system. (Reproduced by permission from Walters et al. (1981), in Lamble (Ed.), *Towards Understanding Receptors*, Elsevier/North Holland Biomedical Press, Amsterdam)

except in cases of faulty vitamin K absorption, since the normal diet covers the need generously.

Vitamin K antagonists, such as **dicoumarol** (a natural product) and **warfarin**, are used as anticoagulants in human therapy (thrombosis, atherosclerosis) and as rat poisons that lead to internal bleeding and death in rodents. **Heparin**, a polysaccharide consisting of 2-*O*-sulfonated glucuronic acid and 2-*N*,6-*O*-disulfonated glucosamine, is also a widely used anticoagulant, but its effect is connected not with Vitamin K but with enzyme inhibition.

## *Selected Readings*

H. F. DeLuca and H. K. Schnoes (1984). Vitamin D: metabolism and mechanism of action. *Annu. Rep. Med. Chem. 19*: 179–190.

D. Hornig (1982). Requirement of vitamin C in man. *Trends Pharmacol. Sci. 3*: 294–296.

S. Lewin (1976). *Vitamin C, its Molecular Biology and Medical Potential*. Academic Press, New York.

L. J. Machlin (Ed.) (1984). *Handbook of Vitamins: Nutritional. Biochemical and Clinical Aspects*. Marcel Dekker, New York.

D. P. McDonnell, D. J. Mangelsdorf, J. W. Pike, M. R. Haussler, and B. W. O'Malley (1987). Molecular cloning of complementary DNA encoding the avian receptor for vitamin D. *Science 235*: 1214–1217.

R. A. Morton (Ed.) (1970). *Fat Soluble Vitamins*. Pergamon Press, Oxford.

C. E. Orfanos, O. Braun-Falco, E. M. Farber, Ch. Grupper, M. K. Polano, and R. Schuppli (Eds.) (1981). *Retinoids: Advances in Basic Research and Therapy*. Springer Verlag, Berlin.

L. Ovesen (1984). Vitamin therapy in the absence of obvious deficiency. What is the evidence? *Drugs 27*: 148–170.

W. H. Sebrell and R. S. Harris (Eds.) (1967–). *The Vitamins*, Vol. 1 (1967), Vol. 2 (1968), Vol. 3 (1971). Academic Press, New York.

L. Stryer (1981). *Biochemistry*, 2nd ed. W. H. Freeman, San Francisco, pp. 896–905.

## 5. NUCLEIC ACIDS AS TARGETS FOR DRUG ACTION

The nucleic acids DNA and RNA have justifiably stood at the center of contemporary biology and biochemistry for the past 25 years. Their remarkable structure and the ever-increasing insight into their intricate functions triggered the major scientific revolution labeled "molecular biology." Since medicinal chemistry and molecular pharmacology are at the confluence of physical chemistry and molecular biology, nucleic acids have been investigated and recognized as the targets of several major groups of drugs. Some antibiotics, numerous antiparasitic agents, many antineoplastic (antitumor) drugs, and most of the antiviral compounds exert their varied actions on different phases of nucleic acid function.

On the basis of molecular mechanisms, drugs that act upon nucleic acids can be classified in the following way:

1. Drugs interfering with DNA replication
   a. Intercalating cytostatic agents: adriamycin, actinomycin, synthetic acridine and quinoline antimalarials
   b. Alkylating agents: nitrogen mustard cytostatic agents
   c. Antimetabolites interfering with DNA synthesis
   d. Antibacterial agents interfering with DNA topoisomerase

2. Drugs interfering with transcription and translation
   a. Tetracycline, aminoglycoside, and chloramphenicol antibiotics; the cytostatic platinum complexes and bleomycin

3. Drugs interfering with mitosis
    a. *Vinca* and *Podophyllum* alkaloid cytostatics and maytansine

4. Antiviral drugs
    a. DNA polymerase inhibitors
    b. Thymidylate kinase inhibitors.

## 5.1. Drugs Interfering with DNA Replication

### 5.1.1. Intercalating Cytostatic Agents

Intercalating drugs bind strongly to the DNA of chromatin in the cell nucleus by slipping between two base pairs of the double helix and forming charge-transfer complexes with the nucleotides. This interaction is not random, and it has been suggested that some compounds (e.g., daunomycin, proflavine) intercalate from the major groove whereas others (actinomycin D and ethidium) do so only from the minor groove of the helix (see Tsai, 1978). There are even indications that some antibiotics are selective for certain base sequences.

Intercalation has been studied thoroughly on oligonucleotide models by Sobell and co-workers, who cocrystallized intercalating drugs with dinucleotides and elucidated the structure of the resulting complexes by x-ray crystallography. Figure 6.27 shows a drawing of the complex formed by iodo-UpA and **ethidium** (6-72), a trypanocide. The ethidium molecule shown forms a charge-transfer stack, and also interacts with the phosphate anions, forming salt bonds with the latter through its amino groups. Above (and below) this intercalated ethidium molecule lies another one which is simply stacked, and does not interact with the phosphate in an ionic bond. The phenyl and ethyl groups of the interacting ethidium molecule lie outside (in the minor groove of) the double helix; only the planar tricyclic ring interacts with the nucleotides. As a result, the base pairs of the helix are twisted by $10°$ and separated by 0.67 nm, and the helix unwinds by $-26°$ at the intercalation site. Since this distorts the double helix, the replication and transcription of genes are compromised.

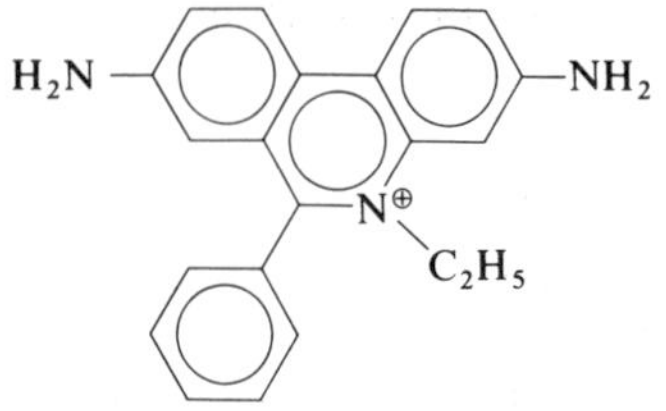

Ethidium
6-72

***Antitumor Antibiotics.*** **Actinomycin D**, produced by the fungus *Actinomyces antibioticus*, is an effective tumor-inhibiting antibiotic (6-73). It is a phenoxazone derivative with two cyclic pentapeptide side chains pointing up and down from the plane of the heterocyclic nucleus (Fig. 6.28A). The peptides form hydrogen bonds

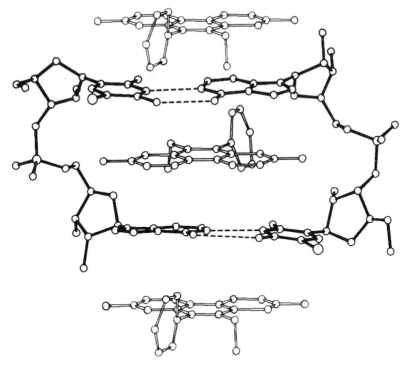

**Fig. 6.27.** The ethidium–iodo-UpA intercalated complex. (Reproduced by permission from Tsai et al. (1977), *J. Mol. Biol. 114*: 301)

**Fig. 6.28.** (**A**) Model of actinomycin D. The ring is shown by the dark lines and is perpendicular; the peptides point up and down. (**B**) The actinomycin–DNA complex. (Reproduced by permission from Stryer (1981), W. H. Freeman, San Francisco)

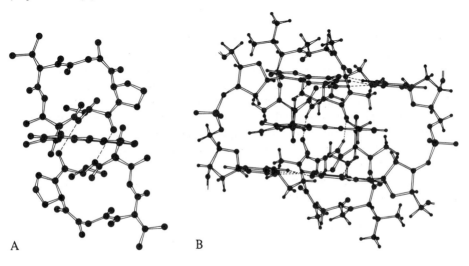

A                  B

with each other. Actinomycin D forms a strong complex with double-stranded DNA ($K_D = 5 \times 10^{-6}$ M) at G–C pairs, binding to the 2-amino group of the guanine. The flat heterocyclic ring is intercalated, as shown in Fig. 6.28B. The antibiotic selectively inhibits the ribosomal RNA chain-elongation step during transcription. Although a very effective drug in certain malignancies (Wilms' tumor, testicular tumors, disseminated cancers), it is a very toxic compound. Some new experimental derivatives are reported to be less toxic.

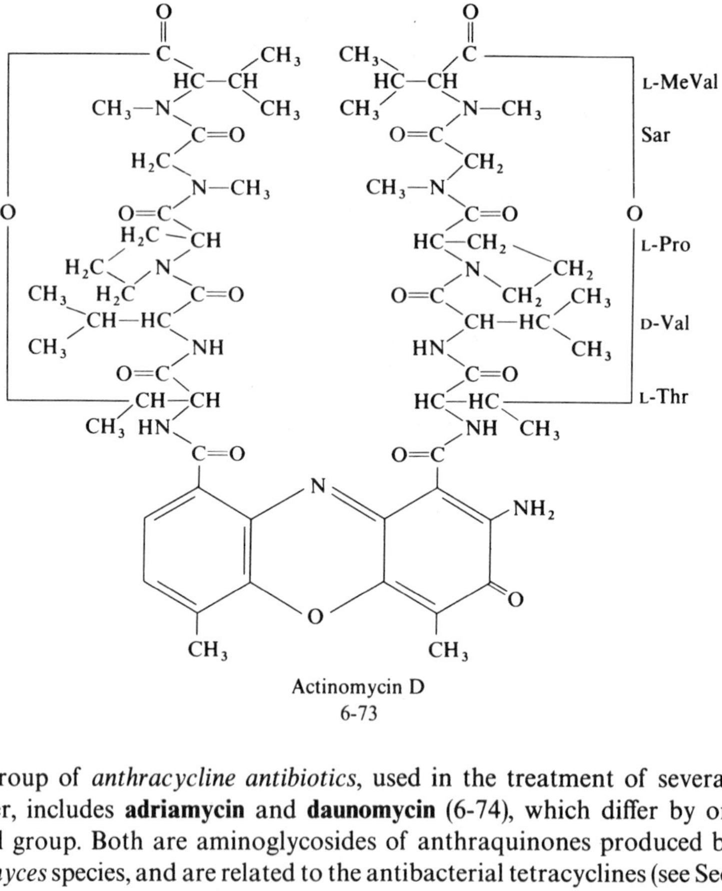

Actinomycin D
6-73

The group of *anthracycline antibiotics*, used in the treatment of several forms of cancer, includes **adriamycin** and **daunomycin** (6-74), which differ by only one hydroxyl group. Both are aminoglycosides of anthraquinones produced by some *Streptomyces* species, and are related to the antibacterial tetracyclines (see Sec. 5.2.2). The four-membered ring system intercalates into DNA, entering from the major groove. The sugar moiety of the drug is ion-bonded through its amino group to the phosphate backbone of DNA. Daunomycin is used only in acute leukemia, but adriamycin is effective in solid tumors also. The 4-desmethoxy derivatives of both compounds are much more potent. The principal drawback of these active cancer chemotherapeutic agents is their severe cardiotoxicity, through the inhibition of cardiac $Na^+$, $K^+$-ATPase. Calcium, however, reverses this toxicity.

***Antimalarials.*** Antimalarial drugs, which are based on 9-aminoacridine and 4-aminoquinoline, and derived from the alkaloid quinine, can also intercalate with

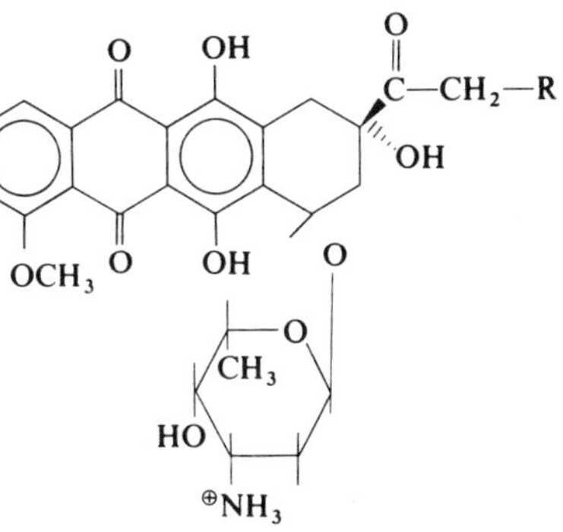

6-74   Daunomycin  (R = H)
       Adriamycin  (R = OH)

DNA, and at least part of their activity is thought to result from the blockade of DNA replication in the malaria parasite. The aromatic rings of these drugs can form charge-transfer complexes with the G–C base pair.

Unlike the slowly acting antifolate antimalarials, the aminoquinolines and aminoacridines are fast-acting schizontocidal agents (see Fig. 6.16), which are active against the parasite in its erythrocytic phase. The most important representatives of this class of drugs are shown in Fig. 6.29.

The natural, parent drug of this family is **quinine** (6-75), an alkaloid produced by the bark of the *Cinchona* tree. It was a well-known substance among South American Indians, and the Spanish Jesuits introduced it to Europe in the seventeenth century. Quinine and some of its epimers are schizontocidal in erythrocytes, but do not kill schizonts in the liver and therefore cannot produce a radical cure. Quinine is also more toxic than some of its synthetic congeners, and is used mainly in combination with **pyrimethamine** (6-55). However, in parts of the world where inexpensive drugs are needed, quinine is still used alone.

The first synthetic drug of the quinine family was **quinacrine** (6-76), introduced in Germany in 1932. Eventually it was found that quinoline derivatives are more active and less toxic than acridines, and **chloroquine** (6-77) was prepared, which is perhaps the most widely used antimalarial even today. It has the same side chain as quinacrine but lacks the nonessential methoxy group. Chloroquine and a number of related derivatives are effective schizontocides in the erythrocyte and can also be used as prophylactics. Their toxicity is low and there are as yet few *Plasmodium* strains that are resistant to them. Among the 8-aminoquinolines, **primaquine** (6-78) is used as an active agent against exoerythrocytic forms of the parasite. It can effect radical cures and also serves as preventive medication in malaria-infested areas.

During World War II and after, as well as during the American involvement in the Viet Nam War, a large-scale antimalarial drug program was mounted in the United States. These protracted efforts notwithstanding, the problem of malaria—after some false hopes—is far from being solved. One of the superior drugs emerging from this program is **mefloquine** (6-79), a quinine-type alcohol. The bridged quinuclidine ring of quinine is replaced in this compound by the simple piperidine, and the —CF$_3$ groups increase lipophilic character and prevent the phototoxicity of

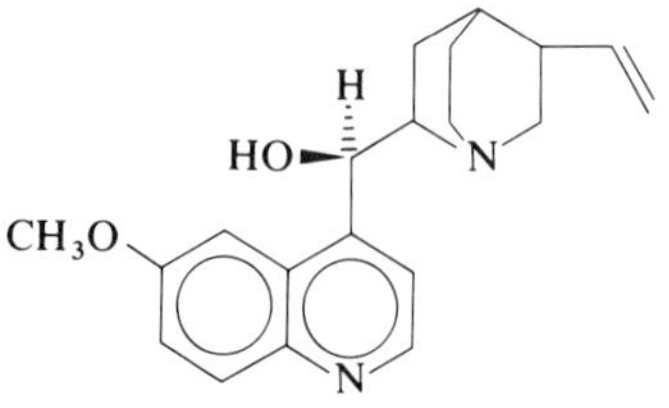


Quinine
6-75

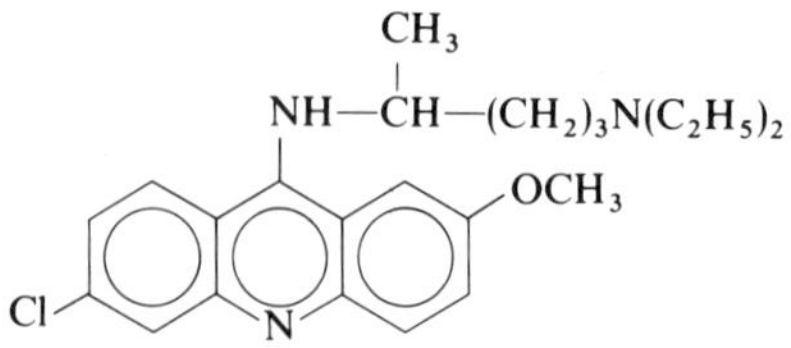


Quinacrine
6-76

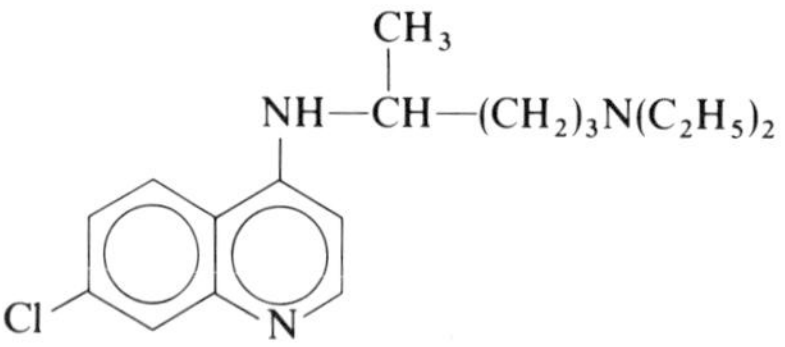


Chloroquine
6-77

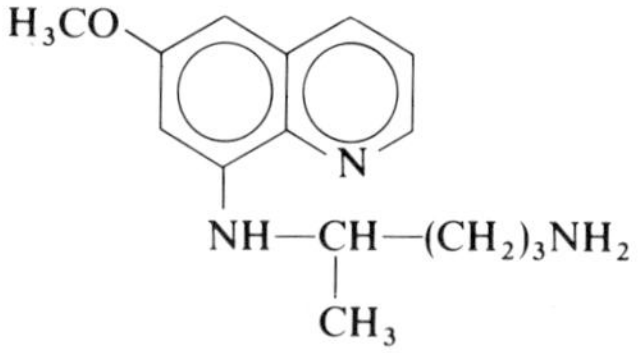


Primaquine
6-78

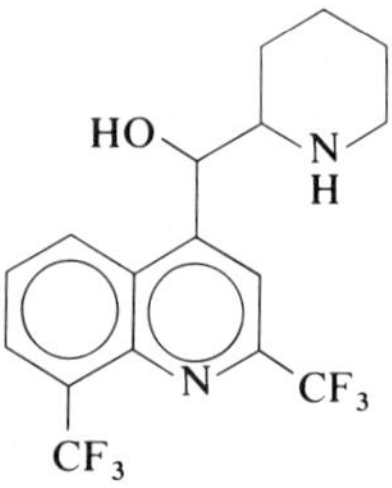


Mefloquine
6-79

**Fig. 6.29.** Some schizontocidal antimalarials.

the otherwise highly active 2-substituted quinolines. Another advantage of mefloquine is its activity against chloroquine-resistant strains of the malaria parasite. It is remarkable and unusual that, unlike the isomers of stereospecific quinine, mefloquine diastereomers are all active compounds. The antifolate antimalarials are discussed in Sec. 3.5.

### *5.1.2. Alkylating Agents*

These antitumor agents are compounds that form carbonium ions or other reactive electrophilic groups, and were discussed as affinity-labeling agents in Chap. 3, Sec. 3.2. Such compounds will also bind covalently to DNA, and either cross-link the two strands of the helix or otherwise interfere with replication or transcription. Since these processes are more prevalent in rapidly dividing malignant cells than in normal tissues, alkylating agents can control, and in some cases even eliminate tumors. However, their selectivity is limited and they have many and serious side effects.

*Nitrogen Mustards.* Alkylating agents were developed from sulfur mustard, the infamous "mustard gas" of World War I, a lethal vesicant and cell poison. Its nitrogen analogue, the nitrogen mustard **mechlorethamine** (6-80 Fig. 6.30), was first used as an antitumor agent in 1942 with some success, and numerous derivatives were subsequently developed. The rationale for this, if any, was to use carrier molecules that are natural products, in the hope that they could direct the active, nitrogen-mustard component of the compound to a selective metabolic site in a tumor. **Melphalan** (6-81; or **phenylalanine mustard**) and **uracil mustard** (6-82) are just two examples of many such compounds. Although they did not fulfill expectations regarding selectivity, they are nevertheless useful oral drugs when employed in conjunction with tumor surgery. Ovarian and breast carcinomas, lymphadenoma, and multiple myeloma are the malignancies most successfully treated with these drugs, especially in combination with mitosis inhibitors (see Sec. 5.3).

**Fig. 6.30.** Representative alkylating antitumor agents.

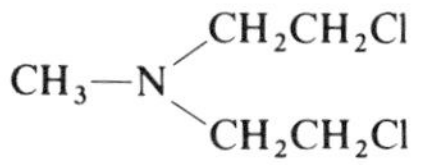

Mechlorethamine
6-80

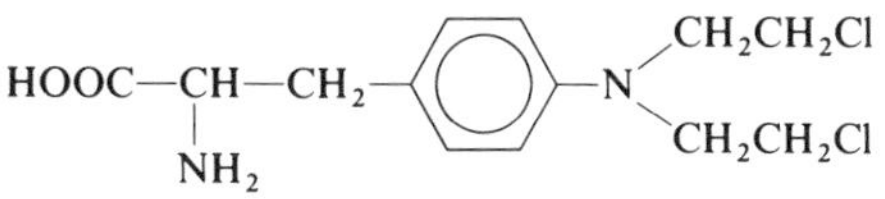

Mephalan
Phenylalanine mustard
6-81

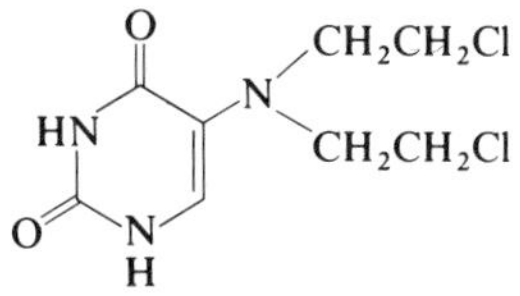

Uracil mustard
6-82

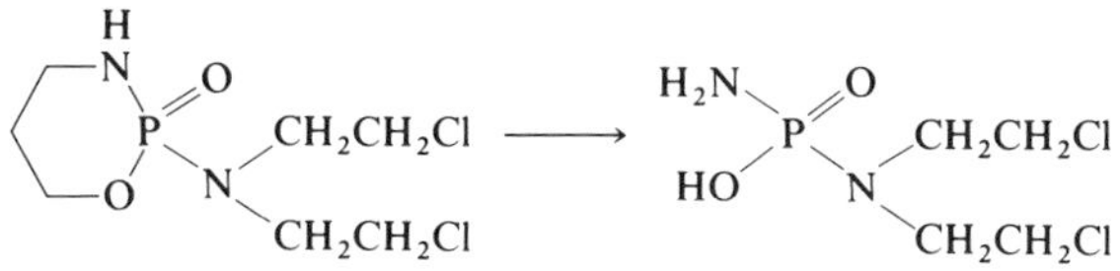

Cyclophosphamide
6-83

Phosphoramidate
mustard
6-84

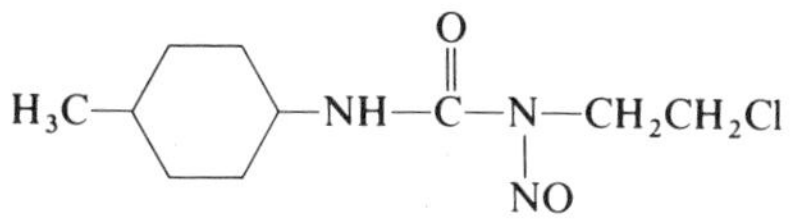

Semustin
6-85

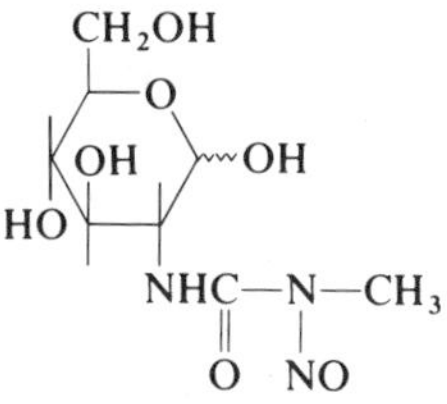

Streptozotocin
6-86

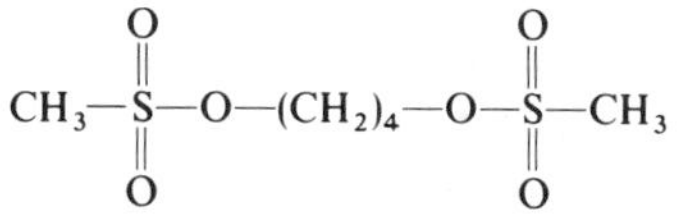

Busulfan (Myleran)
6-87

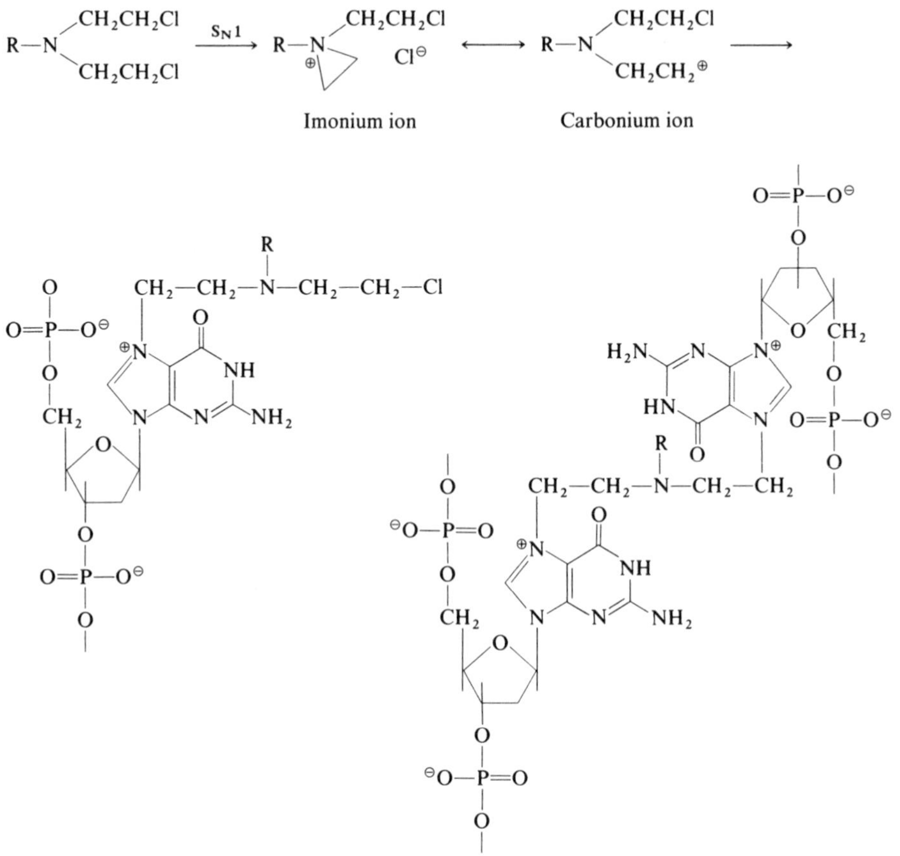

**Fig. 6.31.** Cross-linking of DNA by a nitrogen mustard through two guanines.

A prodrug, **cyclophosphamide** (6-83), was synthesized in the hope of exploiting the high level of phosphoramidase enzymes in certain tumor cells. Cyclophosphamide is nontoxic but is metabolized in the liver, not the tumor, to form the active drug, the **phosphoramide mustard** (6-84). While not without side effects, cyclophosphamide is a relatively successful drug in a number of carcinomas and lymphomas.

The mode of action of these compounds is nonspecific, because the active species, the resonance-stabilized carbonium ion (Fig. 6.31), reacts with any nucleophilic center, including water. Consequently, there is a tremendous waste of drug on the way to the site of action, through hydrolysis alone; this waste is slowed with the aromatic compounds like melphalan. The principal target of the nitrogen mustards seems to be the 7-nitrogen of guanine in DNA, which cross-links the two strands. This action prevents unwinding, causes deguanylation and base-mispairing, and compromises the template function of DNA. Linking within the same strand and binding to nucleoprotein or the phosphate anion are also possible effects and can also lead to functional damage in rapidly proliferating cells, like miscoding and point mutations.

*Nitrosoureas.* The nitrosoureas, represented by **semustin** (6-85), are more recent discoveries. These drugs combine the N—NO group with a monofunctional mustard. The compounds are effective against some brain tumors and the Lewis lung carcinoma, both of which tend to respond poorly to chemotherapy. **Streptozotocin** (6-86) is a naturally occurring glucosamine nitrosourea derivative that shows antileukemic activity as well as antibiotic effects. The nitrosoureas can carbamoylate proteins (e.g., on lysine) by forming isocyanates, whereas the chloroethyl carbonium ion formed could potentially cross-link the strands of DNA. However, these mechanisms are somewhat hypothetical at present (cf. Pratt and Ruddon, 1979).

*Methanesulfonate Esters.* Methanesulfonate esters such as **busulfan** (6-87) produce clinical remission in chronic myelogenous leukemia. Busulfan acts through an $S_N2$ nucleophilic displacement and presumably cross-links DNA, since the methanesulfonate ion in an excellent leaving group.

### 5.1.3. Antimetabolites

Antimetabolite inhibitors of DNA synthesis act by the competitive or allosteric inhibition of a number of different enzymes in purine or pyrimidine biosynthesis. Actually, some such compounds interfere with as many as 10–12 different enzymes—although admittedly to a different degree—and a discussion of such drugs would therefore not have been practical in the section dealing with specific enzymes. Instead, we shall consider their interference with the *de novo* synthesis of DNA and RNA.

We have seen how fluoropyrimidines interfere with pyrimidine synthesis through thymidylate synthetase (Sec. 3.6). Purine synthesis (Fig. 6.32) can also be blocked by **6-mercaptopurine** (6-88) and **6-thioguanine** (6-89). Both require conversion to the mononucleotide in a "lethal synthesis"—a mechanism distinguished from the formation of suicide substrates (Sec. 3.1) in that the enzyme that transforms the inactive pro-drug to the active inhibitor is different from the enzyme that is being blocked. $K_{cat}$ inhibitors are formed and bound by the same enzyme.

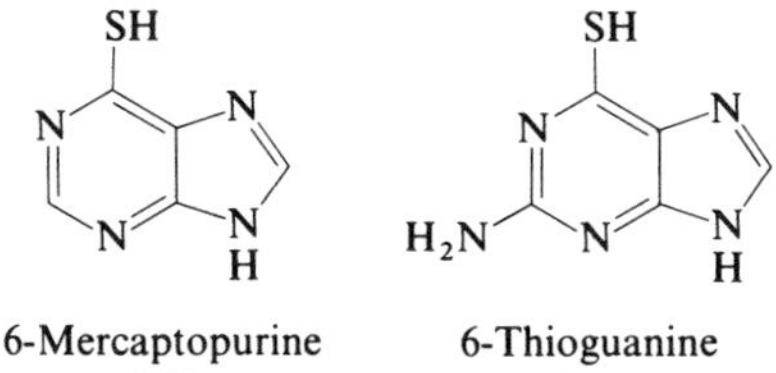

6-Mercaptopurine
6-88

6-Thioguanine
6-89

Both thiopurines primarily block the amidotransferase in the first step of purine synthesis (Fig. 6.32) as pseudo-feedback inhibitors. Additionally, the transformations of inosinic acid to AMP and GMP are also inhibited (for details, see the review of Patterson and Tidd in Sartorelli and Johns, 1975). Both 6-mercaptopurine and 6-thioguanine are used in acute leukemia.

In **cytarabine** (6-90) (**Ara-C**, 1-$\beta$-D-arabofuranosyl-cytosine), the ribose moiety of cytidine is replaced by the epimeric arabinose. This drug inhibits DNA polymerases

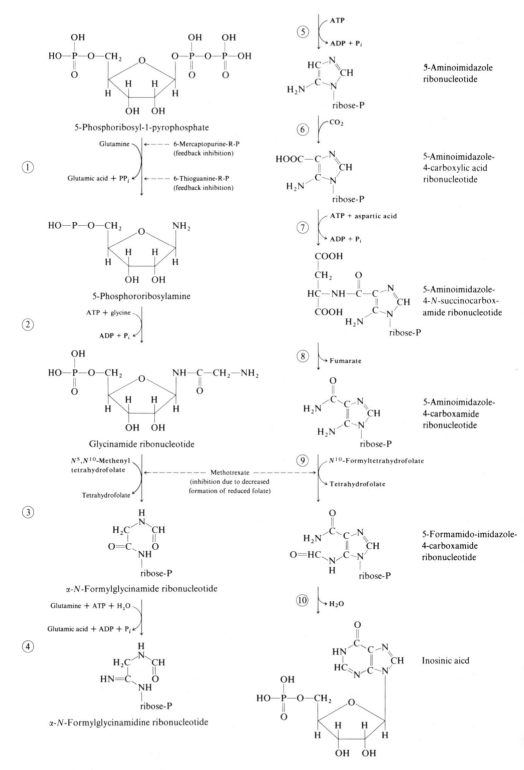

**Fig. 6.32.** Schematic diagram of purine synthesis, including points of attack by antitumor agents. (Modified from Pratt and Ruddon (1979), Oxford University Press, New York)

after its bioconversion to ara-CTP, competing with CTP. The adenine analogue **Ara-A** (6-79) acts in a similar fashion. Both drugs are used in leukemias only; they are inactive against solid tumors.

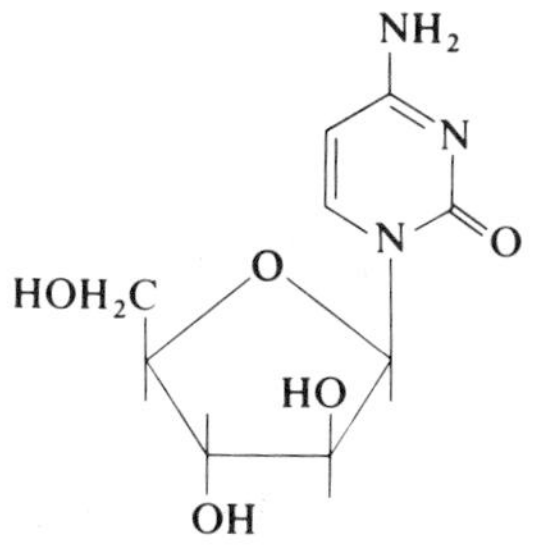

Cytarabine
6-90

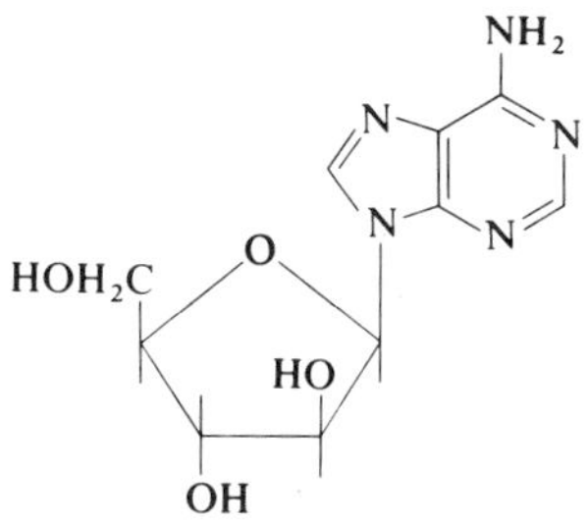

Ara-A
6-91

### 5.1.4. DNA Topoisomerase Inhibitors

DNA is a topologically constrained molecule, because the ends of the circular double helix are fixed in space, allowing the formation of higher-order structures called supercoils. During replication, the double helix must unwind, and unwinding introduces additional positive supertwist. *Topoisomerase I* removes this supertwist by breaking and resealing single strands to produce a relaxed DNA. Since the reaction is thermodynamically favorable, no ATP is needed. *Topoisomerase II* (*gyrase*) catalyzes the passing of two DNA strands through breaks, and thus produces a negative supertwist which further promotes parental strand separation at the replication fork; ATP is needed for this reaction. Since the replicating fork rotates at a speed of about 100 revolutions per second, interference with these enzymes stops replication rapidly. In bacteria, the relaxation and supercoiling effect of the two enzymes oppose each other and thereby maintain the proper superhelical state of the chromosome, which varies depending on the growth phase of the cell. Mammalian gyrase also seems to be regulated by the growth state of the cell (Ross, 1985; Chen and Liu, 1986).

It appears that many intercalating antitumor agents (adriamycin, actinomycin D, ethidium bromide), as well as some agents that do not intercalate, stabilize the enzyme–DNA complex, and thus make the enzyme–induced breaks permanent, leading to DNA cleavage. In this way, the gyrase enzyme is subverted into a lethal factor. Although these investigations are in an early stage, it seems that topo-isomerase I might also be a potential target of chemotherapy.

In addition to antitumor drugs, *quinolone antibacterial agents* also act by inhibiting bacterial DNA gyrase. Although these compounds have been known since 1962, the first prototype, **nalidixic acid** (6-92) and its analogues, were not very active. Recently synthesized compounds, which incorporate a fluorine and a piperazine side chain—for example, **ciprofloxacin** (6-93) and analogues (see Wentland and Cornett, 1985)—are highly active and relatively nontoxic anti-bacterial agents.

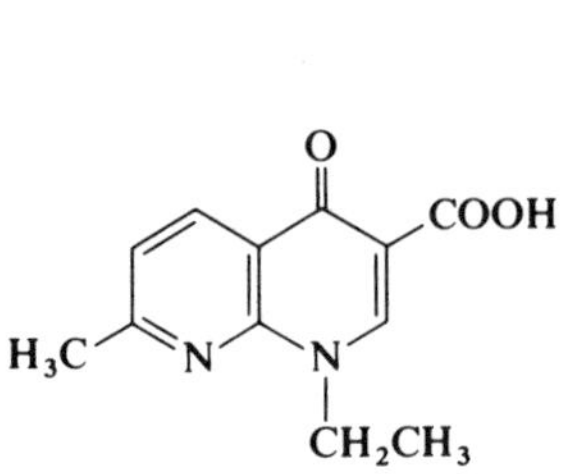

Nalidixic acid
6-92

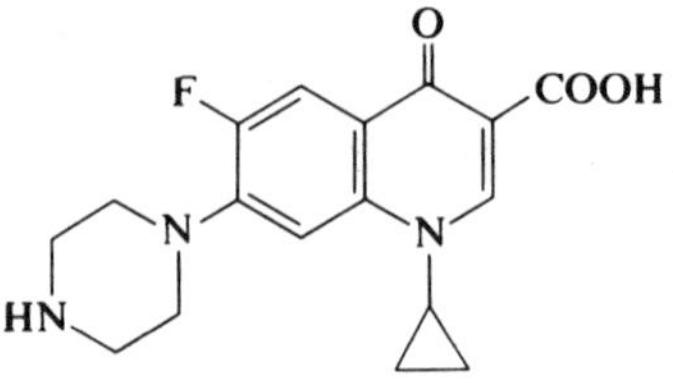

Ciprofloxacin
6-93

## 5.2. Drugs Interfering with Transcription and Translation

The intercalating drugs and nucleic acid synthesis inhibitors discussed in the preceding sections interfere indirectly with every phase of nucleic acid action because the DNA is rendered nonfunctional. Other drugs, discussed below, affect the regulation of protein synthesis even when the genome, the DNA structure, is intact. Such activity can be the result of interference either with transcription of messenger RNA or with translation of the mRNA to protein. There are cytostatic and

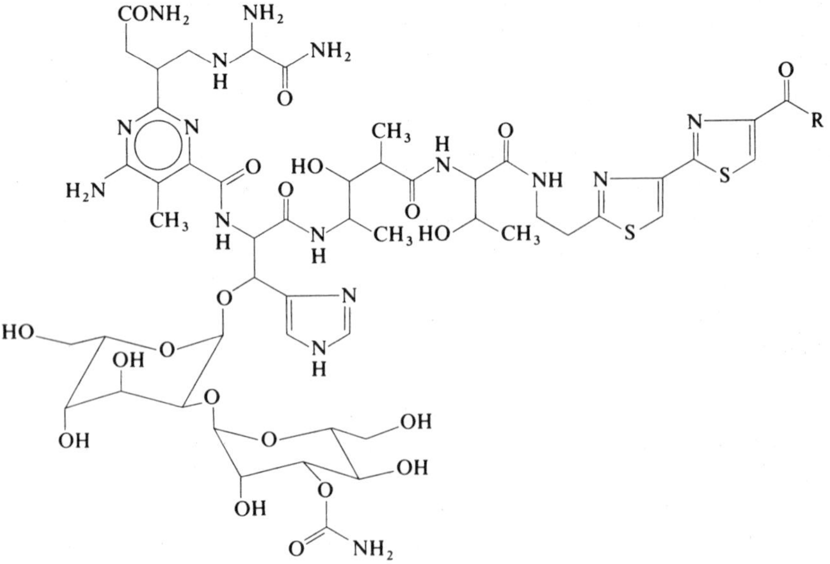

Bleomycinic acid     R = OH

Bleomycin A$_2$     R = NHCH$_2$CH$_2$CH$_2$—S$^+$(CH$_3$)$_2$

Bleomycin B$_2$     R = NHCH$_2$CH$_2$CH$_2$CH$_2$NHC(=NH)NH$_2$

6-94

antibacterial antibiotics, as well as some synthetic compounds, among the drugs acting through such interference mechanisms.

### 5.2.1. Antitumor Agents

Among the *cytostatic* agents in this group are the **bleomycins** (6-94), a very complex group of glycopeptides. Among many biochemical effects, they seem to cause the scission and breakage of viral, bacterial, and animal DNA. They also seem to inhibit DNA ligase, an important DNA replication and repair enzyme. Therefore, bleomycin interferes with the transcription as well as replication of nucleic acids. Since there are no bleomycin-degrading enzymes in the skin, squamous-cell carcinomas of the neck and head respond well to this drug. However, toxic effects also prevail in skin tissue. The unique activity and toxicity of the bleomycins can be utilized in the combination treatment of malignancies, in which drugs with different modes of action and nonoverlapping toxicities are used, often with remarkable success.

Another group of agents that interfere with DNA transcription are the **platinum complexes**. The *cis*-**dichlorodiammine-platinum (II) complex (cisplatin**, 6-95), but not the *trans* isomer, is very active against testicular tumors, epidermoid carcinomas, and ovarian tumors. It binds strongly to DNA by intrastrand binding (i.e. not crosslinking) to oligoguanine sequences, unwinds the duplex, and reduces the length of the DNA molecule. The *trans* isomer is selectively removed from the DNA.

$$H_3N \diagdown \diagup Cl$$
$$Pt$$
$$H_3N \diagup \diagdown Cl$$

Cisplatin
6-95

### 5.2.2. Antimicrobials

*Antibiotics* of different origin can interfere with various phases of transcription and translation of the genetic message in microorganisms.

*Tetracyclines.* The tetracyclines are closely related to the anthracycline glycoside antitumor agents discussed in Sec. 5.1. Their structures are summarized in Table 6.8. Available since the early 1950s, they are broad-spectrum antibiotics, active against a wide variety of microorganisms, including some that either are not sensitive or are resistant to $\beta$-lactam antibiotics. Tetracyclines interfere with protein synthesis by inhibiting the binding of aminoacyl-tRNA to the 30 $S$ subunit of the ribosome in the microorganisms. The release of completed peptides from the ribosome is also blocked.

*Antituberculotics.* Among the *aminoglycoside antibiotics*, **streptomycin** (6-96) and **kanamycin** (6-97) are very important. Streptomycin contains a diguanidine derivative of 1,3-diaminoinositol (streptidine), a specific hexofuranose (streptose) carrying a 3-aldehyde group, and *N*-methyl-2-glucosamine. Kanamycin A is somewhat simpler, being derived from a 1,3-diamino-2-deoxyinositol, a 3-glucosamine,

**Table 6.8.** Structures of tetracycline antibiotics

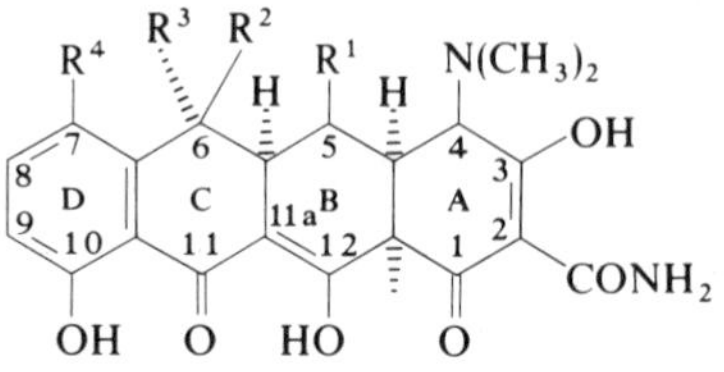

| Name | $R^1$ | $R^2$ | $R^3$ | $R^4$ | Other names |
|---|---|---|---|---|---|
| Tetracycline | H | OH | $CH_3$ | H | Achromycin |
| 7-Chlortetracycline | H | OH | $CH_3$ | Cl | Aureomycin |
| 5-Oxytetracycline | OH | OH | $CH_3$ | H | Terramycin |
| 6-Demethyl-7-chloro tetracycline | H | OH | H | Cl | Declomycin, demeclocycline |
| 6-Demethyl-6-deoxy-5-hydroxy-6-methylene tetracycline | OH | | $=CH_2$ | H | Methacycline |
| 6-Deoxy-5-oxytetra cycline | OH | H | $CH_3$ | H | Doxycycline |

and a 6-glucosamine. Both antibiotics have a fairly wide antibiotic spectrum, but their principal use is as antituberculotic agents. The aminoglycosides decrease the fidelity of translation by binding to the 30 $S$ subunit of the ribosome. This permits the formation of the peptide initiation complex, but prohibits any subsequent addition of amino acids to the peptide. This effect is due to the inhibition of polymerization as well as the failure of tRNA and mRNA codon recognition.

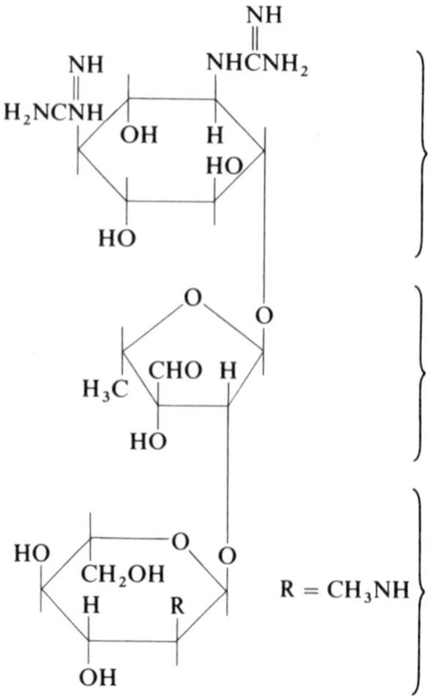

$R = CH_3NH$    N-Methyl-L-glucosamine

Streptomycin

6-96

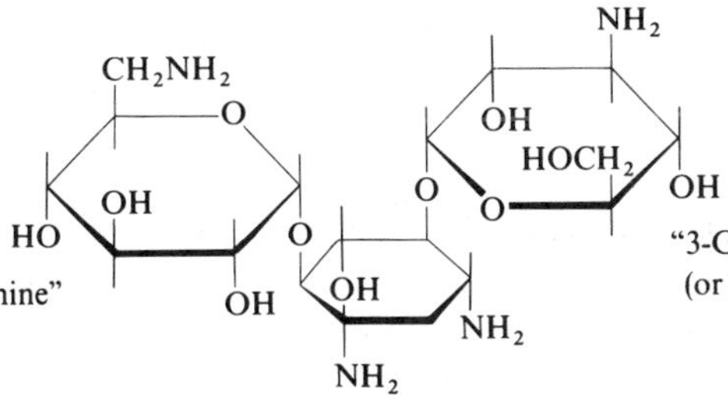

"6-Glucosamine"

"3-Glucosamine"
(or kanosamine)

Kanamycin A
6-97

Streptomycin and its dihydro derivative in (in which a —CH$_2$OH replaces the aldehyde on streptose) are ototoxic, damaging the auditory nerve. Kanamycin is less toxic. Bacterial strains develop resistance to both antibiotics.

Other *nonantibiotic* chemotherapeutic agents active against *Mycobacterium tuberculosis* are **isoniazid** (6-98) an **ethambutol** (6-99).

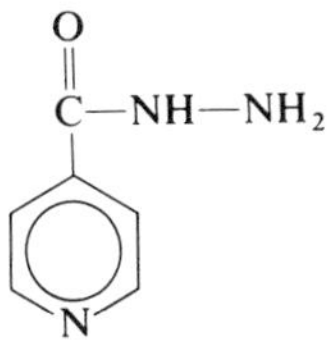

Isoniazid
6-98

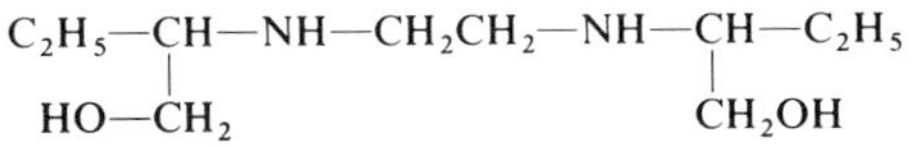

Ethambutol
6-99

Another antibiotic, **rifampicin** (6-100), a semisynthetic derivative of rifamycin, is the drug of choice in the treatment of tuberculosis as well as leprosy, either alone or in combination with other drugs. Rifampicin is much safer than other antituberculotics since it inhibits DNA-directed RNA polymerase in bacteria but not in mammals.

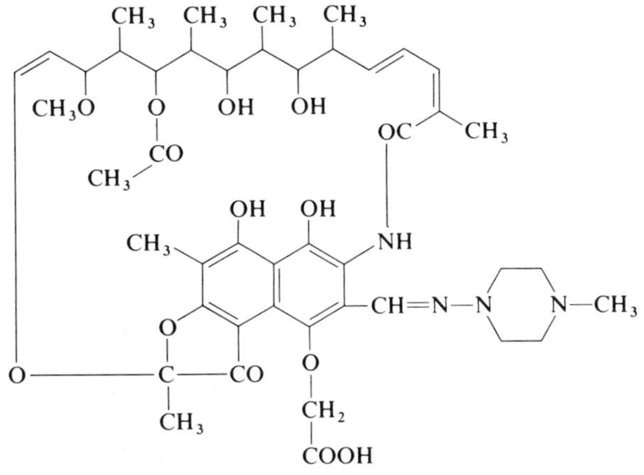

Rifampicin
6-100

These drugs are partly responsible for the fact that the once dreaded tuberculosis, which generally killed young persons, became quite rare in developed nations and is now curable.

**Other Antibiotics. Chloramphenicol** (6-101), like all of the antibiotics discussed in this section, is a product of a *Streptomyces* species. Its structure is remarkably simple, and it is obtained synthetically rather than by fermentation. It is a broad-spectrum antibiotic, but because it can cause fatal blood dyscrasias its use is largely restricted to microorganisms that cannot be well controlled by other antibiotics. Thus, it is the drug of choice against typhoid. Chloramphenicol binds to the 50 *S* subunit of the ribosome and inhibits the enzyme peptidyl transferase. This blocks peptide-bond formation between the amino acid-tRNA on the aminoacyl site and the growing peptide chain on the peptidyl site of the ribosome, interrupting translation.

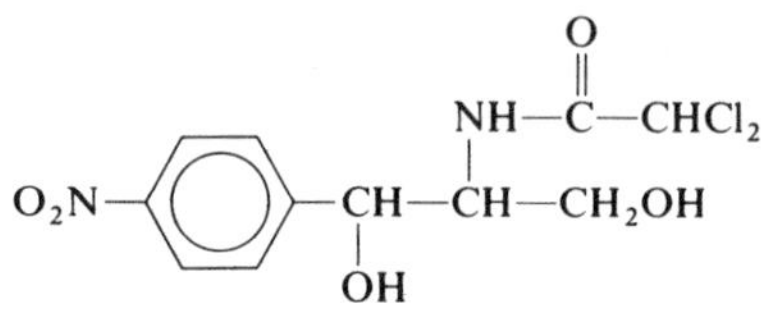

Chloramphenicol
6-101

   **Erythromycin** (6-102) acts in the same way, and can actually compete with chloramphemicol for the same binding site. It is a nontoxic macrocyclic lactone, widely used against penicillin-resistant *Staphylococcus* strains, and is the drug of choice to cure "Legionnaires' disease," caused by *Legionella pneumophila.*

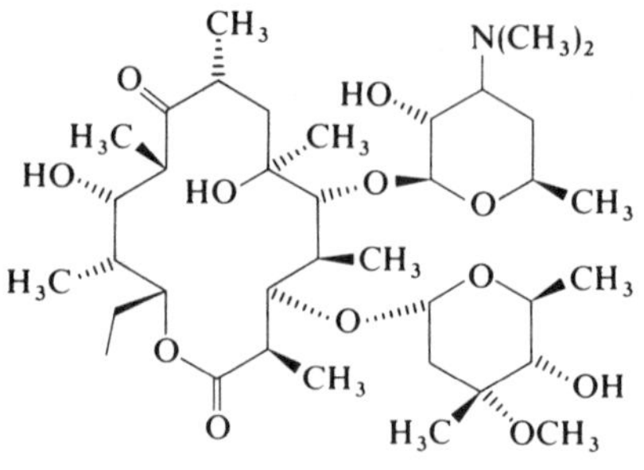

Erythromycin
6-102

### 5.3. Cytostatic Agents Interfering with Mitosis

These antitumor agents stop cell division at the metaphase stage, where the daughter chromosomes normally begin to migrate toward the poles of the cell. They are pulled toward the poles by *microtubules*, which are assembled at one of their ends and disassembled at the other. The microtubules are hollow tubes built from 13 dimers of two different kinds of protein ($\alpha$ and $\beta$) in a helical manner. Their outer diameter is about 24 nm. Some cytostatic agents bind to this $\alpha\beta$ dimer, preventing its incorporation into the growing microtubule.

The best-known drugs acting in this fashion are the *Vinca* alkaloids **vincristine** and **vinblastine** (6-103). They are very complex indole derivatives which nevertheless have been synthesized. Both are quite effective in various leukemias and in Hodgkin's lymphoma, but show considerable neurotoxicity. **Maytansine** (6-104) isolated from an African plant, also shows promise as a useful cytostatic.

For a recent review on antitumor agents, see Doyle and Kaneko (1985).

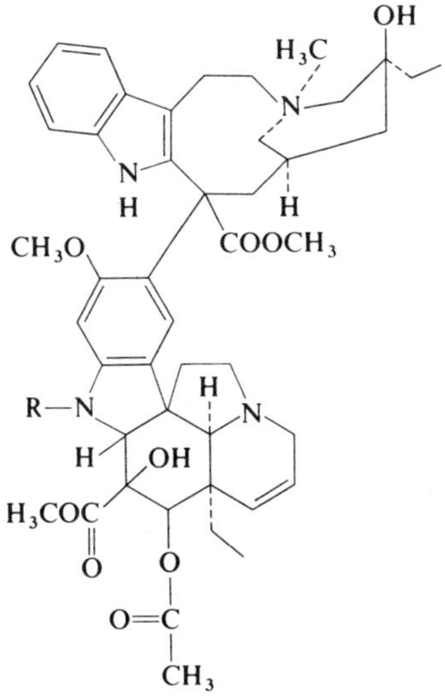

| | R |
|---|---|
| Vincristine | CHO |
| Vinblastine | CH$_3$ |

6-103

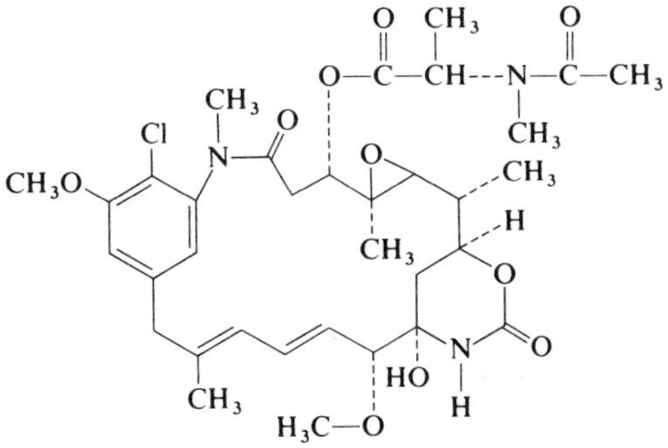

Maytansine
6-104

## 5.4. Antiviral Agents

Viruses are on the borderline of living and inanimate matter: they can reproduce only as extreme cell parasites because they do not possess a complete synthetic machinery. They can also be crystallized, and the virus particle, or virion, shows a high degree of symmetry. Viruses consist of infectious DNA or RNA (but not both) in a protective protein coat which shows helical or spherical symmetry and which may show appendages, such as the tail and tail fibers of the T4 phage. The assembly of such complex structures is no longer a self-assembly, but is directed by enzymes and assisting proteins which form a framework. The replication of viruses is also very different from that of higher organisms. In some, such as the DNA viruses, transcription and translation occur, whereas in others, like the single-stranded RNA viruses, the RNA is its own messenger and activates an RNA-directed DNA polymerase. The latter viruses are known as "retroviruses" because of the reversal of the normal DNA–RNA–protein sequence; the DNA produced in this reverse way then becomes the template for viral RNA and protein. After this the virus uses the enzymes, nucleotides, and amino acids of the infected host cell to build the virion. Because of this diversion of starting material and synthetic capacity, the infected cell may die, with virions being released to infect other cells (lytic pathway); or the viral DNA may join the infected cell by recombination, which then continues to produce virions (lysogenic viruses). In some instances, the host cell is transformed into a malignant, cancerous cell by the virus, obviously a process of great interest. The recent discovery of cancer genes (oncogenes) may shed more light on the process of viral carcinogenesis.

An excellent introduction to virology and virus biochemistry can be found, among other sources, in Stryer (1981).

A high proportion of human and animal diseases are caused by viruses, from the common cold to poliomyelitis, rabies, some leukemias, and many others. The selective chemotherapy of viral diseases is therefore of utmost interest, because not all are amenable to immunotherapy. The development of antiviral drugs was nevertheless a very slow and frustrating field for many years, and only in the past few years has it gained momentum. It is not nearly as well advanced as bacterial chemotherapy.

In the host cell, viruses induce the formation of enzymes which they themselves cannot produce. The most important group of such enzymes are the DNA polymerases, but thymidine kinase is also essential. Interference with these enzymes by either enzyme inhibitors or fraudulent antimetabolites is the basis of the activity of many antiviral drugs. In this respect, antiviral compounds and cytostatics used in the treatment of malignant tumors have much in common, and indeed overlap each other in their activity.

### 5.4.1. Antibiotics

Among the antibiotics, **rifampin** (6-100) is effective against DNA viruses like HSV and the smallpox virus (believed to be extinct). It also prevents the virus-induced transformation of cells to malignant forms. **Bleomycin** (6-94), primarily an antitumor agent, also shows some promise as an antiviral agent.

### 5.4.2. Enzyme Inhibitors and Antimetabolites

Viral *DNA polymerase* inhibitors are compounds already encountered as antitumor agents in Sec. 5.1.3. **Ara-A** or **vidarabine** (6-91) is quite active in some herpes simplex virus type 1 (HSV-1) infections responsible for "cold sores" on the lip, keratitis of the cornea, and encephalitis. Its analogue, **Ara-C** (6-90, **cytarabine**), is primarily an antineoplastic drug, but its 2'-fluoro-5-iodo derivative has shown good activity against HSV-1.

*Thymidylate synthase* is involved in pyrimidine biosynthesis (see Fig. 6.17 and Sec. 3.6), and this can be exploited by antimetabolites that are mistaken for true nucleotide metabolites. The enzyme-mediated incorporation of such antimetabolites into the viral DNA and RNA will destroy the virus, eliminating its infectious properties. This is in contrast to the mode of action of the antitumor agent 5-

fluorouracil, which is a suicide substrate of thymidylate kinase and inactivates the enzyme, thereby interrupting the dTTP supply of the tumor cell.

**Idoxuridine** (6-105) and **trifluridine** (6-106) are phosphorylated to their active form in virus-infected cells, and show specificity for two reasons: their higher affinity for the viral enzyme and the higher phosphorylase levels in infected than in normal cells. Both compounds are used locally on lesions of HSV-1 and HSV-2 (the latter of which causes genital herpes, now reaching epidemic proportions) with fair success. They are rather toxic if administered parenterally, as are all moderately selective antimetabolities. A recently discovered compound, **acyclovir** (6-107), shows a unique specificity and lack of toxicity in HSV-1, HSV-2, and varicella (chickenpox, shingles) infections. A guanine derivative, it lacks the pentose of similar compounds, and is phosphorylated on the alcoholic OH by the viral thymidylate kinase only. Consequently, it is not activated in uninfected cells; additionally, it is a viral DNA polymerase inhibitor but does not readily block the polymerase of the cell itself. Therefore, it is a spectacularly nontoxic drug [$LD_{50}$ (mouse) = 1000 mg/kg, i.p.] and is not degraded metabolically. The new hydroxymethyl derivative of acyclovir **DHPG (dihydroxy-propoxymethylguanine,** 6-108) is even more efficiently phosphorylated, and is active against HSV-1, HSV-2, cytomegalovirus, and Epstein–Barr virus. Another antiherpes compound under investigation is the remarkably simple **phosphonoformate** (6-109; see Dolin, 1985; de Clerq and Walker, 1986).

6-105    Idoxuridine
(6-106    Trifluridine)

6-107    Acyclovir (R = —H)
6-108    DHPG (R = —CH₂OH)

Phosphonoformate
6-109

***Therapy of AIDS.*** The *acquired immune deficiency syndrome (AIDS)* has become a serious public health problem around the world, and commands much concern in medical and lay circles alike. Its causative agent is a human T-cell lymphotropic virus (HTLV-III, also called HIV), that destroys helper/inducer T cells of the immune system and causes extremely high mortality by allowing opportunistic infections and malignancies. Although available studies are, by necessity, still limited (see the reviews by Gupta, 1986; and Mitsuga and Boder, 1987), a compound designed along traditional lines, **3′-azido-3′-deoxythymidine (AZT, retrovir,** 6-110) is used at present in a desperate attempt to control (but not cure) the disease. It seems that novel combinations of antiviral and immunorestorative therapies will be required to make urgently needed progress in this field.

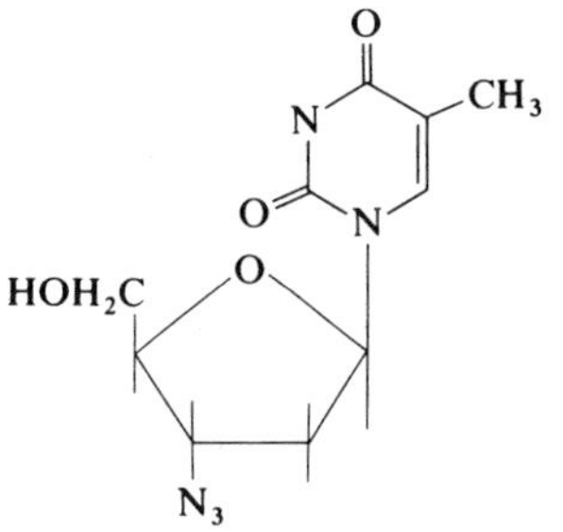

Retrovir
(Azidothymidine)
6-110

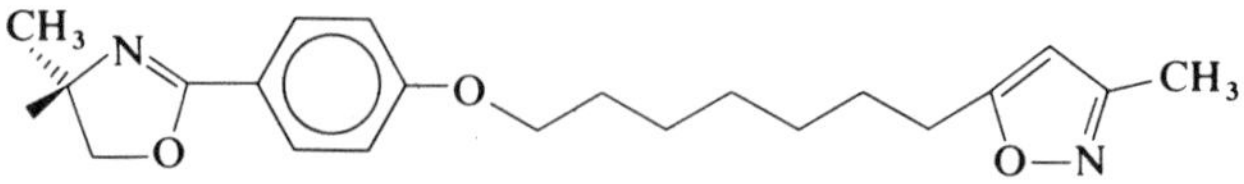

WIN 52084
6-111

### 5.4.3. Antivirals Acting by Other Mechanisms

On a more trivial front, some progress was reported in controlling human *rhinovirus*, the causative agent of the common cold. The compound **WIN 52084** (6-111) prevents the uncoating of the virion (shedding its protein coat), a necessary step in penetrating the host cell. The minimal inhibitory concentration is only 60 nM. The interaction with the virion was shown at atomic resolution, a remarkable feat of molecular pharmacology (Smith et al., 1986).

Among drugs with uncertain modes of action, **ribavirin** (6-112) is not even a nucleotide, since the purine ring is replace by a triazole. It is active against HSV-1 and -2, hepatitis, and perhaps influenza viruses. It seems to have multiple effects on viral replication, blocking RNA synthesis and mRNA capping.

**Amantadine** (1-amino-adamantane; 6-113) and related compounds are available commercially for the prophylactic treatment of influenza A. Amantadine is not active against many strains of the influenza virus, which is a disadvantage when one considers the great variability of the virus. It seems to prevent the injection of viral RNA into the host cell by altering the surface characteristics of the latter. Fortuitously, it is also an antiparkinsonism drug, acting as a cholinergic blocking agent—a totally unrelated effect.

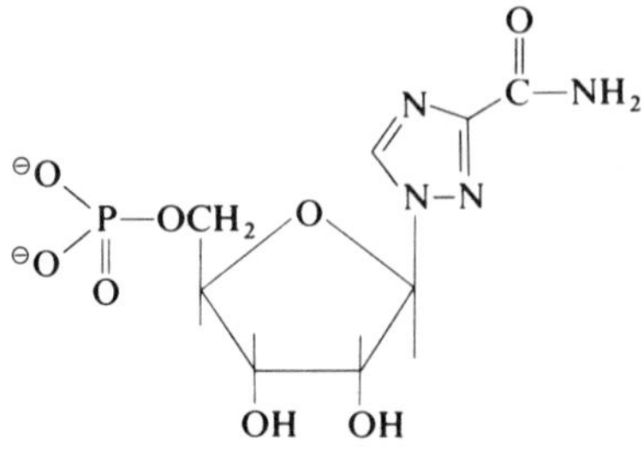

Ribavirin
6-112

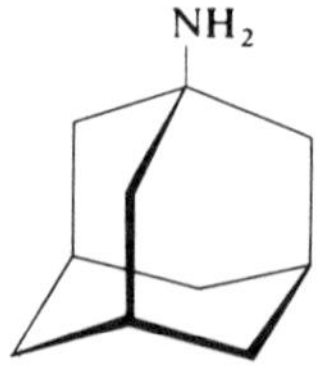

Amantadine
6-113

An antiviral and anticancer compound very much in the news media is **interferon**, a peptide consisting of about 150 amino acids. It is produced by most cells upon viral infection or a challenge by interferon-inducing agents, and protects cells against viral infection by alterating the plasma membrane (i.e., in a nonspecific manner). It also activates an endonuclease that destroys viral mRNA, and a protein kinase that inactivates a protein synthesis initiation factor, IF-2. Interferon can be isolated very laboriously and at tremendous expense from leukocytes and other cells, but the gene that encodes its synthesis has recently been transferred into bacteria by recombinant DNA techniques. It is to be hoped that the ready availability of interferon will advance antiviral and antitumor therapy as well as allow further insight into cellular regulatory mechanisms and immunological processes. A discussion of the interferon therapy is beyond the scope of this book, but some recent references are mentioned in the bibliography that follows (Stringfellow, 1980; Pestka et al., 1981; Came and Carter, 1984).

### *Selected Readings*

J. R. Brown (1978). Adriamycin and related antimicrobial agents. *Progress in Medicinal Chemistry* (G. P. Ellis and G. B. West, Eds.), Vol. 15. Elsevier/North Holland, Amsterdam, pp. 125–164.

P. E. Came and W. A. Carter (Eds.) (1984) *Interferons and Their Application, Handbook Experimental Pharmacology*, Vol. 71. Springer, Berlin.

G. L. Chen and L. F. Liu (1986). DNA topoisomerases as therapeutic targets in cancer chemotherapy. *Annu. Rep. Med. Chem. 21*: 257–262.

E. de Clerq and R. T. Walker (1986). Chemotherapeutic agents for herpesvirus infections. In: *Progress in Medicinal Chemistry* (G. P. Ellis and G. B. West, Eds.), Vol. 23. Elsevier, Amsterdam, pp. 230–255.

S. T. Crooke and A. W. Prestayko (Eds.) (1981). *Cancer and Chemotherapy*, Vol. 3: *Antineoplastic Agents*. Academic Press, New York.

R. Dolin (1985). Antiviral chemotherapy and chemoprophylaxis. *Science 227*: 1296–1303.

T. W. Doyle and T. Kaneko (1985). Antineoplastic agents. *Annu. Rep. Med. Chem. 20*: 163–172.

J. C. Drach (1980). Antiviral agents. *Annu. Rep. Med. Chem. 15*: 149–161.

E. F. Gale, E. Cundliffe, P. E. Reynolds, M. H. Richmond, and M. J. Waring (1981). *The Molecular Basis of Antibiotic Action*, 2nd ed. Wiley, New York.

J. S. Glasby (Ed.) (1979). *Encyclopedia of Antibiotics*. Wiley, New York.

S. Gupta (1986). Therapy of AIDS and AIDS-related syndromes. *Trends Pharmacol. Sci.* 7: 393–397.

H. Mitsuga and S. Boder (1987). Strategies for antiviral therapy in AIDS. *Nature 325*: 773–778.

J. A. Montgomery (1976). Current status of cancer chemotherapy. In: *Drug Research* (E. Jucker, Ed.), Vol. 20. Birkhäuser, Basel, pp. 465–490.

S. Neidle (1979). The molecular basis of action of some DNA-binding drugs. In: *Progress in Medicinal Chemistry* (G. P. Ellis and G. B. West, Eds.), Vol. 16. Elsevier/North Holland, Amsterdam, pp. 151–221.

S. Pestka, S. Maeda, and T. Staehelin (1981). The human interferons. *Annu. Rep. Med. Chem. 16*: 229–241.

W. B. Pratt and R. W. Ruddon (1979). *The Anticancer Drugs*. Oxford University Press, New York.

W. A. Remers (1979). *The Chemistry of Antitumor Antibiotics*, Vol. 1. Wiley, New York.

W. E. Ross (1985). DNA topoisomerases as targets for cancer therapy. *Biochem Pharmacol.* *34*: 4191–4195.

A. C. Sartorelli and D. G. Johns (Eds.) (1974, 1975). *Antineoplastic and Immunosuppressive Agents*, Parts 1 and 2. Springer, New York.

T. J. Smith, M. J. Kremer, M. Luo, G. Vriend, E. Arnold, G. Kramer, M. G. Rossman, M. A. McKinley, G. D. Diana, and M. J. Otto (1986). The site of attachment in human rhinovirus-14 for antiviral agents that inhibit uncoating. *Science 233*: 1286–1293.

D. A. Stringfellow (Ed.) (1980). *Interferon and Interferon Inducers*. Marcel Dekker, New York.

L. Stryer (1981). *Biochemistry*, 2nd ed. Freeman, San Francisco.

D. L. Swallow (1978). Antiviral agents. In: *Drug Research* (E. Jucker, Ed.), Vol. 22. Birkhäuser, Basel, pp. 267–326.

C.-C. Tsai (1978). Stereochemistry of drug–nucleic acid interactions and its biological implications. *Annu. Rep. Med. Chem. 13*: 316–326.

M. P. Wentland and J. B. Cornett (1985). Quinolone antibacterial agents. *Annu. Rep. Med. Chem. 20*: 145–154.

L. M. Werbel and D. F. Worth (1980). Antiparasitic agents. *Annu. Rep. Med. Chem. 15*: 120–129.

J. D. Williams and A. M. Geddes (Eds.) (1976–). *Chemotherapy*, 8 vols. Plenum Press, New York.

### Periodicals of General Interest

*Anti-Cancer Drug Design*. Macmillan, Houndmills, UK.

# 7

# Drug Distribution and Metabolism

The preceding chapters have dealt with the interaction of drugs with various receptor systems, and the biochemical and pharmacological effects of this interaction. In the time frame of drug action, this pharmacodynamic phase is preceded by the pharmacokinetic phase, comprising the physicochemical events that allow the drug to reach its site of action. A solid drug must first dissolve; any dissolved drug must be absorbed at the site of administration and then be transported throughout the organism. This process, however, is very inaccurate in terms of drugs reaching their various targets, and depends on a great many parameters. Besides the specific receptor binding that produces the intended effect, a drug will bind nonspecifically, become trapped in depot sites, and move to other systems, giving rise to unwanted side effects and toxicity.

Together with the pharmacokinetic and pharmacodynamic phases of drug action, the metabolism of the drug molecule must be considered. Drugs are exposed to many enzyme systems involved in the normal housekeeping of cells. These systems recognize a foreign, "xenobiotic" molecule and subject it to biotransformation, often in nonspecific ways designed to eliminate xenobiotic agents. Additionally, the metabolites resulting from such transformations may have pharmacological activity of their own. Metabolism can also provide the activation necessary to convert an inert pro-drug into the active compound.

The activity of a drug is terminated by its excretion, either before or after biotransformation. This may happen in a number of ways, some of which can have further consequences if elimination occurs through the placenta or milk. These correlations are shown in Fig. 7.1.

Pharmacokinetics and drug metabolism are just as important in understanding the total pharmacological activity of a drug as are molecular and biochemical pharmacodynamics. The generally rapid progress in methodology and the demands of drug regulatory agencies have resulted in greatly increased activity in these fields. In addition, it has been recognized that rational drug design is not possible without the proper consideration of drug distribution and metabolism, and the ways in which their modification can improve the overall efficacy of drug action.

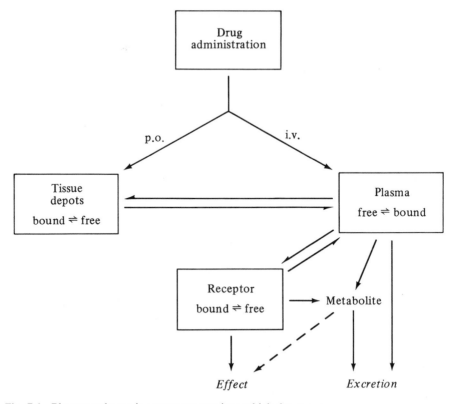

**Fig. 7.1.** Pharmacodynamic compartments into which drugs enter.

## 1. DRUG DISTRIBUTION

Drugs are administered orally (enterally) or by the nonoral (parenteral) route. Absorption of the drug is the first event that will influence the *in vivo* activity of a drug. Parenterally administered drugs are usually in solution and can be absorbed quickly, but oral administration is normally in the form of a solid, which introduces a number of variables that will determine dissolution, absorption, bioavailability, and the rate at which a drug reaches its target. Dissolution and partition in the biophase are physical phenomena; thus, parameters discussed in Chap. 1 also pertain to this field.

Because of all the factors that pertain to the final pharmacological activity of a drug, drugs that are chemically identical are not necessarily equivalent in terms of bioavailability and rate of action. The parameters governing the ultimate therapeutic result of drug administration are discussed in the following sections.

### 1.1. Drug Dissolution

The rate of drug dissolution is the rate-limiting step in drug action when a drug is taken in solid form or as a suspension. The dissolution rate will be determined by:

1. The water solubility of the drug
2. The pH of the medium
3. The $pK_a$ of the drug
4. The form, specific area, and packing of the drug crystals or particles
5. The pharmaceutical formulation (i.e., the nature of the binder, adjuvants and coating of the tablet or capsule)

The dissolution rate can be described by the Noyes–Whitney equation:

$$dC/dt = kS(C_s - C)$$

where $dC/dt$ is the dissolution rate, $S$ is the surface area of the solid, $C_s$ is the solubility of the drug, and $C$ is the concentration at time $t$. The constant $k$ is proportional to the viscosity of the medium, because dissolution is diffusion limited and depends on the thickness of the diffusion layer, a thin unstirred film of saturated solution on the surface of the crystals.

The surface area of the drug can be increased by "micronization"—that is, reduction of particle size by milling to less than 10 $\mu$m. The dissolution rate can thereby be increased by an order of magnitude, but sometimes results opposite to the desired effect are encountered: micronization may augment aggregation of the powder, or the particles may compact during tablet manufacturing. Sometimes micronization enhances the decomposition of the drug in the stomach, as in the case of oral penicillin preparations. Use of amorphous instead of crystalline forms of the drug can also increase solubility considerably, because there is no crystal lattice requiring energy input to achieve solvation and dissolution. Hydrophobic compounds can be treated with wetting agents to facilitate the access of the aqueous phase to the crystal surface.

The pH of the solvent is of obvious importance in drug dissolution. A weak acid like aspirin ($pK_a = 3.5$) is poorly soluble in the acid gastric fluid, but dissolves rapidly in the alkaline intestine. Drugs that increase the stomach pH (antacids, $H_2$ antihistamines) will therefore hasten aspirin dissolution in the stomach. However, it is not necessarily true that the pH of the unstirred diffusion layer is identical to that of the bulk solvent; thus such predictions are not always accurate.

Soluble salts dissolve more rapidly than the corresponding free acids or bases and usually have a faster onset of activity. Sometimes insoluble salts are prepared on purpose, to reduce solubility and achieve a depot effect, or to mask bitter taste in pediatric dosage forms; chewable Al-aspirin and the antimalarial cycloguanil pamoate (8-41) are examples.

## 1.2. Gastrointestinal Drug Administration

The most common and convenient mode of drug administration is the oral route. After dissolution, the drug has to clear the semipermeable membrane barriers between the lumen of the gastrointestinal tract and the systemic circulation. Aqueous pores allow free passage of water, monovalent ions, and small hydrophilic molecules; hydrophobic compounds can pass through the lipid phase of the membrane. Many nutritionally important molecules and drugs cross the membranes by *passive diffusion*; that is, they follow a concentration gradient from a higher to a lower concentration, and the rate of diffusion is proportional to the concentration difference, as described by Fick's law. Since solutes are swept away on the circulation side by the blood flow, the concentration is always lower there. Many other compounds, be they nutrients or drugs, cannot diffuse freely. These have to be transported by carriers that operate on a port–antiport system; thus an "import–export" balance must be maintained. The $Na^+$–$K^+$-ATPase (Chap. 6, Sec. 3.3) is such a system, but there are many others. In *active transport*, molecules are carried against a concentration gradient, and input of energy is required. In *facilitated transport*, a substance is carried along its concentration gradient, but at a much faster rate than passive diffusion would allow. Since the number of these carriers is limited, the transport systems are saturable, and are structure- and stereospecific. Because of carrier capacity limitation, the bioavailability of a drug decreases with increasing dose. Carbohydrates, many amino acids, ions, vitamins, pyrimidines, and drugs are transported by active or facilitated systems.

The gastrointestinal system consists of the stomach, duodenum, jejunum, ileum, colon, and rectum. These component parts are different anatomically, histologically, and physiologically. The *stomach* is lined with a smooth epithelium, and has a pH of 1–2.5, because of dilution and neutralization of the gastric secretion of pH < 1. Nonionized and lipophilic compounds are readily absorbed from the stomach, but weak acids absorb slowly and incompletely. Since gastric emptying is a relatively fast process ($t_{1/2}$ = 20–60 minutes), the role of the stomach in drug absorption is limited. Many drugs (antimuscarinics, $H_2$ antihistamines, opiates) reduce gastric motility and emptying.

The *small intestine* is the most important drug absorption site, because it is long and has a very large surface, due to mucosal invagination that carry villi with microvilli, fingerlike projections that are heavily vascularized. The pH increases from 5 to 7 between the duodenum and the end of ileum; thus weak acids as well as bases are rapidly absorbed. The number of active transport sites is also very high in the small intestine. Residence time is relatively long, approximately 4–6 hours.

The *large intestine* (colon) has a much smaller surface area and is not an efficient absorption site. It is the most alkaline part of the digestive tract (pH is 8–8.5), and is utilized when enteric-coated tablets are administered. In this formulation, tablets are coated with an acid-resistant polymer that passes the stomach intact and protects acid-sensitive drugs from decomposition. The coating dissolves in the alkaline large intestine, and therefore the drug must be absorbed from that site. Prodrugs that are activated by bacterial reduction in the small intestine are also absorbed here. The distal end of the large intestine, the *rectum*, plays an insignificant role in absorption of oral medication, but is utilized in nonoral (parenteral) administration of drugs.


### 1.3. Parenteral Drug Administration

Parenteral (nonoral) drug administration is often more efficient and faster than oral medication, but can also be more inconvenient. Drug absorption is less of a problem than it is with oral drugs, and topical applications are possible.

The most common and rapid method is *intravenous injection* into a peripheral vein. It produces an almost immediate response, with predictable and reliable serum drug levels. Injection should not be too fast, to avoid high local drug levels ("bolus") or precipitation of insoluble material causing an embolism. Drugs that have a short half-life, a narrow therapeutic index ($LD_{50}/ED_{50}$), or a need for prolonged administration are given in an intravenous infusion. When prodrugs are administered by this route, the bioavailability is not always ensured and therefore has to be checked.

Compounds that cannot reach the CNS—that do not pass the blood–brain barrier (see Sec. 1.5) from the systemic circulation—are injected directly into the cerebrospinal fluid by lumbar puncture. Such *intrathecal* and epidural administration of antibiotics and antineoplastic drugs is sometimes necessary, and very small doses of epidural opiates can provide sustained (36–48 hours) pain relief with minimal danger of habituation in postoperative and metastatic cancer pain.

*Intramuscular injections* are less reliable than intravenous delivery in terms of bioavailability, rate of absorption, and local effects. Local drug precipitation, pain, and delayed absorption are not uncommon, and the site of injection may make a difference in the rate of absorption, as vascularization of the deltoid muscle is much higher than, for instance, that of the *gluteus maximus*, which lies under a considerable fat layer. On the other hand, intramuscular depot preparations that dissolve slowly are useful in sustained-release medication, such as antimalarials or hormones.

*Subcutaneous injection* suffers from the same drawbacks as intramuscular injections, but the rate of absorption can be more easily regulated; the effect of local anesthetics can, for instance, be prolonged by addition of a vasoconstrictor (e.g., epinephrine) to the solution. Insulin is normally injected subcutaneously, and the rate of absorption can be regulated by using preparations of different solubility (amorphous or crystalline Zn-complexes). Far more efficacious, however, is the use of a continuous micropump, which administers insulin at varying rates, depending on the need. After meals more insulin is required, and thus in nondiabetics the pancreas provides insulin in such a pulsatile fashion.

Mucosal administration can be useful in some cases. *Sublingual* tablets are used for rapid absorption of nitroglycerin in angina pectoris attacks, and for some androgens. Although the oral cavity is utilized, this is not a gastrointestinal route because the drug is absorbed directly through the highly vascularized oral mucous membrane. *Vaginal* administration is used for contraceptives, prostaglandin $E_2$ abortifacients, estrogens, and antibacterial or antifungal drugs in treating local infections. *Intranasal* administration is widely practiced for decongestant $\alpha$-adrenergic agonists and for some neurohormones (gonadoliberin analogues) that are easily hydrolyzed.

Dermatologic diseases can be treated by *topical application* of drugs to the skin, because systemic effects can be minimized this way. Glucocorticoids, antineoplastics for skin cancer, and antifungal agents are frequently applied in this way. Depending on the vehicle in which the drug is dissolved, compounds can penetrate the intact skin; dimethylsulfoxide is very efficient in such transdermal applications. The eye can also absorb drugs directly; thus drugs that decrease intraocular pressure in glaucoma, and those controlling pupil size are administered as eyedrops. Bioavailability and the amount of drug retained are problematic, however.

*Rectal administration* of drugs is practiced where oral administration is difficult: in children, and in unconscious adults or persons suffering from frequent vomiting as a side effect of cancer chemotherapy or kidney failure. Absorption is reasonably good through the rectal mucosa but slower than from the small intestine, although there are exceptions. First-pass hepatic deactivation (presystemic metabolism) can be avoided at least partially.

### 1.4. Bioavailability of Drugs

As discussed above, the rate and completeness of drug absorption determines the effective dose reaching the site of action. We have seen, that many factors influence absorption: those include permeability, solubility, $pK_a$, mode of administration, pharmaceutical formulation, drug metabolism before or after reaching the site of action, and even the position of the patient (gastric emptying is faster if the patient is lying on the right side). Consequently, chemically identical drugs are not necessarily bioequivalent. Determination of bioavailability is a compulsory part of drug licensing, but secondary manufacturers are not obliged to show therapeutic equivalence with the original preparation. Thus, the problem of bioavailability is still acute, and a surprisingly large number of drugs show large differences between brands. Gibaldi (1984) discusses these in detail. Intravenous injections are normally bioequivalent, except in the case of prodrugs, which are subject to activation by patients with different metabolism (e.g., different serum hydrolase levels).

A very interesting facet of bioavailability is the choice of time of drug administration relative to the circadian rhythm of humans (the timing or periodicity of physiological variables); the field is called *chronopharmacology* or *chronotherapy* (for reviews, see Scheving and Pauly, 1976; Halberg and Halberg, 1984). It has been recognized that most physiological phenomena are time dependent and show a rhythmicity over a period of approximately, but not precisely, 24 hours—the circadian (literally, "around the day") cycle. As examples, the *acrophase* (highest point) in body temperature is at 4 P.M.; blood pressure, at 5–6 P.M.; ACTH production, at 4 A.M.; testosterone level, at 1 A.M.; and growth hormone, at 11 P.M. Cell division also shows circadian rhythm, with an acrophase in bone marrow at 10 P.M., but in skin cells at 1 A.M. It it thus not surprising that the toxicity of drugs is also subject to great variations around the clock. Thus drug tolerance and therapeutic effect are optimal when the administration of an antileukemic drug is timed to the acrophase of lowest toxicity but highest mitosis rate. This necessitates the determination of the circadian rhythm of the patient for these parameters in advance, as there are considerable individual differences. Observation of optimal timing may make a difference between life and death, and the traditional "three times daily" drug administration is physiologically meaningless.

### 1.5. Drug Distribution

Drug distribution is the process by which the adsorbed drug reaches tissues remote from the point of absorption. Drugs reach the circulation directly or indirectly and are diluted into the whole blood volume in minutes. Because peripheral capillaries have large pores (fenestrae), drugs having a molecular weight less than about 600,000 diffuse rapidly into the interstitial fluid occupying the intercellular space. The combined volume of blood and interstitial fluid is called the *central compartment*, and is about 140–190 ml/kg body weight. Drugs bind reversibly to serum proteins, which act as transport molecules. Body fluids low in proteins (cerebrospinal fluid, pericardial fluid) show the same drug concentrations as that of the free drug in plasma. Equilibration between free and bound drug occurs as the free drug enters cells in the *peripheral compartment*. Some drugs (e.g., phenobarbital, acetazolamide, salicylates) bind to erythrocytes; others, like the antileukemic hydroxyurea, bind to leukocytes. Drugs bound to proteins can be displaced by other drugs: aspirin, for instance can dangerously increase the level of free oral antidiabetic agents, triggering hypoglycemic shock. In some diseases such as rheumatoid arthritis, the total plasma protein levels are depressed, altering drug distribution.

Lipids can be an important drug depot for lipophilic molecules. General anesthetics, barbiturates, and phenothiazine neuroleptics can form stores in fatty tissue, and can easily enter neurons and the CNS. The exchange between fat depots and serum is slow, and is determined by partition coefficients.

The capillaries in the CNS are different from those in the periphery because they have no fenestrae, are not porous, and possess a cellular sheath. Therefore, they are not permeable to water-soluble substances, and thus form the *blood–brain barrier* (BBB) (see Goldstein and Betz, 1986). This can impede treatment of infections and neoplasms in the CNS. The BBB can be circumvented either by intrathecal drug administration, by hyperosmotic shock, or by use of lipophilic drug carriers (Chap. 8, Sec. 4.2.2).

The *placenta* requires special consideration, because there is a maternal–fetal equilibration due to the fact that placental capillaries on both sides are normal membranes; thus lipophilic drugs can easily be detected in the fetal circulation after about an hour of maternal ingestion. Furthermore, maternal–fetal equilibration is complicated by the fact that fetal metabolizing enzymes are not developed, and drug half-lives can be 20–40 times longer in the fetus or newborn than in the mother. Thus, drug administration during pregnancy should be severely restricted because of the many unknown effects and possible teratogenic (malformation) damage, of which the thalidomide disaster is the most recent example.

### 1.6. Pharmacokinetic Variability

Individual variability in drug response can be as high as ten fold, given identical doses per kilogram of body weight. There are many reasons for this, but some drugs show greater variability of effect than others. Drugs that show high hepatic clearance (elimination rate/arterial concentration) and presystemic metabolism ($\beta$ blockers, thymoleptics) will also show great individual variability, which can be lessened by parenteral administration.

Adjustment for *body weight* has to be made for children, and also for adults if peak serum levels are an important consideration. Obese patients present a problem because the ratio of lean body mass versus total weight is difficult to determine. The central and peripheral compartments are proportional to lean mass, but fat tissue can act as a drug sink for apolar drugs.

*Neonates and children* are not simply small adults, and can often tolerate larger mg/kg doses than adults. Pediatric dosage is often calculated not on body weight, but on body surface, proportional to that of an average adult ($1.7 m^2$). The reason for this is that the proportion of extracellular fluid of a newborn is almost twice that of an adult. However, the different and less developed drug metabolism of neonates adds another dimension to this clinical problem, and drug half-lives are usually much higher. Older children, on the other hand, show higher levels of drug metabolism, and require higher doses.

In the *aged patient*, the situation is the reverse because both organ function and drug metabolism decrease and therefore drug half-life increases. Consequently, lower drug doses should be administered to the elderly.

*Pharmacogenetics* is the study of the influence of genetic differences in drug distribution and metabolism. Drug half-lives are the same in identical twins but not in fraternal twins. Racial and geographic differences are widespread, and have been studied extensively (see Gibaldi, 1984). Isoniazid acetylation is an example: 80–100% of Eskimos and Orientals acetylate this compound rapidly, whereas Egyptians and many European groups show only a 20–40% incidence of rapid acetylation. Genetically determined enzyme deficiencies in drug metabolizing enzymes may place such individuals in jeopardy of a potential drug overdose.

## 1.7. Drug Elimination

The effects of drugs are diminished in several ways: redistribution between compartments, storage, excretion of the unchanged drug, and excretion of metabolites. The first two factors were discussed in the preceding sections. Unaltered drugs can be eliminated through various organs, but the most important routes of elimination, both for unchanged durgs and for metabolites, are the kidneys and the liver.

Three processes determine the amount of drug eliminated in urine: glomerular filtration, tubular secretion, and reabsorption. The *glomeruli* filter about 10% of the 1.2–1.4 liters of blood they receive every minute, rejecting cells and large proteins, but allowing the passage of water and small drug molecules. Thus only the drugs not bound to plasma proteins will be removed. *Tubular secretion* is a process involving active transport and therefore is not affected by protein binding; both free and bound drugs are transported. One transport system secretes organic acids (e.g., penicillin, salicylates, thiazide diuretics), whereas the other secretes only cations— that is, bases and quaternary salts like hexamethonium, catecholamines, or histamine. Drugs may inhibit each other's secretion because the transport systems have a limited capacity, and energy input is necessary for transport.

*Tubular reabsorption* returns a considerable proportion of substances (including drugs) in the ultrafiltrate into the circulation, especially the physiologically or nutritionally important solutes like glucose, salts, amino acids, and lipid-soluble compounds. The reabsorption of weak acids or bases depends on the pH of urine, and acidification or alkalization of urine with $NH_4Cl$ or $NaHCO_3$ can promote or retard drug elimination and thus influence half-life and pharmacological effect. Since the pH of urine changes in a circadian rhythm (and is lowest during sleep), drug elimination is another function influenced by chronopharmacological considerations (see above). Tubular reabsorption, including most of the water, occurs by passive diffusion because the solute concentration is much higher on the urine side than on the plasma side. However, some ions (e.g., $Li^+$) and glucose are reabsorbed via an active transport system. Some drugs (e.g., sulfonamides, methotrexate) can become so concentrated in urine that they crystallize (a condition known as crystalluria), leading to kidney damage. More than a minimum urine flow ($>190$ ml/hr) and alkaline urine are recommended when these drugs are administered.

The net effect of all these processes can be expressed by the *renal clearance* (RCL):

$$RCL = (dA_u/dt)/C$$

where $A_u$ is the amount of drug excreted in unit time (e.g., one hour) versus drug concentration ($C$) in plasma. Many of the above-mentioned factors influence clearance and therefore modify this simple correlation (see Gibaldi, 1984, for details).

*Biliary excretion* is the result of drug secretion by liver cells, which has much in common with renal excretion. Biliary clearance of 500 ml/min can be achieved. Compounds having a molecular weight less than about 400 are excreted in urine; larger molecules are cleared by the liver. Very large molecules have to be broken down before elimination of their metabolites can proceed. Bile is excreted into the duodenum, where a proportion of drugs (e.g., antibiotics, cardiac glycosides, vitamins, bile acids) is reabsorbed by the *enterohepatic cycle*, but information on this process in humans is limited. Reabsorption can be inhibited by adsorbents; for example; the insoluble ion-exchange resin cholestyramine, used as a cholesterol-lowering agent, binds bile acids and excretes them in feces (Chap. 5, Sec. 1.3).

*Secretion into milk* is not significant, but in view of the immature renal and hepatic functions of infants it is recommended that nursing mothers abstain from taking drugs, or take them just after nursing to avoid high serum and milk concentrations at feeding time. In some cases, nursing should be completely abandoned.

*Drug interactions* are an important part of clinical pharmacology and pharmacokinetics, but are beyond the scope of this book. Among other texts, Hansten (1979) and Gibaldi (1984) deal with the topic.

### Selected Readings

J. Blanchard, R. J. Sawchuk, and B. B. Brodie (Eds.) (1978). *Principles and Perspectives of Drug Bioavailability.* S. Karger, New York.

H. L. Fung, B. J. Aungst, and R. A. Morrison (1979). Pharmacokinetics and drug design. *Annu. Rep. Med. Chem. 14*: 309–320.

M. Gibaldi (1984). *Biopharmaceutics and Clinical Pharmacokinetics*, 3rd ed. Lea and Febiger, Philadelphia.

M. Gibaldi and D. Perrier (1982). *Pharmacokinetics*, 2nd ed. Marcel Dekker, New York.

G. W. Goldstein and A. L. Betz (1986). The blood–brain barrier. *Sci. Am. 225* (3):74–83.

F. Halberg and E. Halberg (1984). Chronopharmacology and further steps towards chronotherapy. In: *Pharmacokinetic Basis for Drug Treatment* (L. Z. Benet, N. Massoud, and J. G. Gambertoglio, Eds.). Raven Press, New York.

P. D. Hansten (1979). *Drug Interactions*, 4th ed. Lea and Febiger, Philadelphia.

S. Oie and L. Z. Benet (1980). Altered drug disposition in disease states. *Annu. Rep. Med. Chem.* 15: 277–287.

S. Pang and J. R. Gillette (1980). Drug absorption, distribution and elimination. In: *Burger's Medicinal Chemistry*, 4th ed. (M. E. Wolff, Ed.), Part 1. Wiley, New York, pp. 55–105.

L. A. Scheving and J. E. Pauly (1976). Chronopharmacology—its implications for clinical medicine. *Annu. Rep. Med. Chem.* 11: 251–260.

J. B. Stenlake (1979). *Foundations of Molecular Pharmacology*, Vol. 2. The Athlone Press, London, pp. 159–212.

## 2. DRUG METABOLISM

The conclusion of the previous discussion on drug elimination could lead to the assumption that a lipophilic drug can be recycled endlessly through glomerular or enterohepatic reabsorption. Most drugs, however, undergo biotransformation, with several possible outcomes:

Phase I: New polar functional groups are introduced or unmasked by oxidation, reduction, or hydrolysis.

Phase II: The original, unaltered drug or the polar metabolite is conjugated to glucuronic acid, sulfate, mercapturic acid, or acetate, becoming even more polar, and is excreted rapidly.

Only two types of compounds avoid this fate: inert anesthetic gases (e.g., Ar, Xe) and highly ionized compounds. The latter simply cannot penetrate cells to become exposed to the intracellular metabolizing enzymes.

It must be understood that metabolism does not necessarily lead to inactive compounds. Often a drug metabolite is a drug in its own right and the precursor is an inactive pro-drug; or it may be capable of covalent binding, and by becoming attached to DNA, may act as a mutagen or carcinogen. The last section of this chapter deals with that eventuality. Since contemporary drug therapy is increasingly aimed at chronic illnesses, and since patients are exposed to drugs for long periods (often years), the secondary effects of drug metabolism have attained increased importance.

### 2.1. Oxidation

Oxidative reactions of many kinds take place in the chief metabolizing organ, the liver, and are catalyzed by nonspecific enzymes. These enzymes are bound to the smooth endoplasmic reticulum which, upon homogenization, gives rise to the microsomal fraction consisting of very small vesicles that sediment only at $100,000 \times g$ acceleration.

The hepatic microsomal membrane contains the mixed-function oxidase system, which catalyzes the reaction

$$R—H + O_2 + NADPH + H^+ \rightarrow R—OH + NADP^+ + H_2O$$

NADPH is necessary to reduce half of the oxygen molecule to water. The oxygen carrier is cytochrome P-450 which, in turn, requires a cytochrome reductase flavoprotein that uses NADP as a coenzyme. The redox system, shown in Fig. 7.2, is organized in the following way:

The resting state of cytochrome P-450, shown at the top of the diagram, is a six-coordinate iron system. The Fe atom is bound to a histidine and a cysteine of the protein. The substrate molecule (R—H) binds reversibly to the cytochrome, and the complex undergoes a reduction to the ferrous state. A second enzyme, the flavoprotein cytochrome P-450 reductase, is necessary for this reaction, which ultimately derives the electron required for the reduction from NADPH through a flavoprotein, $FADH_2$. The reduced complex is now capable of reacting with molecular oxygen. The resultant peroxide then probably undergoes an additional reduction to form a peroxide anion. The involvement of cytochrome $b_5$ is suspected in this latter process, but other mechanisms are possible. The peroxide anion may then dissociate to give $H_2O_2$, or it may rearrange into an oxene, a neutral hexavalent oxygen derivative, as shown in Fig. 7.2. This admittedly hypothetical intermediate, whose existence is assumed on the basis of spectroscopic evidence, then leads to the oxidized R—OH end product with the regeneration of cytochrome P-450 (Ortiz de Montellano, 1986).

The substrates to be oxidized are of a wide variety of structural types, and the oxidation can occur on carbon, nitrogen, or sulfur. Some examples are given in Fig. 7.3.

*Aliphatic hydroxylation* usually occurs on the terminal or adjacent carbon of the molecule, whereas alicyclic rings are oxidized at the least hindered or most activated position—for example, next to a carbonyl group. Such oxidations are also important in the biosynthesis of corticosteroids from progesterone.

*Aromatic compounds* are detoxified by oxidation to their corresponding phenols, followed by coupling and elimination. This oxidation often involves arene oxides (epoxides) as intermediates. The arene oxides are highly reactive electrophiles and, besides undergoing cleavage to phenol, can also react cytotoxically or mutagenically with cellular constituents (see Fig. 1.10). Some peculiar rearrangements, called NIH shifts (after the U.S. National Institutes of Health) and involving 1,2-hydride migration, are also observed during aromatic oxidation (Fig. 7.3, No. 4).

Carbons attached to heteroatoms are also oxidized, and *N-, O-,* and *S-dealkylation* occurs frequently. The transformation of codeine to morphine, or of imipramine to desipramine, are among the more notable dealkylations; these happen to lead to more active metabolites.

*N-oxidation* and *deamination* are also possibilities, and must be distinguished from reactions catalyzed by monoamine oxidase (MAO, Chap. 6, Sec. 3.7). Amphetamines and opiates undergo such oxidation, which may also result in dealkylation.

### 2.2. Reduction and Hydrolysis

Nitro, azo, and carbonyl groups are subject to reduction, resulting in the formation of more polar hydroxy and amino groups. There are several reductases in the liver,

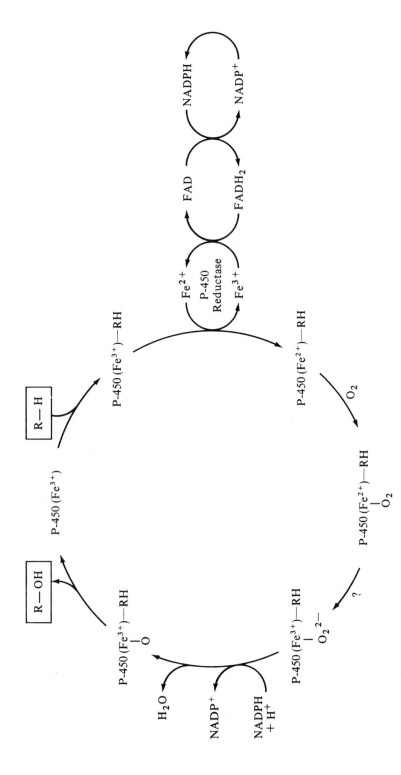

**Fig. 7.2.** Cytochrome P-450 redox system, the microsomal oxidation mechanism.

1. Side-chain oxidation

$$R{-}CH_2{-}CH_3 \longrightarrow R{-}\underset{\displaystyle OH}{\underset{|}{CH}}{-}CH_3 + R{-}CH_2CH_2{-}OH$$

2. Ring hydroxylation

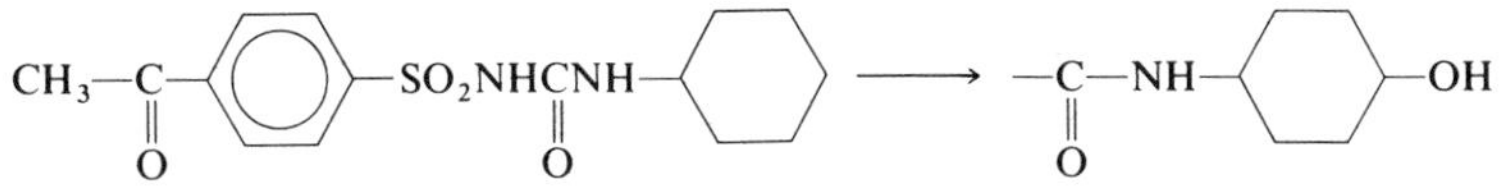
Acetohexamide
(*trans*)

3. Olefin oxidation

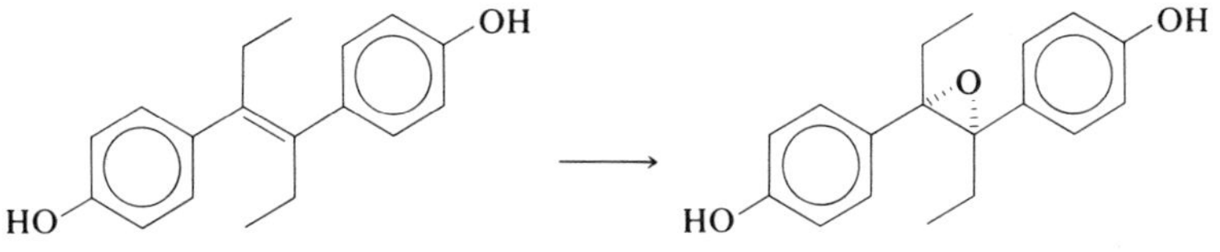
Diethylstilbestrol
(carcinogen)

4. Aromatic oxidation

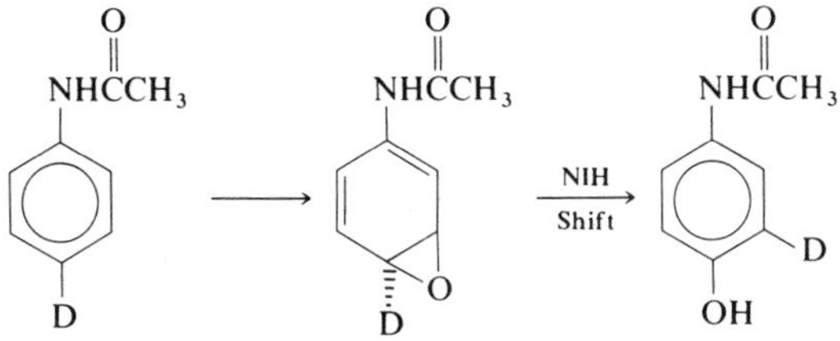
NIH
Shift
Acetanilide

5. *N*-Dealkylation

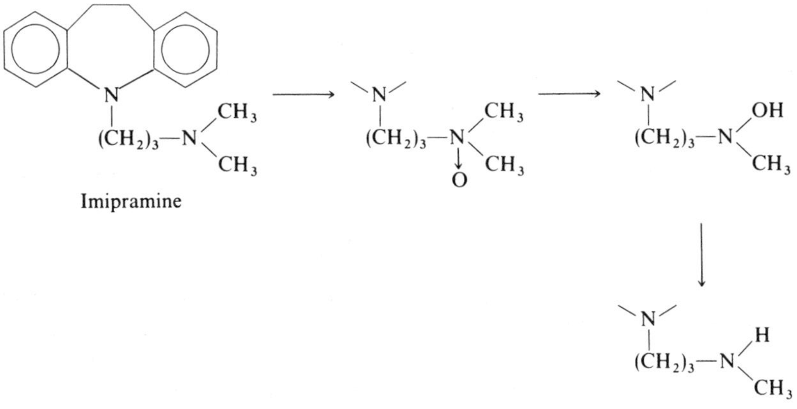
Imipramine
Desipramine

**Fig. 7.3.** Metabolic oxidations catalyzed by cytochrome P-450.

6. *S*-Dealkylation

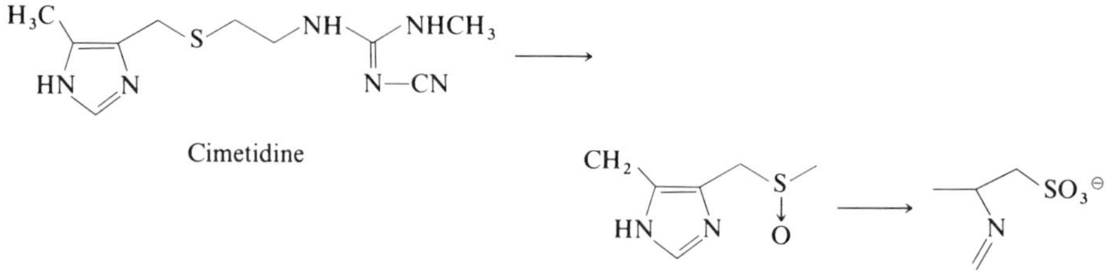

Cimetidine

7. Deamination

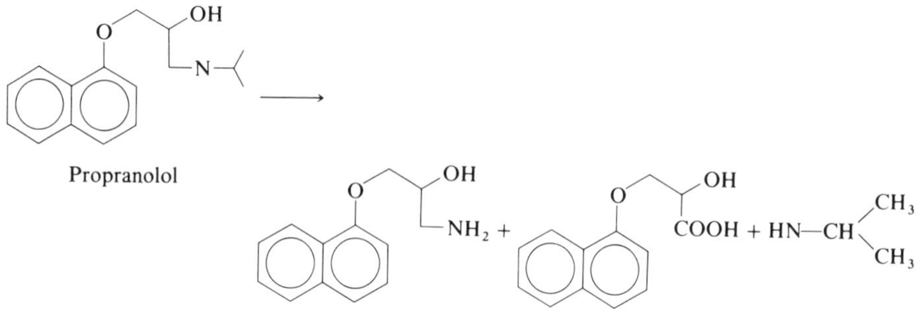

Propranolol

**Fig. 7.3** (*continued*)

which depend upon NADH or NADPH, that catalyze such reactions (Kappus, 1986). Perhaps the best-known reduction is the reductive cleavage of prontosil to sulfanilamide, the first known instance of a prodrug (Fig. 7.4).

Halogenated compounds like the general anesthetic halothane undergo reductive dechlorination; the C—F bond is stable. The halogens are also removed by a minor oxidative reaction to yield trifluoroacetate.

Another process that results in more polar compounds is the *enzymatic hydrolysis* of esters and amides, and there are many enzymes promoting such reactions. Some of these are rather nonspecific, like serum pseudocholinesterase. Naturally there is a great difference in the rates of ester or amide hydrolysis, of which medicinal chemists have taken advantage in designing stable ester analogues. The synthesis of lidocaine, an amide analogue of procaine discussed in Chap. 6, Sec. 1.3, is an example.

## 2.3. Conjugation Reactions

Also known as phase II reactions, conjugation completes the degradation of the drug undergoing oxidative or reductive metabolism, which does not necessarily produce a compound sufficiently hydrophilic or inactive to be excreted readily. The common conjugation reactions are summarized in Fig. 7.5. The most important is glucuronidation, the formation of the glucuronic acid glycoside of a drug via UDP-glucose in liver microsomes. Phenols, alcohols, amines, and amides all form *O*- or *N*-glucuronides, regardless of whether these functional groups are produced meta-bolically or not. Many endogenous substances, like steroids, are also excreted in this way. Glucuronides are normally nontoxic, highly water soluble, and excreted in the urine or bile.

Sulfate conjugation is less common in humans, but is seen with steroids and phenols.

Glutathione conjugation occurs in the cytoplasm and is important in the elimination of polycyclic phenols and halides. It can lead to a mercapturic acid endproduct (Fig. 7.5).

**Fig. 7.4.** Reductive and hydrolytic drug metabolism.

Reduction

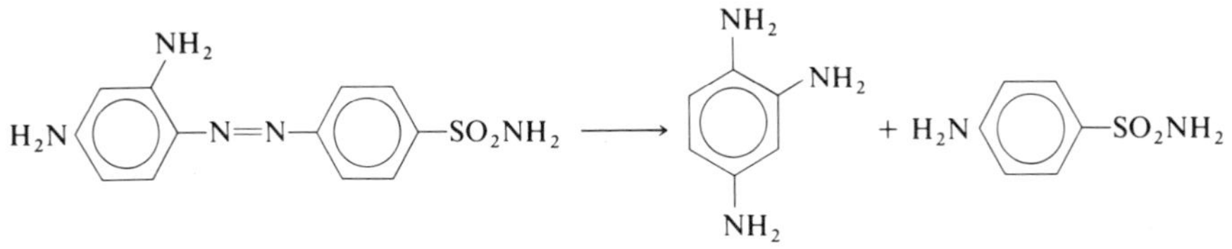

Prontosil rubrum

Sulfanilamide

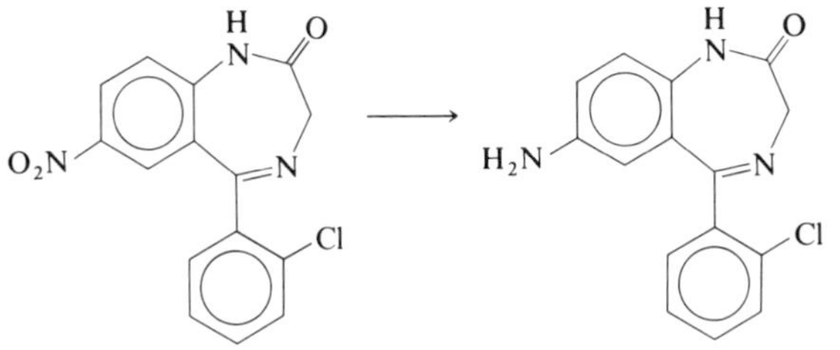

Clonazepam

$$F_3C-CH-Br \longrightarrow F_3C-CH_2Br + Cl^\ominus + F_3C-COOH$$
$$\quad\quad |$$
$$\quad Cl$$

Halothane

Hydrolysis

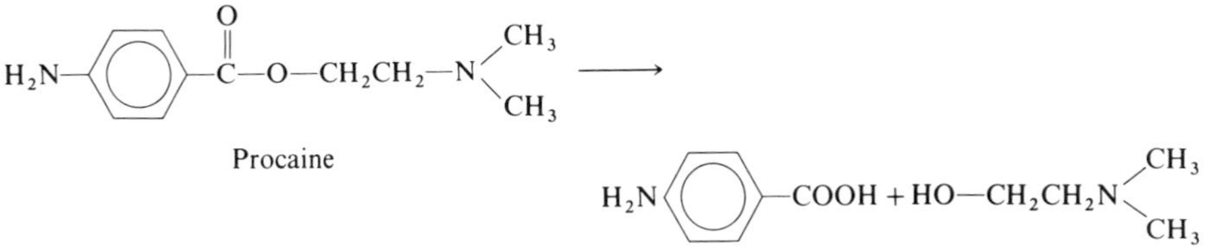

Procaine

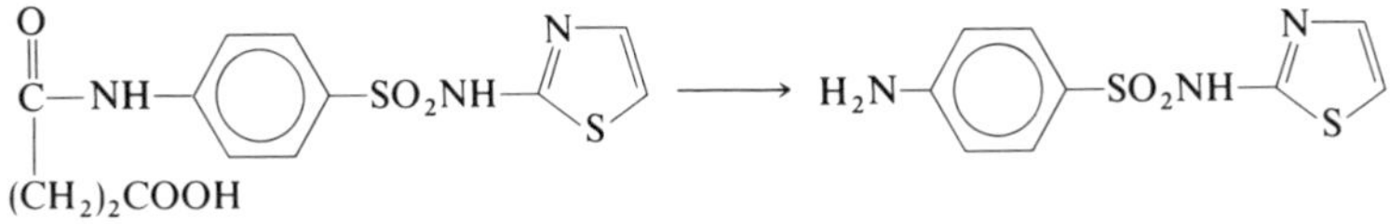

Succinyl-sulfathiazole

Sulfathiazole

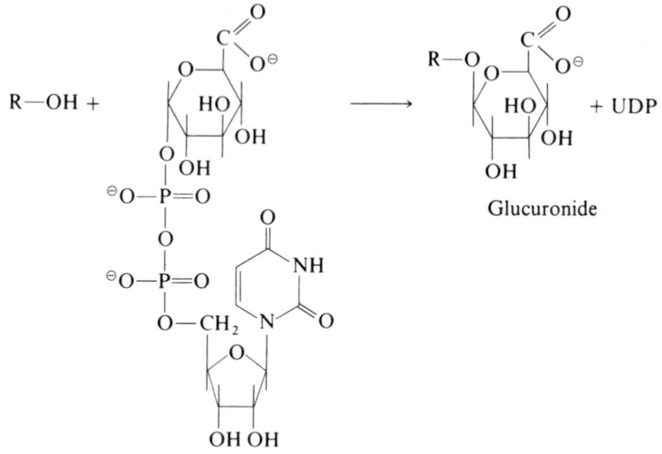

R—OH +

Uridine diphosphate-glucuronate
(UDP-glucuronate)

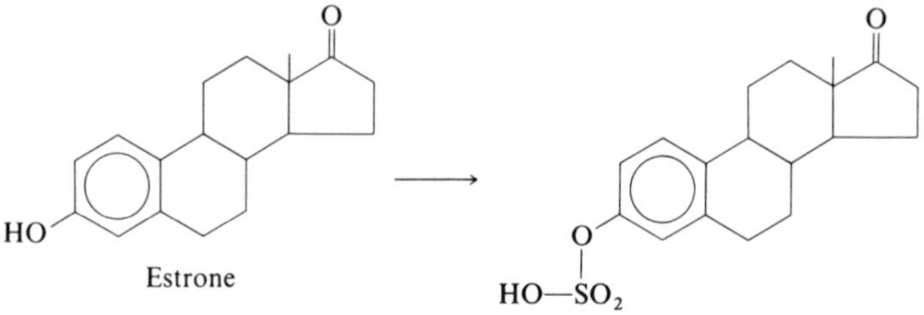

Glucuronide

+ UDP

Estrone

Estrone sulfate

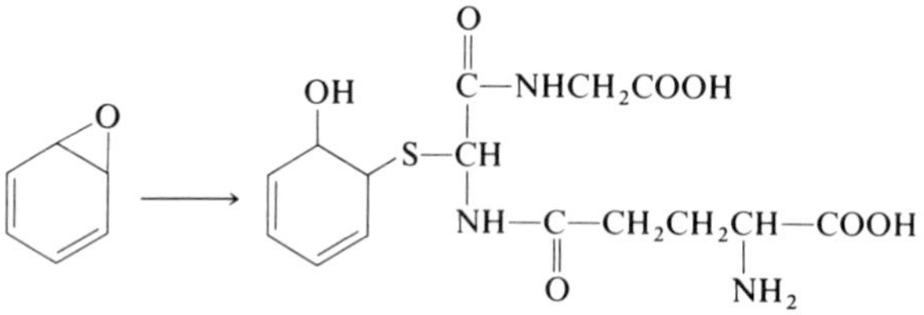

Glutathione conjugate

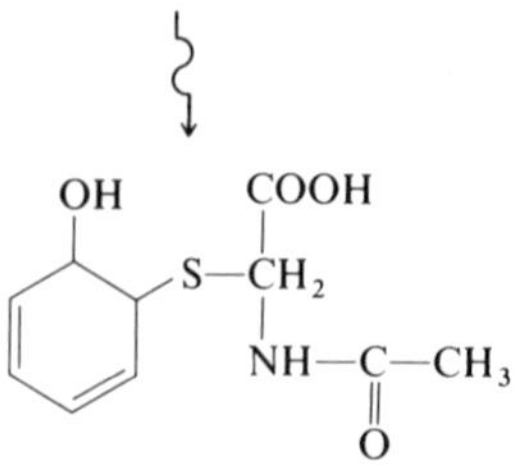

Mercapturic acid conjugate

**Fig. 7.5.** Conjugation reactions.

Acetylation is seen in the metabolism of aromatic amines (e.g., sulfanilamides), hydrazides, and histamine.

### 2.4. Toxic Effects of Drug Metabolism

Biotoxification can be reversible or irreversible. Photosensitization and allergic reactions are usually reversible, but the covalent reaction of a metabolite with biomolecules can lead to carcinogenesis, mutagenesis, or teratogenic effects (developmental alterations in the offspring). All of these must be avoided, if possible, by appropriate drug modification. Ariëns (1984) and Mitchell and Horning (1984) deal with this topic.

There are two pathways that are especially dangerous in terms of producing intermediates capable of permanent cell damage. The first is the formation of arene oxides from polycyclic hydrocarbons, which is the cause of the well-known carcinogenicity of these compounds and their reputation as dangerous environmental pollutants. Figure 7.6 shows the reactions of benzo[a]pyrene that are mediated by cytochrome P-450. The arene oxides that are formed can be opened nonenzymatically by hydrolysis, or by the scavenger enzyme epoxide hydrase, supposedly a protective enzyme. The diol then undergoes a second stereospecific epoxidation (see Chap. 1, Sec. 5), yielding the 9,10-epoxide *trans* to the 7-OH group (Weinstein et al., in Ts'o, 1978) and its isomer, both more mutagenic than the original hydrocarbon. The two epoxides are quite stable: the *trans* epoxide has a half-life of 8 minutes, and then easily reacts with the 2-NH$_2$ group of guanine in DNA, producing single-strand breaks (Guenther, 1984).

**Fig. 7.6.** Oxidative carcinogen formation from benzopyrene. For details of the oxidation, see Fig. 1.10.

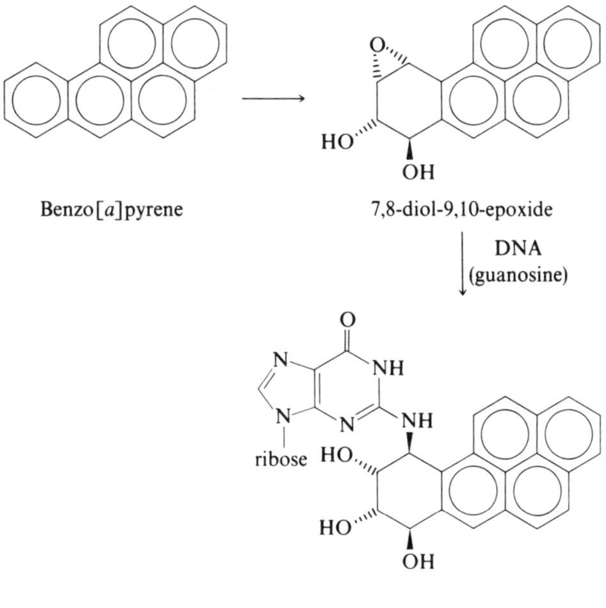

Benzo[a]pyrene                7,8-diol-9,10-epoxide

DNA
(guanosine)

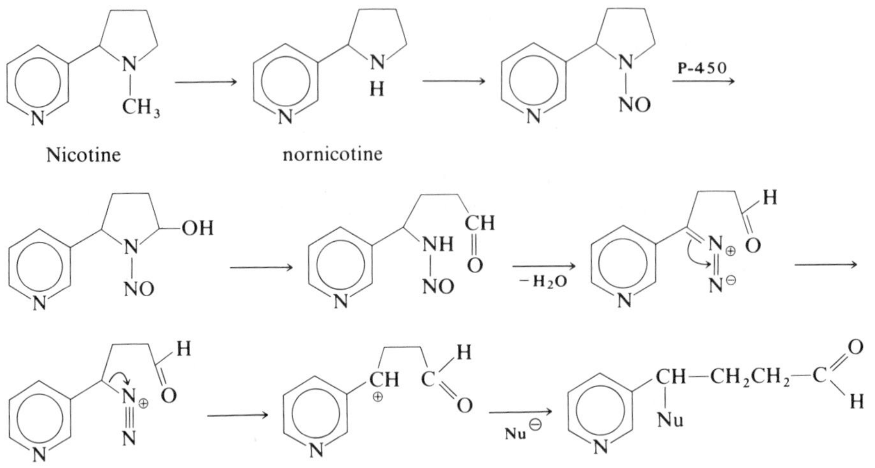

**Fig. 7.7.** Nicotine toxification by nitrosation and hydroxylation. (After Hoffman, in Ts'o and Gelboim, 1978)

Such "third generation" carcinogens can also be formed from other aromatic compounds; one way of inhibiting their production is the introduction of a *p*-fluoro substituent into the ring.

The metabolic toxification of proven carcinogens, as shown in Fig. 7.7, is another interesting pathway to carcinogenesis. As shown in the figure, nicotine is nitrosated nonenzymatically, and the *N*-nitroso derivative hydroxylated by the P-450 enzyme. The resulting diazonium ion is a nucleophile, capable of reacting with DNA (Hoffman et al., in Ts'o, 1978).

An especially insidious case of metabolic toxification was recently uncovered when it was discovered that young women who had many years previously been exposed *in utero* to diethylstilbestrol taken by their mothers, developed vaginal adenocarcinoma (see Williams, 1981).

Further complications in the metabolic picture arise if one considers that some drugs and environmental pollutants (barbiturates, rifampicin, spironolactone, halogenated insecticides, and some other compounds) are *enzyme inducers*, increasing the synthesis of liver microsomal enzymes, including oxygenases. This can result in increased drug toxicity through enhanced metabolite production, or in increased drug activity through the production of pharmacologically active degradation products. When a drug itself induces enzyme activity tolerance can develop. Phenobarbital is one example of such a drug, although barbiturates induce activity in many enzymes. The mechanism of enzyme induction is unknown.

Drug metabolism—and drug activity—are subject to many other factors. Individual variability can be as high as ten fold; congenital abnormalities (e.g., Down's syndrome) or ethnic factors can influence enzyme availability. For instance, whereas 60% of Swedes have low levels of *N*-acetyltransferase, all Canadian Eskimos show a high enzyme concentration. Newborns have not developed many enzyme systems, just as drug metabolism undergoes a decrease in old people. For a recent review, see Neuman (1987).

## Selected Readings

E. J. Ariëns (1984). Domestication of chemistry by design of safer chemicals: structure–activity relationships. *Drug Metab. Rev. 15*: 425–504.

M. J. Coon, A. H. Connery, R. W. Estabrook, H. V. Gelboin, J. R. Gillette, and P. J. O'Brien (Eds.) (1980). *Microsomes, Drug Oxidation and Chemical Carcinogenesis*, 2 vols. Academic Press, New York.

J. Edelson, D. P. Benzinger, and J. E. Peterson (1981). Drug metabolism. *Annu. Rep. Med. Chem. 16*: 319–327.

T. Guenther (1984). Cellular processing of carcinogen–DNA adducts. *Trends Pharmacol. Sci. 5*: 365.

W. B. Jacoby (Ed.) (1981). Detoxication and drug metabolism: conjugation and related systems. In: *Methods in Enzymology* (S. P. Colowick and N. O. Kaplan, Eds.), Vol. 77. Academic Press, New York.

T. Jenner and B. Testa (1981). *Concepts in Drug Metabolism*, 2 vols. Marcel Dekker, New York.

H. Kappus (1986). Overview of enzyme systems involved in bioreduction of drugs and in redox cycling. *Biochem. Pharmacol. 35*: 1–6.

L. K. Low and N. Castagnoli Jr. (1980). Drug biotransformation. In: *Burger's Medicinal Chemistry*, 4th ed., Part 1 (M. E. Wolff, Ed.). John Wiley, New York, pp. 107–226.

B. H. Migdalof, K. J. Kripalani, and S. M. Singhvi (1979). Drug metabolism. *Annu. Rep. Med. Chem. 14*: 188–197.

J. R. Mitchell and M. G. Horning (Eds.) (1984). *Drug Metabolism and Toxicity*. Raven Press, New York.

S. D. Nelson (1982). Metabolic activation and drug toxicity. *J. Med. Chem. 25*: 753–765.

H. G. Neuman (1987). Toxication mechanisms in drug metabolism. *Adv. Drug. Res. 15*: 1–28. Academic Press, New York.

P. R. Ortiz de Montellano (Ed.) (1986). *Cytochrome P-450. Structure, Mechanism, and Biochemistry*. Raven Press, New York.

B. Testa (1986). Chiral aspects of drug metabolism. *Trends Pharmacol. Sci. 7*: 60–64.

B. Testa and P. Jenner (1976). *Drug Metabolism: Chemical and Biochemical Aspects*. Marcel Dekker, New York.

J.-P. Tillement, G. Houin, R. Zini, S. Urien, E. Albengres, J. Barr, M. Lecomte, P. D'Athis, and B. Sebille (1984). The binding of drugs to blood plasma macromolecules: recent advances and therapeutic significance. In: *Adv. Drug. Res.* (B. Testa, Ed.), Vol. 13. Academic Press, London, pp. 59–94.

P. Ts'o and H. Gelboin (Eds.) (1978). *Polycyclic Hydrocarbons and Cancer*. Academic Press, New York.

C. Walsh (1980). Scope and mechanism of enzymatic monooxygenation reactions. *Annu. Rep. Med. Chem. 15*: 207–216.

D. A. Williams (1981). Drug metabolism. In: *Principles of Medicinal Chemistry*, 2nd ed. (W. O. Foye, Ed.). Lea and Febiger, Philadelphia.

### Periodicals of General Interest

*Drug Metabolism and Disposition*. Am. Soc. Pharmacol. Exp. Ther.
*Drug Metabolism Reviews*. Marcel Dekker, New York.

# 8
# Principles of Drug Design

## 1. ANALOGUE SYNTHESIS VERSUS RATIONAL DESIGN

Drug design today is more of a hope than an achievement. It means the application of previously recognized correlations of biological activity with physicochemical characteristics in the broadest sense, in the hope that the pharmacological success of a not yet synthesized compound can be predicted. Few drugs in use today were discovered entirely in this way. The cholinesterase inhibitor antidote pyridine aldoxime methiodide (PAM, **pralidoxime**; 6-38), the antiulcer drug **cimetidine** (4-181), and some antimetabolites active against leukemias are examples. One of the principal difficulties in this approach is that the available—and very sophisticated—methods for predicting drug action cannot foretell toxicity and side effects, nor do they help in anticipating the transport characteristics or metabolic fate of the drug *in vivo*. These are, of course, as important in producing a therapeutically successful drug as the abstract *in vitro* or cellular effect of the pharmacon.

Very often our best efforts are frustrated by basic ignorance of the biology or biochemistry underlying a disease, and we are reduced to what Lewis Thomas (1974), in one of his incisive essays, calls "halfway technology" in reference to the complex and costly management of diseases whose basic causes are not understood. The treatment of rheumatoid arthritis, most malignant tumors, and all mental diseases falls into this category, and contrasts glaringly with the simplicity of dealing with most infectious diseases of bacterial origin and even some viral diseases like poliomyelitis.

Although some practicing medicinal chemists and molecular pharmacologists still regard efforts at rational drug design with some condescension and ill-concealed impatience, a slow but promising development gives renewed hope that progress in this area will not be less rapid than in the application of biology and physical chemistry to human and animal pathology. The explosive development of computer-aided drug design (see Sec. 3) in three dimensions promises to lead to the era of true rational drug design.

Until the early 1960s, drug design was an intuitive endeavor based on long experience, keen observation, serendipity, sheer luck, and a lot of hard work. The probabilities of finding a clinically useful drug were not good; it was estimated that anywhere from 3000 to 5000 compounds were synthesized in order to produce one practical drug. With today's more strict drug safety regulations, the proportions are even worse and the costs skyrocket, retarding the introduction of new drugs to a dangerous extent. The classical method usually applied in drug development was molecular modification—the design of analogues of a proven active "lead" compound. The guiding principle was the paradigm that minor changes in a molecular structure lead to minor, quantitative alterations in its biological effects. Although this may be true in closely related series, it depends on the definition of "minor" changes. The addition of two very small hydrogen atoms to the $\Delta^8$ double bond of ergot alkaloids eliminates their uterotonic activity, but replacement of the $N-CH_3$ substituent by the large phenethyl group in morphine increases the activity less than ten fold. Extension of the side chain of diethazine by only one carbon atom led to the serendipitous discovery of chlorpromazine and modern psychopharmacology. Burger (1983) provides a retrospective account of classical drug design.

There are two conclusions to be drawn from these random examples. First, a merely structural change in an organic molecule is meaningless in structure–activity relationship (SAR) studies as long as its physicochemical consequences remain unexplored and the molecular basis of its action remains unknown. Structure, in the organic chemical sense, is only a repository, a carrier of numerous parameters of vital importance of drug activity, as amply illustrated in the first chapter of this book.

The second conclusion to be drawn from the above examples—and innumerable others—is that the discovery of qualitatively new pharmacological effects is often a discontinuous jump in an otherwise monotonous series of drug analogues and is hard to predict, even with fairly sophisticated methods.

Despite the great success of the classical methods of drug design, their unpredictability and the tremendous amount of wasted effort expended have necessitated the development of more rational methods with a much higher predictive capability, in an effort to elevate drug design from an art to a science. The approach involving the design of analogues of an active lead compound remains unchanged, and the expertise of the medicinal chemist is as much in demand as ever; however, the intuitive process of selecting structural modifications for synthesis becomes circumspect in this approach, and models based on multiple regression analysis and pattern recognition methods, using very powerful computer techniques, are employed as aids. It is obviously much faster and cheaper to calculate the required properties of novel compounds from a large pool of data on their analogues than to synthesize and screen all such new compounds in the classical fashion. Only promising candidates are investigated experimentally. The results gained this way are incorporated into the data base, expanding and strengthening the theoretical search. Eventually, sufficient material accumulates to aid in making a confident decision about whether the "best" analogue has been prepared or whether the series should be abandoned (Gross, 1983).

Although a beginning has been made, drug design is far from being either automatic or foolproof. The choice of a proper lead compound—a necessity in quantitative drug design—is still based on experience, serendipity, and luck, given our basic ignorance of molecular phenomena at the cellular level. Now, however, we can at least have the confidence that the discovery of new drugs and the development of existing ones will be able to keep pace with the progress of biomedical research.

## Selected Readings

E. J. Ariëns (Ed.) (1972–). *Drug Design*, 10 vols. Academic Press, New York.
A. Burger (1983). *A Guide to the Chemical Basis of Drug Design*. Wiley, New York.
R. Franke (1984). *Theoretical Drug Design Methods*. Elsevier, Amsterdam.
F. Gross (1983). *Decision Making in Drug Research*. Raven Press, New York.
J. A. Keverling Buisman (Ed.) (1982). *Strategy in Drug Research*. Elsevier, New York.
H. J. Smith and H. Williams (Eds.) (1983). *Introduction to the Principles of Drug Design*. Wright, Bristol, England.
L. Thomas (1974). The technology of medicine. In: *The Lives of a Cell*. Penguin Books, New York.

## 2. DISCOVERY OF "LEAD" COMPOUNDS

Lead (as in *guide*) compounds are still a *sine qua non* of drug design, as we are just beginning to reach the level of sophistication at which an intimate knowledge of receptor mechanisms allows us to design a novel drug without a previous model (see Sec. 3.4). There are several well-tested methods for uncovering leads other than by *de novo* design.

*Random screening*, while seemingly wasteful, still has its place in developing lead compounds in areas in which theory lags. Screening for antitumor activity has been carried on for more than 20 years by the U.S. National Cancer Institute and Sloan-Kettering Memorial Cancer Institute, with tens of thousands of compounds being tested annually on tumors *in vivo* and *in vitro* on bacteria, and on viruses. More recently, a computerized prescreening method has been applied to this process (see Hodes, 1979), saving time and expense, and hence the screening is not as random as it used to be. A successful random search for antibacterial action was conducted by several pharmaceutical companies in the 1950s. They tested soil samples from all over the world, which resulted in the discovery of many novel structures and some spectacularly useful groups of antibiotics, notably the **tetracyclines**. In fact, microbial sources have supplied an enormous number of new drug prototypes, sometimes of staggering complexity. Recently, the large-scale automated testing of microbial mutants has been realized and combined with recombinant DNA techniques to speed up the efficient discovery and production of new antibiotics.

*"Weak" drugs* are compounds that have shown some biological activity, but not enough to make them drug candidates. Skillful modification of their structure and physicochemical properties may considerably increase the activity of the resultant derivatives; quite a few examples of such improvements have been discussed in this book.

Sometimes, the *side effects* of a drug may constitute such "weak" activity. The keen observation of pharmacological activity has led to the development of several groups of drugs in this way. A recent example is the development of the **sulfonylurea** hypoglycemic drugs and **chlorothiazide** diuretics from the antibacterial sulfanilamides. In such instances, molecular modification leads to the abolition of the former main effect and enhancement of the former side effect. In some cases, both effects might be useful, as was true for the sedative antihistaminic agents; the medicinal chemist must then separate the two effects rather than suppressing one or the other. Within this same category one could include drugs that were developed through investigations of drug metabolism, since a drug metabolite is frequently much more active than the parent compound. Such manipulations can decrease the toxicity of a drug or otherwise improve its effects; these are discussed in Sec. 4, on prodrugs.

Although ethnopharmacology, the scientific investigation of folk medicine and "old wives' tales," has led to some *bona fide* drugs (e.g., **reserpine**, **quinine**, **ephedrine**), it has not proven to be a good or efficient source of leads. However, *natural products* have always been and still are an inexhaustible source of drug leads as well as drugs (Krogsgaard-Larsen et al., 1984), and vitamins, hormones, and neurotransmitters are all natural products as well as being drugs in cases of deficiency disease. Renewed interest in natural products and the novel structures they provide is especially noticeable in marine pharmacology, a practically virgin territory. Several research institutes (e.g., the Medical University of South Carolina) as well as well-established groups (notably the Scripps Institute of Oceanography and the University of Hawaii) are producing some very promising results in this field. Papers by Sarett (1979), Mitscher and Al-Shamma (1980), de Souza et al. (1982), and Kaul and Daftari (1986) and the books by Faulkner and Fenical (1977) and Scheuer (1981–) provide entry to this vast new field of natural products and the drugs it may provide. The isolation of prostaglandins from a coral was one of the more startling recent discoveries in marine pharmacology.

An extension of natural products chemistry is the *biochemical information* derived from the study of metabolic pathways, enzyme mechanisms, and cell physiological phenomena; this research has revealed exploitable differences between host and parasite (including malignant cells), and between normal and pathological function in terms of these parameters. The large and fertile area of *antimetabolite* (metabolic inhibitors) and *parametabolite* (metabolic substitutes) chemistry is based on such stratagems, and has found use in the field of enzyme inhibition (covered in depth in Chap. 6, Sec. 3) as well as in conjunction with nucleic acid metabolism (Chap. 6, Sec. 5). The design of drugs based on biochemical leads remains a highly sophisticated endeavor, light-years removed from the random screening of sulfonamide dyes in which it has its origin.

### Selected Readings

N. J. de Souza, B. N. Ganguli, and J. Reden (1982). Strategies in the discovery of drugs from natural sources. *Annu. Rep. Med. Chem. 17*: 301–310.

D. J. Faulkner and W. Fenical (Eds.) (1977). *Marine Natural Product Chemistry.* Plenum Press, New York.

L. Hodes (1979). Computer-aided selection of novel antitumor drugs for animal screening. In: *Computer Assisted Drug Design* (E. C. Olson and R. E. Christoffersen, Eds.). American Chemical Society, Washington, DC.

P. N. Kaul and P. Daftari (1986). Marine pharmacology: bioactive molecules from the sea. *Annu. Rev. Pharmacol. Toxicol. 26*: 117–142.

P. Krogsgaard-Larsen, S. B. Christensen, and H. Kofod (Eds.) (1984). *Natural Products and Drug Development*. Munksgaard, Copenhagen.

L. A. Mitscher and A. Al-Shamma (1980). New developments in natural products of medicinal interest. *Annu. Rep. Med. Chem. 15*: 255–266.

L. H. Sarett (1979). The impact of natural product research on drug discovery, In: *Drug Research* (E. Jucker, Ed.), Vol. 23. Birkhäuser, Basel, pp. 51–62.

P. Scheuer (Ed.) (1981–). *Marine Natural Products*, 5 vols. Academic Press, New York.

## 3. PHARMACOPHORE IDENTIFICATION

All drugs have pharmacological activity as a result of stereoelectronic interaction with a receptor. The receptor macromolecule recognizes the arrangement of certain functional groups in three-dimensional space and their electron density. It is the recognition of these groups rather than the structure of the entire drug molecule that results in an interaction, normally consisting of noncovalent binding. The collection of relevant groups responsible for the effect is called the *pharmacophore*, and their geometric arrangement is called the *pharmacophoric pattern*, whereas the position of their complementary structures on the receptor is the *receptor map* (see Gund, 1979; Humblet and Marshall, 1980). Over the years, many attempts have been made to define the pharmacophores and their pattern on many drugs. The first attempts were rather naive and simplistic, but the recent use of crystallographic methods, statistical pattern recognition techniques, and the development of the idea of "conformational space" (see Humblet and Marshall, 1980) have contributed greatly to the evolution of sophisticated methods of practical significance. What elevated this discipline from a rut of many years' duration was the recognition that the "minimum energy conformation" of a drug would fit a receptor only in the context of a rigid "lock and key" type interaction (as proposed by Emil Fisher for enzymes, around the turn of the century)—a plainly obsolete analogy in view of the Koshland induced-fit theory and Burgen's "zipper" hypothesis. The situation is patently more complex, for the "conformational space" of Marshall represents a population of conformers having energies within about 25 kJ of the minimum energy value and capable of undergoing conformational perturbations when reacting with the receptor. Only up-to-date computer methods, including computer graphics, permit calculation of the "*excluded volume maps*" *of a receptor*—the sum total of the maps of all active compounds capable of binding to that receptor. Refinement of this pseudo-electron-density surface can be achieved by considering compounds not capable of binding, even though possessing the required pharmacophores. Excluded volume maps of inactive compounds ($V_I$) can be subtracted by computer graphics and yield the *receptor essential volume* of the active site. At the same time, the "favorable" volume ($V_F$) is defined as the overlap of the excluded volumes of all active and inactive compounds; $(V_I - V_F)$ represents the volume responsible for negative interactions with the receptor (Sufrin et al., 1981). The great advantage of this approach is its capacity for accommodating noncongeneric drug series (i.e., allowing the comparison of compounds with similar activities but dissimilar structures). The reviews of Gund (1980) and Humblet and Marshall (1980) provide further entry into this fascinating field. Naturally, these methods do not take into account the pharmacokinetic and metabolic aspects of drug action, and they can therefore be only a part in the intricate process of rational drug design.

### 3.1. Structure Modifications

Classical structure modifications have been the mainstay of drug synthesis since the earliest days. As emphasized earlier at several points, structural modifications expressed in organic chemical terms are really only symbols for modification of the physicochemical properties of various structures, as will be discussed in Sec. 3.2. Nevertheless, the medicinal chemist usually thinks in terms of structure, since that is the language of organic synthesis. It is therefore appropriate to deal with such an approach, provided one keeps in mind that it is somewhat obsolete because it is twice removed from the arena of drug–receptor interactions.

#### 3.1.1. Variation of Substituents

The variation of substituents can follow many directions. It can be used to increase or decrease the polarity, alter the $pK_a$, and change the electronic properties of a molecule. Exploration of *homologous series* is one of the most often used strategies in this regard, because polarity changes that are induced are very gradual. The case of the antibacterial action of aliphatic alcohols has been discussed in detail in Chap. 1 Sec. 2; an increase in chain length leads to increased activity, with a sudden cutoff point at $C_6$–$C_8$, due to insufficient solubility of these homologues in an aqueous medium because of their high lipophilicity. In aliphatic amines, micelle formation commences at about $C_{12}$, and effectively removes the compound from potential interaction with receptors requiring monomeric ligands. On the other hand, local anesthetics depend on lipid solubility in the membrane, and the duration of anesthesia produced by the nupercaine derivatives varies between 10 and 600 minutes for a series of alkyl substituents ranging from —H to —n-pentyl. A classical example of the effect of alkyl chains on activity involves **hexamethonium** (4-27) and **decamethonium** (4-32), the first of which is a ganglionic blocking agent and the second a neuromuscular blocker. Another well-known example is the profound qualitative change in action between **promethazine** (4-175), an $H_1$-antihistaminic drug in which two —$CH_2$— groups separate the ring and side-chain nitrogens, and **promazine** (and **chlorpromazine**, 4-112), which has three methylene groups and predominantly exhibits tranquilizing properties. Higher homologues can, on occasion, become antagonists of the lower members of a series.

### 3.1.2. Introduction of Double Bonds

The introduction of double bonds changes the stereochemistry of a molecule and decreases the flexibility of carbon chains. The $E$ and $Z$ isomers will show very different binding properties. The $\Delta^1$-double bond of **prednisone** (5-38) increases its antirheumatic activity over that of its parent compound, **cortisol** (5-35), by about 30 fold.

### 3.1.3. Variations in Ring Structure

Variations in ring structure are endless in drug synthesis, and are often used in the service of some other change, or are introduced simply for patent-right purposes. Inspection of some of the bewildering variations of rings in the older $H_1$ antihistamines (Fig. 8.1) reveals them to be simply variations on the ethylenediamine structure (8-1 to 8-6) differing only quantitatively in their effect. Among the antihistamines shown in Fig. 8.1, the closure of the diphenylmethane group of **cyclizine** (8-4) to the phenothiazine ring in **parathiazine** (8-5) gave rise to the first phenothiazine compounds.

**Fig. 8.1.** Ring variations in ethylenediamine $H_1$-antihistaminics.

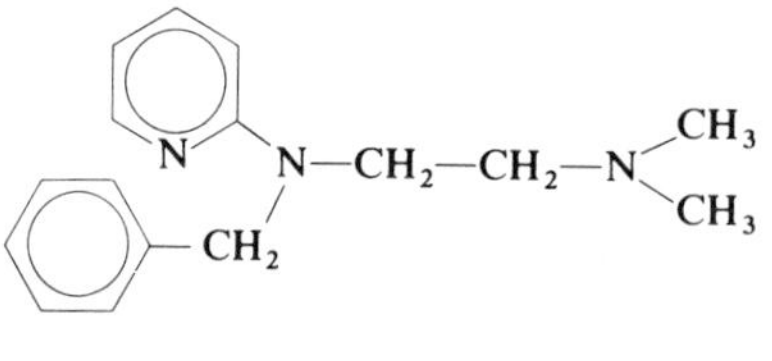

Tripelennamine
8-1

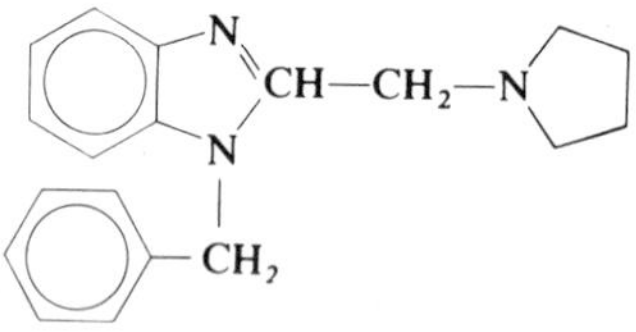

Clemizole
8-2

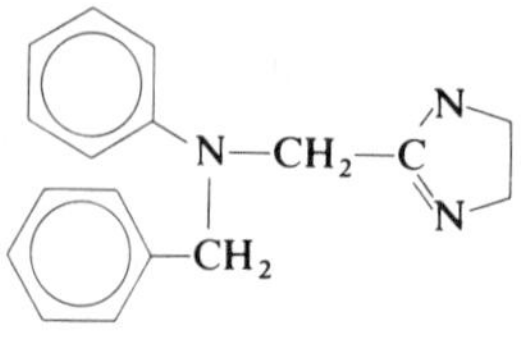

Antazoline
8-3

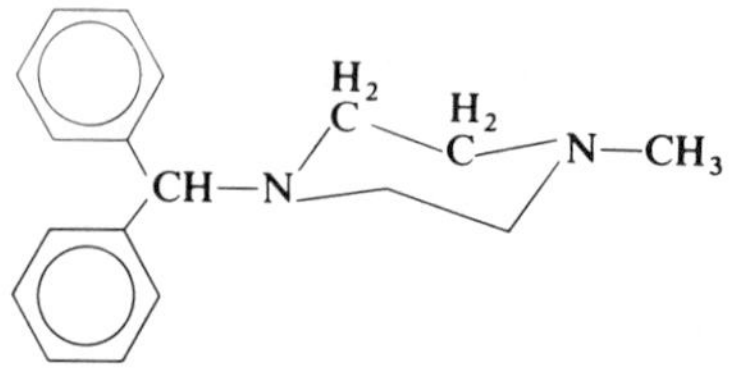

Cyclizine
8-4

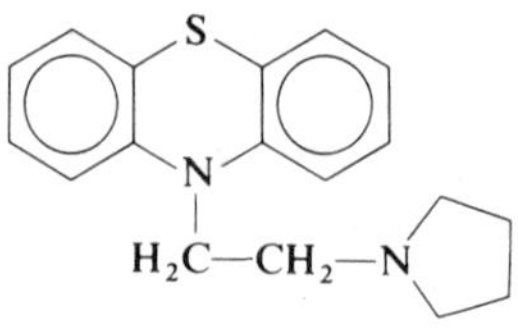

Parathiazine
8-5

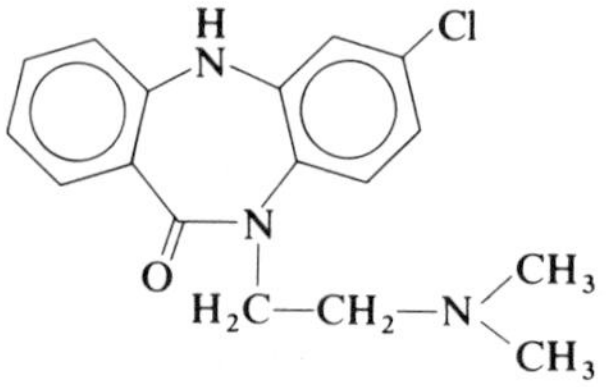

Tarpane
8-6

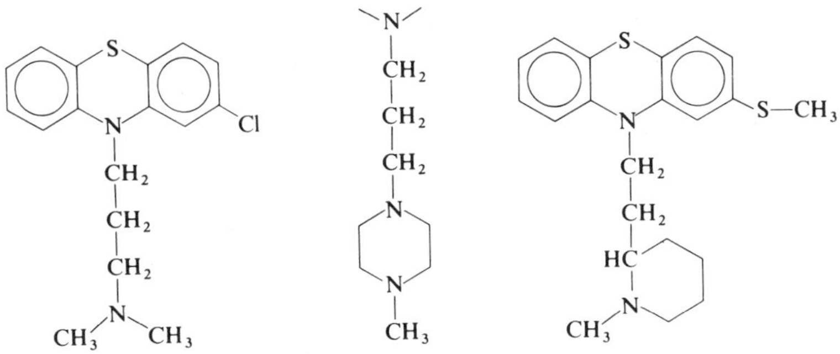

Chlorpromazine
8-7

Prochlorperazine
8-8

Thioridazine
8-9

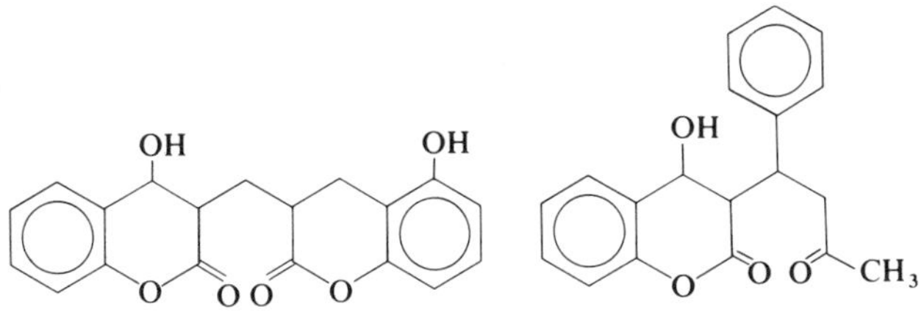

Dicumarol
8-10

Warfarin
8-11

**Fig. 8.2.** Examples of ring closure and ring opening in drug modification.

Sometimes, however, relatively minor changes in ring structure lead to profound qualitative changes. The most famous example of this occurs in the transition from the neuroleptic DA-blocking phenothiazine drugs (Table 4.7) to the antidepressant dibenzazepines such as imipramine (4-54, Fig. 4.25), in which the replacement of —S— by $—CH_2—CH_2—$ changes the molecular geometry while also being isosteric (see next section).

Ring opening or closure usually leads to subtle changes in activity, provided that nothing else changes. Three examples (among many possibilities) come to mind: incorporation of the *N*-methyl substituents of **chlorpromazine** (8-7) into a closed piperazine ring in **prochlorperazine** (8-8) tremendously increases the antiemetic effect while the neuroleptic activity declines. Of course, this may be due to the introduction of a new basic center (Fig. 8.2). In **thioridazine** (8-9), the neuroleptic effect increases with the introduction of a closed ring, but extrapyramidal side effects become noticeable.

A ring-opening modification leading to increased activity is shown in Fig. 8.2, in the anticoagulants **dicumarol** (8-10) and **warfarin** (8-11).

### *3.1.4. Structure Pruning and Addition of Bulk*

As noted earlier, the pharmacophore of a drug is usually confined to a few functional groups or parts of the whole molecule, which can be a large one. In the case of such complex natural products as alkaloids, which may be difficult or impractical to synthesize (e.g., **tubocurarine**), the first design attempt is usually directed at simplification of the molecule, pruning away those structural elements that are not part of the pharmacophore and do not serve to hold crucial binding groups in their appropriate positions. The most successful dissection of a molecule can be seen in the case of **morphine**, in Fig. 8.3. Starting with morphine (8-12), the oxygen bridge (i.e., the furan ring) is first removed, resulting in **levorphanol** (8-13), a morphinan. By eliminating ring C, the benzomorphan series is obtained. Its most successful member is **pentazocine** (8-14), which retains only the two methyl groups from ring C, and has a low addiction liability. The simplest (and, incidentally, oldest) modification of the morphine molecule is seen in **meperidine** (8-15), a phenylpiperidine that has many congeneric analogues. **Fentanyl** is designed along similar lines and has some tremendously active analogues, like sufentanil (8-16) (see Chap. 5, Sec. 3.7.3). Even in the **methadone** (8-17) molecule, the remnants of the piperidine ring are discernible. On the basis of these and other analogues, the opiate pharmacophore consists of:

**Fig. 8.3.** Successive "pruning" of the morphine molecule.

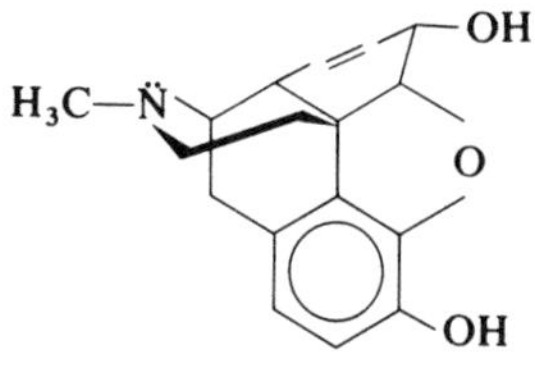

Morphine
8-12

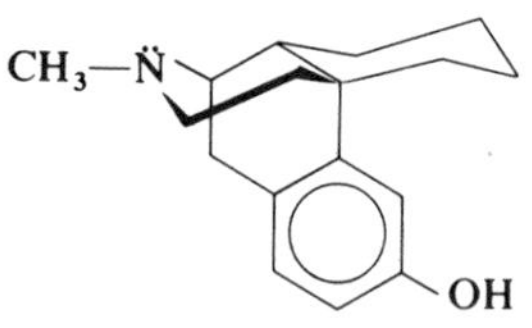

Levorphanol
8-13

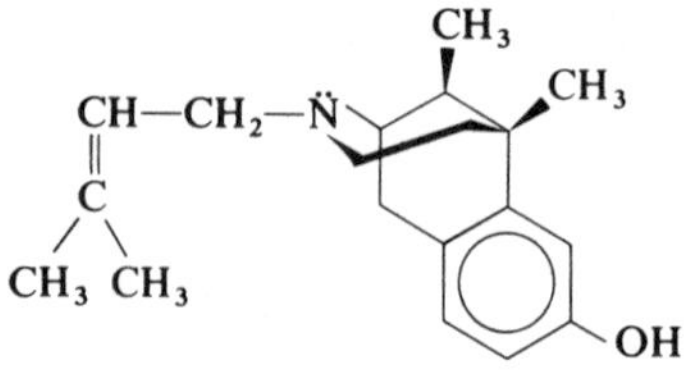

Pentazocine
8-14

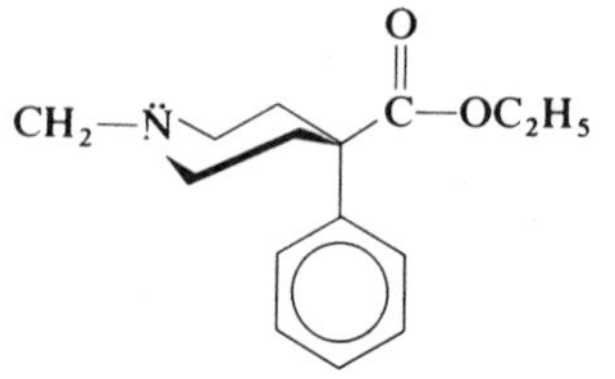

Meperidine
8-15

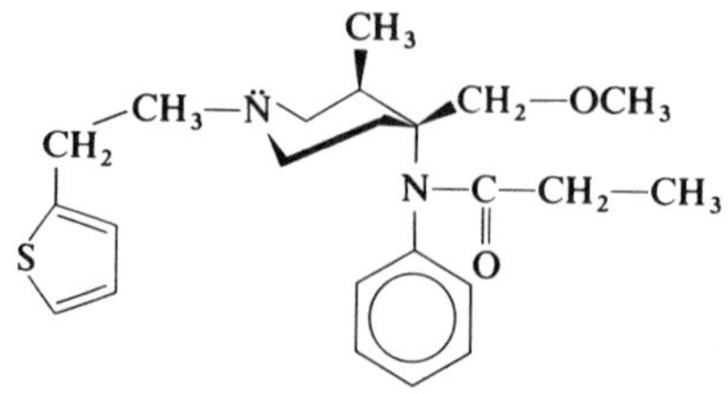

Sufentanil
8-16

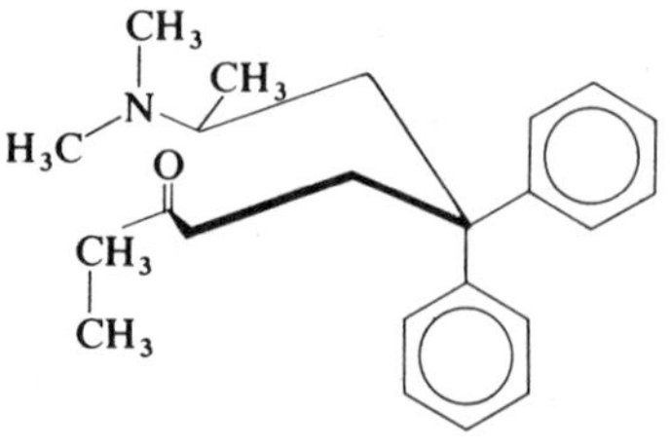

Methadóne
8-17

1. A nonbonding N electron pair
2. A phenyl ring three carbons removed from the N
3. A quaternary carbon next to the phenyl ring

Basically, the same criteria apply to the enkephalins.

The addition of bulky substituents to a drug molecule often results in the emergence of antagonists, since it permits the utilization of auxiliary binding sites on the receptor. This trend is especially noticeable among the neurotransmitters. For example, the anticholinergics, $\beta$-adrenergic blocking agents, and some serotonin antagonists show this correlation (Fig. 8.4, structures 8-18 to 8-23).

Large substituents often prevent enzymatic attack on a drug, thereby prolonging its useful life. This technique was used to impart resistance to $\beta$-lactamase to the semisynthetic penicillins (see Table 6.3), as shown in the case of **diphenicillin** (8-25) in Fig. 8.4. The need for the proximity of the phenyl group to the lactam is quite interesting; o-phenylbenzyl penicillin (8-26) is inactive as an enzyme inhibitor because the phenyl group no longer hinders access of the enzyme to the lactam bond.

### 3.2. Physicochemical Alterations

Alteration of the physicochemical characteristics within a drug series is, of course, a result of structural modification; it is just our point of view that changes. It is rather difficult to change only a single parameter with any specific modification, with the potential exception of lipophilicity, which increases with the addition of "inert" hydrocarbon groups. The degree of lipophilicity—so important in drug action and quantitative SAR (QSAR) investigations—is otherwise subject to change together with the Hammet $\sigma$ constant, a descriptor of the electron-donor or -acceptor capability of a substituent. A plot showing this correlation, constructed by Craig, is shown in Fig. 1.19. When planning a drug series, maximum variability can be obtained by choosing substituents from each quadrant of the plot and, for further refinement of the SAR, observing the general trend in activity when only the $\sigma$ or $\pi$ parameter is varied. Correlations like this are valuable in noncomputerized SAR correlation methods such as the Topliss scheme (see Chap. 1, See. 9, and also this chapter, Sec. 3.3).

### 3.2.1. Isosteric Variations

The *isosteric replacement* of atoms or groups in a molecule is widely used in the design of antimetabolites or drugs that alter metabolic processes. Isosteric groups, according to Erlenmeyer's definition are isoelectronic in their outermost electron shell. However, since their size and polarity may vary, the term isostere is somewhat misleading. Isosteres are classified according to their valence (i.e., number of electrons in the outer shell):

Class I: halogens; OH; SH; $NH_2$; $CH_3$
Class II: O, S, Se, Te; NH; $CH_2$

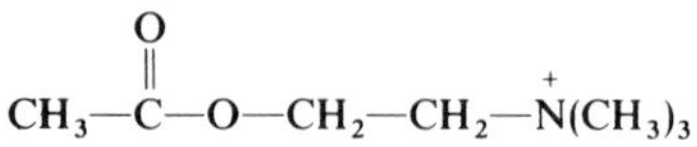

agonist

Acetylcholine
8-18

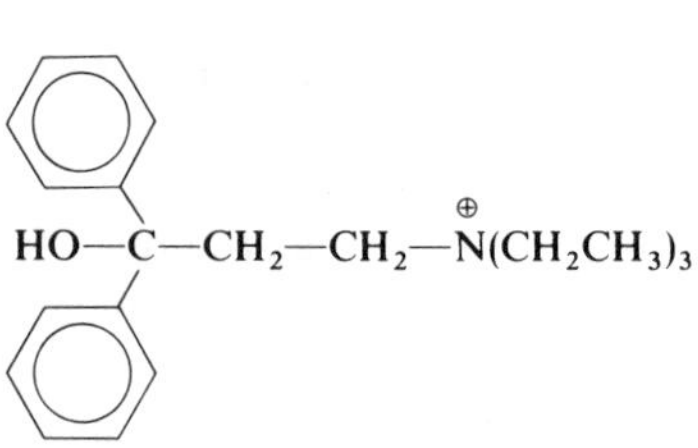

antagonist

Tridihexethyl bromide
8-19

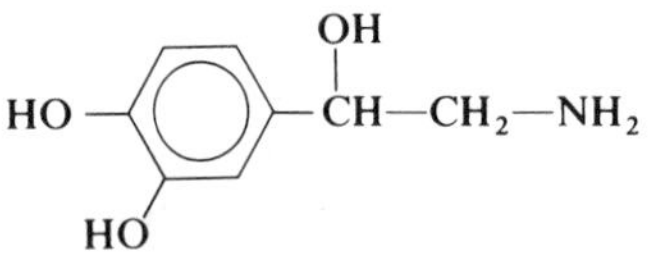

agonist

Norepinephrine
8-20

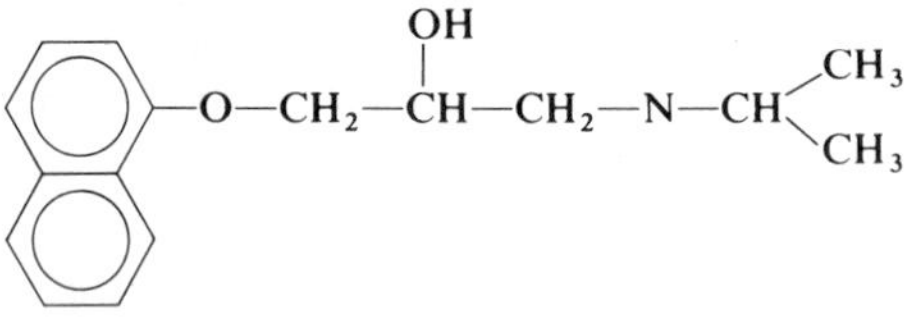

antagonist

Propranolol
8-21

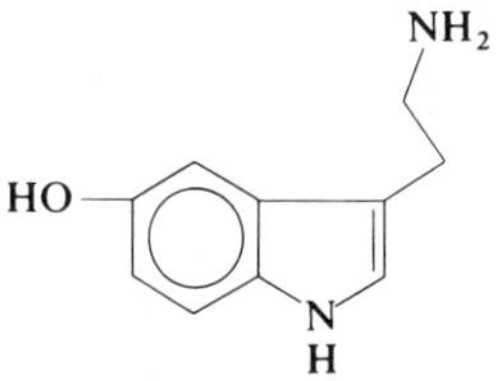

agonist

Serotonin
8-22

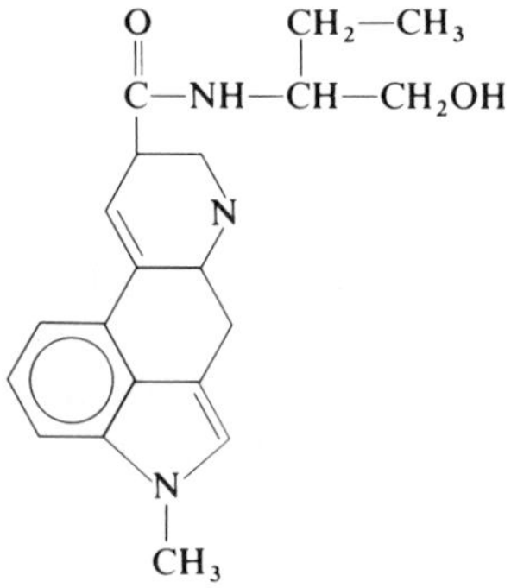

antagonist

Methysergide
8-23

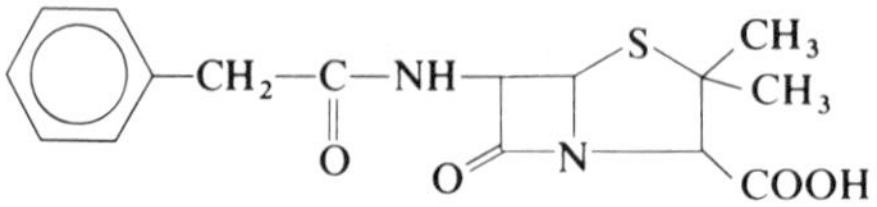

Benzylpenicillin
8-24

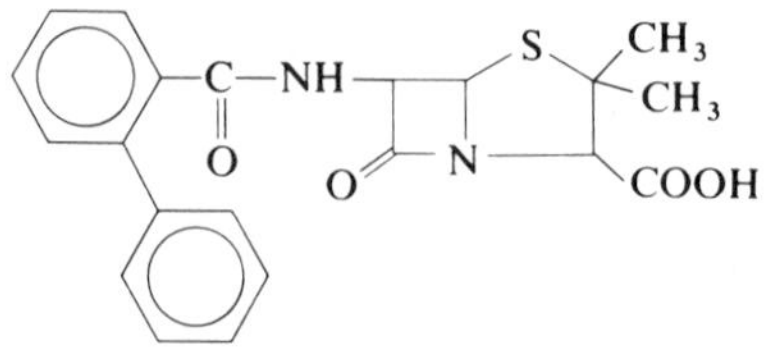

lactamase inhibitor

Diphenicillin
8-25

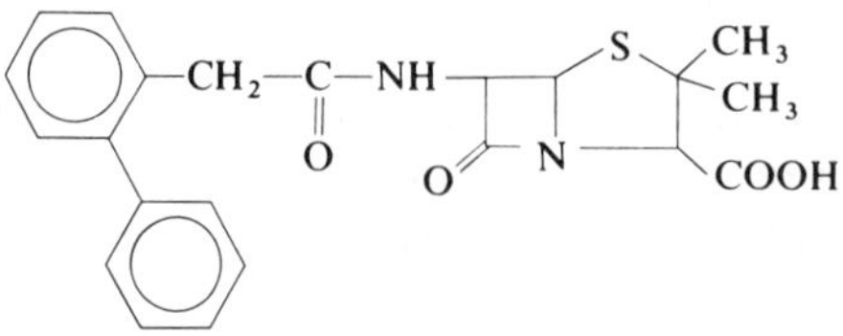

no inhibitor

*o*-Phenylbenzyl-penicillin
8-26

**Fig. 8.4.** Effect of bulky substituents on pharmacological activity.

Class III: N, P, As, CH
Class IV: C, Si, $N^+$, $P^+$, $S^+$, $As^+$
Class V: —CH=CH—, S, O, NH (in rings)

Thus, for instance, the exchange of OH for SH in hypoxanthine gives the antitumor agent **6-mercaptopurine** (6-88) (cf. Chap. 6, Sec. 5.1.3). Fluorine, the smallest halogen, replaces hydrogen well, giving, for instance, **fluorouracil** (Fig. 6.19), which is also an antitumor antimetabolite. Interchanges of —CH— and nitrogen are common in rings, as seen in the antiviral agent **ribavirin** (6-112).

The oldest example of the use of "nonclassical" isosteres involves the replacement of the carboxamide in folic acid by sulfonamide, to give the sulfanilamides. Diaminopyrimidines, as antimalarial agents (Chap. 6, Sec. 3.5.2, Fig. 6.18), are also based on folate isosterism, in addition to the exploitation of auxiliary binding sites on dihydrofolate reductase. This concept of nonclassical isosteres or *bioisosteres*— that is, moieties that do not have the same number of atoms or identical electron structure—is really the classical structure modification approach; Lipinski (1986) describes many new examples.

### 3.2.2. Conformational Variations

Conformational variations in a drug molecule are difficult to study (cf. Martin, 1978, pp. 364–366). When "preferred conformations" were believed to be a useful concept, it was thought that rigid analogues of a flexible drug might determine which of several possible conformers "fit" the receptor. Besides the fact that rigid analogues are often difficult to synthesize, they may bind in an orientation different from that of the flexible parent compound; moreover, flexibility, in terms of induced fit, may be required for the proper binding of a drug. It seems possible however, to calculate the relative conformational energies of various molecules from steric, solvation, torsional, and H-bonding forces. The utility of this concept remains to be seen, but recent progress in interactive computer graphics allows the study of models and the manipulation of conformations on three-dimensional representations, which literally permits the tailoring of compounds to a known, given receptor (see Sec. 3.4).

### Selected Readings

P. Gund (1979). Pharmacophoric pattern searching and receptor mapping. *Annu. Rep. Med. Chem. 14*: 299–308.

P. Gund (1980). Three-dimensional molecular modeling and drug design. *Science 208*: 1425–1431.

C. Humblet and G. R. Marshall (1980). Pharmacophore identification and receptor mapping. *Annu. Rep. Med. Chem. 15*: 267–276.

C. A. Lipinski (1986). Bioisosterism in drug design. *Annu. Rep. Med. Chem. 21*: 283–291.

Y. C. Martin (1978). *Quantitative Drug Design*. Marcel. Dekker, New York.

J. R. Sufrin, D. A. Dunn and G. R. Marshall (1981). Steric mapping of the L-methionine binding site of ATP: L-methionine S-adenosyltransferase. *Mol. Pharmacol. 19*: 307–313.

J. G. Topliss and Y. C. Martin (1975). Utilization of operational schemes for analog synthesis in drug design. In: *Drug Design* (E. J. Ariëns, Ed.), Vol. 5. Academic Press, New York, pp. 1–21.

### 3.3. Numerical Techniques in Drug Design

Chapter 1 discussed the quantitative approaches to structure–activity correlations, including methods that in part are retrospective as well as predictive, since a "training set" of compounds of known pharmacological activity must first be established (Austel, 1984). The purpose of such methods is to increase the probability of finding active compounds among those eventually synthesized, thus keeping synthetic and screening efforts within reasonable limits in relation to the success rate.

To utilize the wealth of information contained in large training sets, researchers must employ the methods of multivariate statistics because the raw data simply overwhelm the unassisted human mind. The general aim of these methods is to project into two-dimensional space a multidimensional (and thus incomprehensible) data matrix, and to order independent variables according to their contribution to variance (principal component analysis, correspondence analysis).

Logically, one must begin with a discussion of *pattern recognition* and *cluster analysis*, the two most recent quantitative methods, which make use of sophisticated statistics and computer software. These methods are outlined in detail by Stuper, Brugger, and Jurs (1979), and by Mager (1980).


### *3.3.1. Pattern Recognition*

Pattern recognition can be used to deal with a large number of compounds, each characterized by many parameters—for example, the presence or absence of certain functional groups; spectral lines; the number of certain atoms per molecule; and the radius of gyration (Wolff and Parsons, 1983). First, however, these raw data must be processed by scaling and normalization—the conversion of diverse units and orders of magnitude from many sources—so that the chosen parameters become comparable. Feature selection methods exist for weeding out irrelevant "descriptors" and obtaining those that are potentially most useful. By using "eigenvector" or "principal component" analysis algorithms, these multidimensional data are then projected two-dimensionally onto a plot whose axes are the two principal components or two (transformed and normalized) parameters accounting for most of the variance; these are the two eigenvectors with the highest values. Previously unrecognized relational patterns between large numbers of compounds characterized by multidimensional descriptors will thus emerge in a new, comprehensible, two-dimensional plot. The projection of unknowns onto this eigenvector plot will determine their relationship to active and inactive compounds. Decisions are made without attention to any statistics of the data, by reducing it to a geometric problem, called the "nearest neighbor" classification. Figure 8.5 shows such a plot with four classes and two unknowns. It is obvious that unknown $X$ can be classified easily, and it is apparent why unknown $Y$ leads to an unreliable decision. For example, Stuper et al. (1979) described their training set of 140 tranquilizers and 79 sedatives belonging to 25 types of compounds. These drugs were characterized by 69 original descriptors, which were reduced to about 38. The predictive capacity of their classification ranged from 90% to 92%. The duration of sleep induced by bar-

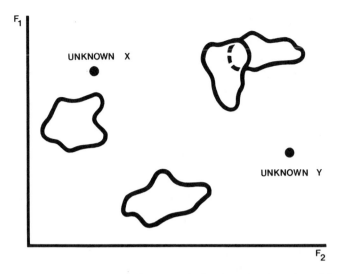

**Fig. 8.5.** Schematic representation of cluster analysis, plotted as function of factors $F_1$ and $F_2$: Unknown $X$ is easy to associate with a cluster; unknown $Y$ has an ambiguous position.

biturates was treated in an analogous manner, resulting in a 93–94% predictive capacity among 160 compounds.

### 3.3.2. Cluster Analysis

Cluster analysis is similar in concept to pattern recognition. It can define either the similarity or dissimilarity of observations or can reveal the number of groups formed by a collection of data. The distance between clusters of data points is defined either by the distance between the two closest members of two different clusters, or by the distances between the centers of clusters. Other useful statistical methods are outlined by Martin (1978).

### 3.3.3. Regression Analysis

Regression analysis is currently the most widely used correlative method in drug design, because it simplifies problems within a set of compounds by using a limited number of descriptors, notably the Hansch hydrophobic constant $\pi$, Hammett constants, or other electronic characteristics of substituents, and the Taft steric constant $E_s$. The extrathermodynamic linear relation of these parameters, in its parabolic form, is expressed in the equation:

$$\log 1/C = -k_1\pi^2 + k_2\pi + \sigma + k_3E_s + k_4$$

where $C$ is the molar drug concentration necessary for a constant effect, and $k_1, k_2, k_3$, and $k_4$ are constants which are found by regression analysis. Further details of this method are given in Chap. 1, Sec. 9, and a "do-it-yourself," step-by-step explanation of the procedure is given by Martin (1978). This correlation is used in the *Hansch method* and its variants.

### 3.3.4. Artificial Intelligence Methods

Methods of artificial intelligence have been applied to create an "expert system" that can read molecular structures and manipulate them to generate potential descriptors monitored by statistical analysis (Klopman and Contreras, 1985). The descriptors are molecular fragments similar to those used by the Free–Wilson approach (Chap. 1, Sec. 9.2). The main advantages of the system are its capability of handling very large data bases that are beyond human comprehension or overview, and the automatic generation of descriptors avoiding investigator prejudice. It has been used to evaluate and design benzodiazepine anticonvulsants.

## 3.4. Four-Dimensional Drug Design Using Computer Graphics

The enormous progress in computer hardware and software, elucidation of macromolecular structure and ligand–receptor interactions, crystallography, and molecular modeling since about 1980 has brought us to the threshold of a breakthrough in drug design. We are now in a position to design lead compounds *de novo*, based on the structure of the receptor macromolecule. This new generation of drugs will contain *four dimensions* of information: three for the spacial stereostructure of the molecule; and the fourth, expressing electronic charge distribution. However, as Cohen (1985) warns in his extensive review of the topic, computerized drug design is still only an instrument that reduces empiricism in an experimental science; the inherent approximations of innumerable conformers and molecular parameters of drug and receptor, and the methodological inaccuracies and difficulties of comparison, will never allow the elimination of insight and trial.

The *tools* of the modeler are multifaceted. Interatomic distances and other parameters are obtainable from data banks: the Cambridge Crystallography Data bank contains over 30,000 small molecules, but only about 50 proteins. X-ray crystallographic data of drug–receptor interactions are also available (Stezowski and Chandrasekhar, 1986). There are many powerful computer programs available; some are listed in Table 8.1; others are discussed by Cohen (1985) and by Hrib (1986). Characteristically, a program like SCRIPT takes the two-dimensional formula of a compound drawn with a joystick , generates all conformers and draws their perspectivic formula, derives the minimal conformational potential, and draws a stereoscopic view of the minimum energy conformers. It also defines bond distances and angles, shows a space-filling model, and allows the rotation of all the models on the computer screen to explore the molecule from all angles (Cohen, 1985).

### 3.4.1. Docking Experiments

The most important aspect of computer-assisted modeling is the capability to perform three-dimensional color-coded docking experiments, a tool now used by all major drug companies. Starting with the x-ray structure of the macromolecule, a space-filling molecular model is created, including hydrate envelopes around it. By separately generating the three-dimensional model of a hypothetical drug as

**Table 8.1.** Some computer programs of interest to medicinal chemists

| Name | Functions | Location | Reference |
|---|---|---|---|
| ADAPT | Structure–activity relationships by linear regression, learning machine, pattern recognition | | Stuper et al., 1979 |
| CIS | Chemical Information System: 30,000 mass spectra; 4000 $^{13}$C-NMR spectra; Substructure Search System: 34,000 x-ray crystal structures, 27,000 powder diffraction spectra; Mathematical Modeling System | U.S. National Institutes of Health (NIH) | Chu, 1980 |
| HANSCH | Hansch method; linear regression | Pomona College | Martin, 1978 |
| LHASA | Interactive structural graphics, organic synthesis | Harvard U. | Gund, 1977 |
| MMMS | Molecular modeling, drug design, quantum chemical calculations | Merck, Sharp & Dohme, Inc. | Gund, 1980 |
| MMS-X | Drug design, receptor mapping, conformational analysis | Washington U., St. Louis | Gund, 1980 |
| MOLPAT | Pharmacophoric pattern search | | Gund, 1979 |
| OCCS | Interactive structural graphics, organic synthesis | Harvard U. | Gund, 1977 |
| PROPHET | Model building in three dimensions, tabulation, graphing, and statistical analysis of pharmacological data | NIH | Chu, 1980 |
| SAS | Statistical Analysis System | | Martin, 1978 |
| SUMEX | Structure elucidation, synthesis (SECS) | Stanford U. network | Gund, 1977 |
| | 3D computer graphics | | Meyer, 1980 |

described above, a modeler can manipulate the two by modern fast computers and can directly examine the fit of the ligand in the active site; the investigator can change the substituents, conformation, and rotamers of the drug on the screen, and can repeat the docking. The molecules can be color-coded; thus, one can zoom in and out of the picture, and literally "sit" in the active site by "peeling" away layers of the macromolecule to obtain an unimpeded view of the drug macromolecule fit. It is indeed a wondrous experience to watch a skilled computer modeler at work, performing a "direct" drug design. It is also possible to add the fourth dimension of electronic surface charges to the picture. The method is, as would be expected, expensive in terms of programing and hardware. Because the technique is in its infancy, no major results have been reported to date, but Marshall (1987) provides an excellent review.

It is, however, instructive to review two studies along these lines; these studies resulted in the design of active molecules that could not have been conceived by classical drug design using existing leads.

The first study, conducted by Olson and his co-workers (1981), concerned dopaminergic ligands. As shown in Fig. 8.6, a three-dimensional receptor model was constructed which was based on a number of dopaminergic agonists and antagonists. This model allowed $\pi-\pi$ or $n-\pi$ interactions, a hydrogen bond, and an optional lipophilic site (L). The pyrroloisoquinoline **Ro 22-1319** (Fig. 8.7) and also

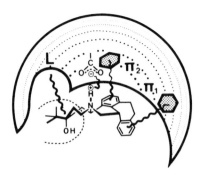

**Fig. 8.6.** Interaction of (+)-butaclamol with the receptor model of Olson et al. (1981). (Adapted by permission from Cohen (1985), Academic Press, New York)

**Fig. 8.7.** Interaction of different classes of ligands with the receptor model of Olson et al. (1981), shown in Fig. 8.6. (Reproduced by permission from Olson et al. (1981), American Chemical Society, Washington, DC; and Cohen (1985), Academic Press, New York)

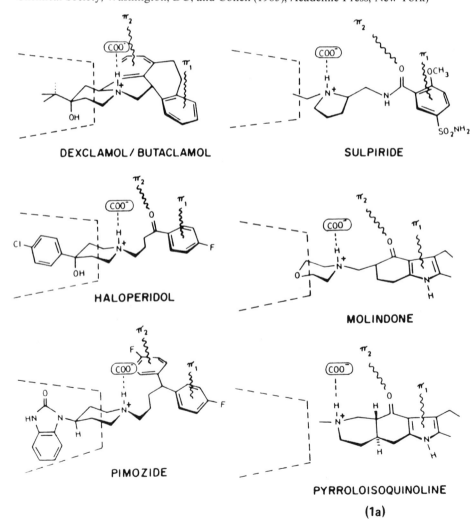

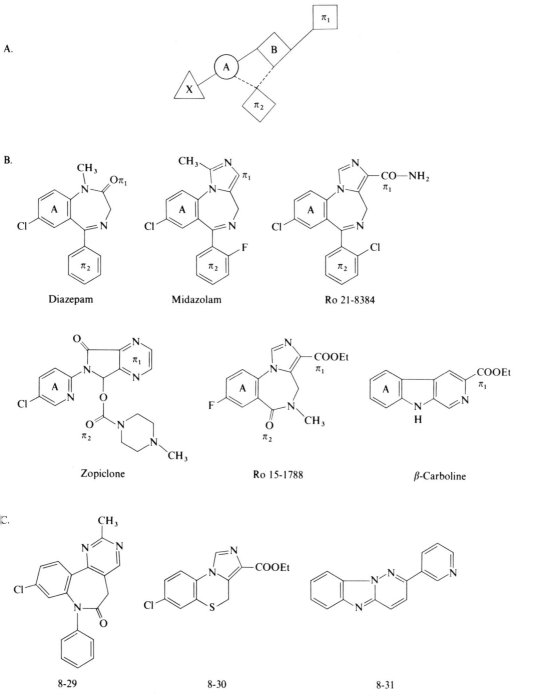

**Fig. 8.8.** (A) The Fryer and Gilman model of the benzodiazepine binding scheme. (B) Six known compounds have been demonstrated to fit this model. (C) Three new compounds, designed to fit the binding scheme, have proved to be highly active. (Modified by permission from Cohen (1985), Academic Press, New York)

compounds 8-27 and 8-28 (which were designed on this basis; see Fig. 8.7) proved to be $D_2$ antagonists and powerful neuroleptics with extrapyramidal effects.

8-27                                    (+)Ro   14-8625
                                             8-28

The second study (Fryer and Gilman; cited by Cohen, 1985) analyzed the 3D structure of a number of benzodiazepines (Fig. 8.8), and proposed a complex binding model. Compounds 8-29, 8-30, and 8-31—designed on the basis of this model—proved to be highly active anxiolytics. All three structures constitute new lead compounds.

### Selected Readings

V. Austel (1984). Design of test series in medicinal chemistry. *Drugs of the Future 9*: 349–365.

J. M. Blaney, E. C. Jorgensen, M. L. Conolly, T. E. Ferrin, R. Langridge, S. J. Oatly, J. M. Burridge, and C. F. Blake (1982). Computer graphics in drug design: molecular modeling of thyroid hormone–realbumin interactions. *J. Med. Chem. 25*: 785–790.

A. S. V. Burgen and G. C. K. Roberts (Eds.) (1986). *Molecular Graphics and Drug Design. Topics in Molecular Pharmacology*, Vol. 3. Elsevier, Amsterdam.

R. C. Chu (1980). The quantitative analysis of structure-activity relationships. In: *The Basis of Medicinal Chemistry: Burger's Medicinal Chemistry*, 4th ed. (M. E. Wolff, Ed.). Part 1. Wiley, New York, pp. 393–418.

N. C. Cohen (1983). Towards the rational design of new leads in drug research. *Trends Pharmacol. Sci. 4*: 503–506.

N. C. Cohen (1985). Drug design in three dimensions. *Adv. Drug Res.* (B. Testa, Ed.), Vol. 14. Academic Press, New York, pp. 41–145.

P. Gund (1977). Computer-assisted organic synthesis analysis. *Annu. Rep. Med. Chem. 12*: 288–297.

P. Gund (1979). Pharmacophoric pattern searching and receptor mapping. *Annu. Rep. Med. Chem. 14*: 299–308.

P. Gund (1980). Three-dimensional molecular modeling and drug design. *Science 208*: 1425–1431.

A. J. Hopfinger (1985). Computer-assisted drug design. *J. Med. Chem. 28*: 1133–1139.

N. J. Hrib (1986). Recent development in computer-assisted organic synthesis. *Annu. Rep. Med. Chem. 21*: 303–311.

G. L. Kirschner and B. R. Kowalski (1978). The application of pattern recognition to drug design. In: *Drug Design* (E. J. Ariëns, Ed.), Vol. 8. Academic Press, New York, pp. 73–131.

G. Klopman and R. Contreras (1985). Use of artificial intelligence in structure–activity relationships of anticonvulsant drugs. *Mol. Pharmacol. 27*: 86–93.

P. P. Mager (1980). The MASCA model of pharmacochemistry I. Multi-variate statistics. In: *Drug Design* (E. J. Ariëns, Ed.), Vol. 9. Academic Press, New York, pp. 187–236.

P. P. Mager (1984). *Multidimensional Pharmacochemistry: Design of Safer Drugs.* Academic Press, New York.

G. R. Marshall (1987). Computer-aided drug design. *Annu. Rev. Pharmacol. Toxicol. 27*: 193–213.

Y. C. Martin (1978). *Quantitative Drug Design.* Marcel. Dekker, New York.

E. F. Meyer, Jr. (1980). Interactive graphics in medicinal chemistry. In: *Drug Design* (E. J. Ariëns, Ed.), Vol. 9. Academic Press, New York, pp. 267–289.

G. L. Olson, H. C. Cheung, K. D. Morgan, L. Todaro, L. Berger, A. B. Davidson, and E. Boff (1981). A dopamine receptor model and its application in the design of a new class of rigid pyrrolo[2,3-*g*]isoquinoline antipsychotics. *J. Med. Chem. 24*: 1026–1034.

J. J. Stezowski and K. Chandrasekhar (1986). X-ray crystallography of drug molecule–macromolecule interactions as an aid to drug design. *Annu. Rep. Med. Chem. 21*: 293–302.

A. J. Stuper, W. E. Brugger, and P. C. Jurs (1979). *Computer Assisted Studies of Chemical Structure and Biological Function.* Wiley, New York.

D. D. Wolff and M. L. Parsons (1983). *Pattern Recognition Approach to Data Interpretation.* Plenum Press, New York.

## 4. PRO-DRUGS AND "SOFT" DRUGS

The knowledge of pharmacodynamic processes and drug metabolism *in vivo* can be utilized to improve a wide variety of drug characteristics. This knowledge is the basis of a drug design system in which an active, lead compound is not modified irreversibly, but in such a way that it will be regenerated by metabolic processes. The reversibly modified compound—usually inactive in itself—is called a pro-drug because it releases the active compound as a metabolite.

The purpose of a pro-drug can be to:

1. Increase or decrease the metabolic stability of a drug
2. Interfere with transport characteristics
3. Mask side effects or toxicity
4. Improve the flavor of a drug

### 4.1. Regulation of Drug Stability

The regulation of drug stability can take two directions: A pro-drug can increase the *in vivo* stability of an active compound and prolong its action, or can automatically limit the duration of and prevent potential toxicity.

There are many examples of drug stabilization. Among local anesthetics, **procaine** (8-32) is an ester and is therefore very easily hydrolyzed by esterases. By conversion of the ester into an amide, in **lidocaine** (8-33), the duration of action is increased by several fold. Lidocaine is also used intravenously as an antiarrhythmic agent. In that application, it must pass through the liver—the principal drug-metabolizing organ—in which it loses an *N*-ethyl group to become a convulsant and emetic. To minimize these unwanted and toxic effects, **tocainide** (8-34)—whose α-methyl group prevents degradation, and which lacks the vulnerable *N*-ethyl groups—was prepared. This compound is not a pro-drug in the strict sense, but rather represents a molecular modification.

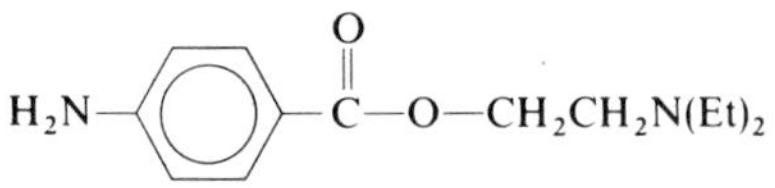

$H_2N$—⬡—$\overset{\overset{\displaystyle O}{\|}}{C}$—O—$CH_2CH_2N(Et)_2$

Procaine
8-32

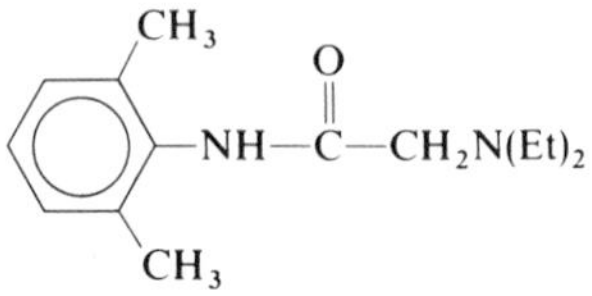

⬡—NH—$\overset{\overset{\displaystyle O}{\|}}{C}$—$CH_2N(Et)_2$ (with $CH_3$ groups at ortho positions)

Lidocaine
8-33

⬡—NH—$\overset{\overset{\displaystyle O}{\|}}{C}$—$\overset{\overset{\displaystyle}{\underset{\underset{\displaystyle CH_3}{|}}{CH}}}$—$NH_2$ (with $CH_3$ groups at ortho positions)

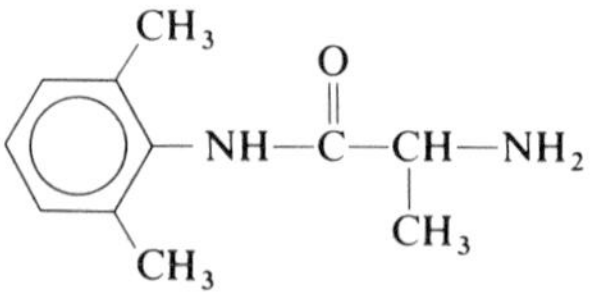

Tocainide
8-34

Replacement of a "vulnerable moiety" like a methyl group (to use the terminology of Ariëns and Simonis, 1974) by a less readily oxidized chlorine was used to transform the short-acting **tolbutamide** (8-35), an oral antidiabetic, into the long-acting **chlorpropamide** (8-36), with a half-life six fold greater than its parent.

Drug stability can be improved by other means than pro-drug creation. In Chap. 4, Sec. 4.2, the inhibition of DOPA decarboxylation achieved by the enzyme inhibitor **carbidopa** (4-93) was mentioned briefly.

A *decrease* in stability is often a desirable modification. For example, **succinylcholine (suxamethonium**; 8-37)—a neuromuscular blocking agent used in surgery—has a self-limiting activity, since the ester is hydrolyzed in about 10 minutes, preventing the potential for overdose, which could be fatal with more stable curarizing agents.

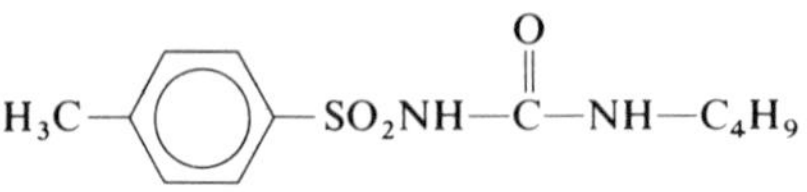

$H_3C$—⬡—$SO_2NH$—$\overset{\overset{\displaystyle O}{\|}}{C}$—$NH$—$C_4H_9$

Tolbutamide
8-35

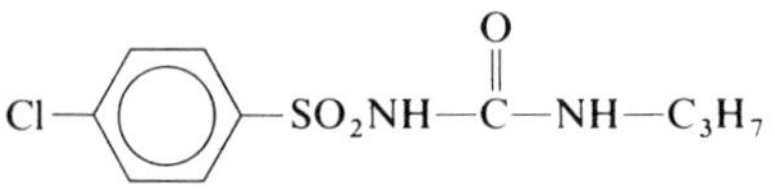

$Cl$—⬡—$SO_2NH$—$\overset{\overset{\displaystyle O}{\|}}{C}$—$NH$—$C_3H_7$

Chlorpropamide
8-36

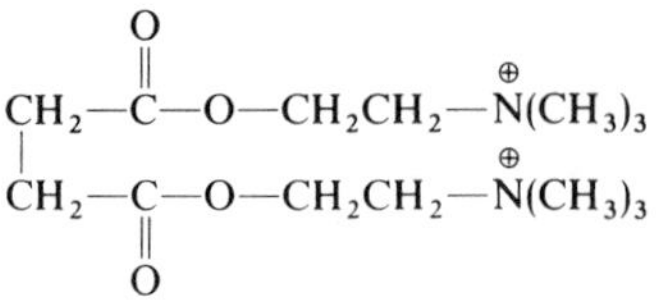

$CH_2$—$\overset{\overset{\displaystyle O}{\|}}{C}$—$O$—$CH_2CH_2$—$\overset{\oplus}{N}(CH_3)_3$
$|$
$CH_2$—$\underset{\underset{\displaystyle O}{\|}}{C}$—$O$—$CH_2CH_2$—$\overset{\oplus}{N}(CH_3)_3$

Succinylcholine
8-37

⬡—NH—$\overset{\overset{\displaystyle O}{\|}}{C}$—$CH_3N(Et)_2$ (with $CH_3$ at top and $\underset{\underset{\displaystyle O}{\|}}{C}$—$OCH_3$ at bottom)

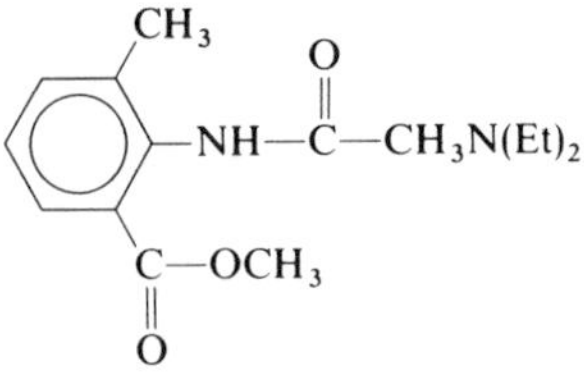

Tolycaine
8-38

An ester group can be introduced into a local anesthetic, such as **tolycaine** (8-38), to prevent the drug from reaching the CNS if it is injected intravascularly by accident or abuse. The ester group is fairly stable in the tissues, but is very rapidly hydrolyzed in the serum to the polar carboxylic acid, which cannot penetrate the blood–brain barrier.

## 4.2. Interference with Transport Characteristics

Interference with transport characteristics can serve many purposes. The introduction of a hydrophilic "disposable moiety" can restrict a drug to the gastrointestinal tract and prevent its absorption. Such a type of drug is represented by the intestinal disinfectant **succinyl-sulfathiazole** (8-39). On the other hand, lipophilic groups can ensure peroral activity, such as in the case of the penicillin derivative **pivampicillin** (8-40), which enters the circulation and then slowly releases the antibiotic in its free acid form, producing high blood levels of the latter.

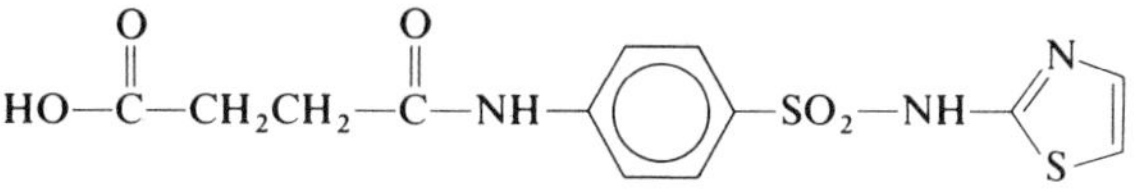

Succinyl-sulfathiazole
8-39

Lipophilic groups that are not easily hydrolyzed are used extensively for *depot preparations*, which liberate the active drug molecule slowly, for a period of days or weeks. Steroid hormone palmitates and pamoates, and antimalarial esters or insoluble salts (e.g., **cycloguanil pamoate**; 8-41) can deliver the active drugs for a long time—cycloguanil, for several months. This can be a great convenience for the patient, especially in areas with remote medical facilities.

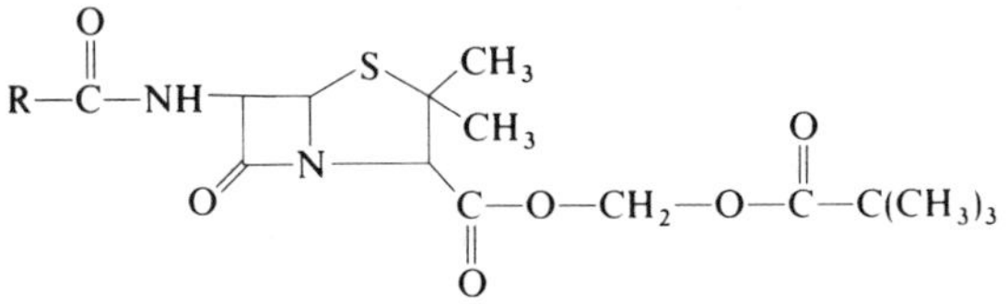

Pivampicillin
8-40

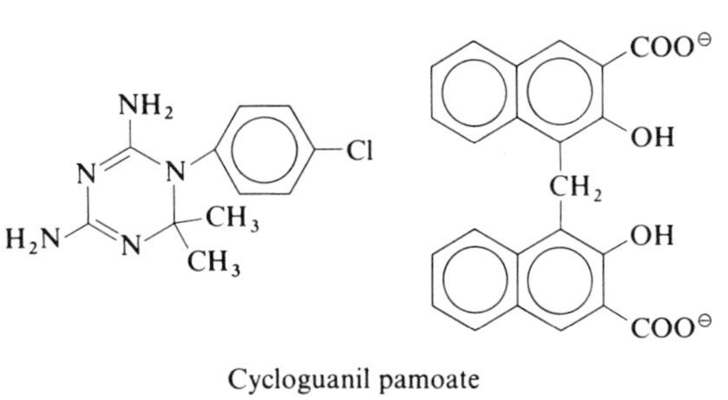

Cycloguanil pamoate
8-41

Drug designers have attempted for many years to use selective drug-transport moieties, and have met with moderate success. The idea is to attach a drug, such as an antitumor agent, to a natural product that will accumulate selectively in a specific organ and act as a "Trojan horse" for the drug. The attachment of alkylating agents to estrogens has been tried in the treatment of ovarian cancer, and amino acids have also been used as drug carriers. A recent ingenious application of the carrier concept is the utilization of *antibodies*—which can, at least in principle, be tailored to any site—as drug carriers (Kadin and Otterness, 1980). In this regard, antitumor agents such as **adriamycin** and **methotrexate** have been linked covalently to leukemia antibodies and melanoma antibodies with some initial success. The large-scale preparation of antibodies is, of course, a major difficulty in this approach; however, the new monoclonal antibodies hold great promise.

### 4.2.1. Soft Drug Design

"Soft drug" design is a novel concept introduced by Bodor (1982, 1984), and is in a sense the opposite of prodrug design. Whereas a prodrug is inactive and must be activated enzymatically, a soft drug is active but is deactivated enzymatically in a predictable and controllable way. Thus, instead of relying upon an unpredictable, multistep metabolic pathway, this technique involves the design of a simple, one-step process that does not allow for metabolites that might cause DNA damage (such as reactive epoxides). The key feature of a soft drug is its controllable and predictable disposition. Endogenous substances such as neurotransmitters and hormones are natural soft drugs, since very rapid and efficient ways have evolved for their disposition without the production of any reactive intermediates.

An interesting example of a soft drug is the soft antimuscarinic agent shown as compound (8-42). It has a $pA_2$ of 9.3, and is thus more active than atropine ($pA_2 = 8.5$). Used topically to control hyperhydrosis (uncontrollable sweating), it has none of the systemic antimuscarinic effects, such as mydriasis or dry mouth, because it is rapidly deactivated to a carboxylic acid, an aldehyde, and a tertiary amine; so much so, in fact, that its slow infusion in low doses does not produce any effect. In contrast, a "hard" drug like atropine accumulates under such conditions, producing normal activity. Alkylating agents, disinfectants, and hydrocortisol derivatives have been successfully designed on similar principles.

$$\text{CH}-\overset{\displaystyle \underset{\displaystyle O}{\|}}{\text{C}}-\text{O}-\text{CH}_2-\overset{\displaystyle \text{CH}_3}{\underset{\displaystyle \text{CH}_3}{\text{N}^{\oplus}}}$$

8-42

### 4.2.2. Site-Specific Drug Delivery Systems

The soft drug idea was carried a step further by the Bodor group (Bodor, 1984; Simpkins et al., 1986) and extended into site-specific drug delivery systems. Such compounds are really pro-"soft" drugs: the administered compound undergoes predictable enzymatic transformations that result in the delivery and sequestration of a soft drug at the site of action; this soft drug then undergoes further predictable degradations, providing a sustained release of the active drug. All the intermediates are nontoxic, and even the active site receives the drug at a low, relatively nontoxic level. One example of the several successful applications of this principle is the delivery of the cytostatic **trifluridine** to the CNS by use of the $N$-methyl-dihydro-pyridine–$N$-methyl-pyridinium redox system for transport purposes (Fig. 8.9). Trifluridine is too polar to cross the blood–brain barrier in unaltered form. At-tachment of the dihydropyridine ester allows passage of the compound into the CNS, where it is oxidized to the pyridinium compound, which is highly polar and therefore remains in the CNS, because it cannot cross the blood–brain barrier in the reverse direction either. The **pyridinium-trifluridine** is a soft prodrug from which the active trifluridine is slowly liberated. The distribution of the reduced and oxidized forms in the blood and brain is also shown in Fig. 8.9. Other strategies for drug delivery across the BBB are discussed by Pardridge (1985).

### 4.3. Masking of Side Effects or Toxicity

Masking of the side effects or toxicity of drugs was historically the first application of the pro-drug concept. This concept goes back to the turn of the century, and in fact many pro-drugs were not at the time really recognized as such. For instance, castor oil is a laxative because it is hydrolyzed intestinally to the active ricinoleic acid. However, the classical example is **prontosil** (8-43), which undergoes a reduction to sulfanilamide. The analgesic **phenacetin** (8-44) acts in the form of its hydrolysis product, $p$-acetaminophenol. Another classical example of side-effect masking occurs in **aspirin** and its many analogues—the result of a considerable effort to eliminate the gastric bleeding caused by salicylic acid (see Chap. 5, Sec. 4.2).

Prontosil rubrum
8-43

Phenacetin
8-44

Selective bioactivation (toxification) is illustrated in the case of the insecticide **malathion** (8-45). This acetylcholinesterase inhibitor is desulfurized selectively to the toxic malaoxon (8-46), but only by insect and not mammalian enzymes. Malathion is therefore relatively nontoxic to mammals ($LD_{50}$ = 1500 mg/kg, rat; p.o.). Higher organisms rapidly detoxify malathion by hydrolyzing one of its ester groups to the inactive acid, a process not readily available to insects. This makes the compound doubly toxic to insects since they cannot eliminate the active metabolite.

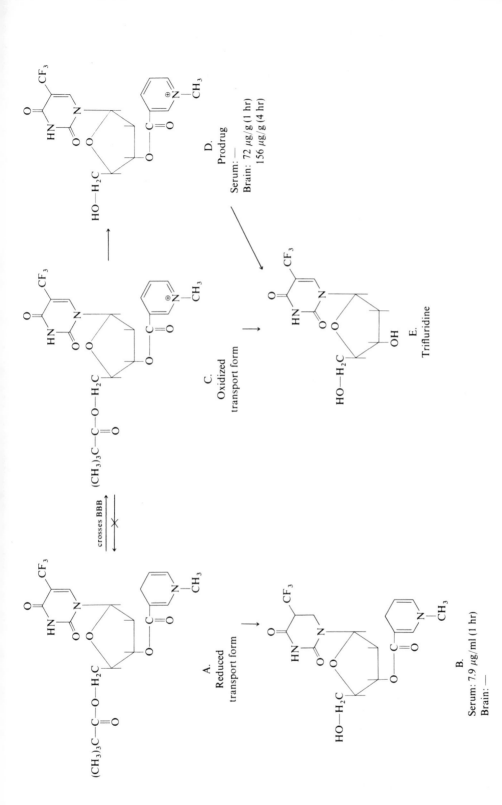

**Fig. 8.9.** Trifluridine in a soft drug modification. The fully esterified hydrophobic dihydropyridine derivative (A) is able to cross the blood–brain barrier (**BBB**) reversibly. It can either undergo partial hydrolysis peripherally (**B**), or be oxidized in the CNS to the pyridinium compound (**C**) which is then hydrolyzed to the prodrug (**D**). Both C and D are hydrophilic salts that cannot cross the **BBB**; thus they remain confined to the CNS and slowly hydrolyze to trifluridine (**E**). Concentrations in serum and brain are shown under the key compounds.

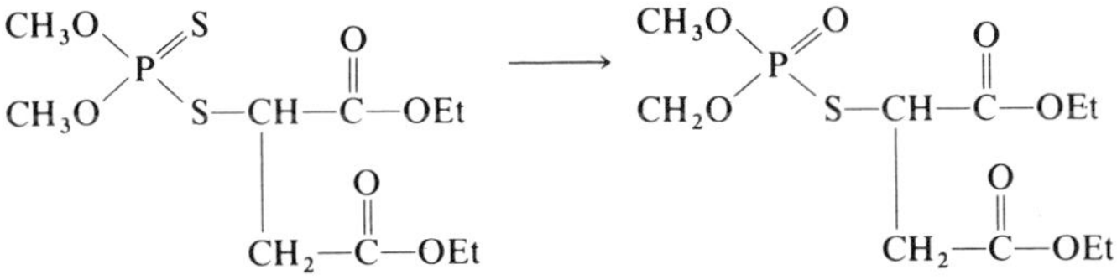

Malathion
8-45

Malaoxon
8-46

## 4.4. Improvement of Taste

Taste improvement is quite an important aspect of drug modification, especially in pediatric medicine. The extremely bitter taste of some antibiotics such as **clindamycin** or **chloramphenicol** can be masked successfully by preparing esters or pamoate salts of these drugs (cf. 8-41), which are very insoluble and therefore have no taste.

## 5. INNOVATIONS IN DRUG DELIVERY

Novel drug delivery systems can also have a profound effect on pharmacokinetics, even if they do not involve the use of pro-drugs in the classical sense. The reviews by Chandrasekaran et al. (1978), Henderson (1983), Beck and Pope (1984), and Johnson and Lloyd-Jones (1987), give an overview of this new field, based partly on the rapid progress in polymer technology. Novel polymers have permitted the development of membranes with controlled diffusion rates. For example, **pilocarpine**, used in the treatment of glaucoma, can be applied in a steady-release ocular insert that lasts for a week. The intrauterine release of **progesterone** as a contraceptive has also been achieved, with a single insert lasting a year. The great advantage of this is that the constant release rate of 65 $\mu$g/day means that much less drug is released than with the use of oral contraceptive tablets. The transdermal delivery of **scopolamine** as an antiemetic for motion sickness represents another successful application of microporous membrane technology. Here the drug is applied in a plastic strip similar to a Band-Aid, usually behind the ear. Low-density lipoproteins and liposomes (drug-filled lipid–cholesterol vesicles measuring a fraction of a micrometer) are also being used to protect drugs from enzymatic destruction during transport in the bloodstream.

Osmotic minipumps—cylinders measuring about 25 × 5 mm—are widely used to deliver constant amounts of drug solutions to experimental animals. They last for 1 or 2 weeks but require surgical implantation. The osmotic compartment swells in contact with tissue fluid and squeezes the drug reservoir, displacing the drug solution in a continuous flow. The rate of delivery is specified by the size of the opening in the container and the swelling rate of the osmotic "syringe" (cf. Bruck, 1983).

The great advantage of these systems is the uniform drug delivery they permit, as opposed to the enormous drug level fluctuations inherent with the traditional oral or injected parenteral modes of drug administration. Patient compliance and convenience of use are also ensured. Although these interesting developments in bioengineering are not, strictly speaking, in the realm of drug design or even medicinal chemistry, they can nevertheless contribute substantially to the success of drug therapy.

## Selected Readings

E. J. Ariëns and A.-M. Simonis (1974). Design of bioactive compounds. In: *Medicinal Chemistry, Topics in Current Chemistry*, Vol. 52. Springer, New York, pp. 1–61.

L. R. Beck and V. Z. Pope (1984). Controlled-release delivery system for hormones. A review of their properties and current therapeutic use. *Drugs 27*: 528–547.

N. Bodor (1982). Soft drugs: strategies for design of safer drugs. *Strategy in Drug Design* (J. A. Keverling-Buisman, Ed.). Elsevier, Amsterdam.

N. Bodor (1984). Novel approaches to the design of safer drugs: soft drugs and site-specific delivery systems. *Adv. Drug Res.* (B. Testa, Ed.), Vol. 13. Academic Press, New York, pp. 255–331.

S. D. Bruck (Ed.) (1983). *Controlled Drug Delivery*, Vol. 1: *Basic Concepts*; Vol. 2: *Clinical Applications*. CRC Press, Boca Raton, Fla.

S. K. Chandrasekaran, F. Theeuwes, and S. I. Yum (1978). The design of controlled drug delivery systems. In: *Drug Design* (E. J. Ariëns, Ed.), Vol. 8. Academic Press, New York, pp. 133–167.

Y. W. Chien (1982). *Novel Drug Delivery Systems*. Marcel Dekker, New York.

N. L. Henderson (1983). Recent advances in drug delivery system technology. *Annu. Rep. Med. Chem. 18*: 275–284.

P. Johnson and G. Lloyd-Jones (Eds.) (1987). *Drug Delivery Systems*. UCH Publishers, New York.

S. D. Kadin and J. G. Otterness (1980). Antibodies as drug carriers and toxicity reversal agents *Annu. Rep. Med. Chem. 15*: 233–244.

W. M. Pardridge (1985). Strategies for delivery of drugs through the blood–brain barrier. *Annu. Rep. Med. Chem. 20*: 305–313.

M. J. Poznansky and R. L. Juliano (1984). Biological approaches to the controlled delivery of drugs: a critical review. *Pharmacol. Rev. 36*: 277–336.

J. E. Shaw (1980). Drug delivery systems. *Annu. Rep. Med. Chem. 15*: 302–315.

J. N. Simpkins, J. McCormack, K. S. Estes, M. E. Brewster, E. Sheck, and N. Bodor (1986). Sustained brain-specific delivery of estradiol causes long-term suppression of luteinizing hormone secretion. *J. Med. Chem. 29*: 1809–1812.

A. A. Sinkula (1975). Prodrug approach in drug design. *Annu. Rep. Med. Chem. 10*: 306–316.

## Periodicals of General Interest

*Drug Design and Delivery*. Harwood, London.

# Appendix: Drugs Arranged by Pharmacological Activity*

**ABORTIFACIENTS**
**Carboprost**
*Prostin/15 M*

**Dinoprost tromethamine**
*Prostin F$_{2\alpha}$*

**Prostaglandin E$_2$ (dinoprostone)**
*Prostin E$_2$*


**ACETYLCHOLINESTERASE (AChE) INHIBITORS**
**Ambenonium**
*Mysuran, Mytelase*

**Echothiophate**
*Phospholine, Tammelin*

**Edrophonium**
*Enlon, Tensilon*

**Neostigmine**
*Prostigmin*

**Physostigmine**
*Eserine, Physostol*


**AChE INHIBITOR ANTIDOTE**
**Pralidoxime (PAM)**
*Protopam*


**ACNE PRODUCTS.** *See* **ANTIACNE AGENTS**


**ADRENERGIC VASOCONSTRICTORS**
[*see also* **ANTIHYPERTENSIVES; NASAL DECONGESTANTS; VASOCONSTRICTORS**]
Cyclopentamine (cyclopentadrine)
*Clopane, Cyklosal*

**Epinephrine**
*Adnephrine, Hemostatin, Nephridine, Vasotonin*

**Mephentermine**
*Vialin, Wyamine*

**Naphazoline**
*Niazol, Rhinantin, Vasocon*

**Nordefrin**
*Corbasil, Levonordefrin*

Tetrahydrozoline
*Tetryzolin, Visine*

Xylometazoline


**ADRENOCORTICOSTEROIDS.** *See* **GLUCOCORTICOIDS; MINERALOCORTICOIDS**


**ALDOSE REDUCTASE INHIBITORS**
**Sorbinil**

**Tolrestat**

---

* Drugs shown in boldface are discussed in the text. Drugs shown in italics are proprietary names.

## ANABOLIC STEROIDS
**Ethylestrenol**
*Maxibolin*

**Fluoxymesterone**
*Androfluorone, Halotestin, Ora-Testryl*

**Methandrostenolone**
*Danabol, Dianabol, Methandroid, Nerobol,
  Stenolon*

**Nandrolone**
*Androlone, Durabolin, Nandralin*

**Oxandrolone**
*Anavar*

**Oxymetholone**
*Androl*

Stanozolol
*Tevabolin, Winstrol*


## ANALEPTICS
[*see also* **ANOREXICS; STIMULANTS
  (CENTRAL)**]
**Bemegride**
*Eukraton, Malysol, Megimide*

**Caffeine**
*NōDōz, Vivarin*

Doxapram
*Dopram*

**Nikethamide**
*Coramine*


## ANALGESICS
[*see also* **ANTIINFLAMMATORY
  AGENTS**]
**Acetaminophen (*N*-acetyl-*p*-aminophenol,
  APAP)**
*Panadol, Tylenol*

Anidoxine

**Aspirin (acetylsalicylic acid, ASA)**
*Aspro, Bufferin, Ecotrin, Empirin*

Bicifadine

**Buprenorphine**
*Buprenex, Temgesic*

**Butorphanol**
*Satadol, Torate, Torbutrol*

**Cinanserin**

**Diflunisal**
*Dolobid, Flovacil*

**Diprenorphine**

**Flufenisal**

Methotrimeprazine
*Levoprome*

Naproxen
*Naprosyn*

## ANALGESICS (NARCOTIC
  AGONIST)
**Anileridine**
*Alidine, Leritine, Nipecotan*

**Codeine**

**Etonitazene**

**Fentanyl**
*Sublimaze*

**Hydromorphone**
*Dilaudid*

**Ketobemidone**
*Cliradon, Cymidon, Ketogan*

**Levorphanol**
*Levo-Dromoran*

**Meperidine**
*Demerol, Dolantin, Mephedine*

**Methadone**
*Amidon, Dolophine, Fenadone, Polamidon*

**Morphine**
*MSIR, Roxanol*

**Oxycodone**

**Oxymorphone**
*Numorphan*

Propoxyphene (dextropropoxyphene)
*Darvon, Dolene*

**Sufentanil**
*Sufenta*

## ANALGESICS (NARCOTIC AGONIST–ANTAGONIST)

**Buprenorphine**
*Buprenex, Temgesic*

**Butorphanol**
*Stadol, Torate, Torbutrol*

**Nalbuphine**
*Nubain*

**Pentazocine**
*Algopent, Fortalin, Talwin*

## ANDROGENS

**Fluoxymesterone**
*Android-F, Halotestin Ora-Testryl*

**Methyltestosterone**
*Android, Metandren*

**Nandrolone**
*Durabol, Strabolene*

**Testosterone**
*Homosteron, Oreton-F, Virosterone*

## ANESTHETICS (INHALATION)

**Cyclopropane (trimethylene)**

**Ethyl ether (diethyl ether; ether)**

**Enflurane**
*Ethrane*

**Halothane**
*Fluothane*

**Isoflurane**
*Forane*

**Methoxyflurane**
*Penthrane*

**Nitrous oxide ("laughing gas")**

## ANESTHETICS (INTRAVENOUS)

**Alfaxalone (alphaxalone)**
*Althesin, Aurantex*

Etomidate
*Amidate*

**Fentanyl**
*Sublimaze*

Ketamine
*Ketalar*

**Methohexital**
*Brevital*

**Midazolam**
*Versed*

**Sufentanil**
*Sufenta*

Thiamylal
*Surital*

**Thiopental**
*Pentothal*

## ANESTHETICS (LOCAL/TOPICAL)

**Benzocaine (ethyl aminobenzoate)**
*Anesthesin, Orthesin, Parathesin*

**Cocaine**

Dibucaine
*Nupercainal*

Dyclonine
*Dyclone*

**Lidocaine**
*Xylocaine*

Pramoxine
*Tronothane, Prax*

**Procaine (*p*-aminobenzoate)**
*Anesthesol, Ethocaine, Novocaine*

**Tetracaine**
*Pontocaine*

**Tolycaine**
*Baycain*

## ANGIOTENSIN CONVERTING ENZYME INHIBITORS

**Teprotide**

## ANOREXICS

**Benzphetamine**
*Didrex*

**Fenfluramine**
*Obedrex, Ponderal, Pondimin*

**Phendimetrazine**
*Dacarate, Phenazine, Statobex*

**Phentermine**
*Ionomin*

**Quipazine**

**ANTERIOR PITUITARY
  SUPPRESSANTS**
**Danazol**
*Cyclomen, Danocrine, Danol*

**ANTHELMINTICS**
Mebendazole
*Vermox*

Niclosamide
*Niclocide*

Oxamniquine
*Vansil*

Piperazine
*Antepar, Vermizine*

Praziquantel
*Biltricide*

Pyrantel
*Antiminth*

**Pyrvinium**
*Povan Filmseals*

**Quinacrine**
*Atabrine*

Thiabendazole
*Mintezol*

**ANTIACNE AGENTS**
**Benzoyl peroxide**

**Etretinate**
*Tegison*

**Isotretinoin (13-*cis*-retinoic acid)**
*Accutane*

**Retinoic acid**
*Aberel*

**Tretinoin (*trans*-retinoic acid, vitamin A
  acid)**
*Retin-A*

**ANTIADRENERGICS**
[*see also* **ANTIHYPERTENSIVES**]
**Alprenolol**
*Aprobal, Aptine*

**Atenolol**
*Tenormin*

**Bretylium**
*Bretylan, Bretylel, Darenthin, Ornid*

Cetamolol

**Dihydroergotamine**
*Diergotan, Ergotex, Ikaran*

**Labetalol**
*Normodyne, Trandate*

Metoprolol
*Lopressor*

**Phentolamine**

**Timolol**
*Blocadren*

**ANTIANDROGENS**
**Buserelin**
*Suprefact*

**Cyproterone**
*Cyproteron*

**Flutamide**
*Sebatrol*

**ANTIANGINAL AGENTS**
[*see also* **CALCIUM CHANNEL
  BLOCKERS**]
Amyl nitrite

Dipyridamole
*Persantine*

Isosorbide dinitrate
*Isonate, Isordil, Sorbitrate*

**Nitroglycerin**
*Nitro-Dur, Nitrostat*

**Pentaerythritol tetranitrate (P.E.T.N.)**
*Pentol, Pentritol*

**ANTIANXIETY AGENTS.** See
  **TRANQUILIZERS**

**ANTIARRHYTHMICS**
[*see also* **CALCIUM CHANNEL BLOCKERS**]
**Amiodarone**
*Cordarone*

**Bretylium**
*Bretylan, Bretylel, Darenthin, Ornid*

**Digoxin**
*Lanoxin*

**Diltiazem**
*Cardizen*

**Disopyramide**
*Norpace*

**Flecainide**
*Tambocor*

**Lidocaine**
*Xylocaine*

Mexiletine
*Mexitil*

**Procainamide**
*Procan, Pronestyl*

**Propranolol**
*Inderal*

**Quinidine**
*Cardioquin, Quinidex*

**Tocainide**
*Tonocard*

**Verapamil**
*Calan, Isoptin*

**ANTIARTHRITICS.** *See*
**ANTIINFLAMMATORY AGENTS;**
**ANTIRHEUMATICS**

**ANTIASTHMATICS**
[*see also* **ANTIHISTAMINES;**
**BRONCHODILATORS**]
**Albuterol**
*Proventil, Ventolin*

**Cromoglycate**
*Cromolyn, Intal*
**Isoproterenol**
*Isuprel*

Neocromil

Oxatomide

**Terbutaline**
*Brethine, Bricanyl*

**ANTIBACTERIALS**
[*see also* **ANTIBIOTICS**]
**Ciprofloxacin**

Furazolidone
*Furoxone*

**Nalidixic acid**
*NegGram*

Sulfacytine
*Renoquid*

Sulfadiazine
*Microsulfon*

**Sulfamerazine**
*Debenal, Mesulfa*

**Sulfamethoxine (sulfadimethoxine)**
*Agribon, Dibosol, Madribon*

Sulfamethoxazole
*Ganatol*

**Sulfanilamide**
*Deseptyl, Prontosil album, Streptocid album*

**Sulfapyridine**

Sulfasalazine
*Azaline, Azulfidine*

**Sulfathiazole**
*Cibazol, Eleudron, Norsulfasol, Sulfamul*

**Sulfisoxazole**
*Gantrisin*

**ANTIBERIBERI AGENTS**
**Thiamine (vitamin B$_1$)**
*Betalin, Biamine*

**ANTIBIOTICS**
[*see also* **ANTIBACTERIALS**]
Amoxicillin
*Amoxil, Augmentin, Azlocillin, Azlin*

**Ampicillin**
*Omnipen, Penbritin, Polycillin, Totacillin*

**Carbenicillin**
*Geocillin, Geopen, Pyopen*

**Cefazolin**
*Ancef, Kefzol*

**Cefotaxime**
*Claforan*

**Cephalexin (Cefalexin)**
*Ceporex, Keflex, Keforal, Oracef*

**Cephaloridine**
*Ceflorin, Ceporin, Keflodin, Loridine*

**Cephalosporin C**

**Cephalothin**
*Cepovenin, Coaxin, Keflin, Microtin*

**Chloramphenicol**
*Chloromycetin*

**Chlortetracycline**
*Aureomycin, Biomycin*

**Clindamycin**

**Cloxacillin**
*Bactopen, Cloxapen, Tegopen*

**Cyclacillin**
*Cyclapen-W*

**Demeclocycline**
*Declomycin, Deganol*

**Dicloxacillin**
*Dycill, Dynapen, Pathocil, Veracillin*

**Diphenicillin**
*Ancillin*

**Doxycycline**
*Doxy-Lemmon, Liomycin, Vibramycin, Vivox*

**Erythromycin**
*Erythrocin, Ilosone, Ilotycin, Robimycin*

**Fenbecillin (Phenbecillin)**
*Penspek*

**Gentamycin**
*Garamycin*

**Gramicidin A**

**Imipenem-cilastatin**
*Primaxin*

**Kanamycin**
*Kantrex Klebcil*

**Lincomycin**
*Lincocin*

**Methacycline**
*Declomycin, Rondomycin*

**Methicillin**
*Staphcillin*

**Metronidazole**
*Flagyl, Metryl*

**Mezlocillin**
*Mezlin*

**Minocycline**
*Minocin*

**Moxalactam**
*Moxam*

**Nafcillin**
*Nafcil, Naftopen Unipen*

**Neomycin**
*Mycifradin*

**Nocardicin A–G**

**Oxacillin**
*Bactocill, Cryptocillin, Prostaphlin*

**Oxytetracycline**
*Oxymycin, Terramycin*

**Penicillin G**

**Penicillin V (phenoxymethyl penicillin)**
*Beepen, Pen-Vee, V-Cillin*

**Phenethicillin**
*Chemipen, Darcil, Maxipen, Synapen, Syncillin*

**Piperacillin**
*Pipracil*

**Polymyxin B**

**Rifampin (rifampicin)**
*Rifadin Rimactane*

**Streptomycin**

**Tetracycline**
*Achromycin, Steclin, Sumycin*

**Thienamycin**

Tobramycin
*Nebcin*

**Trimethoprim (TMP)**
*Proloprim, Trimpex*

**Trimethoprim w. sulfamethoxazole (co-trimoxazole, TMP-SMZ)**
*Bactrim, Septra*

Vancomycin
*Vancocin*

**ANTICANCER AGENTS.** *See* **ANTITUMOR AGENTS**

**ANTICHOLESTEROLEMICS.** *See* **ANTIHYPERCHOLESTER-OLEMICS/ANTIHYPERLIPO-PROTEINEMICS**

**ANTICHOLINERGICS/ ANTISPASMODICS**
[*see also* **ANTIPARKINSON AGENTS**]
**Atropine**
*Atropisol, Eumydrin, Lyopine, Metropine*

**Belladonna**

**Belladonna levorotatory alkaloids**
*Bellafoline*

**Benztropine**
*Cogentin*

**Homatropine**
*Bufopto Homatrocel, Homapin, Homatrisol, Mesopin*

**L-Hyoscyamine**
*Anaspaz, Levsin*

Mepenzolate

**Methantheline**
*Banthine*

Methscopolamine
*Cantil*

**Oxyphencyclimine**
*Antulcus, Caridan, Daricon*

**Oxyphenonium**
*Antrenyl*

**Pirenzepine**
*Bisvanil, Gasteril, Ulcosan*

**Propantheline**
*Ketaman, Norpanth, Pantheline, Pro-Banthine*

**Scopolamine (hyoscine)**
*Hyosol, Isopto Hyoscine*

**Tridihexethyl**
*Pathilon*

**ANTICHOLINERGICS (OPHTHALMIC)**
**Atropine**

**Cyclopentolate**
*Cyclogyl, Mydrilate*

**Scopolamine**

**ANTICOAGULANTS**
Anisindione
*Miradon*

**Dicumarol (bishydroxy-coumarin)**

**Heparin**

Phenprocoumon
*Liquamar*

**Warfarin**
*Athrombin, Coumadin, Panwarfin*

**ANTICONVULSANTS**
**Acetazolamide**
*Diamox*

**Carbamazepine**
*Tegretol*

**Clonazepam**
*Clonopin, Rivotril*

**Clorazepate**
*Tranxene*

**Diazepam**
*Valium*

**Ethosuximide**
*Succimal, Zarontin*

Methsuximide
*Celontin*

**Phenytoin**
*Dilantin*

**Primidone**
*Mysoline*

**Progabide**

**Trimethadione**
*Tridione*

**Valproic acid**
*Depakene*


## ANTIDEPRESSANTS
**Amitriptyline**
*Elavil, Emitrine, Laroxyl, Saroten, Triptisol*

**Amoxapine**
*Asendin*

**Bupropion (amfebutamon)**
*Wellbatrin*

**Citalopram**
*Nitalapram*

**Clomipramine (chlorimipramine)**
*Anafranil*

**Clorgiline (clorgyline)**

**Deprenyl**
*Eldepryl, Jumex, Selegiline*

**Desipramine**
*Norpramin, Pertofrane*

**Doxepin**
*Adapin, Sinequan*

**Fluoxetine**
*Prozac*

**Fluvoxamine**

**Imipramine**
*Deprinol, Janimine, Presamine, Tofranil*

**Iprindole**
*Galatur, Prondol Tertran*

**Isocarboxazid**
*Marplan*

**Mianserin**
*Athymil, Norval, Tolvin*

**Nomifensine**
*Alival, Merital*

**Nortriptyline**
*Aventyl, Norzepine, Pamelor*

**Pargyline**
*Eutonyl*

**Phenèlzine**
*Nardil*

**Pirandamine**

**Proniazide**
*Marsilid*

**Protriptyline**
*Vivactil*

**Quipazine**

**Tandamine**

**Tranylcypromine**
*Parnate*

**Zimelidine**
*Normud, Zelmid*


## ANTIDIABETIC AGENTS
[*see also* **ALDOSE REDUCTASE INHIBITORS**]
**Acetohexamide**
*Dimelin, Dymelor, Ordimel*

**Buformin**
*Biguanal, Gliporal, Krebon*

**Chlorpropamide**
*Catanil, Diabinese, Millinese, Stabinol*

**Glipizide**
*Glucotrol*

**Glyburide, (glibenclamid, glybenzcyclamide)**
*Diabeta, Euclamin, Micronase*

**Glymidine**
*Gondafon, Lycanol, Redul*

**Insulin injection**
*Insular, Velosulin*

**Insulin zinc suspension**
*Lente Insulin, Semitard*

**Metformin**
*Diabefagos, Flumamine, Glucophage,
  Metiguanide*

**Pirogliride**

**Tolbutamide**
*Orinase*

## ANTIDIURETIC AGENTS
**Vasopressin**
*Pitressin*

## ANTIDOTES
[*see also* **EMETICS**]
**Charcoal, activated**
*Activated Charcoal*

**Deferoxamine** (*Use:* Fe intoxication)
*Desferal Mesylate*

**Dimercaprol** (*Use:* As, Au, Hg poisoning)
*BAL In Oil*

**Edetate calcium disodium (Ca-EDTA)**
  (*Use:* Pb poisoning)
*Calcium Disodium Versenate*

**Naloxone** (*Use:* narcotic overdose)
*Narcan*

**Naltrexone** (*Use:* narcotic overdose)
*Naltrexone, Trexan*

**Physostigmine** (*Use:* anticholinergic
  overdose)

**Pralidoxime (PAM)** (*Use:* organophosphate pesticide poisoning)
*Protopam*

## ANTIEMETIC/ANTIVERTIGO
  AGENTS
**Chlorpromazine**
*Hibernal, Largaktil, Prompapar, Thorazine*

**Cyclizine**
*Marezine, Nautazine, Valoid*

**Dexamethasone**
*Decadron, Dexone*

**Dimenhydrinate**
*Dramamine, Marmine*

**Diphenhydramine**
*Benadryl, Nordryl*

**Metoclopramide**
*Maxolone, Reglan*

**Nabilone**
*Cesamet*

**Perphenazine**
*Trilafon*

**Prochlorperazine**
*Compazine*

**Promethazine**
*Baymethazine, Phenergan*

**Scopolamine**
*Hyoscine, Hyosol, Skopolate, Transderm-
  Scōp, Triptone*

**Sulpiride**
*Dogmatil, Miradol, Mirbanil*

**Triflupromazine**
*Vesprin*

## ANTIEPILEPTICS. See
  ANTICONVULSANTS

## ANTIESTROGENS
**Clomiphene**
*Clomid*

**Tamoxifen**
*Nolvadex, Serophene*

## ANTIFUNGAL AGENTS
**Amphotericin B**
*Fungizone*

**Butoconazole**
*Femstat*

**Clotrimazole**
*Lotrimin, Mycelex, Trimysten*

**Flucytosine (5-fluorocytosine, 5-FC)**
*Ancobon*

**Ketoconazole**
*Nizoral*

**Miconazole**
*Micatin, Monistat*

**Nystatin**
*Mycostatin, Nilstat*

**Naftifine**
*Exoderil*

**Tridemorf**

**Undecylenic acid**
*Desenex*

**ANTIHERPESVIRUS AGENTS**
[*see also* **ANTIVIRAL AGENTS**]
**Acyclovir (acycloguanosine)**
*Zovirax*

**Idoxuridine (IDU)**
*Herplex, Stoxil*

**Phosphonoformate**

**ANTIHISTAMINES (H₁)**
**Antazoline**
*Antistine, Azalone*

**Astemizole**
*Hismanal*

Brompheniramine
*Bromphen, Dimetane*

**Carbinoxamine**
*Ciberon, Clistin*

**Chlorpheniramine**
*Aller-Chlor, Chlor-Trimeton, Chlor-
  Tripolon, Histalen*

**Clemizole**
*Allercur, Histacuran*

Clobenzepam
*Tarpan*

**Cyclizine**
*Marezine, Nautazine*

**Cyproheptadine**
*Nuran, Periactinol, Peritol*

**Diphenhydramine**
*Benadryl, Fenylhist, Nordryl*

**Isothipendyl**
*Adantol, Nilergex, Udantol*

**Mianserin**
*Athymil, Norval, Tolvin*

**Oxatomide**

**Promethazine**
*Baymethazine, Phenergan*

*Pyrathiazine* (**parathiazine**)
*Pyrrolazote*

**Pyrilamine (mepyramine)**
*Neo-Antergan*

**Terfenadine**
*Aldaban, Seldane*

**Trimeprazine**
*Temaril*

**Tripelennamine**
*PBZ, Pelamine, Pyribenzamine,
  Pyrinamine*

**Triprolidine**
*Actidil, Myidil*

**ANTIHISTAMINES (H₂).** *See*
  **ANTIULCER AGENTS**

**ANTIHYPERCHOLESTEROLEMICS/
  ANTIHYPERLIPOPROTEINEMICS**
**Butoxamine**

**Cholestyramine**
*Cuemid, Questran*

**Colestipol**
*Colestid*

**Clofibrate**
*Atromid-S*

**Dextrothyroxine**
*Choloxin*

Gemfibrozil
*Lopid*

**Metformin (*N,N*-dimethylguanide,
  DMGG)**
*Diabex, Diabefagos, Flumamine,
  Glucophage*

**Mevastin (compactin)**

**Mevinolin (monacolin K)**

**Nicotinamide (niacinamide)**

## ANTIHYPERTENSIVES
[see also **CALCIUM CHANNEL
    BLOCKERS; DIURETICS;
    GANGLIONIC BLOCKERS**]
Acebutolol
*Sectral*

**Atenolol**
*Tenormin*

**Amiflamine**

**Benzpindolol**

**Betaxolol**
*Kerlone*

**Bretylium**
*Bretylan, Darenthin, Ornid*

**Captopril**
*Capoten*

**Clonidine**
*Catapres*

**Clorgyline**

Cyclothiazide
*Aquirel, Doburul, Fluidil*

**Deprenyl (deprenalin)**
*Eldepryl, Jumex, Selegiline*

**Diazoxide**
*Hyperstat*

**Enalapril**
*Vasotec*

**Guanabenz**
*Wytensin*

**Guanadrel**
*Hylorel*

**Guanethidine**
*Ismelin*

**Hydralazine**
*Alazine, Apresoline*

**Ketanserin**

**Labetalol**
*Normodyne, Trandate*

**Mecamylamine**
*Inversine*

**Methyldopa**
*Aldomet, Dopamet, Medopren*

**Pargyline**
*Eutonyl*

**Phenoxygenzamine**
*Dibenzyline*

Phentolamine
*Regitine*

**Pindolol**
*Visken*

**Piperoxan**
*Benodaine*

**Prazosin (furazosin)**
*Hypovase, Minipress, Sinetens*

**Propranolol**
*Inderal*

**Prorenoate**

**Reserpine**
*Rauwolfia, Rauval*

**Sotalol**
*Beta-Cardone, Sotacor, Sotalex*

**Teprotide**

**Tianeptine**
*Stablon*

**Timolol**
*Blocadren*

**Tolazoline**
*Priscoline*

**Trimethaphan**
*Arfonad*


## ANTIINFECTIVES (TOPICAL)
**Benzalkonium chloride (BAC)**
*Benirol, Germicin, Zephiran*

**Cetrimonium bromide**
*Bromat, Cetab, Cetylamine, C.T.A.B.*

**Chlorhexidine**
*Hibiclens, Hibitane, Lisium*

**Fenticlor**
*Novex*

**Glutaraldehyde**
*Cidex*

**Hexachlorophene**
Bilevon
*pHisoHex, Septisol*

**4-Hexylresorcinol**
*Sucrets*

Iodine

Thimerosal
*Mersol, Merthiolate*

**ANTIINFLAMMATORY AGENTS**
[*see also* **ANTIRHEUMATICS**]
**Aspirin (acetylsalicylic acid, ASA)**
*A.S.A., Aspro, Bufferin, Ecotrin*

**Diflunisal**
*Dolobid*

**Flufenamic acid**
*Ansatin, Arlef, Meralen, Paraflu*

**Flufenisal**

**Indomethacin**
*Indameth, Indocin, Liometacen, Osmosin*

**Indoprofen**
*Flosin, Isindone*

**Mefenamic acid**
*Ponstel*

**Naproxen**
*Anaprox, Apranax, Naprosyn*

**Oxyphenbutazone**
*Oxalid*

**Phenylbutazone**
*Azolid, Butazolidin, Butazone*

**Sulindac**
*Clinoril*

**Suprofen**
*Suprol*

**ANTILEPROTICS**
**Acedapsone**
*Hansolar, Rodilone*

**Dapsone (DDS)**
*Disulone, Novophone*

**Rifampin (rifampicin)**
*Rifadin, Rifoldine, Rimactan*

**ANTIMALARIALS**
**Chlorguanide (chloroguanide)**
*Diguanyl, Guanatol, Paludrine*

**Chloroquine**
*Aralen, Nivaquine B, Sanoquin*

**Cycloguanil**
*Camolar*

**Hydroxychloroquine**
*Plaquenil*

**Mefloquine**
*Lariam*

**Primaquine**

**Pyrimethamine**
*Daraprim*

**Quinacrine**
*Atabrine*

**Quinine**
*Legatrin*

**Sulfadoxine w. pyrimethamine**
*Fonsidar*

**Trimethoprim**
*Proloprim, Trimpex*

**ANTIMUSCARINICS.** *See*
    **ANTICHOLINERGICS/**
    **ANTISPASMODICS**

**ANTINEOPLASTIC AGENTS.** *See*
    **ANTITUMOR AGENTS**

**ANTIPARKINSON AGENTS**
**Amantadine**
*Amazolon, Midantan, Symmetrel*

**Benztropine**
*Cogentin*

Biperidan
*Akineton*

**Bromocriptine**
*Parlodel*

**Carbidopa**
*Lodosyn, Sinemet*

**Diphenhydramine**

Ethopropazine
*Parsidor*

**Levodopa (L-dopa)**
*Dopar, Doparkine, Larodopa, Ledopa,
  Levopa*

**Lisuride**
*Cuvalit, Lysenyl*

**ANTIPELLAGRA AGENTS**
**Niacinamide (Nicotinamide)**

**ANTIPSYCHOTICS**
**Acetophenazine**
*Tindal*

**Benperidol**
*Anquil, Frenactyl, Glianimon*

**Butaclamol**
*Lodine*

**Chlorpromazine**
*Hibernal, Largactil, Prompapar, Thorazine*

**Chlorprothixene**
*Taractan*

**Clozapine**
*Leponex*

**Flupentixol (flupenthixol)**
*Emergil, Fluanxol, Metamin*

**Fluphenazine**
*Permitil, Prolixin*

**Lithium**
*Eskalith*

Perphenazine
*Trilafon*

**Haloperidol**
*Haldol*

**Pifluthixol**

**Pimozide**
*Opiran, Orap*

**Prochlorperazine**
*Compazine*

**Promazine**
*Sparine*

**Spiperone (spiroperidone)**
*Spiropitan*

**Sulpiride**
*Dogmatil, Guastil, Miradol*

**Thioridazine**
*Mellaril, Millazine*

**Thiothixene**
*Navane*

**Triflupromazine**
*Vesprin*

**Trifluoperazine**
*Stelazine, Suprazine*

**ANTIPYRETICS.** *See*
  **ANTIINFLAMMATORY AGENTS**

**ANTIRACHITIC AGENTS**
**Calcifediol (25-[OH]-$D_3$)**
*Calderol*

**Calcitriol (1,25[OH]-$D_3$)**
*Rocaltrol*

**Cholecalciferol ($D_3$)**
*Delta-D*

**Calcitonin**
*Calcimar*

**Dihydrotachysterol (DHT)**
*Hytakerol*

**Ergocalciferol**
*Calciferol, Drisdol, Deltalin*

**ANTIRHEUMATICS**
[*see also* **ANTIINFLAMMATORY
  AGENTS**]
**Auranofin**
*Ridaura*

**Aurothioglucose**
*Solganal*

**Cortisone**
*Cortone, Cortistab*

**Dexamethasone**
*Dalanone, Decadron, Deronil, Hexadrol*

Fenopren
*Nalfon*

Gold sodium thiomalate
*Myochrysine*

**Hydroxychloroquine**
*Plaquenil*

**Hydrocortisone**
*Cortef, Hydrocortone,*

**Ibuprofen**
*Advil, Brufen, Medipren, Motrin, Nuprin*

Ketoprofen
*Orudis*

**Meclofenamate**
*Meclomen*

**Methylprednisolone**
*Depo-Medrol, Medrol*

**Naproxen**
*Anaprox, Apranax, Naprosyn*

Paramethasone
*Haldrone*

Piroxicam
*Feldene*

**Prednisolone**
*Cortalone, Decortin H, Dicortil, Sterane,
    Sterolone*

**Prednisone**
*Deltasone, Meticorten, Orasone*

Tolmetin
*Tolectin*

**Triamcinolone**
*Aristocort, Cenocort, Kenacort, Trilone*

**ANTISHOCK AGENTS**
[*see also* **HYPERTENSIVES**]
Dobutamine
*Dobutrex*

**Dopamine**
*Intropin*

**Epinephrine**

**Ephedrine**
*Ephetonin, Racephedrine*

**Isoproterenol**
*Isuprel*

Mephentermine
*Wyamine*

**Metaraminol**
*Aramine*

**Methoxamine**
*Pressomin, Vasoxine, Vasoxyl*

**Norepinephrine (levarterenol)**
*Levophed*

**Phenylephrine**
*Neo-Synephrine*

**ANTITHYROIDS**
**Methimazole**
*Basolan, Tapazole*

**Propylthiouracil (PTU)**
*Propacil*

**Sodium iodide I 131**

**ANTITUBERCULOTICS**
Aminosalicylate (*p*-aminosalicylate)
*P.A.S., Teebacin*

Capreomycin
*Capastat*

**Cycloserine**
*Seromycin*

**Ethambutol**
*Mycambutol, Tibutol*

**Ethionamide**
*Trecator*

**Isoniazid**
*Laniazid, Nydrazid, Rimifon*

**Kanamycin**
*Kanicin, Kantrex, Klebcil*

**Rifampin (rifampicin)**
*Rifadin, Rimactine*

**Streptomycin**
*Strycin*

**ANTITUMOR AGENTS**
**Asparaginase**
*Elspar*

**Azacitidine (5-azacytidine, 5-AZC)**
*Ladakamycin*

**Aminopterin (4-amino-PGA)**

**Bleomycin (BLM)**
*Blenoxane*

**Busulfan**
*Myleran*

**Carmustine (BCNU)**
*BiCNU*

**Chlorambucil**
*Keukeran*

**Cisplatin (CDDP)**
*Platinol*

**Cyclophosphamide**
*Cytoxan, Neosar*

**Cytarabine (cytosine arabinoside, Ara-C)**
*Cytosar-U*

**Dacarbazine (DTIC, imidazole carboxamide)**
*DTIC-Dome*

**Dactinomycin (actinomycin D, ACT)**
*Cosmegen*

**Daunorubicin (daunomycin, rubidomycin)**
*Cerubidine, Daunoblastina, Ondena*

**Doxorubicin (adriamycin, ADR)**
*Adriacin, Adriblastina*

**Etoposide (VP-16-213)**
*VePesid*

**Floxuridine (FUDR)**

**Fluorouracil (5-FU)**
*Adrucil*

**Flutamide**
*Sebatrol*

**Fosfestrol (diethylstilbestrol diphosphate)**
*Cytonal, Honvan, Honvol, Stilphostrol*

**Hydroxyurea**
*Hydrea*

**Interferon $\alpha_{2a}$ (leukocyte interferon, IFLrA, LeIF, rIFN-A)**
*Roferon-A*

**Interferon $\alpha_{2b}$ (fibroblast interferon, FIF)**
*Intron A*

**Lomustine (CCNU)**
*Belustine, CeeNu*

**Maytansine**
*NSC 153858*

**Mechlorethamine (nitrogen mustard, $HN_2$)**
*Cloramin, Embichen*
*Mustargen, Mustine*

**Mephalan (phenylalanine mustard, PAM, L-PAM, L-sarcolysin)**
*Alkeran*

**Mercaptopurine (6-MP)**
*Purinethol*

**Methotrexate (amethopterin, MTX)**
*Folex, Mexate*

**Semustin**
*Methyl-CCNU*

**Streptozocin (streptozotocin)**
*Zanosar*

**Thioguanine (TG)**
*Lanvis*

**Uracil Mustard**
*Uramustil*

**Triethylenethiophosphoramide (TSPA, TESPA)**
*Thiotepa*

**Vidarabine (ara-A)**
*Vira-A*

**Vinblastine (vincaleukoblastine, VLB)**
*Exal, Velban, Velbe*

**Vincristine (VCR; leurocristine, LCR)**
*Kyocristine, Oncovin*

**ANTITUSSIVES**
**Codeine**

**Dextromethorphan**
*Romilar*

**Diphenhydramine**
*Benylin, Noradryl, Valdrene*

**ANTIULCER AGENTS**
**Cimetidine**
*Tagamet*

**Famotidine**
*Pepdul*

**Nileprost**

**Omeprazole**

**Oxyphencyclimine**
*Antulcus, Caridan, Daricon, Naridan*

**Pirenzepine**
*Bisvanil, Gasteril, Leblon, Ulcosan*

Proglumide
*Gastridene, Midelid, Milid, Promid*

**Propantheline**
*Giquel, Ketaman, Pantheline, Pro-Banthine*

**Ranitidine**
*Ranidil, Zantac*

**Tridihexethyl**
*Claviton, Pathilon*

**ANTIVIRAL AGENTS**
[*see also* **ANTIHERPESVIRUS AGENTS**]
**Acyclovir (acycloguanosine)**
*Zovirax*

**Amantadine**
*Midantan, Symmetrel*

**Azidothymidine**
*AZT, Retrovir*

**Idoxuridine (IDU)**
*Dendrid, Herplex, Stoxil*

**Ribavirin**
*Viramid, Virazole*

**Trifluridine (trifluorothymidine)**
*Viroptic*

**Vidarabine (adenine arabinoside, ara-A)**
*Vira-A*

**BRONCHODILATORS**
[*see also* **ANTIASTHMATICS**]
**Albuterol (salbutamol)**
*Aerolin, Proventil, Ventolin*

**Aminophylline (theophylline ethylenediamine)**
*Amoline, Somophyllin, Truphylline*

**Ephedrine**

**Epinephrine (adrenaline)**
*Bronkaid, Primatene*

**Ethylnorepinephrine**
*Bronkephrine*

**Isoproterenol**
*Aerolone, Aludrine, Isuprel, Norisodrine, Proternol*

Metaproterenol
*Alupent, Metaprel*

**Methoxyphenamine**
*Orthoxine*

**Soterenol**

**Terbutaline**
*Brethine, Bricanyl*

**Theophylline**
*Bronkodyl, Elixophyllin, Somophyllin*

**CALCIUM CHANNEL BLOCKERS**
**Cinnarizine**
*Aplactan, Cerepar, Denapol, Glanil, Mitronal*

**Diclofurine**

**Diltiazem**
*Cardizem*

**Fendiline**
*Cordan, Difmecor, Sensit*

**Flunarizine**
*Flugeral, Gradient, Issium, Sibelium*

**Nifedipine**
*Adalat, Procardia*

**Verapamil**
*Calan, Isoptin*

**CARDIAC DEPRESSANTS.** *See* **ANTIARRHYTHMICS**

**CHELATING AGENTS.** *See* **ANTIDOTES**

## CARDIAC STEROIDS (CARDENOLIDES, CARDIOTONICS)
[see also **INOTROPIC AGENTS**]
**Deslanoside (desacetyllanatoside C)**
*Cedilanid-D*

**Digitoxin**
*Crystodigin*

**Lanatoside C**
*Allocor, Cedilanid, Digilanide C*

**Ouabain (G-strophanthin)**
*Astrobain, Gratibain*

## CHOLINERGIC AGONISTS
**Acetylcholine**
*Acecoline, Miochol, Pragmoline*

**Ambenonium**
*Mytelase*

**Bethanechol**
*Duvoid, Mictone, Urabeth, Urecholine*

**Carbachol**
*Bufopto Carbacel, Carbamiotin, Carcholin,
   Doryl, Lentin*

**Edrophonium**
*Enlon, Tensilon*

**Guanidine**

**Methacholine**
*Mecholin, Mecholyl*

**Neostigmine**
*Prostigmin*

**Pilocarpine**
*Pilocar*

**Pirenzepine**
*Bisvanil, Gasteril, Leblon*

**Pyridostigmine**
*Mestinon, Regonol*

## CONTRACEPTIVES, FEMALE (ESTROGEN/PROGESTIN)
**Ethinyl estradiol/Ethynodiol diacetate**
*Demulen*

**Ethinyl estradiol/Megestrol**
*Ovex, Planovin*

**Ethinyl estradiol/Norethindrone acetate**
*Norlestrin*

**Ethinyl estradiol/Norethindrone**
*Brevicon, Norinyl, Ovcon*

**Ethinyl estradiol/Norgestrel**
*Ovral*

**Mestranol/Ethynodiol diacetate**
*Ovulen*

**Mestranol/Megestrol**
*Delpregnin*

**Mestranol/Norethindrone**
*Norinyl, Ortho-Novum*

**Mestranol/Norethynodrel**
*Enovid*

## CONTRACEPTIVES, MALE
**Gossypol**

## DIURETICS
**Acetazolamide**
*Acetamox, Cidamex, Diamox, Diluran*

Amiloride
*Midamor*

Bumetanide
*Bumex*

**Chlorothiazide**
*Diachlor, Diuril, Lyovac*

**Ethacrynic acid**
*Crinuryl, Edecrin, Taladren*

**Furosemide**
*Lasix*

**Glycerin (glycerol)**
*Glyrol, Osmoglyn*

**Hydrochlorothiazide**
*Esidrix, Mictrin*

**Mannitol**
*Osmitrol*

**Polythiazide**
*Drenusil, Nephril, Renese*

**Prorenoate**

**Spironolactone**
*Alatone, Aldactone*

Triamterene
*Dyrenium*

**Urea**
*Ureaphil*

**EMETICS**
**Apomorphine**

**Ipecac syrup**

**ESTROGENS**
**Chlorotrianisene**
*Merbentul, Tace*

**Diethylstilbestrol (DES)**

**Estradiol**
*Estrace, Estraval, Follidrin, Ovocyclin,*
  *Progynon*

**Estrogens, Conjugated**
*Conestron, Estrifol, Genisis, Premarin*

**Estrogens, Esterified**
*Estratab, Menest*

**Estropipate (piperazine estrone sulfate)**
*Ogen*

**Estrone**
*Bestrol, Estrol, Kestrone, Menformon,*
  *Ovex, Theelin*

**Ethinyl Estradiol**
*Estinyl, Feminone*

**Hexestrol**
*Hexanoestrol, Retalon, Synthovo*

**Mestranol**
*Norquen, Ovastol; Enovid, Norinyl, Ovulen*
  [constituent]

**Quinestrol**
*Estrovis*

**GANGLIONIC BLOCKERS**
[*see also* **ANTIHYPERTENSIVES**]
**Hexamethonium**
*Bistrium, Esametina, Gangliostat,*
  *Vegolysen*

**Mecamylamine**
*Inversine, Mekamine, Mevasine, Versamine*

**Pempidine**
*Pempidil, Perolysen, Tenormal, Viotil*

**Trimetaphan**
*Arfonad*

**GESTAGENS.** *See*
  **CONTRACEPTIVES; PROGESTINS**

**GLUCOCORTICOIDS**
**Bethamethasone**
**Corticotropin** [injection]

**Cortisone**
*Cortone, Cortistab*

**Dexamethasone**
*Dalanone, Decadron, Deronil, Hexadrol*

**Diflucortolone**
*Nerisone, Texmeten*

**Fluocinolone**
*Coriphate, Fluonid, Synotic*

**Hydrocortisone**
*Cortef, Hydrocortone*

**Prednisolone**
*Cortalone, Decortin H, Dicortil,*
  *Sterane, Sterolone*

**Prednisone**
*Deltasone, Meticorten, Orasone*

**Triamcinolone**
*Aristocort, Cenocort, Trilone*

**GLUCOSE-ELEVATING AGENTS**
Diazoxide
*Proglycem*

**Glucagon**

**GERMICIDES.** *See* **DISINFECTANTS**

**HISTAMINE ANTAGONISTS.** *See*
  **ANTIHISTAMINES (H$_1$);**
  **ANTIULCER AGENTS**

**HYPERTENSIVES**
[*see also* **ADRENERGIC**
  **VASOCONSTRICTORS**]
**Ephedrine**

**Mephentermine**
*Wyamine*

**Metaraminol**
*Aramine*

**Methoxamine**
*Vasoxyl*

**Norepinephrine (levarterenol)**
*Levophed*

**Phenylephrine**
*Neo-Synephrine*

## HYPNOTICS/SEDATIVES (BARBITURATE)
**Anobarbital**
*Amytal*
*Barbamyl, Dorminal*

**Barbital**
*Medinal, Veronal*

**Butabarbital**
*Butisol*

**Hexobarbital**
*Citopen, Evipan, Somnalert, Sombulex*

**Pentobarbital**
*Nembutal*

**Phenobarbital**
*Luminal*

## HYPNOTICS/SEDATIVES (NONBARBITURATE)
Acetylcarbromal
*Paxarel*

**Brotizolam**
*Lendormin*

**Chloral hydrate**
*Noctec*

**Ethchlorvynol**
*Placidyl*

**Flunitrazepam**
*Rohypnol, Roipnol*

**Flurazepam**
*Dalmane*

Glutethimide
*Doriden*

**Lorazepam**
*Ativan*

**Midazolam**
*Versed*

Paraldehyde
*Paral*

**Temazepam**
*Restoril*

**Triazolam**
*Halcion*

**Zopiclone**
*Imovance*

## HYPOGLYCEMIC AGENTS. *See* ANTIDIABETIC AGENTS

## INTRANASAL STEROIDS
**Beclomethasone**
*Beconase*

**Dexamethasone**
*Decadron*

**Flunisolide**
*Nasalide*

## INOTROPIC AGENTS
[*see also* **CARDIAC STEROIDS**]
**Amrinone**
*Inocor*

**Milrinone**

## LEPROSTATICS. *See* ANTILEPROTICS

## MIGRAINE PROPHYLACTICS
**Deoxycorticosterone (desoxycorticosterone, DOCA)**
*Cortesan, Decortin, Percorten*

**Dihydroergotamine**
*D.H.E.*

**Ergotamine**
*Ergomar, Ergostat, Femergin*

**Fludrocortisone**
*Alflorone, Florinef, Fludrocortone*

**Methysergide**
*Deseril, Sansert*

**Propranolol**
*Deralin, Inderal*

**MINERALOCORTICOIDS**
**Aldosterone**
*Aldocortin, Electrocortin*

**MITOTIC INHIBITORS.** *See*
**ANTITUMOR AGENTS**

**MUSCLE RELAXANTS.** *See*
**NEUROMUSCULAR BLOCKERS;**
**RELAXANTS (SKELETAL**
**MUSCLE)**

**NARCOTIC ANALGESICS.** *See*
**ANALGESICS (NARCOTIC)**

**NARCOTIC ANTAGONISTS**
**Cyclorphan**

**Levallorphan**
*Lorfan*

**Nalorphine**
*Anarcon, Nalline*

**Naloxone**
*Narcan*

**Naltrexone**
*Trexan*

**Oxilorphan**

**NASAL DECONGESTANTS**
**Guanabenz**
*Wytensin*

**Methoxamine**
*Vasoxyl*

**Naphazoline**
*Privine*

**Oxymetazoline**
*Afrin, Nafrine, Nasivin, Sinerol*

**Phenylephrine**
*Adrianol, Fenilfar, Neo-Synephrine*

**Tetrahydrozoline**
*Tyzine*

**Xylometazoline**
*NeoSynephrine II, Otrivin*

**NEUROLEPTICS.** *See*
**ANTIPSYCHOTICS**

**NEUROMUSCULAR BLOCKERS**
**Decamethonium**
*Syncurine*

Gallamine
*Flaxedil*

Metocurine
*Metubine*

**Pancuronium**
*Pavulon*

**Succinylcholine**
*Anectine*
*Sucostrin*

**Tubocurarine**
*Delacurarine, Tubarine*

**OVULATION STIMULANTS**
**Clomiphene**
*Clomid, Serophene*

**Tamoxifen**
*Nolvadex*

**OXYTOCICS (UTEROTONIC**
**AGENTS)**
**Carboprost**
*Prostin*

**Ergocryptine**

**Ergonovine**
*Basergin, Ergotrate, Ermetrine*

**Ergotamine**
*Ergomar, Femergin, Gynergen*

**Methylergonovine**
*Methergine*

**Oxytocin**
*Pitocin, Syntocinon*

**Prostaglandin $E_2$ (dinoprostone)**
*Prostin $E_2$*

**PLATELET ANTIAGGREGATORY AGENTS**
Ciprostene

Oxagrelate

**PROGESTINS**
[see also **CONTRACEPTIVES**]
**Danazol**
Cyclomen, Danocrine, Danol

**Ethisterone**
Gestoral, Progestoral, Prolidon

**Hydroxyprogesterone**
Duralutin, Hylutin

**Medroxyprogesterone**
Depo-Provera, Deporone, Oragest

**Megestrol**
Megace, Ovaban

**Nafarelin**

**Norethindrone**
Aygestin, Conludag, Norlutate, Norlutin

**Progesterone**
Amen, Corlutin, Prolutin, Provera

**PSYCHOTHERAPEUTIC AGENTS.**
See **ANALEPTICS;
ANTIDEPRESSANTS;
ANTIPSYCHOTICS; STIMULANTS
(CENTRAL); TRANQUILIZERS**

**RELAXANTS (MUSCLE)**
**Baclofen**
Baclon, Lioresal

Cinflumide

**Progabide**

**RELAXANTS (SKELETAL MUSCLE)**
**Baclofen**
Lioresal

Chlorzoxazone
Paraflex

Cyclobenzaprine

**Dantrolene**
Dantrium

**Diazepam**
Valium

**RESPIRATORY DRUGS.** See
**ANTIASTHMATICS;
ANTIHISTAMINES;
ANTITUSSIVES;
BRONCHODILATORS;
INTRANASAL STEROIDS;
NASAL DECONGESTANTS**

**SEDATIVES.** See
**HYPNOTICS/SEDATIVES**

**SEROTONIN ANTAGONISTS**
Altanserin

**Ketanserin**

**Ritanserin**

**SEROTONIN INHIBITORS**
**Cinanserin**

**Fenclonine**

Fonazine
Banistyl, Bonpac, Promaquid

**Mianserin**
Athymil, Norval, Tolvin

**STIMULANTS (CENTRAL)**
[see also **ANALEPTICS;
ANOREXIANTS**]
**Amphetamine**
Benzedrine

**Dextroamphetamine**
Dexaminex, Dexedrine

**Fenfluramine**
Obedrex, Ponderal, Pondimin

Methylphenidate
Ritalin

Pemoline
Cylert

**STIMULANTS (GASTRIC SECRETORY)**
Ceruletide
Ceosunin, Cerulen

**Histamine**

**THYMOLEPTICS.** *See*
   **ANTIDEPRESSANTS**

**THYROID HORMONES**
**Levothyroxine (L-thyroxine, T$_4$)**
*Euthyrox, Levothroid, Synthroid*

**Liotrix**
*Euthroid, Thyrolar*

**Lyothyronine (3,5,3′-triiodothyronine, T$_3$)**
*Cytobin, Cytomel, Triothyrone*

**Thyroglobulin**
*Proloid*

**THYROID INHIBITORS**
**Methimazole**
*Basolan, Mercazol, Tapazole*

**Methylthiouracil**
*Alkiron, Basecil, Methicil, Thimecil*

**Propylthiouracil**
*Propacil, Propycil*

**TRANQUILIZERS**
**Chlordiazepoxide**
*Lipoxide, Librium*

**Chlorpromazine**
*Hibernal, Largactil, Prompapar, Thorazine*

Clorazepate
*Tranxene*

**Diazepam**
*Valium*

**Doxepin**
*Adapin, Sinequan*

Droperidol
*Inapsine*

**Fluprazine**

**Lorazepam**
*Ativan*
*Lorax, Lorsilan*

Meprobate
*Equanil, Meptrospan, Miltown*

Mesoridazine
*Serentil*

**Midazolam**
*Versed*

**Oxazepam**
*Bonare, Quilibrex, Serax, Serepax*

**Prazepam**
*Centrax*

**Prochlorperazine**
*Compazine*

**Reserpine**
*Sandril, Serpalan, Rau-Sed, Serpasil,*
  *Serolfia*

**Suriclone**

**Thioridazine**
*Mellaril, Millazine*

**Trifluoperazine**
*Modalina, Stelazine*

**Zopiclone**
*Imovance*

**UTEROTONICS.** *See* **OXYTOCICS**

**VASOCONSTRICTORS**
[*see also* **ADRENERGIC**
   **VASOCONSTRICTORS**]
**Angiotensin amide**
*Hypertensin*
Felypressin

Midodrine
*Gutron*

**VASODILATORS**
**Amyl nitrite**

**Betahistine**
*Aequamen, Betaserc, Medan, Serc*

Darodipine

**Flunarizine**
*Flugeral, Gradient*

**Pindolol**
*Decreten, Visken*

## VASODILATORS (CORONARY)

**Clonitrate**
*Dilate*

**Diltiazem**
*Anginyl, Cardizem, Tildiem*

**Dipyridamole**
*Anginal, Cardoxin, Persantine*

**Droprenilamine**
*Valcor*

## VASODILATORS (PERIPHERIAL)

Inositol niacinate
*Dilexpal, Esantene, Linodil*

**Nicotinyl alcohol**
*Roniacol, Rontinol*

Nylidrin
*Adrin, Arlidin, Dilatal, Rydrin*

**Tolazoline**
*Artonil, Priscoline, Vasodil*

## *Selected References*

*Compendium of Pharmaceutical Specialties*, 21st ed. (1986). Canadian Pharmaceutical Association, Ottawa, Ontario.

C. R. Craig and R. E. Stitzel (1986). *Modern Pharmacology*, 2nd ed. Little, Brown, Boston.

R. F. Doerge (Ed.) (1982). *Wilson and Giswold's Textbook of Organic Medicinal and Pharmaceutical Chemistry*, 8th ed. J. B. Lippincott, Philadelphia.

*Drug Facts and Comparisons.* (1987). 1987 ed. Facts and Comparisons, St. Louis.

E. E. J. Marler (1985). *Pharmacological and Chemical Synonyms*, 8th ed. Elsevier, Amsterdam.

H. R. Patterson, E. A. Gustafson, and E. Sheridan (1982). *Current Drug Handbook 1982–1984.* W. B. Saunders, Philadelphia.

*United States Adopted Names, and the USP Dictionary of Drug Names* (M. C. Griffith, Ed.), 1961–1982 Cumulative List. (1987). U.S. Pharmacopoeial Convention, Rockville, MD.

M. Windholz (Ed.) (1983). *The Merck Index*, 10th ed. Merck & Co., Rahway, NJ.

# Index